Studies in Computational Intelligence

Volume 660

Series editor

Janusz Kacprzyk, Polish Academy of Sciences, Warsaw, Poland
e-mail: kacprzyk@ibspan.waw.pl

About this Series

The series "Studies in Computational Intelligence" (SCI) publishes new developments and advances in the various areas of computational intelligence—quickly and with a high quality. The intent is to cover the theory, applications, and design methods of computational intelligence, as embedded in the fields of engineering, computer science, physics and life sciences, as well as the methodologies behind them. The series contains monographs, lecture notes and edited volumes in computational intelligence spanning the areas of neural networks, connectionist systems, genetic algorithms, evolutionary computation, artificial intelligence, cellular automata, self-organizing systems, soft computing, fuzzy systems, and hybrid intelligent systems. Of particular value to both the contributors and the readership are the short publication timeframe and the worldwide distribution, which enable both wide and rapid dissemination of research output.

More information about this series at http://www.springer.com/series/7092

Nilanjan Dey · V. Santhi
Editors

Intelligent Techniques in Signal Processing for Multimedia Security

 Springer

Editors
Nilanjan Dey
Department of Information Technology
Techno India College of Technology
Kolkata
India

V. Santhi
School of Computing Science and
   Engineering
VIT University
Vellore
India

ISSN 1860-949X   ISSN 1860-9503   (electronic)
Studies in Computational Intelligence
ISBN 978-3-319-83137-4   ISBN 978-3-319-44790-2   (eBook)
DOI 10.1007/978-3-319-44790-2

Printed on acid-free paper

This Springer imprint is published by Springer Nature
The registered company is Springer International Publishing AG
The registered company address is: Gewerbestrasse 11, 6330 Cham, Switzerland

Preface

Through the recent years, innovative computational intelligence procedures and frameworks have been established as convenient techniques to address the various security issues. Computational intelligence approaches are widely used for modeling, predicting, and recognition tasks in several research domains. One of these is denoted by multimedia that finds enormous applications in security, entertainment, and health. Advancement in sensor technology, signal processing, and digital hardware lead to a variety of signal types including text, audio, animation, and video. Technological advancement in this media production and computer hardware/communications as well make information processing/transmission relatively simple and quick. However, the rate of exposure to various threats increases for multimedia data. Thus, providing security for multimedia data becomes an urgent challenging task.

Recently, various security mechanisms are developed to provide security for multimedia data, such as cryptography, steganography, and digital watermarking. Cryptographic techniques are used to protect digital data during the transmission from the sender to the receiver, but once it gets decrypted at the reception end, there is no control over the data for providing further security. In contrast, steganography is applied for secret communication between trusted parties with the limitation of payload. Moreover, digital watermarking uses the identity of the multimedia data owner. Thus, signal processing provides a major role in providing security for multimedia data.

This book consists of 21 chapters, including a brief discussion about the multimedia content security chapter followed by eight chapters that reported various aspects in the multimedia security domain. These chapters handled the intelligent security techniques for high-definition multimedia data as well as the morphing and steganography security techniques for information hiding, the digital watermarking as a solution for the multimedia authentication in addition to the representation of reversible watermarking with real-time implementation, and the comparison between singular value decomposition and randomized singular value decomposition–based watermarking. Furthermore, the concept and challenges of the biometric-based security systems were carried out followed by the evaluation of different cryptographic approaches in wireless communication network. In addition, this book contains another set of chapters that deal with the personal authentication and recognition systems in several applications, including personal authentication system based on hand images, surveillance system security using gait recognition, and face recognition under restricted constraints such as dry/wet face conditions, and the three-dimensional face identification using developed approach was also included. Thereafter, a proposed attendance recording system based on partial face recognition algorithm was conducted followed by the recognition of the human emotion that applied in surveillance video applications. Moreover, the concept of the security based on watermarking in the healthcare and medical applications and the pixel repetition-based high-capacity/reversible data hiding method for e-healthcare applications were included. Afterward, two chapters introduced several security issues for different types of multimedia contents, namely secured compressed data transmission over Global System for Mobile Communication (GSM) voice channel using wavelet transform and early tamper detection using joint stego-watermark approach. Finally, since biometrics has a great significant role in the security and authentication, the last chapter proposed a multifingerprint unimodel-based biometric authentication to support the cloud computing.

This book offered a precise concept and proposed approaches for the security in various applications combined with computational intelligence for multimedia security technologies. Based on the efforts done and the included contributions, it is expected very good endorsement from almost all readers of this book—from the graduated to postgraduate students' levels and researchers, professionals, and engineering. We are the editors wishing this book will stimulate further research to develop several intelligent techniques in signal processing for multimedia security applications in the various domains based on algorithmic and computer-based approaches.

Actually, this volume cannot be in this outstanding form without the contributions of the promising group of authors to whom we introduce our appreciation. Moreover, it was impossible to achieve this quality without the impact of the anonymous referees who assisted us during the revision and acceptance process of the submitted chapters. Our thankfulness is extended to them for their diligence in reviewing the chapters as well. Special thanks are directed to our publisher, Springer, for the endless support and guidance.

We hope this book presents promising concepts and outstanding research results supporting further development of security based on intelligent techniques.

Kolkata, India   Nilanjan Dey
Vellore, India   V. Santhi

Contents

Part I
Overview on Multimedia Data Security and Its Evaluation

Security of Multimedia Contents: A Brief

Amira S. Ashour and Nilanjan Dey

Abstract Through the past few years, an explosion in the use of digital media has been increased. A new infrastructure of digital audio, image, animation, video recorders/players, electronic commerce, and online services are speedily being deployed. These multimedia forms are applied for further transmission between users in the various applications through the Web or wireless communication that leads to security threats. Generally, the multimedia authenticity/security solution is based on using hiding techniques including digital watermarking and cryptography. This chapter introduced an extensive brief overview of the multimedia security problem in the prevailing research work in the various applications. It provided evolution of the different hiding techniques along with their properties and requirements. Finally, some challenges are addressed for future research direction.

Keywords Multimedia · Multimedia security · Watermarking techniques · Cryptography · Hiding techniques

1 Introduction

Recently, a great revolution occured in the digital information has led to profound changes in the society. The digital information advantages also generate new challenges/opportunities for innovation. Accompanied by the powerful software, new devices including high-quality printers/scanners, digital cameras, and digital voice recorder are used worldwide to create, deploy, and enjoy the multimedia

A.S. Ashour
Department of Electronics and Electrical Communications Engineering,
Faculty of Engineering, Tanta University, Tanta, Egypt
e-mail: amirasashour@yahoo.com

N. Dey (✉)
Department of Information Technology, Techno India College
of Technology, Kolkata, India
e-mail: neelanjan.dey@gmail.com

© Springer International Publishing Switzerland 2017
N. Dey and V. Santhi (eds.), *Intelligent Techniques in Signal Processing for Multimedia Security*, Studies in Computational Intelligence 660,
DOI 10.1007/978-3-319-44790-2_1

3

content. Internet and wireless network compromise ubiquitous ways to deliver and exchange the information. Multimedia is the combination of several items including text, audio, images, animation, and/or motion through video. These items use different techniques for digital formatting. The multimedia information volume is massive compared to traditional textual data. Multimedia is concerned with the computer-controlled integration of these contents to represent, store, transmit, and process each information type digitally. It can be distinguished from the traditional moving pictures/movies by the production scale, where multimedia is smaller and less exclusive. In addition, it can be interactive with the audience through text entry, voice commands, touch screen, and/or video capture. The prime multimedia system's characteristic is the use of several kinds of media to provide content and functionality. An elementary multimedia system embraces input/output devices, memory, backing storage, and a variety of software [1].

The multimedia hardware and software components availability assists the existing applications enhancement toward being more user-friendly with developing new multimedia applications. Generally, the different applications are crucial for the whole domain of multimedia computing and communications. Multimedia applications can comprise numerous types of media using programming and enhanced user interaction. Such applications include patient monitoring systems, distance learning, electronic magazines, video on demand, remote robotic agents, and interactive distributed virtual reality games. Figure 1 illustrated a classification for the multimedia applications.

Figure 1 depicted the three main categories of the multimedia applications, which include the following: (i) Information systems that provide information to the user/users, (ii) remote representation systems to represent the user at a remote location either in a passive or active way, and (iii) entertainment systems through multimedia telecommunication. Such wide variety of applications results in huge and different types of information contents that require security during its transmission.

The majority of these applications are accessing content and multimedia systems through Web or wireless telecommunication and transmission. However, since the digital multimedia contents (audio, images, and video) are easily copied and multiplied without information loss, a security problem arises which requires secure information management to maintain secured and authenticated data. Recently, researchers are interested in developing security approaches to prevent misappropriation, illegal copying, and misrepresentation of the multimedia contents in the

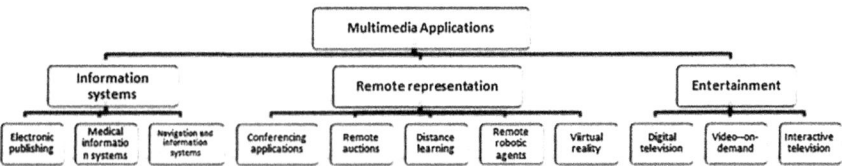

Fig. 1 Categories for the multimedia applications domain

various applications [2–6]. The three main basic threats that applied to any multimedia system are namely (i) confidentiality of the unauthorized information revealing, (ii) the availability of unauthorized withholding of information or resources, and (iii) integrity of the unauthorized data modification.

Several challenging topics for the multimedia contents transmission include the security, fair use of the multimedia data content, and the required fast delivery of the multimedia content to a variety of users/devices. Attempts to solve such problems improve the understanding of this complex technology as well as offer new techniques to be explored. Typically, multimedia security is based mainly on digital watermarking and cryptography. Digital watermarking [7] is the process of hiding a message related to the actual content of the digital data inside the signal itself for ownership rights protection. The information is embedded invisibly in the digital data as the original (host) data to form watermarked data in the digital watermarking. This embedded information is decodable from the watermarked data, even it is processed, redistributed, or copied. The watermarked information should be robust to maintain intact facing to the attacks [8], to remain hidden with any unauthorized user, and to be non-interfering with the watermarked document and fragile use. Furthermore, the cryptography is a very convenient approach for multimedia content protection. Multimedia contents are encrypted before the content provider (original user) transmits them through the Web using encryption key. Digital watermarking and cryptography systems have potential applications including authentication [9], copyright protection [10], authorized access control [11], distribution tracing [12], and in the medical domain [13].

Consequently, the objective of this chapter is to provide a brief overview of the current problems and developments in the security domain of multimedia contents. The various multimedia securities based on digital watermarking and steganography are introduced. In addition, the common challenges were introduced for opening new future directions. The structure of the remaining sections is as follows. Section 2 introduces the types of the hiding techniques in the multimedia applications. Section 3 highlights the common procedures for the data hiding. Afterward, the multimedia data hiding applications were introduced in Sect. 4 followed by common challenges of the multimedia data security systems. Finally, the conclusion is depicted in Sect. 5.

2 Hiding Techniques

Recently, the digital information revolution and the great advancement in the transmission and networking facilitate the multimedia transfer between the end users. Digital audio, image, animations, and videos are definitely transmitted over networked information systems. These multimedia offers various benefits; nevertheless, they also suffers from problems of security and illegal reproduction and distribution of its valuable contents. However, simple editing and impeccable reproduction for the multimedia contents in the digital domain, the ownership protection, authentication, annotation, access control, and the unauthorized tampering prevention are required. In order to achieve these requirements, data hiding schemes that embed secondary data in the multimedia contents have been proposed for a variety of applications for sending the information in multimedia communication to achieve supplementary functionalities and to enhance the performance. The basic requirements for any data hiding technique include the robustness, imperceptibility, and the ability to hide large number of bits. Moreover, in practice other encountered problems during data hiding, including the binary images perceptual models and the image/video uneven embedding capacity, are to be considered.

The simple editing, perfect reproductions, and multimedia content distribution through the internet, attack the focus toward solving the problems of copyright infringement, unauthorized tampering, and illegal distribution. This necessitates the developing of techniques that associate invisible data with the multimedia sources through embedding, such as the digital watermarking, whose traditional counterpart is not necessarily imperceptible. Since embedding data in multimedia sources is broaden, various techniques, including data hiding, steganography, and digital watermarking, are conducted in several applications [14–27]. However, the data hiding and digital watermarking are the most common techniques. They refer to embedding secondary data in the primary sources of the multimedia. The embedded data is called watermark(s) that can be used for several purposes with different security, robustness, and embedding capacity requirements. The main advantage of data hiding versus other techniques as a solution for the security is its capability to associate secondary data with the primary media in a continuous way, which is desirable in many applications.

There are several categorizations for the data hiding techniques. The straightforward categorization is based on the type of primary multimedia sources. Under this classification, the data hiding systems have two categories, namely data hiding for perceptual and data hiding for non-perceptual sources. The perceptual multimedia sources include audio, binary image, color or grayscale image, video, and 3-D graphics. However, the non-perceptual data includes text and executable codes that require lossless processing, transmission, and storage, while the perceptual data has a perceptual tolerance range. The perceptual property allows data embedding and lossy compression either with a controllable amount of perceptual degradation. Though, several data hiding techniques can be applied to the multimedia contents (audio, image, video, and 3-D graphics) [15]. Nevertheless, dimensionality and causality lead to quite different techniques, and resources are required for processing 1-D data contents (e.g., audio) from that for 2-D (e.g., images) and for 3-D (e.g., video).

In addition, another classification for the data hiding techniques in terms of perceptibility is perceptible hiding and imperceptible hiding. Generally, the perceptible watermarks are employed with image and video, where a visual, such as a logo, is overlaid on an image/video. Braudaway et al. [28] proposed a visible watermarking system through modifying the luminance of the original image in consistent with a binary watermark pattern. Based on the local luminance, the modification amount is adaptive to provide a consistent perceptual contrast [29]. Similarly, the same visible watermarking was applied in [30] for video contents.

Digital watermarking is considered as a special case of information hiding that hides information in the multimedia contents being transferred as well as store the information inside it. This embedded information is further being extracted or detected. Typically, watermarking is performed to hide copyright protection data inside multimedia content. It is possible to hide the information within digital audio, image, and video files in the sense that it is perceptually undetectable. Several schemes are applied to the hide information with preserving being recovered if the host signal is edited, compressed, or converted from digital to analog and back.

Digital watermarking can be considered as a means to identify the owner or distributor of the multimedia data. Commonly, the watermark techniques are classified [31] according to (i) the human perception into visible or invisible; (ii) the robustness into fragile, semi fragile, or robust; (iii) the multimedia content's types into audio, text, image, or video; iv) its characteristics into blind, semi-blind, or non-blind; and (v) the working domain into spatial domain or transform domain.

Generally, multimedia data embedding offers a tool for embedding significant control, descriptive, or reference information in a given multimedia content. This information can be used to track the use of a specific content, such as (i) clip for pay-per-use applications in internet electronic commerce of digital media and (ii) audio or visual object within a given content. Providing different access levels to the data is one of the significant applications of the data embedding process.

3   Data Hiding Common Procedures

The most common data hiding techniques include the cryptography, steganography, and watermarking. Since in cryptography the cipher data is a scrambled output of the plaintext (in the case of text data), the attacker can guess that encryption has been performed and hence can employ decryption techniques to acquire the hidden data. In addition, the cryptography techniques often require high computing power to perform encryption. Thus, the steganography is often preferred in some situations compared to the cryptography, where it is the process of sensitive data masking in any cover multimedia, such as audio, still images, and video over the internet. This makes the attacker unable to recognize that the data is being transmitted, since it is hidden from the naked eye and impossible to distinguish from the original media.

Generally, there are four steps for the steganography, namely (i) the selection of the cover media where the data will be hidden, (ii) the identification of the secret message/information that is required to be masked in the cover image, (iii) the determination of the function that will be used to hide the data in the cover media and its inverse to retrieve the hidden data, and (iv) the optional key to authenticate or to hide and unhide the data.

Naseem et al. [39] proposed an Optimized Bit Plane splicing algorithm to hide the data in the images. Along with hiding the data pixel by pixel and plane by plane, the proposed method involved hiding the data based on the intensity of the pixels. The pixels intensity was categorized into different ranges based on the pixel intensity. Moreover, the bits were hidden randomly in the plane instead of hiding them adjacent to each other, and the planes were transmitted sporadically. Thus, it was difficult to guess and intercept the transmitted data. Dey et al. [40] proposed a new technique to hide data in stego-images which achieved improvement over the Fibonacci decomposition method. The authors exploited prime numbers in order to hide data in the images. In order to increase the bit planes number of the image, the original bit planes were converted to some other binary number system using prime numbers as the weighted function. Thus, the bits number to represent each pixel increases that can be used hide data in higher bit planes. The results depicted that the proposed method generated the stego-image which is virtually indistinguishable from the original image.

The most effective data hiding technique is the watermarking. It is defined as the process of embedding watermark in the original audio, image, or video. Afterward, the watermark can be detected and extracted from the cover that contains information such as copyright and authorship. Consequently, any watermarking approach contains (i) watermark, (ii) encoder for embedding the watermark in the data, and (iii) decoder for extraction. Generally, there are main characteristics of the digital watermarking techniques, namely (i) the robustness, which is defined as the watermark system's ability to resist the image processing operations, (ii) the transparency, (iii) imperceptibility which indicates the similarity between the watermarked image and the host image, and (iv) the capacity that expresses the fixed information bits amount.

4 Multimedia Data Hiding Applications

Data hiding is a technique that is used to convey information while retaining the original appearance. It is useful in the multimedia communication applications to achieve additional functionalities with better performance. There are several criteria to measure the performance of multimedia data hiding system, namely data payload, robustness, the embedding effectiveness, and the fidelity [32]. The data payload ratio is the ratio between the watermark size and its carrier size. Consequently, proper detection algorithms are required for the extraction of the watermark. The embedding effectiveness is defined as the probability of immediate detection after embedding. Temporarily, the perceptual similarity between the host un-watermarked data and watermarked data when presented to a user is referred to as the fidelity watermarking system. The application domain of the multimedia security is another criterion to categorize data hiding techniques has various applications that can be classic applications, such as ownership protection, fingerprinting, authentication (tampering detection), annotation, and copy/access control. A brief explanation of these applications is as follows:

- Ownership Protection: In this application, a watermark indicating the ownership is embedded in the multimedia source. The watermark can handle the different attacks so that the owner can show the presence of this watermark in case of argument to prove the ownership. Little ambiguity/false alarm detection should be achieved. The total embedding capacity that denotes the bits number to be embedded and extracted with small probability of error, should not be high in most scenarios.
- Fingerprinting: In this application, the watermark is used to trace the creator or receivers of a particular copy of multimedia source based on the fingerprint. The robustness against destroying and the capability to carry a non-trivial number of bits are required.
- Authentication: In this application, beforehand a set of secondary data is embedded in the multimedia source which was used later to determine whether the host media is tampered or not. Copying a valid authentication watermark in an unauthorized/tampered media content must be inhibited, while no interest is directed to the robustness against making the watermark undetectable or removing it. In addition, it is required to locate the tampering and to distinguish any changes from the changes that may occur due to the content tampering, which requires high embedding capacity. Without the original un-watermarked copy, the detection can be performed that known as non-coherent (blind) detection.
- Annotation: In this application, the embedded watermark conveys large number of bits without the use of original unmarked copy in detection. A convinced robustness degree against lossy compression is desired.
- Copy/Access Control: In this application, the embedded watermark represents certain copy control or access control policy. A recording/playback system is used for watermarking detection, where definite software/hardware actions are carried out.

Moreover, other applications for multimedia data hiding system can be depicted based on the content's type. Accordingly, and due to the increased number of digital binary in the daily life, such as the captured handwritten signatures by the electronic signing pads, insurance information, social security records, and the financial documents, the unauthorized use of a signature become a critical issue. Secret data hiding in binary images is more problematic compared to other multimedia formats, where binary images entail only one bit representation to indicate black and white. Thus, several researchers are interested to develop data hiding techniques that applied for the authentication and tamper detection of binary images. Hidden data is inserted into an image in image authentication watermarking in order to detect any malicious image alteration. Few number of cryptography-based secure authentication techniques were available for binary images, where a message authentication code (or digital signature) of the whole image was computed and the resulting code is inserted into the image itself. Meanwhile, other data hiding techniques were employed with the binary images as follows.

Yang and Kot [33] proposed a new blind data hiding approach for binary images authentication to preserve the connectivity of pixels in a local neighborhood. The flippability of a pixel was defined by imposing three transition criteria in a moving window centered at the pixel. In addition, the watermark can be extracted without denoting the original image, where the embeddability of a block was invariant in the watermark embedding process. The host image uneven embeddability was handled by embedding the watermark only in the embeddable blocks. The locations were selected to guarantee high visual quality of the watermarked image. The authors compared the use of different types of blocks in terms of their abilities to increase the capacity. Venkatesan et al. [34] deployed a novel authentication watermarking method for binary images by using the prioritized sub-blocks through pattern matching scheme to embed the code. Before embedding, the shuffling was applied to equalize the uneven embedding capacity. It detected any alteration with sustaining good visual quality for all types of binary images. The security of the algorithm lies only on the used private keys.

Jung and Yoo [35] suggested a novel technique for data hiding in binary images using optimized bit position to replace a secret bit, where blocks were sub-divided. For a specified block, the parity bit was decided to be changed or not in order to embed a secret bit. The binary image quality of the resulting stego-image was improved by finding the best position to insert a secret bit for each divided block, while preserving low computational complexity.

Additionally, researchers were focused on data hiding for images and videos. Wu et al. [36] proposed a multi-level data hiding scheme based on a new classification of embedding schemes to convey secondary data in high rate. The authors studied the detection performance of spread spectrum embedding. The experimental results illustrated the effectiveness of multi-level data hiding. The authors used spread spectrum embedding and odd-even embedding. Min et al. [37] applied multilevel embedding for data in image and video. The multilevel embedding allowed the amount of embedded information that can be reliably extracted to be adaptive with respect to the actual noise conditions. Furthermore, for multilevel embedding to video, the authors proposed strategies to handle uneven embedding capacity from region to region within a frame as well as from frame to frame. Control information was also embedded for accurate extraction of the user data payload and to combat such distortions as frame jitter. The proposed procedure can be used for various applications such as copy control, access control, robust annotation, and content-based authentication.

Paruchuri and Cheung [38] proposed exhaustive search and fast Lagrangian approximation novel compression-domain video data hiding approaches to determine the optimal embedding strategy in order to minimize the output perceptual distortion and the output bit rate. The hidden data was embedded in selective discrete cosine transform (DCT) coefficients that were found in most video compression standards. The coefficients were selected based on minimizing a cost function that combines the distortion and bit rate through a user controlled weighting. The results established that the exhaustive search produced optimal results, while the fast Lagrangian approximation was significantly faster and amenable to real-time implementation.

5   Challenges and Future Directions

The most critical challenges that the data hiding techniques can face and/or can improve in multimedia security applications can be addressed as follows:

- The flippability model can be improved for different types of binary images including texts, drawings, and dithered images.
- The method for recovering binary image from high-quality printing/scanning can be improved by using grayscale information from the scanned image.
- A balance between robustness, imperceptibility, and capacity can be studied.
- It is required to compromise the payload size with the imperceptibility.
- The computational cost of inserting and detecting watermarks should be minimized.
- Overcome the different types of attacks, such as the protocol attacks, cryptographic attacks, security attack, removal attack, and the image degradation attacks.
- Reliability and confidentiality should be considered.
- Due to the multimedia contents non-stationarity nature, such as audio, digital image, and video, the number of bits that can be embedded varies significantly. This unevenly distributed embedding capacity adds difficulty to data hiding.

6   Conclusion

Internet and wireless network offer advanced ways for information delivery and exchange. The multimedia data security with fast delivery of multimedia content becomes a challenging topic. The solutions to these problems require multimedia data hiding security and communication development. Several techniques have been proposed for several applications for ownership protection, authentication, and access control. Data hiding in the multimedia applications achieved its functionality and enhanced performance. Generally, the data hiding can be considered to be a communication problem where the embedded data is the signal to be transmitted. The various embedding approaches target at different robustness-capacity trade-off.

The current chapter introduced the different embedding algorithms based on the application under concern, such as the data hiding algorithms for binary images and data hiding applications in image/video communication. Watermarking has a substantial concern with multimedia security, confidentiality, and integrity. Authentication is to trace the multimedia content origin, while the integrity is to detect whether the changes have been done. Recently, attempts are conducted to solve the trade-offs between the robustness and capacity of the multimedia applications for information protection, safety, and management.

References

1. Qiao L (1998) Multimedia security and copyright protection. Ph.D. Thesis, University of Illinois at Urbana-Champaign
2. Dey N, Biswas S, Das P, Das A, Chaudhuri SS (2012) Feature analysis for the reversible watermarked electrooculography signal using low distortion prediction-error expansion. In: 2012 International conference communications, devices and intelligent systems (Codis), pp 624–27
3. Dey N, Maji P, Das P, Biswas S, Das A, Chaudhuri S (2012) Embedding of blink frequency in electrooculography signal using difference expansion based reversible watermarking technique. Buletinul Ştiinţific Al Universităţii "Politehnica" Din Timişoara, Seria Electronică Şi Telecomunicaţii Transactions on Electronics and Communications 57(71)
4. Dey N, Biswas D, Roy A, Das A (2012) DWT-DCT-SVD based blind watermarking technique of gray image in electrooculogram signal. In: 2012 12th international conference on intelligent systems design and applications (ISDA), pp 680–685
5. Dey N, Mukhopadhyay S, Das A, Chaudhuri S (2012) Analysis of P-QRS-T components modified by blind watermarking technique within the electrocardiogram signal for authentication in wireless telecardiology using DWT. Int J Image Graph Signal Process (IJIGSP) 7:33–46
6. Nandi S, Roy S, Dansana J, Karaa WB, Ray R, Chowdhury SR, Chakraborty S, Dey N (2014) Cellular automata based encrypted ECG-hash code generation: an application in inter human biometric authentication system. Int J Comput Netw Inf Secur 6(11):1
7. Kumari G, Kumar B, Sumalatha L, Krishna V (2009) Secure and robust digital watermarking on grey level images. Int J Adv Sci Technol 11
8. Miyazaki A, Okamoto A (2002) Analysis and improvement of correlation-based watermarking methods for digital images. Kyushu University
9. Jin C, Zhang X, Chao Y, Xu H (2008) Behavior authentication technology using digital watermark. In: International conference on multimedia and information technology (MMIT '08), pp 7–10
10. Jin C, Wang S, Jin S, Zhang Q, Ye J (2009) Robust digital watermark technique for copyright protection. In: International symposium on information engineering and electronic commerce (IEEC '09), pp 237–240
11. Pan W, Coatrieux G, Cuppens-Boulahia N, Cuppens F, Roux C (2010) Watermarking to enforce medical image access and usage control policy. In: 2010 sixth international conference on signal-image technology and internet-based systems (SITIS), pp 251–260
12. Li M, Narayanan S, Poovendran R (2004) Tracing medical images using multi-band watermarks. In: 26th conference of the IEEE annual international engineering in medicine and biology society (IEMBS '04), vol 2, pp 3233–3236
13. Kishore P, Venkatram N, Sarvya C, Reddy L (2014) Medical image watermarking using RSA encryption in wavelet domain. In: 2014 First international conference on networks & soft computing (ICNSC), pp 258–262
14. Petitcolas FAP, Anderson RJ, Kuhn MG (1999) Information hiding—a survey. In: Proceedings of IEEE, pp 1062–1078
15. Hartung F, Kutter M (1999) Multimedia watermarking techniques. In: Proceedings of IEEE, pp 1079–1107

16. Dey N, Nandi B, Das P, Das A, Chaudhuri SS (2013) Retention of electrocardioGram features insiGnificantly devalorized as an effect of watermarkinG for. Advances in biometrics for secure human authentication and recognition, 175
17. Biswas S, Roy AB, Ghosh K, Dey N (2012) A biometric authentication based secured ATM banking system. Int J Adv Res Comput Sci Softw Eng, ISSN, 2277
18. Bose S, Chowdhury SR, Sen C, Chakraborty S, Redha T, Dey N (2014) Multi-thread video watermarking: a biomedical application. In: 2014 International conference on circuits, communication, control and computing (I4C). IEEE, pp 242–246
19. Bose S, Chowdhury SR, Chakraborty S, Acharjee S, Dey N (2014) Effect of watermarking in vector quantization based image compression. In 2014 International conference on control, instrumentation, communication and computational technologies (ICCICCT). IEEE, pp 503–508
20. Dey N, Samanta S, Yang XS, Das A, Chaudhuri SS (2013) Optimisation of scaling factors in electrocardiogram signal watermarking using cuckoo search. Int J Bio-Inspired Comput 5 (5):315–326
21. Dey N, Mukhopadhyay S, Das A, Chaudhuri SS (2012) Analysis of P-QRS-T components modified by blind watermarking technique within the electrocardiogram signal for authentication in wireless telecardiology using DWT. Int J Image Graph Signal Process 4 (7):33
22. Dey N, Maji P, Das P, Biswas S, Das A, Chaudhuri SS (2013) An edge based blind watermarking technique of medical images without devalorizing diagnostic parameters. In: 2013 international conference on advances in technology and engineering (ICATE). IEEE, pp 1–5
23. Chakraborty S, Samanta S, Biswas D, Dey N, Chaudhuri SS (2013) Particle swarm optimization based parameter optimization technique in medical information hiding. In: 2013 IEEE international conference on computational intelligence and computing research (ICCIC). IEEE, pp 1–6
24. Dey N, Biswas S, Das P, Das A, Chaudhuri SS (2012) Lifting wavelet transformation based blind watermarking technique of photoplethysmographic signals in wireless telecardiology. In: 2012 World congress on information and communication technologies (WICT). IEEE, pp 230–235
25. Dey N, Pal M, Das A (2012) A session based watermarking technique within the NROI of retinal fundus images for authencation using DWT, spread spectrum and Harris corner detection. Int J Mod Eng Res 2(3):749–757
26. Dey N, Samanta S, Chakraborty S, Das A, Chaudhuri SS, Suri JS (2014) Firefly algorithm for optimization of scaling factors during embedding of manifold medical information: an application in ophthalmology imaging. J Med Imaging Health Inform 4(3):384–394
27. Chakraborty S, Maji P, Pal AK, Biswas D, Dey N (2014) Reversible color image watermarking using trigonometric functions. In: 2014 International conference on electronic systems, signal processing and computing technologies (ICESC). IEEE, pp 105–110
28. Braudaway GW, Magerlein KA, Mintzer F (1996) Protecting publicly-available images with a visible image watermark. In: SPIE conference on optical security and counterfeit deterrence techniques, vol 2659, pp 126–133
29. Jain AK (1989) Fundamentals of digital image processing. Prentice Hall, Upper Saddle River
30. Meng J, Chang S-F (1998) Embedding visible video watermarks in the compressed domain. In: International conference on image processing (ICIP)
31. Serdean C (2002) Spread spectrum-based video watermarking algorithms for copyright protection. Ph.D. thesis, University of Plymouth
32. Cox I, Miller M, Bloom J, Fridrich J, Kalker T (2007) Digital watermarking and steganography. Morgan Kaufmann Publishers Inc., San Francisco, CA, USA. ISBN: 978-0123725851
33. Yang H, Kot AC (2007) Pattern-based data hiding for binary image authentication by connectivity-preserving. IEEE Trans Multimed 9(3):475–486

34. Venkatesan M, MeenakshiDevi P, Duraiswamy K, Thyagarajah K (2008) Secure authentication watermarking for binary images using pattern matching. IJCSNS Int J Comput Sci Netw Secur 8(2):241–250

35. Jung KH, Yoo KY (2009) Data hiding method with quality control for binary images. J Softw Eng Appl 2(1):20

36. Wu M, Hong HY, Gelman A (1999). Multi-level data hiding for digital image and video. In: Photonics East'99. International society for optics and photonics, pp 10–21

37. Min WU, Heather Y, Bede L (2003) Data hiding in image and video: part II—designs and applications. IEEE Trans Image Process 12(6):696–705

38. Paruchuri JK, Cheung SCS (2008) Joint optimization of data hiding and video compression. In: IEEE International symposium on circuits and systems, 2008. ISCAS 2008. IEEE, pp 3021–3024

39. Naseem M, Hussain IM, Khan MK, Ajmal A (2011) An optimum modified bit plane splicing LSB algorithm for secret data hiding. Int J Comput Appl (0975–8887), 29(12)

40. Dey S, Abraham A, Sanyal S (2007) An LSB data hiding technique using prime numbers. In: Third international symposium on information assurance and security, 2007. IAS 2007, pp 101–108

A Survey on Intelligent Security Techniques for High-Definition Multimedia Data

S.D. Desai, N.R. Pudakalakatti and V.P. Baligar

Abstract Multimedia security has advanced tremendously over the decades due to the change in variety and volume of data. In the current security context, intelligent systems for multimedia security are very much in demand. Various applications such as biometric, e-commerce, medical imaging, forensics, aerospace, and defense require high-end data security systems. Conventional cryptography, watermarking, and steganography fall in short to provide security for high-resolution 2D/3D image and high-definition video. Persistent demand exists for designing new security algorithms for 3D graphics, animations, and HD videos. Traditional encryption method does not suffice the current need, as its securing ability is limited when it gets decoded. Steganography techniques are reported for securing text, audio, and video content, but observed to be few in number compared to image steganography techniques. Watermarking techniques for securing video content, text, and animations are reported in the literature but seem to be few in numbers as compared to image watermarking techniques. Majority of the literature is observed to apply digital watermarking as security means for video and image data. However, digital watermarking for 3D graphics is a current research topic. On the other hand, video watermarking techniques shall be broadly classified based on domain and human perception. Usually, video watermarking techniques do not alter video contents. But current trend shows that security techniques are designed based on video content. This kind of security methods is claimed to be far superior as they concentrate not only on watermarking but also on synchronization of watermark. In this chapter, we present a comprehensive review of multimedia security techniques emphasizing on their applicability, scope, and shortcomings especially when applied to high-definition multimedia data. Problematic issues of intelligent

S.D. Desai (✉)
Department of IS&E, BVB CET, Hubli 580031, India
e-mail: sd_desai@bvb.edu

N.R. Pudakalakatti · V.P. Baligar
Department of CS&E, BVB CET, Hubli 580031, India
e-mail: nehapud.np@gmail.com

V.P. Baligar
e-mail: vpbaligar@bvb.edu

© Springer International Publishing Switzerland 2017
N. Dey and V. Santhi (eds.), *Intelligent Techniques in Signal Processing for Multimedia Security*, Studies in Computational Intelligence 660,
DOI 10.1007/978-3-319-44790-2_2

techniques in signal processing for multimedia security and outlook for the future research are discussed too. The major goal of the paper was to provide a comprehensive reference source for the researchers involved in designing multimedia security technique, regardless of particular application areas.

Keywords Robustness · Multimedia · Security systems · Cryptography · Steganography · Watermarking

1 Introduction

In the recent years, the exponential growth of digital media and the ease with which digital content is exchanged over the Internet have created security issues. Hence, there is a need for good security systems. In this section, we present a comprehensive categorization and classification of security system for multimedia. Figure 1 represents the categorization and classification of security systems. Security systems shall be broadly categorized based on the information encrypting or information hiding process as shown in Fig. 1. Three categories such as digital watermarking, steganography, and cryptography fall under the security systems. All these three categories have common purposes but different approaches.

When a digital signal or pattern is embedded in the host such as image, audio, video, or text, it is called as digital watermarking. It is mainly used to identify ownership of copyright for such signals. Extraction is said to be major factor in watermarking because any kind of distortion during retrieval is unacceptable [1–5]. Imperceptibility and robustness are the two most desirable properties of a digital watermarking [6–9]. Digital watermark should be imperceptible to the human eye, and illegal removal of the watermark should be prevented. The watermark should be robust toward various intentional and unintentional attacks [10]. Nowadays, digital watermarking has been successfully used in wide range of application as it provides a high level of security. Applications related to medical, defense, and agriculture are unlimited.

Current applications are as follows:

- Content identification and management,
- Content protection for audio and video content,
- Broadcast monitoring,
- Locating content online, and
- Audience measurement.

Steganography is the art and science of hiding vital information. It aims at hiding the message so that there is no knowledge of the existence of the message. Due to the increase in popularity of IP telephony, it is attracting the attention of research community as a perfect carrier medium. Good qualities of steganography are hiding capacity, imperceptibility, and irrecoverability.

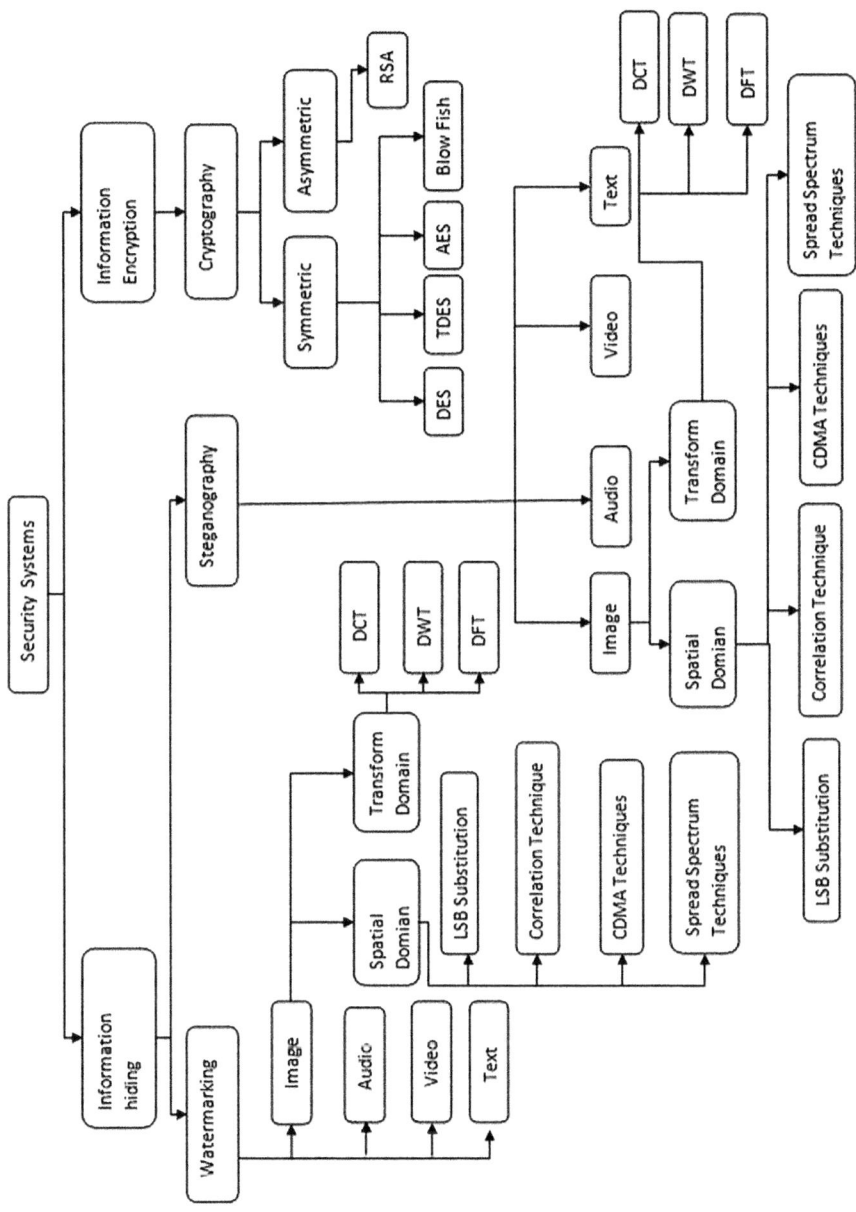

Fig. 1 Classification of security algorithms

Current applications are as follows:

- Documents protection against forgery,
- Medical imaging,
- Protection of data alteration, and
- Media database systems.

On the other hand, cryptography or information encryption scrambles messages so that they cannot be understood. Confidentiality, integrity, access control, and authentication are that main purposes of the cryptography.

Current applications are as follows:

- ATM encryption,
- Online banking, and
- E-mail privacy.

Presently, there are many embedding/extraction schemes proposed, but only few techniques have a commercial value. Both watermarking and steganography belong to the same category of information hiding, but the objective of both the techniques is totally opposite. In watermarking, important information is external data. The internal data are additional data for protecting external data. But in steganography, external data are not important as it is just a carrier medium for internal data (secret data). Generally, the watermarking and steganography are broadly classified into two main categories: first one as spatial domain and second one as transform domain. Spatial domain techniques are said to be less robust compared to transform domain techniques. Cryptography is broadly classified based on encryption algorithms.

In the recent years, audio and video technologies have better resolution compared to the previously used standards. HDM videos are defined by 3 attributes such as number of lines in the vertical display resolution, the scanning system, and the number of frames per second. The most common HD video modes are presented in Table 1.

Table 1 Common HD modes

Video Mode	Frame size in pixels ($W * H$)	Pixels per image	Scanning type	Frame rate
720p	1,280 × 720	921,600	Progressive	23.976, 24, 25, 29.97, 30, 50, 59.94, 60, 72
1080i	1,920 × 1,080	2,073,600	Interlaced	25 (50 fields/s), 29.97 (59.94 fields/s), 30 (60 fields/s)
1080p	1,920 × 1,080	2,073,600	Progressive	24 (23.976), 25, 30 (29.97), 50, 60 (59.94)
1440p	2,560 × 1,440	3,686,400	Progressive	24 (23.976), 25, 30 (29.97), 50, 60 (59.94)

HD audio also known as high-resolution audio in a trend in the audio market. In fact, there is no standard definition for HD audio. It basically describes audio signals with bandwidth and dynamic range as compared to compact disk digital audio. Some of the well-known HD audio formats are FLAC, ALAC, WAV, AIFF, and DSD.

In preceding sections, we discuss in detail about different security systems, comparison among them, classification, and algorithms used.

2 Digital Watermarking

Presently, digital watermarking is a widely used technique for data encryption. Digital watermarking is applied on image, audio, video, and text. Image watermarking is predominantly being used other than audio, video, and text. Watermarking is of two types: (i) blind watermarking and (ii) non-blind watermarking. When original data are not needed during extraction of the watermark, it is said to be blind. On the other hand, when original data and key are required during extraction process, it is said to be non-blind. Based on the human perception, digital watermarking can be classified as visible and invisible watermarking. In visible watermarking, embedded watermark is visible to human eye. Those which fail can be classified as invisible watermarks.

2.1 Digital Image Watermarking

Image watermarking has a lot of attention in the research community compared to all other watermarking techniques. There are two main categories of digital image watermarking techniques, which are based on the embedding position, spatial domain, and transform domain.

2.1.1 Spatial Domain

The spatial domain image is represented by pixels. In this technique, watermark embedding is achieved by directly modifying the pixel value of the host image.

Techniques in spatial domain generally share the following characteristics:

- Watermark is applied in pixel domain.
- Simple operations are applied when combining with host signal.
- No transforms are applied.
- Watermark is detected by correlating excepted pattern with received signal [11].

The methods used in the spatial domain are the least significant bit (LSB), correlation-based, and spread-spectrum techniques. LSB is easiest, and the most commonly method used in spatial domain [11].

A. LSB

The LSB technique works by replacing some of the information in a given pixel with information from the data in the image. The LSB embedding is performed on the least significant bit because it minimizes the variation in colors that the embedding creates [12]. LSB substitution suffers from drawbacks. Any addition of noise would likely to defeat the watermark. Once the algorithm is discovered, the embedded watermark could be easily modified by the hackers [13].

Advantages of LSB are as follows:

- Easiest method of watermarking.
- Can insert lot of data if image is simple.
- Enhances security when used with more sophisticated approach such as pseudorandom generator.

Disadvantages of LSB are as follows:

- Less robust.
- Highly sensitive to signal processing operations and easily corrupted.
- Less secured because if small portion of watermarked image is detected, then the whole message can be extracted.

The mean square error (MSE) and the peak signal-to-noise ratio (PSNR) are the two metrics used to compare image quality and are described as follows [11]:

$$\text{MSE}(x, y) = \frac{1}{N} \sum_{i=1}^{N} (x_i - y_i)^2 \tag{1}$$

$$\text{PSNR} = 10 \log_{10} \frac{L^2}{\text{MSE}} \tag{2}$$

where x and y are two finite-length discrete signals, where N is the number of signal samples (pixels, if the signals are images), x_i and y_i are the values of the ith samples in x and y, and L is the dynamic range of allowable image pixel intensities [11].

A.1. Watermark embedding and extraction using LSB

Select an image as a cover image or a base image in which watermark will be inserted. n represents the number of least significant bits to be utilized to hide most significant bits of watermark under the base image. LSB embedding and extraction are shown in Figs. 2 and 3.

Fig. 2 LSB embedding

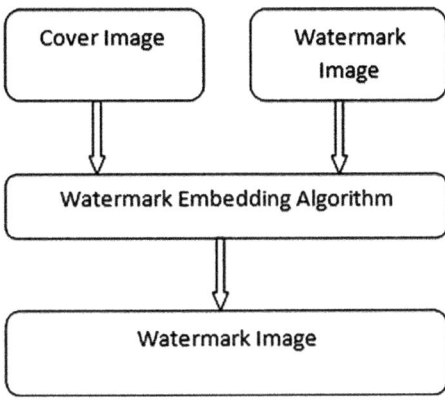

Fig. 3 LSB extraction

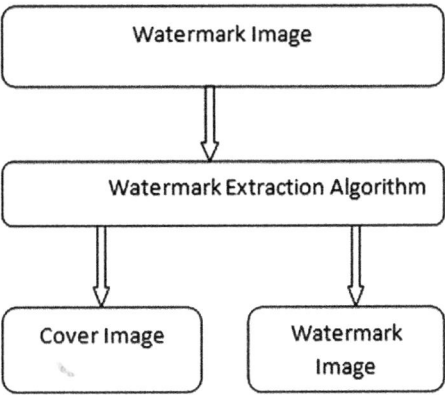

Algorithm 1: LSB method

Embedding
Input: Cover Image and Watermark.
Output: Watermark Image.
    1· For each pixel in base, watermarked and watermarked image
    2: Cover image: Set n least significant bits to zero.
    3: watermark: Shift right by 8-n bits.
    4: Watermarked image: add values from base and watermark.
    5: End

As an example, let us consider the first pixel of the base image along with first pixel of watermark image. The final first pixel of watermarked image is calculated as follows:

Explanation:	
Consider 1st pixel of cover image:	01101011
Set $n = 1$ and LSB to zero:	01101010
Consider first pixel of watermark and shift right by $8\text{-}n$ bits:	1
Watermarked Image:	$\begin{array}{r} 01101010 \\ +\qquad 1 \\ \hline 01101011 \end{array}$

Extraction
Input: Watermark Image.
Output: Base Image and Watermark.
      1:   For each pixel in watermarked image and extracted image
      2:   Watermarked image: Shift left by 8-n bits.
      3:   Append n bits of zeroes at MSB
      4:   Extracted image: Set to the shifted value of watermarked
           image.

Explanation:	
Consider watermarked image:	01101011
Shift left by 8–1 bits:	1101011
Append n bits of zeroes at MSB($n = 1$) which is the cover image:	01101011

A.2. Results of LSB

Figure 4 represents the original, the watermark, and the watermarked images. This is visible watermarking. The figures are the experimental results obtained by Chopra et al. [11].

B. Correlation-based watermarking

Early researches used LSB method for watermarking. And the only disadvantage was that if small portion of watermarked image was detected, then the whole message can be extracted [14]. In this method, to increase the security and the robustness, pseudorandom noise (PN) sequence and key are used [14]. PN sequence is a sequence of binary numbers that appear to be random but is actually deterministic. Here, linear feedback shift register (LFSR) circuit is used to generate pseudorandom sequence. Periodic shift registers are those shift registers which have nonzero initial state and output is fedback to the input [14].

PN sequences are good tool for watermarking because of the following reasons [14]:

(a)   **(b)**   **(c)**

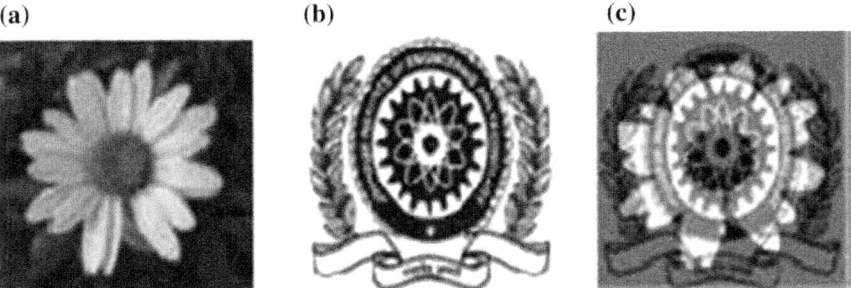

Fig. 4 LSB results. **a** Original, **b** watermark, and **c** watermarked images (visible watermarking)

- Periodic sequences are produced by the generator that appears to be random.
- These sequences are generated by algorithm that makes use of initial speed.
- Unless the key and algorithm are known, it is impossible to generate the sequence.

B.1. Watermark embedding and extraction using correlation technique—threshold-based correlation

Figures 5 and 6 represent the embedding and extraction process of threshold-based correlation technique.

Algorithm 2: Correlation Technique

Embedding
Input: Base Image and Watermark.
Output: Watermark Image.
  1: *PN pattern W(x,y) is added to the original image I(x,y) based on the equation 1.3 shown below*

   $$I_W(x,y) = I(x,y) + k * W(x,y) \qquad\qquad 1.3$$

   where I(x, y) = Original(base) Image
   W(x, y) = PN pattern
   K = Gain Factor
   I_w (x, y) = Watermarked Image
  2: *Apply both watermark and the PN sequence to the product modulator.*
  3: *PN pattern W(x,y) is added to the base image I(x,y) to produce the resultant watermarked Image $I_w(x,y)$.*

Extraction
Input: Watermarked Image.
Output: -Base Image and Watermark.
  1: *Watermarked Image $I_w(x,y)$ is multiplied at the receiver with the PN sequence which is same as that used during embedding process to extract the watermark a(x,y).*

B.2. Results of correlation-based technique

Figure 7 represents the original, the watermark, and the watermarked images. The figures are the experimental results obtained by Gajriya et al. [14]

2.1.2 Transform Domain

The transform domain images are represented by frequencies. In this technique, the transform coefficients are modified instead of directly changing the pixel values.

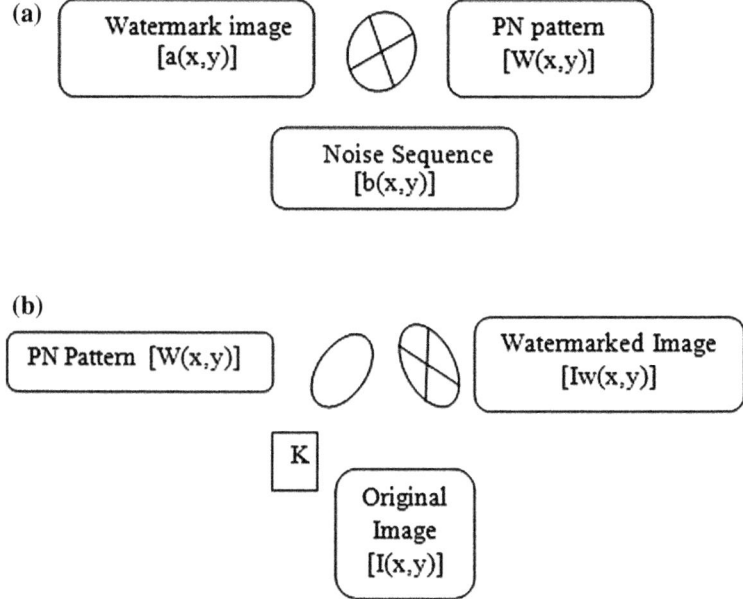

Fig. 5 Threshold-based embedding. **a** Generation of PN sequence and **b** watermarking the original image with PN sequence

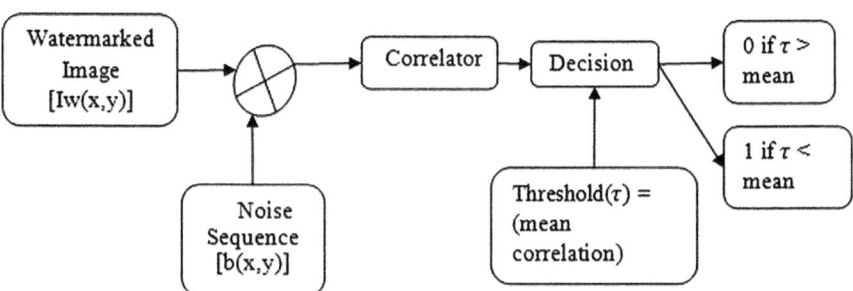

Fig. 6 Threshold-based extraction

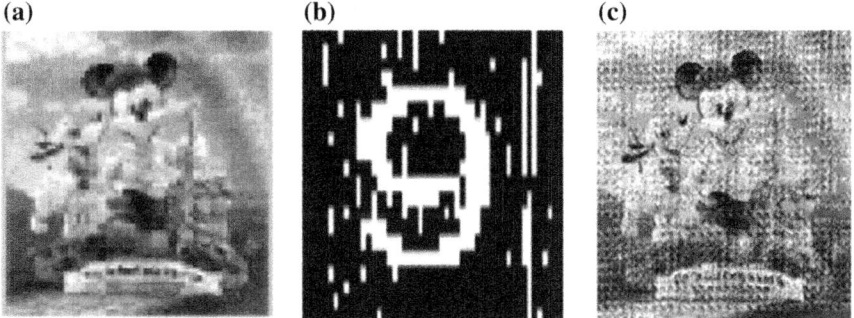

Fig. 7 Results. **a** Original, **b** watermark, and **c** watermarked images

First, the host image is converted into frequency domain using any of the transformation methods. The most commonly used transformation methods are DCT, DWT, and DFT.

A. DWT

DWT is a modern technique frequently used in digital image processing which has excellent multiresolution and spatial localization characteristics [15]. Its multiresolution characteristics hierarchically decompose an image [16]. DWT provides both the frequency and the spatial domain of an image; i.e., it captures both location information and frequency components. The original image is decomposed into four subimages: three high-frequency parts (HL, LH, and HH, named detail subimages) and one low-frequency part (LL, named approximate subimage). Edge information is present in detailed subimages. Compared to the detail subimages, approximate subimage is much more stable, since the majority of the image energy concentrates here. Therefore, the embedding is done in approximate subimages to have better robustness [16].

A.1. Watermark embedding and extraction using DWT

Alpha blending technique is a process of mixing two images to get the final image. It is accomplished in computer graphics by blending each pixel from first image and corresponding pixel in second image [17, 41]. It can be used to create partial or full transparency [16]. Formula of alpha blending technique used for watermark embedding and extraction is given as follows:

$$\text{Watermark Image} = A \times (\text{LL1}) + B \times (\text{WM1}) \tag{4}$$

$$\text{Recovered Image} = (\text{WM} - A \times \text{LL1}) \tag{5}$$

A and B represent scaling factors for cover and watermark image, respectively.

LL1 represents low-frequency approximation of cover image.
WM1 represents watermark image.
WM represents watermarked image.

Advantages of alpha blending technique are as follows:

- Insertion and extraction of watermark becomes simpler.
- This technique can be used to embed invisible watermark into salient features of the image [18].
- It provides high security.
- Image is resistant to several attacks [17].

Figures 8 and 9 represent DWT extraction and embedding.

Algorithm 3: DWT

Embedding
Input: Cover Image and Watermark.
Output: Watermarked Image.
  1: In this first we take cover image(base image) and is decomposed into 4 components using 2D DWT.
  2: The same procedure is applied on the watermark image which is to be embedded into cover image.
  3: Apply DWT for both cover image and watermark image.
  4: Now alpha blending technique is used for inserting a watermark.

Extraction
Input: Watermarked image.
Output: Cover image and watermark.
  1: The watermarked image and cover image is first decomposed into sub images using DWT
  2: Then alpha blending formula is applied to recover the water-marked image.

Fig. 8 DWT embedding

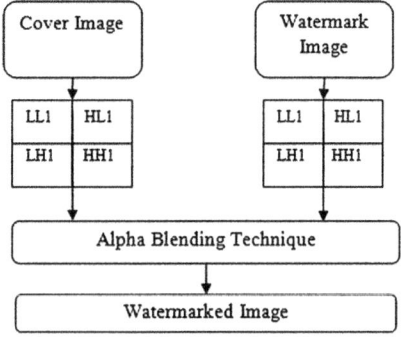

Fig. 9 DWT extraction

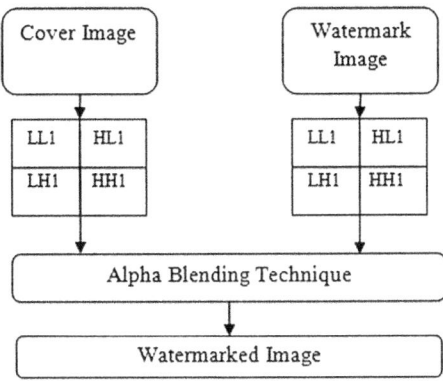

Fig. 10 DWT results. **a** Original, **b** watermark, and **c** watermarked images

A.2. Results of DWT

Figure 10 represents the original, the watermark, and the watermarked images. The figures are the experimental results obtained by Narang and Vashisth [16].

B. DFT

The DFT is a most popular technique used in signal analysis, signal study, and synthesis to define the effect of various factors on signal. The Fourier transform is used in transforming the signal from time domain to frequency domain or from frequency domain to time domain. This transformation is reversible and maintains the same energy [19]. The Fourier transform and the inverse Fourier transform are given in Eqs. 5 and 6 [19]:

$$F(\omega) = \int_{-\infty}^{\infty} f(t)e^{-j\omega t}dt \tag{6}$$

$$F(t) = \frac{1}{2\pi} \int\limits_{-\infty}^{\infty} f(t)e^{j\omega t}d\omega \tag{7}$$

Performance parameters such as PSNR and normalized correlation (NC) are used in measuring the quality of watermarked image. PSNR and NC are described in the Eqs. 7 and 8 [19]:

$$\text{PSNR} = 10\log_{10} XY \operatorname{map} p_{x,y}^2 / \sum \left(p_{x,y} - p_{x,y}\right)^2 \tag{8}$$

$$NC = \sum p_{x,y} - p_{x,y} / \sum p_{x,y}^2 \tag{9}$$

B.1. Watermark embedding and extraction using DFT

Algorithm 4: DFT
Embedding
Input: Cover Image and Watermark.
Output: Watermarked Image.
  1: *Divide the original image, in which watermark is embedding,*
      into the sub-blocks of 256*256.
  2: *Transform the all image blocks into 8*8 matrixes by using*
      DFT transform.
  3: *Arnold Scrambling is used to change the binary watermark*
      and generate two unrelated pseudo-random sequence.
  4: *Modify the corresponding value of amplitude spectrum.*
  5: *Apply Discrete Fourier Transform to each image blocks to*
      produce the image with watermark.

Extraction
Input: Watermarked Image.
Output: Cover Image and Watermark.
  1: *Apply image segmentation process to divide the image into*
      256*256 sub-blocks, which is embedded watermark.
  2: *Transform the all image blocks into 8*8 matrix by using DFT*
      transform.
  3: *Produce the two unrelated pseudo-random sequence.*
  4: *Compare the watermark's amplitude spectrum and the pseudo-*
      random sequence and calculate the relativity between both of
      them and then produce watermark matrix with the help of
      embedding rules.
  5: *Use Arnold transform scrambling to watermark matrix.*

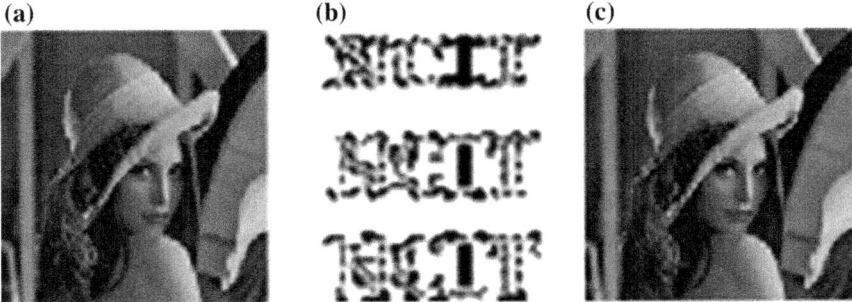

Fig. 11 DFT results. **a** Original, **b** watermark, and **c** watermarked images

B.2. Results of DFT

Figure 11 represents the original, the watermark, and the watermarked images. The figures are the experimental results obtained by Kaushik et al. [19].

C. DCT

DCT is a popular image transformation method which is used in many image processing applications. This transformation allows each transform coefficient to be encoded independently without losing compression efficiency [20]. Due to its good performance, it has been used in JPEG standard for image compression. DCT has been applied in many fields such as data compression, pattern recognition, and image processing [20].

Performance parameters such as PSNR are used in measuring the quality of watermarked image [20]. PSNR is described in Eq. 7:

$$\text{PSNR} = 10\log_{10} 225^2 \Big/ \frac{1}{N \times N} \sum_{i=0}^{N-1} \sum_{j=0}^{N-1} \left(x_{i,j} - x\right)^2 \tag{10}$$

C.1. Watermark embedding and extraction using DCT

Figures 12 and 13 represent watermark embedding and extraction using DCT.

Algorithm 5: DCT

Embedding
Input: Cover Image and Watermark.
Output: Watermarked Image.
    1: Extract every 8 bit data from the watermark bit stream.
    2: Using pseudo random system generate the random number, which
       points to one of n blocks of host image.
    3: Embed extracted the 8-bit watermarking data into the 8 lower-
       band coefficients in the block pointed by previous step.
    4: Repeat step 1 to step 3, until the watermark bit stream is run out.
    5: The proposed employee replace bit to embedded watermark bit
       stream, and it was hidden at position bit 3 in the selected 8-bit
       coefficient. If the watermark bit is "1" then bit 3 to "1" otherwise
       "0".

Extraction
Input: Watermarked Image.
Output: Cover Image and Watermark.
    1: Transform the watermarked image to frequency domain by DCT.
    2: Use the same set of random numbers, which is applied in the
       embedding process.
    3: Apply the random number to find the exact location of the
       DCT block in the watermarked image
    4: Extract 8-bit watermark data from each DCT block by
       means of the inverse embedded. The watermark bit is "1"
       when bit 3 is "1" of selected DCT-block coefficient otherwise
       the watermark bit is "0".
    5: Rearrange the 8-bit data into watermark image.

C.2. Results of DCT

Figure 14 represents the original, the watermark, and the watermarked images using DCT. The figures are the experimental results obtained by Marjuni et al. [20].

2.2 Audio Watermarking

Recently, audio watermarking is used as one of the most popular approaches for providing copyright protection. Digital audio watermarking is different from digital image watermarking. An effective audio watermarking scheme must have the following properties: (1) imperceptibility, (2) robustness, (3) payload, and (4) security [21].

Fig. 12 DCT embedding

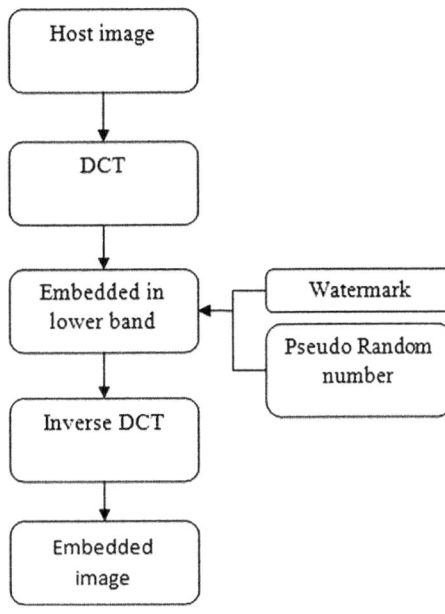

Fig. 13 DCT embedding

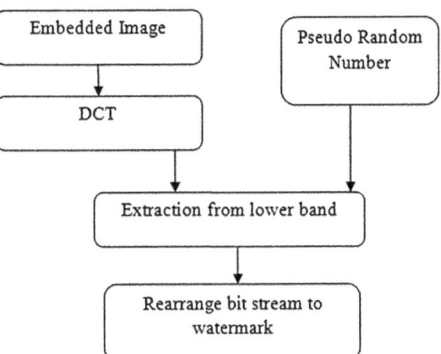

1. Imperceptibility refers to the maintaining of audio quality after adding the watermark.
2. Robustness refers to the ability to extract a watermark from a watermarked audio signal after various intentional and unintentional attacks.
3. Payload refers to the amount of data that can be embedded into the host audio signal.
4. Security refers to the watermark that can only be detected by the authorized person.

Robustness and imperceptibility are the most important requirements for digital audio watermarking. A watermark embedding should be imperceptible by the user;

(a)   (b)   (c)

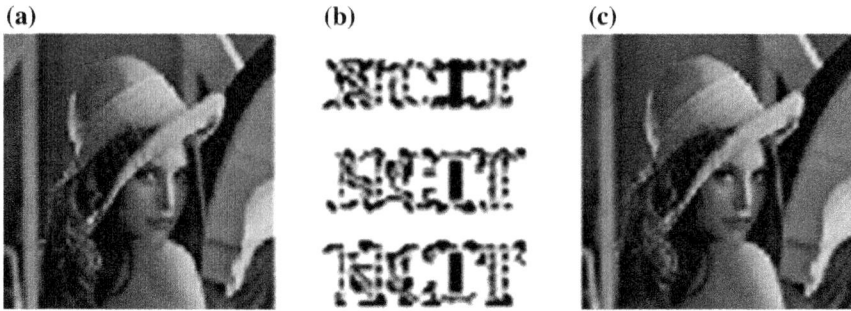

Fig. 14 DCT results. **a** Original, **b** watermark, and **c** watermarked images

that is, it should be highly transparent or invisible to prevent unauthorized detection and removal. On the other hand, watermark should be resistant to several attacks such as noise addition, cropping, resampling, requantization, MP3 compression, shifting, time scale modification (TSM), and pitch scale modification (PSM) [22, 23].

In general, audio watermarking is classified into 2 types: time domain and transform domain [21]. Time domain watermarking schemes are very easy to implement, and they also require less computing resources compared to transform domain watermarking methods. On the other hand, time domain watermarking systems are usually weaker against various attacks. The widely used transform domains for audio watermarking are DCT, DWT, and FFT.

Audio watermarking applications are as follows:

- Copyright protection—the copyright owner will be having knowledge of secrete key to read the embedded watermark.
- Vendor identification—copyright notice can be embedded in the audio signal and can determine vendor of the copyrighted audio.
- Evidence of proprietorship—the original owner can prove his proprietorship by extracting watermark from watermarked image.
- Validation of genuineness—the copyright audio can be proved easily by adding watermark.
- Fingerprinting—information about authenticated customers can be embedded as secret watermarks right before the secure delivery of the data [24].

2.3   Video Watermarking

Video watermarking is a rapidly evolving field in the area of multimedia. Many watermarking techniques have been proposed for video watermarking. Video watermarking scheme should have the following properties: (1) imperceptibility, (2) robustness, (3) payload, and (4) security. An effective and efficient video watermarking should be resistant to several attacks such as JPEG coding, gaussian noise addition, histogram equalization, gamma correction, lossy compression, frame averaging, frame swapping, rotation, and rescaling.

Based on the domain, video watermarking can be broadly classified into two domains. The first one is the spatial domain watermarking where embedding and detection of watermark are performed by manipulating the pixel values of the frame. The second category is the transform domain techniques in which the watermark is embedded by changing its frequency components. The commonly used transform domain techniques are DFT, DCT, and DFT and also principle component analysis (PCA) transform. The transform domain watermarking schemes are relatively more robust than the spatial domain watermarking schemes, particularly in pixel removal, noise addition, rescaling, rotation, and cropping.

Video watermarking applications are as follows [25]:

- Copy control—it prevents copying from unauthorized persons.
- Broadcast monitoring—it monitors the video contents that are broadcasted.
- Fingerprinting—it traces the unauthorized or malicious uses.
- Enhanced video coding—it includes additional information.
- Copyright protection—it proves the authenticity of the owner.

2.4 Text Watermarking

Authentication and copyright protection are the two main applications of digital watermarking. In addition to image, audio and video text is also an important medium. Textual contents over the Internet include newspapers, e-books, and messages. Hence, text has to be protected from intentional and unintentional attacks such as random insertion, deletion or, reordering of words or sentences to and from the text [14, 40]. Text watermarking techniques help to protect the text from illegal copying, forgery, and redistribution. It also helps to prevent copyright violations [26]. Text watermarking developed so far use either image or textual watermark. Watermarking which makes use of both text and image is more secured and provides better robustness to various attacks [27]. Compared to other watermarking techniques, less research works are proposed for text. Text watermarking algorithms can be broadly classified into 4 types:

- Image-based methods,
- Syntactic methods,
- Semantic methods, and
- Structural methods [26, 28].

In the next section, we discuss in detail steganography, its classification, and the algorithms used.

3   Steganography

Steganography refers to the technique of hiding secret messages into media such as text, audio, image, and video, while steganalysis is the art and science of detecting the presence of steganography. There are two main categories of image steganography techniques, which are based on embedding position, spatial domain, and transform domain.

3.1   Image Steganography

In this section, we present spatial and transform domain image steganography methods.

3.1.1   Spatial Domain

A. LSB

   In spatial domain, LSB is most commonly used technique and very easy to implement, but it is less robust compared to other techniques. It embeds the secret into least significant bits of pixel value of the base image (cover image) [29].

   Performance parameters such as PSNR are used to measure the quality of stego-image. The formula is given as follows [39]:

$$\text{PSNR} = 10 \log C_{\text{max}}^2 / \text{MSE} \tag{11}$$

$$\text{MSE} = \text{mean} - \text{square} - \text{error} \tag{12}$$

$$\text{MSE} = \frac{1}{MN\left((S - C)^2\right)} \tag{13}$$

$$C_{\text{max}} = 255. \tag{14}$$

where M and N are the dimensions of the image, S is the resultant stego-image, and C is the cover image.

A.1. Data embedding and extraction using LSB

Figures 15 and 16 represent the watermark embedding and extraction using LSB

Algorithm 6: LSB method

Embedding
Inputs: Cover image, stego-key and the text file
Output: stego image
 1: Extract the pixels of the cover image.
 2: Extract the characters of the text file.
 3: Extract the characters from the Stego key.
 4: Choose first pixel and pick characters of the stego key and place it
    in first component of the pixel.
 5: Place some terminating symbol to indicate end of the key.
 6: Insert characters of text file in each first component of next pixels
    by replacing it.
 7: Repeat step 6 till all the characters has been embedded.
 8: Again place some terminating symbol to indicate end of data.
 9: Obtained stego image.[16].

Extraction
Inputs: Stego-image file, stego-key
Output: Secret text message.
 1: Extract the pixels of the stego image.
 2: Now, start from first pixel and extract stego key characters from
    first component of the pixels. Follow Step3 up to terminating sym-
    bol, otherwise follow step 4.
 3: If this extracted key matches with the key entered by the receiver,
    then follow Step 5, otherwise terminate the program.
 4: If the key is correct, then go to next pixels and extract secret mes-
    sage characters from first component of next pixels. Follow Step 5
    till up to terminating symbol, otherwise follow step 6.
 5: Image extraction algorithm.
 6: Extract secret message[16].

Fig. 15 LSB embedding

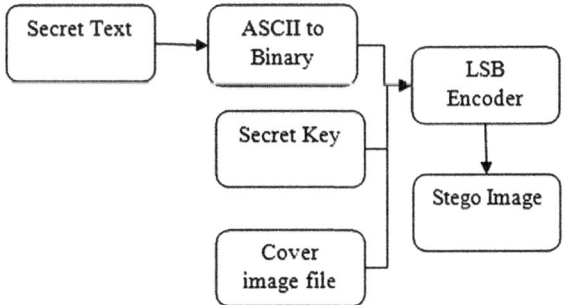

Fig. 16 LSB extraction

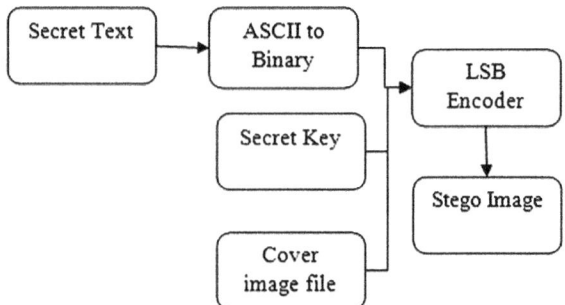

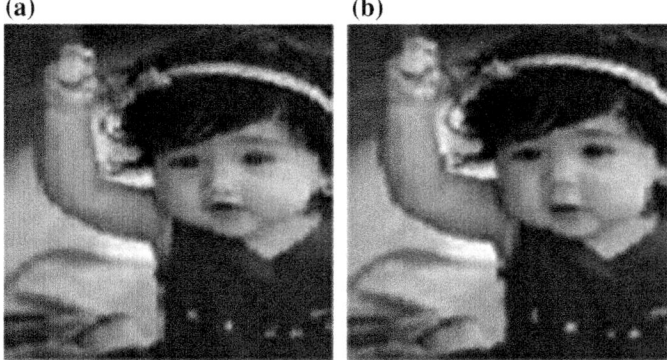

Fig. 17 LSB results. **a** Original image and **b** stego-image

A.2. Results of LSB

Figure 17 represents the original image and stego-image. The figures are the experimental results obtained by Devi [39].

3.1.2   Transform Domain

A. DWT

DWT divides the image into low- and high-frequency components. The original image is embedded in approximate coefficients which are low-frequency components. And additional information about the image is embedded in detailed coefficients which are high-frequency components [31].

A.1. Data embedding and extraction using DWT

Algorithm 7: DWT

Embedding
Input: An m × n carrier image and a secret message/image.
Output: An m × n stego-image.
    1: Read the cover image (Ic)
    2: Calculate the size of Ic
    3: Read the secret image (Im)
    4: Prepare Im as message vector
    5: Decompose the Ic by using Haar wavelet transform
    6: Generate pseudo-random number (Pn)
    7: Modify detailed coefficients like horizontal and vertical coefficients of wavelet decomposition by adding Pn when message bit = 0.
    8: Apply inverse DWT
    9: Prepare stego image to display.

Extraction
Input: An m × n carrier image and an m × n stego-image.
Output: Secret message/image.
    1: Read the cover image (Ic)
    2: Read the stego image (Is)
    3: Decompose the Ic and Is by using Haar wavelet transform
    4: Generate message vector of all ones
    5: Find the correlation between the original and modified coefficients
    6: Turn the message vector bit to 0 if the correlation value is greater than mean correlation value
    7: Prepare message vector to display secret image.

A.2. Results of DWT

Figure 18 represents the original image and stego-image. The figures are the experimental results obtained by Banik [30].

B. DFT

It is used to get the frequency component for each pixel. The DFT of spatial value $f(x, y)$ for the image size $M * N$ is given by [32]

$$F(u, v) = \frac{1}{\sqrt{MN}} \sum_{x=0}^{M-1} \sum_{y=0}^{N-1} f(x, y) e^{-j2\pi\left(\frac{ux}{M} + \frac{vy}{N}\right)} \tag{15}$$

$$F(u, v) = \frac{1}{\sqrt{MN}} \sum_{x=0}^{M-1} \sum_{y=0}^{N-1} F(u, v) e^{-j2\pi\left(\frac{ux}{M} + \frac{vy}{N}\right)} \tag{16}$$

where $u = 0$ to $M - 1$ and $v = 0$ to $N - 1$.

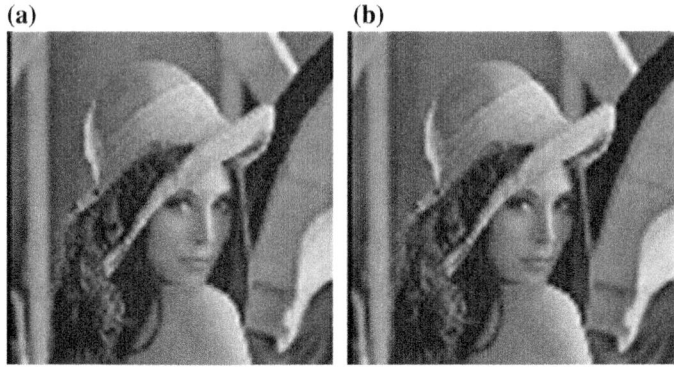

Fig. 18 DWT results. **a** Original image and **b** stego-image

B.1. Data embedding and extraction using DFT

Algorithm 8: DFT

Extraction
Inputs: Source image, Secret Image
Output: Stego image
  1:   Read the source image.
  2:   Read the image.
  3:   Apply DFT. Embedded the image in the real part of transform
        domain excluding 1st pixel.
  4:   Insert the image bit one by one.
  5:   Apply Inverse DFT.
  6:   Repeat steps 3 to 5 for the whole embedding process.
  7:   Stop.

Extraction
Inputs: Stego Image
Output: Source image and Secret image
  1:   Read the noisy embedded image.
  2:   Apply DFT.
  3:   Extract the image from real part of transform domain.
  4:   Repeat steps 2 to 3 for complete decoding of as per image size.
  5:   Apply Inverse DFT.
  6:   Stop

C. DCT

DCT provides high-energy compaction compared to DFT for natural images [33]. DCT is a general 8*8 transform for digital image processing and signal processing [34]. Two-dimensional DCT can be defined as follows:

$$f(x,y) = C(u)C(v) \sum_{u=0}^{N-1} \sum_{v=0}^{N-1} F(u,v) \cos\left[\frac{(2x+1)u\pi}{2N}\right] \cos\left[\frac{(2x+1)v\pi}{2N}\right] \quad (17)$$

$$F(u,v) = \frac{2}{N} C(u)C(v) \sum_{u=0}^{N-1} \sum_{v=0}^{N-1} f(x,y) \cos\left[\frac{(2x+1)x\pi}{2N}\right] \cos\left[\frac{(2x+1)y\pi}{2N}\right] \quad (18)$$

where $F(u,v)$ is cosine transform coefficient, and u and v is generally frequency variables. For $x, y = 0, 1, 2 \ldots N - 1$. N is horizontal and vertical pixel number of pixel block.

C.1. Data embedding and extraction using DCT

Figures 19 and 20 represent DCT embedding and extraction.

Algorithm 9: DCT

Embedding
Input: Cover Image I, Secret Message
Input Parameters: Quantization Matrix (Q)
Output: Stego Image S
    1: Begin
    2: Read the cover image, I.
    3: Divide the cover image, I into blocks of size 8 x 8.
    4: Find the DCT of I.
    5: Obtain the Quantized DCT blocks by dividing the DCT of I by the
       quantization matrix.
    6: Hide the secret message in the Quantized DCT.
    7: Obtain the dequantized matrix and inverse DCT.
    8: Restructure the 8 x 8 blocks into a single array.
    9: Stego image is formed.
    10: End

Extraction
Input: Stego Image S
Input Parameters: Quantization Matrix (Q)
Output: Cover Image I, Secret Message
    1: Begin
    2: Read the stego image, S.
    3: Divide the stego image, S into blocks of size 8 x 8.
    4: Find the DCT of S.
    5: Obtain the Quantized DCT blocks by dividing the DCT of
       S by the quantization matrix.
    6: Extract the secret message from the quantized DCT blocks
       and concatenate DCT LSB to secret message.
    7: Obtain the dequantized matrix and inverse DCT.
    8: Restructure the 8 x 8 blocks into a single array.
    9: Cover image and secret message are obtained[13].
    10: End

Fig. 19 DCT embedding

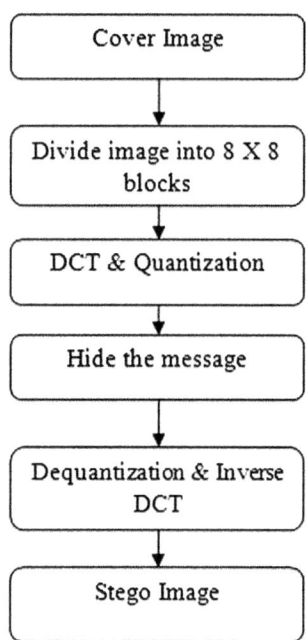

Fig. 20 DCT extraction

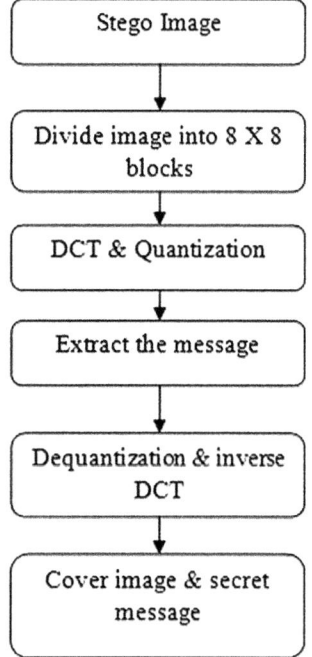

(a)   **(b)**

Fig. 21 DCT results. **a** Original image and **b** stego-image

C.2. Results of DCT

Figure 21 represents the original image and stego-image. The figures are the experimental results obtained by Bansal and Chhikar [33].

In this section, we discuss briefly on cryptography, its classification, purposes, and algorithms used.

4 Cryptography

Cryptography is a technique used to scramble confidential data to make it unreadable for unauthorized persons [35]. Many security system use cryptography for providing security because it consists of different encryption algorithms that makes encrypted data to be unreadable [35]. It is not only used to provide security for data but can also be used for user authentication.

A. Purpose of cryptography

- Confidentiality—confidential means private or secret. It ensures that only the sender and authorized person have access to the message [35].
- Authentication—origin of the message is authenticated [35, 36].
- Integrity—it makes sure that the contents of the message are not modified during transmission [37].
- Non-repudiation—it ensures that the sender of the message does not claim of not sending the message to the authorized person [36].
- Access control—it ensures that the only person has access to the message [37].

Cryptography can be classified into symmetric and asymmetric cryptography [35]. In symmetric cryptography, only one key is used for encryption and decryption. In asymmetric cryptography, 2 keys are used: one for encryption and another for decryption.

4.1 DES

DES stands for data encryption standard, and it was the first standard recommended by National Institute of Standard and Technology [36]. It is a block cipher that uses secrete key for both encryption and decryption [35]. It takes fixed length of plain text and transforms into same length of cipher text. In DES, each block size is of 64 bits and uses 56 bits key. The decryption can only be done using the key which was used to encrypt the message [38].

4.2 TDES

TDES stands for triple DES. It is the enhancement of DES but uses 3 keys each of 56 bits key size. TDES provides key of larger size, hence increasing the computational complexity [38]. Encryption method is same as DES, but it is applied 3 times to increase the encryption level [35]. TDES algorithm which uses three keys requires 2^{168} possible combinations, and the algorithm which uses two keys requires 2^{112} combinations. Hence, TDES is said to be the most strongest encryption algorithm, but one disadvantage is that it is very time-consuming [35].

4.3 AES

It is a block cipher and a symmetric key algorithm means which uses same key for both encryption and decryption [36]. AES works on the principle known as substitution and permutation which makes it faster and more efficient algorithm [38]. AES has fixed block size of 128 bits, and a key size of 128, 192, or 256 bits [35].

4.4 RSA

RSA stands for Ron Rivest, Adi Shamir, and Leonard Adleman, who first publicly described the algorithm in 1977 [36]. It is a asymmetric algorithm and is widely used for secure data transmission.

Two different keys are used for encryption and decryption: (i) public key and (ii) private key. The former one is used for encryption, and the latter one is used to for decryption [35].

In the next section, a comparative study is made on different encryption techniques used so far.

Table 2 Comparative study

Criteria	Watermarking	Steganography	Cryptography
Carrier	Any multimedia data	Any multimedia data	Text files
Visibility	Visible or invisible	Not visible	Visible
Robustness	High	Moderate	Less
Capacity	Depends on the size of hidden data	Usually low	High
Objective	To protect copyright information	To prevent discovery of secret message	To prevent unauthorized access
Failure	Removed	Detected	Deciphered

5 Comparative Study

Watermarking, steganography, and cryptography are well-known security techniques and are widely used to provide security for highly confidential data. All three techniques are useful for real-time encryption and are suitable for different applications. A comparative study on these security techniques is given in Table 2.

6 Conclusions and Future Scope

In the digital world nowadays, the security of digital images become more and more important since the communications of digital products over open network occur more and more frequently. In this chapter, we have surveyed existing work on image encryption

It also presents a background discussion on major techniques of steganography and watermarking. Some important techniques used in steganography and watermarking and few encryption algorithms used in cryptography are discussed. Both steganography and watermarking are divided into spatial domain and transform domain. In spatial domain, LSB is the most important technique used for image encryption. In transform domain, DCT, DWT, and DFT are the commonly used techniques for image encryption. Spatial domain techniques are said to be less robust compared to transform domain techniques. Watermarking is robust to most of the signal processing techniques and geometric distortions compared to steganography and cryptography. PSNR is the most commonly used parameter to measure the quality of encrypted data. Lena and boat images are the most commonly used testing images. Few algorithms on image watermarking and steganography are reviewed and are robust against different kinds of attacks. General description on text, audio, and video is also given.

Nowadays, many companies have already been active in digital watermarking. For example, Microsoft has developed a system that limits unauthorized playback which helps music industry for copyright protection. Future development of multimedia systems in the Internet is conditioned by the development of efficient methods to protect data owner from copyright violations. But many encryption systems do not solve this problem. Hence, different encryption systems must be proposed in order to give copyright protections.

References

1. Dey N, Das P, Roy AB, Das A, Chaudhuri SS (2012) DWT-DCT-SVD based intravascular ultrasound video watermarking. In: 2012 World congress on information and communication technologies (WICT). IEEE, pp 224–229
2. Dey N, Mukhopadhyay S, Das A, Chaudhuri SS (2012) Analysis of P-QRS-T components modified by blind watermarking technique within the electrocardiogram signal for authentication in wireless telecardiology using DWT. Int J Image Graph Signal Process 4(7):33
3. Dey N, Maji P, Das P, Biswas S, Das A, Chaudhuri SS (2013) An edge based blind watermarking technique of medical images without devalorizing diagnostic parameters. In: 2013 International conference on advances in technology and engineering (ICATE), IEEE, pp 1–5
4. Dey N, Biswas S, Roy AB, Das A, Chowdhuri SS (2013) Analysis of photoplethysmographic signals modified by reversible watermarking technique using prediction-error in wireless telecardiology. International conference on intelligent infrastructure the 47th annual national convention at Computer Society of India
5. Chakraborty S, Samanta S, Biswas D, Dey N, Chaudhuri SS (2013) Particle swarm optimization based parameter optimization technique in medical information hiding. In: 2013 IEEE International conference on computational intelligence and computing research (ICCIC). IEEE, pp 1–6
6. Dey N, Das P, Chaudhuri SS, Das A (2012) Feature analysis for the blind-watermarked electroencephalogram signal in wireless telemonitoring using Alattar's method. In: Proceedings of the fifth international conference on security of information and networks. ACM, pp 87–94
7. Dey N, Biswas S, Das P, Das A, Chaudhuri SS (2012) Feature analysis for the reversible watermarked electrooculography signal using low distortion prediction-error expansion. In: 2012 International conference on communications, devices and intelligent systems (CODIS). IEEE, pp 624–627
8. Dey N, Biswas S, Das P, Das A, Chaudhuri SS (2012) Feature analysis for the reversible watermarked electrooculography signal using low distortion prediction-error expansion. In: 2012 International conference on communications, devices and intelligent systems (CODIS). IEEE, pp 624–627
9. Chakraborty S, Maji P, Pal AK, Biswas D, Dey N (2014) Reversible color image watermarking using trigonometric functions. In: 2014 International conference on electronic systems, signal processing and computing technologies (ICESC). IEEE, pp 105–110
10. Chen WY, Huang SY (2000) Digital watermarking using DCT transformation. Department of Electronic Engineering, National ChinYi Institute of Technology, Taichung
11. Chopra D, Gupta P, Sanjay G, Gupta A (2012) LSB based digital image watermarking for gray scale image. IOSR J Comput Eng (IOSRJCE). ISSN:2278-0661
12. Kaur G, Kaur K (2013) Implementing LSB on Image Watermarking using text and image. Int J Adv Res Comput Commun Eng. ISSN:2319-5940
13. Sharma PK (2012) Rajni, "Information security through Image watermarking using Least Significant Bit Algorithm,". Comput Sci Inf Technol 2(2):61–67
14. Gajriya LR, Tiwari M, Singh J (2011) "Correlation based watermarking technique-threshold based extraction,". Int J Emerg Technol 2(2):80–83

15. Awasthi M, Lodhi H (2013) Robust image watermarking based on discrete wavelet transform, discrete cosine transform & singular value decomposition. Adv Electr Electron Eng 3(8):971–976
16. Narang M, Vashisth S (2013) Digital watermarking using discrete wavelet transform. Int J Comput Appl 74(20)
17. Singh AP, Mishra A (2011) Wavelet based watermarking on digital image. Indian J Comput Sci Eng 1(2), 86–91
18. Kashyap N, Sinha GR (2012) Image watermarking using 2-level DWT. Adv Comput Res 4 (1):42–45
19. Kaushik AK (2012) A novel approach for digital watermarking of an image using DFT. Int J Electron Comput Sci Eng 1(1):35–41
20. Marjuni A, Fauzi MFA, Logeswaran R, Heng SH (2013) An improved DCT-based image watermarking scheme using fast Walsh Hadamard transform. Int J Comput Electr Eng 5(3):271
21. Dhar PK, Kim JM (2011) Digital watermarking scheme based on fast Fourier transformation for audio copyright protection. Int J Secur Appl 5(2):33–48
22. Dhar PK, Khan MI, Jong-Myon K (2010) A new audio watermarking system using discrete fourier transform for copyright protection. Int J Comput Sci Netw Secur 6:35–40
23. Lei B, Soon Y, Zhou F, Li Z, Lei H (2012) A robust audio watermarking scheme based on lifting wavelet transform and singular value decomposition. Sig Process 92(9):1985–2001
24. Arnold M (2000) Audio watermarking: features, applications, and algorithms. In: IEEE international conference on multimedia and expo (II), pp 1013–1016
25. Doerr G, Dugelay JL (2003) A guide tour of video watermarking. Signal Process Image Commun 18(4):263–282
26. Jaseena KU, John A (2011) Text watermarking using combined image and text for authentication and protection. Int J Comput Appl 20(4):8–13
27. Jaseena KU, John A (2011) An invisible zero watermarking algorithm using combined image and text for protecting text documents. Int J Comput Sci Eng 3(6):2265–2272
28. Jalil Z, Jaffar MA, Mirza AM (2011) A novel text watermarking algorithm using image watermark. Int J Innov Comput Inf Control 7(3)
29. Gupta S, Gujral G, Aggarwal N (2012) Enhanced least significant bit algorithm for image steganography. IJCEM Int J Comput Eng Manag 15(4):40–42
30. Banik B (2013) A DWT Method for image steganography. Int J Adv Res Comput Sci Softw Eng 3(6):983–989
31. Nag A, Biswas S, Sarkar D, Sarkar PP (2011) A novel technique for image steganography based on DWT and Huffman encoding. Int J Comput Sci Secur 4(6):497–610
32. Ghoshal N, Mandal JK (2012) Image authentication technique in frequency domain based on discrete Fourier transformation (IATFDDFT). arXiv:1212.3371
33. Bansal D, Chhikara R (2014) An improved DCT based steganography technique. Int J Comput Appl 102(14)
34. Kaur B, Kaur A, Singh J (2011) Steganographic approach for hiding image in DCT domain. Int J Adv Eng Technol 1(3):72
35. Patel BK, Pathak M (2014) Survey on cryptography algorithms. Int J Innov Technol 4(7)
36. Elminaam DSA, Kader HMA, Hadhoud MM (2008) Performanceevaluation of symmetric encryption algorithms. IJCSNS Int J Comput Sci Netw Secur 8(12):280286
37. Patel KD, Belani S (2011) Image encryption using different techniques: a review. Int J Emerg Technol Adv Eng 1(1):30–34
38. Mitali VK, Sharma A (2014) A survey on various cryptography techniques. Int J Emerg Trends Technol Comput Sci 3(4):6
39. Devi KJ (2013) A sesure image steganography using LSB technique and pseudo random encoding technique. Doctoral dissertation. National Institute of Technology-Rourkela
40. Kaur M, Mahajan K (2015) An existential review on text watermarking techniques. Int J Comput Appl 120(18)
41. Dey N, Roy AB, Dey S (2012) A novel approach of color image hiding using RGB color planes and DWT. arXiv:1208.0803

Intelligent Morphing and Steganography Techniques for Multimedia Security

Anant M. Bagade and Sanjay N. Talbar

Abstract Data security plays an important role in today's digital world. There is a potential need to do the research in the field of image morphing and steganography for data security. The development of morphing over the past years allows an organization into three categories of morphing algorithms namely geometric, interpolation and specialized algorithms depending upon the pixel mapping procedure. It gives an insight of how an appropriate morphing method is useful for different steganographic methods categorized into spatial domain, transform-based domain, spread spectrum, statistical and Internet Protocol. The geometric transformation morphing methods are more suitable in spatial domain steganography. This chapter includes the review of different morphing and steganography techniques. Hybrid approaches using morphing for steganography have a special status among steganographic systems as they combine both the features of morphing and steganography to overcome the shortcomings of individual methods.

Keywords Image morphing · Steganography · Image data security · Classification of morphing and steganography algorithms

1 Introduction

Image morphing has been proved to be a powerful visual effect tool. Many examples now appear in film and television for transformation of one digital image to another. Gradual inclusion of source image to destination image using 2-D

A.M. Bagade (✉)
Department of Information Technology, Pune Institute of Computer Technology,
Pune, Maharashtra, India
e-mail: ambagade@pict.edu

S.N. Talbar
Department of Electronics and Telecommunication,
SGGS Institute of Engineering and Technology, Nanded, Maharashtra, India
e-mail: sntalbar@sggs.ac.in

© Springer International Publishing Switzerland 2017
N. Dey and V. Santhi (eds.), *Intelligent Techniques in Signal Processing for Multimedia Security*, Studies in Computational Intelligence 660, DOI 10.1007/978-3-319-44790-2_3

geometric transformations is called morphing. Morphing among multiple images involves a series of transformations from one image to another, which is usually called metamorphosis [1]. Cross dissolve is one of the techniques of morphing. Image metamorphosis between two images begins with an animation establishing their correspondence with pairs of primitive features, e.g. mesh nodes, line segments, curves or points.

Morphing is a powerful technique that can enhance many multimedia projects, presentations, education and computer-based training. There are various techniques of morphing and they are dependent on how the correspondence between the two images is established. Existing morphing techniques use points and lines for correspondence in the source and destination images. Morphing algorithms use mesh warping, field morphing, radial basis functions, thin-plate splines, energy minimization and multi-level free form deformation techniques to generate the morph [2].

Today, image data security is a challenging problem. Steganography is a method of secret data hiding in a cover media. The least significant bit is a standard steganographic method, but has some limitations. The limitations are less data hiding capacity, stego-image quality and imperceptibility.

Data security plays an important role in today's digital world. There is a potential need to do the research in the field of image morphing and steganography for data security. The development of morphing over the past years allows an organization into three categories of morphing algorithms namely geometric, interpolation and specialized algorithms depending upon the pixel mapping procedure. It gives an insight of how an appropriate morphing method is useful for different steganographic methods categorized into spatial domain, transform-based domain, spread spectrum, statistical and Internet Protocol. The geometric transformation morphing methods are more suitable in spatial domain steganography.

This chapter includes the review of different morphing and steganography techniques. Hybrid approaches using morphing for steganography have a special status among steganographic systems as they combine both the features of morphing and steganography to overcome the shortcomings of individual methods [3]. Steganography is a method to hide the data inside a cover medium in such a way that the existence of any communication itself is undetectable as opposed to cryptography where the existence of secret communication is known, but is indecipherable. The word steganography originally came from a Greek word, which means 'concealed writing'. Steganography has an edge over cryptography because it does not attract any public attention, and the data may be encrypted before being embedded in the cover medium. Hence, it incorporates cryptography with an added benefit of undetectable communication.

Steganography can be used to exchange secret information in an undetectable way over a public communication channel. Image files are the most common cover medium used for steganography. In most cases, with resolution higher than human perception, data can be hidden in the 'noisy' bits or pixels of the image file. Because of the noise, a slight change in those bits is imperceptible to the human eye, although it might be detected using statistical methods (i.e. steganalysis). One of the most common and naive methods of embedding message bits is LSB replacement in spatial domain where the bits are encoded in the cover image by replacing the least significant bits of pixels [4].

2 Morphing Methods

2.1 Geometric Transformation Methods

Geometric transformation is performed by using motion vector calculus [5], linear and bilinear transformations, meshes, etc. The work on image geometric transformation started in early 1980s. Mesh warping was pioneered at Industrial Light and Magic (ILM) by Douglas Smythe [6, 7] for use in the movie 'Willow' in 1988. It has been successfully used in many subsequent motion pictures [8, 9]. Beier and Neely [8, 10] have developed a model for pixel transformation using field morphing. It transforms the pixels corresponding to the lines drawn on source and destination images. Thin-plate spline transforms the pixels corresponding to the surface positions. Affine, linear and bilinear transformations are used in morphing to generate the morphs [11]. The following section discusses the mesh morphing, field morphing, thin-plate spline and coordinate transformation techniques. These techniques are based on geometric transformation.

2.1.1 Mesh Morphing

Mesh morphing falls into a category of geometric correspondence problem. How the control points of one mesh should be mapped to another mesh is a research problem. Many approaches to the correspondence problem have been described, but there appears to be no general solution. Klein [12] has merged the various mesh connectivities under a projection. This works well for star-shaped, swept or revolutionary objects. There are some general problems arising while re-meshing process. These problems are validity, quality fidelity and correspondence. Lee et al. [13] have given the solution using multi-resolution mesh morphing and 2-pass mesh warping algorithm [7, 14]. The use of meshes for feature specification facilitates straightforward solution for warp generation. Fant's algorithm [15] was used to resample the image in a separable implementation. The advantage of this technique is that controls are given to the animator [9]. The limitation is that if the grid size increases, the computations also increase. Genetic algorithm and fuzzy logic will provide the better solution to solve the correspondence problems in mesh generation.

2.1.2   Field Morphing

Beier and Neely [8] have proposed a technique for morphing based upon fields of influence surrounding two-dimensional control primitives. It is based on the pixel distance with respect to the lines drawn on source and destination images. There are two ways to warp an image. The first is called 'forward mapping', which scans through the source image pixel by pixel and copies them to the appropriate place in the destination image. The second is 'reverse mapping', which goes through the destination image pixel by pixel and samples the correct pixel from the source image. The most important feature of inverse mapping is that every pixel in the destination image gets set to something appropriate. In the forward mapping case, some pixels in the destination might not get painted and have to be interpolated. The problem in the field morphing is that animator has given a control to mark the features on source and destination images. Hence, the accuracy of algorithm depends upon the animator. This problem can be solved by designing an algorithm for automatic feature detection. Thus, the human intervention error will be less, and hence, there will be increase in accuracy.

2.1.3   Thin-Plate Spline-based Image Morphing

Bookstelin [14] has invented the method of thin-plate spline in the year 1989 [16]. It is a conventional tool for surface interpolation over scattered data. It is an interpolation method that finds a minimally bended smooth surface that passes through all given points. The thin-plate spline warping allows an alignment of the points and the bending of the grid showing the deformation needed to bring the two sets on top of each other [17]. Thin-plate spline is applied to coordinate transformation using two splines, one for the displacement in the x direction and another for the displacement in the y direction. When there are fewer than 100 interpolation points and the curvature of underline function is important, then piecewise interpolation is not useful. It is essential to design a smooth interpolation algorithm that can reduce the complexity from $O(n^3)$ to $O(n^2)$. Further complexity can be reduced to $O(n \log n)$ using conjugate gradient method.

2.1.4   Coordinate Transformation Morphing

There are many coordinate transformations for the mapping between two triangles or between two quadrangles. It usually used affine and bilinear transformations for the triangles and quadrangles, respectively. Besides, bilinear interpolation is performed in pixel sense.

2.1.5 Affine Transformation and Triangular Warping

Suppose there are two triangles *ABC* and *DEF*. An affine transformation is a linear mapping from one triangle to another. For every pixel p within triangle *ABC*, it is assumed that the position of p is a linear combination of A, B and C vectors. The transformation is given by the following equations:

$$p = \lambda_1 A + \lambda_2 B + \lambda_3 C \tag{1}$$

$$\lambda_1 + \lambda_2 + \lambda_3 = 1 \quad \lambda_i \geq 0 \tag{2}$$

$$q = T(p) = \lambda_1 D + \lambda_2 E + \lambda_3 F \tag{3}$$

Here, λ_1 and λ_2 are unknown and there are two equations for each of the two dimensions. Consequently, λ_1 and λ_2 can be solved, and they are used to obtain q. The affine transformation is a one-to-one mapping between two triangles [11, 18].

2.1.6 Bilinear Transformation

Suppose there are two quadrangles *ABCD* and *EFGH* as shown in Fig. 1. The bilinear transformation is a mapping from one quadrangle to another. For every pixel p within quadrangle *ABCD*, it is assumed that the position of p is linear combination of vectors A, B, C and D. Bilinear transformation is given by the following equations:

$$p = (1 - u)(1 - v)A + u(1 - v)B + uvC + (1 - u)vD, \quad 0 \leq u, v \leq 1 \tag{4}$$

$$q = (1 - u)(1 - v)E + u(1 - v)F + uvG + (1 - u)vH \tag{5}$$

There are two unknown components u and v because this is a 2-D problem. Hence, u and v can be solved, and they are used to obtain q. Again, the bilinear transformation is a one-to-one mapping between two quadrangles [11].

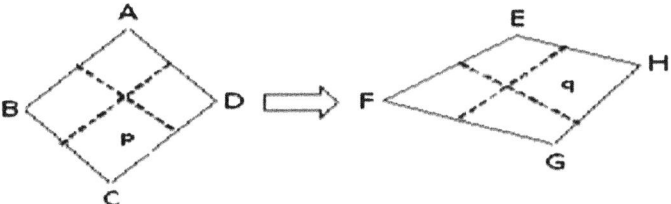

Fig. 1 Example of two quadrangles

2.1.7 Cross Dissolving

Cross dissolving is described by the following equation:

$$C(x, y) = \alpha A(x, y) + (1 - \alpha)B(x, y), \quad 0 \leq \alpha \leq 1 \tag{6}$$

where A and B form the pair of images, and C is the morphing result. This operation is performed pixel by pixel, and each of the colour components, i.e. RGB is dealt individually [11]. Cross dissolving is required to generate the accurate morph from warped images. Cross-dissolve map pixel positions pixel by pixel otherwise interpolation interpolates the part of the image corresponding to destination image results in faster processing.

2.2 Interpolation Methods

2.2.1 Morphing Using Fuzzy Vertex Interpolation

Sederberg and Greenwood [19] have identified a problem of correspondence between the vertices of two polygons. An interpolation should produce a smooth transformation at any intermediate polygon. The most widely used interpolation technique is linear interpolation. The limitation of applying this technique directly to the corresponding vertices of polygons under warping is that the resulting in battening is sometimes significantly distorted as compared with the original polygons. The approach based on physical blending alleviates this difficulty by imposing constraints on the angles of the polygons. This approach is further improved in the intrinsic shape-blending method by interpolating the lengths and orientations of the edges rather than the vertices of the polygon. The above two difficulties can be solved by using Shoemake and Duffs method [20], which produces an inbetweening by interpolating transformation matrices. It is difficult to use the transformation matrix approach to warp non-trivial polygons, since a transformation matrix common to all the vertex correspondence between the two polygons may not exist. The vertex corresponding problem can be solved by using fuzzy affine transformation [21, 22]. The fuzzy affine transformation will give better results than matrix vertex correspondence because of the correct vertex identification for interpolation.

2.2.2 A Level Set Approach for Image Morphing

The goal of level set approach is to produce a blend consisting of a sequence of images that incrementally move the contours or level sets of one image towards those of another. If we construct a blend that consists of incremental movements of contours, it will adhere to the Gestalt principles of proximity and similarity for establishing correspondences between successive images [23]. Thus, the blend will give the appearance of motion because pixels of a specific colour (or grey level) in one image are necessarily nearby in the next which is not a general property of a simple linear interpolation. Geometric active contour and gradient-based level set active contour methods are useful for level set approach in medical images. High computation time is required for the above methods. Computation time can be reduced by improvement in spatial and temporal resolution.

2.2.3 View Morphing

Suppose I_0 is a photograph of an object. Then, move the object in a direction parallel to the image plane of the camera, zoom out, and take a second picture I_1. Linear interpolation of corresponding pixels in parallel views with image planes I_0 and I_1 creates image $I_{0.5}$, representing another parallel view of the same scene. Chen and Williams [24, 25] previously considered this special case arguing that linear image interpolation should produce new perspective views when the camera moves parallel to the image plane. In addition to changes in view point, view morphing accommodates changes in projective shape. By using this, view morphing enables transition between images of different objects that gives strong sense of meta-morphosis in 3D. View morphing relies exclusively on image information.

2.3 Specialized Methods

2.3.1 Layered Morphing Method

Layered morphing consists of two kinds of processes. One is a process to separate a base image into layered images and the other is to perform the morphing of images layer by layer [26]. The first process is to separate a base image into layered images. Suppose there is a base image in which one object is in front of another object keeping a part of it out of sight. In this case, if you perform morphing without separating it into layers, the resulting image cannot present observers a part hidden by another object even when they change their position. This gives an observer a sense of incongruity. Endo et al. [27, 28] have proposed a solution to solve this problem. They have separated the objects into their own layers so that their appropriate parts are seen when an observer changes his/her point of view. The second process is to reconstruct any image seen from other viewpoints than those used to take the base images. The manual process of base image separation can be avoided using fusion and cross-dissolve techniques.

2.3.2 Star Skeleton Approach for Morphing

Star skeleton approach is based on a new polygon representation scheme, which is called as star skeleton. A star polygon is a polygon for which there exists a point, i.e. the star point from which all other points are visible. The star skeleton is appropriate for the shape-blending problem for two reasons [29]. First, it represents all points in the interior of the shape as well as on its boundary. Second, it defines an explicit dependence between interpolated polygon vertices relative to a common structure (the skeleton). Shapira and Rappoport [19] have preferred a star decomposition over a convex decomposition because it has fewer pieces. Star skeleton produces better results than skeletons using a minimal convex decomposition or a triangulation. Star skeleton is useful for multi-polygon shapes and image morphing. The complexity of the algorithm can be reduced using meta-skeleton.

2.3.3 Volumetric Morphing

One class of 3D metamorphosis algorithm deals with sampled or volumetric representation of the objects. Hughes [30] has proposed transforming objects into the Fourier domain and then interpolating low frequencies over time while slowly adding the high frequencies, thus minimizing the object distortion caused by the high-frequency components. Cheung et al. [31] have proposed a similar method for use in simulating a metal forming process. Chen et al. [32] have extended Beier and Neely's feature-based image metamorphosis method to volumetric representation. Two objects are translated into volumes, and each corresponding pair of volume values is interpolated according to the primitive shapes, defined by the user [33]. A wavelet-based approach is superior to Fourier transformation for further improvement in the performance [34, 35].

2.3.4 A Light-Field Morphing Using 2-D Features

The principal advantage of light-field morphing is the ability to morph between image-based objects whose geometry and surface properties may be difficult to model with traditional vision and graphics techniques. Light field is based on ray correspondence and not on surface correspondence. Light-field morphing is a flexible morphing scheme [36–39].

3 Classification of Steganographic Methods

Any steganographic system is expected to work with a high embedding capacity, imperceptibility and stego-image quality. To achieve all these requirements, we have used morphing for hiding image data. Morphing is a process of generating the intermediate images from the source image to the destination image. Source image gradually fades towards the destination image by generating intermediate frames. Intermediate images contain some part of the source and some part of the destination images. This feature helps to hide any source or destination images between the selected intermediate images by using morphed steganography.

Steganographic methods are broadly classified into four categories depending upon their method of data embedding. The first category is spatial domain; the secret data is hidden in cover media in appropriate pixels. The second category is transform domain; secret data is hidden using discrete cosine transform, discrete Fourier transform and discrete wavelet transform, etc. in cover media. Third category is spread-spectrum domain; secret data is hidden in broad spectrum. Fourth category is statistical domain; secret data is hidden in first- and higher-order statics.

3.1 Spatial Domain

Spatial domain embeds the secret data bitwise in cover media using different techniques. Chan and Cheng [40] have proposed the algorithm for enhancing the stego-image quality obtained by simple LSB method. This algorithm identifies the error rate between cover image and stego-image. The nature of error rate is identified and depending upon the error rate the value of error difference is segmented into three intervals. In first interval, error rate is positive to positive. In second interval, error rate is negative to positive, and in third interval, error rate is negative to negative. The new best location for the insertion of values is identified by observing the results of these three intervals. They have compared their results using WMSE and PSNR to state the stego-image quality. The results comparison is made with optimal pixel adjustment process (OPAP), least significant bit (LSB) and optimal least significant bit (OLSB) with respect to their PSNR. The result analysis shows that OPAP method gives better PSNR than LSB and OLSB. They have focused only on quality parameters and not on capacity of stego-data. Yu et al. [41] have presented two contributions with respect to steganography methods. The first contribution includes adaptive LSB steganography, which satisfies various requirements such as high embedding capacity and image security. The second contribution minimizes the degradation of the stego-image by finding the best mapping between secured message and covered message based on chaos genetic algorithm.

Chang et al. [42] have identified the problems in OPAP and PVD methods. Also, they have given the solution to overcome the limitations using tri-way pixel value differencing. The PVD algorithm is proposed using novel approach of tri-way pixel value differencing. PVD method uses only one direction to increase the hiding capacity. Tri-way differencing scheme uses three different directional edges. The proposed scheme uses 2×2 blocks that include four pixels of $P(x, y), p(x + 1, y), P(x, y + 1)$ and $P(x + 1, y + 1)$ where x and y are pixel locations in image. Pixel location is described by nine possible ways. The capacity of PVD method is compared with the proposed method. Security justification is stated by using RS stego-analysis [43].

Yadollahpour and Nair [44] have proposed a method to detect LSB stego-images by using 2-D autocorrelation coefficient. Support vector machine and weighted Euclidean distance are used to classify the stego-image. Correlation coefficient and its variance are used to distinguish between original image and stego-image. Classifier with fusion technique is used for better performance of steganography. Hsieh and Zhao [45] have invented a method for image enhancement and image hiding based on linear fusion. Morphing is a part of linear image fusion. The author has proposed a method using morphing to hide a source or destination images by using linear fusion. Morphed steganography is performed without using any cover media. Source secret image is hidden inside the intermediate image. Linear fusion with steganography could be a better steganographic combination, and a new superior method can be derived for steganography [46, 47]. Wang and Moulin [48] have proposed two methods for perfectly secure steganography. Positive capacity and random coding exponents are used for perfectly secure steganographic systems. The algorithm is robust enough that neither a warden nor unlimited computational resource can detect the presence of a hidden message. Randomized codes are required to achieve perfectly secure steganographic system.

Zhang [49] has proposed a novel steganographic scheme with the allowable modifications $\{-2, -1, 0, +1, +2\}$. The secret data is hidden into the first layer, second layer and third layer of LSB of cover image in a layer by layer manner. Generic model has been designed to hide the data within different parameters. Liao et al. [50] have investigated a steganographic method to improve the embedding capacity and provide imperceptible visual quality. A novel method based on four pixel differencing and modified LSB substitution is presented. This method clarifies whether a block is smooth area or an edge area. The pixels in the edge area can tolerate much more changes without making perceptible distortion than smooth areas. The proposed method increases the hiding capacity with sacrificing the attacks resistance a little for obtaining higher embedding capacity. El-Emam [51] has proposed a new steganographic algorithm with high security. Three layer securities are used to secure data by observing the content in which it was trans-ferred. The proposed algorithm is free from visual and statistical attacks. Wang et al. [52] have presented the algorithm based on the idea that smoother area has less data hiding capacity, and on the other hand, more an edge area more secret data can be embedded. To support this, reminder of the two consecutive pixels is cal-culated by using modulus operation. The proposed method solves the falling-off-boundary problem by readjusting the reminder of the two pixels saying secure against the RS detection attack. Cryptography along with steganography is a better security solution. Narayana and Prasad [53] have introduced two methods that combine the cryptography and steganography. Original text is converted into cipher text by S-DES algorithm using secret key. The encountered data is embedded into another image. Steganography when combined with encryption provides a secure means of secret communication between two parties.

Regalia [54] has designed a matrix embedding capacity under distortion constraints as opposed to any cryptographic measure. Kharrazi et al. [55] have presented how to use information fusion methods to aggregate the outputs of multiple steganalysis techniques. The author frames different rules for steganalysis. The fusion technique increases detection accuracy and other scalability pre- and post-classification. Manjunatha Reddy [56] has presented a novel image steganography method that utilizes a two-way block matching procedure to search for the highest similarity blocks of important image part. Careful design of the two-way block matching procedure and the hop embedding scheme has resulted in both the stego-image and extracted secured image of high quality. F5 [57] is one of the most popular algorithm and is undetectable using the chi-square technique. F5 uses matrix encoding along with permutated straddling to encode message bits. Permutated straddling helps to distribute the changes evenly throughout the stego-image. Another popular algorithm is Hetzl and Mutzel [58], where the authors claim to use exchanging coefficients rather than overwriting them. They use the graph theory techniques where two interchangeable coefficients are connected by an edge in the graph with coefficients as vertices of the graph. The embedding is done by solving the combinatorial problem of maximum cardinality matching. Spatial domain techniques are more suited for morphed steganography purpose. Morphed steganography achieves all steganography properties with high security. Long length stego-keys are generated during the morphing process, and these keys are useful for secure transmission of secret data over covert channel.

3.2 *Transform Domain*

Transform domain method uses discrete cosine transform (DCT), discrete Fourier transform (DFT) and discrete wavelet transform (DWT) for secret data hiding [59–61]. Al-Ataby and Al-Naima [62] have proposed a modified high-capacity image steganography technique that depends on wavelet transform with acceptable level of imperceptibility and distortion. One-dimensional and multi-level 2-D wavelet decomposition techniques are used for steganography. The proposed method allows high payload in the cover image with very little effect on the statistical nature. JSteg [63] was one of the first JPEG steganography algorithm developed by Derek Upham. JSteg embeds message bits in LSB of the JPEG coefficients. JSteg does not embed any message in DCT coefficients with value 0 and 1. This is to avoid changing too many zeros to 1s since number of zeros is extremely high as compared to number of 1s. The data hidden by using this method is easily detected by chi-square method.

Provos and Honeyman [64] is another JPEG steganography programme, which improves stealth by using the Blowfish encryption algorithm to randomize the index for storing the message bits. This ensures that the changes are not concentrated in any particular portion of the image, a deficiency that made JSteg more easily detectable. Similar to the JSteg algorithm, it also hides data by replacing the

LSB of the DCT coefficients. The only difference is that it also uses all coefficients including the ones with value 0 and 1. The maximum capacity of JP Hide and Seek to minimize visual and statistical changes is around 10 %. Hiding more capacity can lead to visual changes to the image, which can be detected by the human eye. Patchwork algorithm developed at MIT, which uses seven random pairs of pixels, increases the brightness of the brighter pixels and decreases the brightness of the others. DCT is a popular method in frequency domain. DCT and DFT are used for JPEG steganography.

3.3   Spread-Spectrum Domain

In spread-spectrum steganography, the message is embedded in noise and then combined with the cover image to produce the stego-image. The power of embedded signal is much lower than the power of the cover image. Marvel et al. [65] have defined spread-spectrum communication as the process of spreading the bandwidth of narrow band signal across a wide band of frequencies. In spread-spectrum steganography, cover media is considered as a communication channel and secret message is a signal to be transmitted through it [66, 67].

3.4   Statistical Domain

Statistical steganalysis, analyses underlying statistics of an image to detect the secret embedded information. Statistical steganalysis is considered powerful than signature steganalysis because the mathematical techniques are more sensitive than visual perception. These types of techniques are developed by analysing the embedding operation and determining certain image statistics that get modified as a result of the embedding process. The design of such techniques needs a detailed knowledge of embedding process. These techniques totally depend upon the mathematical modelling for embedding and extraction of secret data. These techniques yield very accurate results when used against a target steganography technique and watermarking [54, 68, 69].

OutGuess, proposed by Provos [70], was one of the first algorithms to use first-order statistical restoration methods to counter chi-square attacks. The algorithm works in two phases, the embedding phase and the restoration phase. After the embedding phase, using a random walk, the algorithm makes corrections to the unvisited coefficients to match it to the cover histogram. Model-based steganography (MBS1), proposed by Sallee and Pal et al. [71, 72], claims to achieve high embedding efficiency with resistance to first-order statistical attacks. While OutGuess preserves the first-order statistics by reserving around 50 % of the coefficients to restore the histogram, MB1 tries to preserve the model of some of the statistical properties of the image during the embedding process. Statistical restoration refers to the class of embedding data such that the first- and higher-order statistics are preserved after the embedding process. As mentioned earlier, embedding data in a JPEG image can lead to change in the typical statistics of the image, which in turn, can be detected by steganalysis. Most of the steganalysis methods existing today employ first- and second-order statistical properties of the image to detect any anomaly in the stego-image. Statistical restoration is done to restore the statistics of the image as close as possible to the given cover image [73].

3.5 TCP/IP Steganography

The need of the security is to protect the sensitive data from unauthorized access. The Internet Protocol (IP) does not design with security in mind [74]. Within IP, there are many fields that are not being used for normal transmission. Some bits are reserved or option can be set by the user as per their requirement [75]. Harshawardhan and Anant [4, 76] have proposed a mechanism for sending data through 16-bit identification field of the header through chaotic mixing, and the generation of sequence number is used to hide the data and convey the information to the recipient. Ahasan and Kundur [77] have presented a technique by using covert channel communication using IP steganography [78]. Author has given the mechanism for transmission of data through IP header identification and flag fields. They have used a chaotic mixing algorithm and sequence number method to get the original message. The results are shown by impossible sequences, evident sequences and best estimation sequences. Error rate is calculated to measure the performance along with network behaviour. The difference between the secret sequence and the original sequence carries the covert information. Panajotov and Aleksanda [79] have given a steganography in Internet layer. Steganography with IPV4 and IPV6 were presented based on maximum transmission unit. Many covert channels are available on transport layer and application layer. Many covert channels are discovered and many more need to be discovered. This is a continuous race between the hacker and security experts. Bellovin [80] has explained security problems in TCP/IP protocol suite and discussed many protocol-level issues.

4 Comparative Discussions

This chapter introduced morphed steganographic algorithms that are capable of increasing hiding capacity, image quality and imperceptibility. Hiding capacity increases due to the consideration of only pixel difference between selected morphed image and destination image. There is a gain of PSNR with standard algorithms as compared to proposed algorithm. The high value of PSNR results into good quality of stego-image and proves the imperceptibility property of steganography. Our proposed algorithm's PSNR is increased by 15.72 % than existing mentioned algorithms. Hiding capacity is also increased as compared to existing stated algorithms. Owing its ability to increase in hiding capacity and imperceptibility, it is applicable to any image data security applications. Also proposed work introduces a new dimension to network security by investigating the TCP/IP protocol suite for hiding morphed stego-keys using the existence of hidden communication channels. These covert channels find interesting applications in network security and in facilitating various network processes, which are in line with modern concepts. The covert channel exploration in TCP/IP suite, therefore, has much potential in a network environment. Morphed stego-keys play an important role in image security over the Internet. Every time different stego-keys are generated during morphing process and it is difficult to attack on these keys. Covert channel and morphed stego-keys provide a more secure environment. The stated algorithm provides the best security mechanism to MITM attack.

5   Conclusions

There has been an increase of research in the field of morphing and steganography since 1990s. Different strategies using combination of multiple features, different classification methods and morphing using snakes have been considered extensively. Only a few works have been reported in the area of video and voice morphing. Morphing was limited only to the animation and film industry. There is a lot of scope of morphing in the field of image hiding and image data security. The interdisciplinary nature of morphing will open many thrusting areas of research for the researcher. There is a tremendous potential of research in the field of morphing with respect to complexity and correctness of morph formation. Lot of work is not reported till with respect to morphing among multiple images and video sequences. Research is required in the field of applicability of morphing concept in steganography. Cryptography along with steganography would be a better security mechanism. Morphed steganography involves encrypted stego-keys and is useful for security purpose. The applicability of morphing to steganography will open the new doors for the researchers. Few papers have been published in the application area of morphing with respect to steganography. Researchers can focus on the area of image security and steganography using morphing. The research is useful for defence security applications, biometric security and satellite image security. This chapter presents comparative analysis of morphing and steganographic techniques. From survey, it has been observed that there is no method, which you can call efficient. Each method has its own merits and demerits. The nature of acceptability depends upon the application where to use the concept.

References

1. Cohen-Or D, Solomovici A, Levin D (1998) Three-dimensional distance field metamorphosis. ACM Trans Graph 17(2):116–141
2. Arad N, Dyn N, Reisfeld D, Yeshurun Y (1994) Image warping by radial basis functions: applications to facial expressions. J Graph Models Image Process 56(2):161–172
3. Bagade AM, Talbar SN (2014) A review of image morphing techniques. Elixir Electr Eng J 70 (2):24076–24079
4. Kayarkar H, Sugata S (2012) A survey on various data hiding techniques and their comparative analysis. ACTA Techn Corviniensis 5(3):35–40
5. Acharjee S, Chakraborty S, Samanta S, Azar AT, Hassanien AE, Dey N (2014) Highly secured multilayered motion vector watermarking. In: Advanced machine learning technologies and applications. Springer, pp 121–134
6. Fant KM (1986) A nonaliasing real-time spatial transform technique. IEEE Comput Graph Appl 6(1):71–80
7. Wolberg G (1990) Digital image warping. IEEE Computer Society, Los Alamitos, pp 222–240
8. Beier T, Neely S (1992) Feature-based image metamorphosis. Comput Graph 26(2):35–42
9. Lee S, Wolberg G, Chwa K-Y, Shin SY (1996) Image metamorphosis with scattered feature constraints. IEEE Trans Vis Comput Graph 2(4):337–354
10. Beier T, Costa B, Darsa L, Velho L (1977) Warping and morphing of graphics objects. Course notes siggraph
11. Vlad A (2010) Image morphing techniques. JIDEG 5(1):25–28
12. Klein R (1998) Multiresolution representation for surfaces meshes based on vertex decimation method. Comput Graph 22(1):13–26
13. Lee AWF, Dobkin D, Sweldens W, Schröder P (1999) Multiresolution mesh morphing. Comput Graph Interact Techn 99:343–350
14. Bookstein FL (1989) Principal warps: thin-plate splines and the decomposition of deformations. IEEE Trans Pattern Anal Mach Intell 2(6):567–585
15. Alexa M (2003) Differential coordinates for local mesh morphing and deformation. Vis Comp 105–114
16. Lee W-S, Thalmann NM (1998) Head modeling from pictures and morphing in 3D with image metamorphosis based on triangulation modeling and motion capture techniques for virtual environments. In: Lecture notes in computer science, pp 254–267
17. Bremermann H (1976) Pattern recognition by deformable prototypes in structural stability, the theory of catastrophes and applications in the sciences. In: Springer notes in mathematics, vol 25. Springer, pp 15–57
18. Bagade AM, Talbar SN (2010) Image morphing concept for secure transfer of image data contents over internet. J Comput Sci 6(9):987–992
19. Sederberg TW, Greenwood E (1992) A physically based approach to 2D shape blending. Comput Graph 26(2):25–34
20. Shoemake K, Duff T (1992) Matrix animation and polar decomposition. In: Proceedings of graphics interface, pp 258–264
21. Sederberg TW, Gao P, Wang G, Mu1 H (1993) 2D shape blending: an intrinsic solution to the vertex path problem. In: ACM computer graphics (proceedings of SIGGRAPH'93), pp 15–18
22. Zhang Y (1996) A fuzzy approach to digital image warping. IEEE Comput Graph Appl 16 (4):34–41
23. Palmer SD (1999) Vision science-photons to phenomenology. MIT Press, Cambridge, pp 171–185
24. Chen SE, Williams L (1993) View interpolation for image synthesis. In: (Proceedings of SIGGRAPH'93) computer graphics and interactive techniques, pp 279–288
25. Seitz SM, Dyer CR (1996) View morphing: synthesizing 3D metamorphoses using image transforms. In: (Proceedings of SIGGRAPH'96), pp 21–30

26. Shum H-Y, He L-W (1999) Rendering with concentric mosaics. In: (Proceedings of SIGGRAPH'99), pp 299–306
27. Endo T, Katayama A, Tamura H, Hirose M, Tanikawa T, Saito M (1998) Image-based walk-through system for large-scale scenes. In: Proceedings of VSMM'98, pp 269–274
28. Gong M, Yang Y-H (2001) Layered based morphing. Elsevier J Graph Models 63(1):45–59
29. Whitaker RT (2000) A level-set approach to image blending. IEEE Trans Image Process 9 (11):1849–1861
30. Hughes JF (1992) Scheduled Fourier volume morphing. In: ACM Proceedings of SIGGRAPH'92, pp 43–46
31. Cheung KK, Yu K, Kui K (1997) Volume invariant metamorphosis for solid and hollow rolled shape. Proc Shape Model 226–232
32. Chen L-L, Wang GF, Hsiao K-A, Liang J (2003) Affective product shapes through image morphing. In: Proceedings of DPPI'03, pp 11–16
33. Kanai T, Suzuki H, Kimura F (2000) Metamorphosis of arbitrary triangular meshes. IEEE Comput Graph Appl 20(2):62–75
34. Dey N, Roy AB, Das A, Chaudhuri SS (2012) Stationary wavelet transformation based self-recovery of blind-watermark from electrocardiogram signal in wireless telecardiology. Recent Trends Comput Netw Distrib Syst Secur. Springer, Berlin, pp 347–357
35. Dey N, Dey M, Mahata SK, Das A, Chaudhuri SS (2015) Tamper detection of electrocardiographic signal using watermarked bio-hash code in wireless cardiology. Int J Signal Imaging Syst Eng 8(1–2):46–58
36. Gortler SJ, Grzeszczuk R, Szeliski R, Cohen MF (1996) The lumigraph. In: Proceedings of SIGGRAPH'96, pp 43–54
37. Levoy M, Hanrahan P (1996) Light field rendering. In: Proceedings of SIGGRAPH'96, pp 31–42
38. Wang L, Lin S, Lee S, Guo B, Shum H-Y (2005) Light field morphing using 2D features. IEEE Trans Vis Comput Graph 11(1)
39. Zhang Z, Wang L, Guo B, Shum H-Y (2002) Feature-based light field morphing. ACM Trans Graph 21(3):457–464
40. Chan C-K, Cheng LM (2004) Hiding data in images by simple LSB substitution. Elsevier J Pattern Recogn 37(3):469–474
41. Yu L, Zhao Y, Ni R, Li T (2010) Improved adaptive LSB steganography based on chaos and genetic algorithm. EURASIP J Adv Signal Process 2010:1–6
42. Chang K-C, Huang PS, Tu T-M, Chang C-P (2008) A novel image steganographic method using tri-way pixel-value differencing. J Multimedia 3(2):37–44
43. Liao X, Wen Q, Zhang J (2011) A steganographic method for digital images with four-pixel differencing and modified LSB substitution. Elsevier J Vis Commun Image Represent 22:1–8
44. Yadollahpour A, Naimi HM (2009) Attack on LSB steganography in color and grayscale images using autocorrelation coefficients. Eur J Sci Res 31(2):172–183
45. Hsieh C-H, Zhao Q (2006) Image enhancement and image hiding based on linear image fusion. In: Ubiquitous intelligence and computing, Lecture notes in computer science, vol 4159, pp 806–815
46. Bagade AM, Talbar SN (2014) A high quality steganographic method using morphing. KIPS J Inf Process Syst 10(2):256–270
47. Yang C-H, Wang S-J (2006) Weighted bipartite graph for locating optimal LSB substitution for secret embedding. J Discrete Math Sci Cryptogr 9(1):152–154
48. Wang Y, Moulin P (2008) Perfectly secure steganography: capacity, error exponents, and code constructions. IEEE Trans Inf Theory 54(6):2706–2722
49. Zhang X (2010) Efficient data hiding with plus–minus one or two. IEEE Signal Process Lett 17(7):635–638
50. Liao X, Wen Q, Zhang J (2011) A steganographic method for digital images with four-pixel differencing and modified LSB substitution. Elsevier J Vis Commun Image Represent 22 (1):1–8
51. El-Emam NN (2007) Hiding a large amount of data with high security using steganography algorithm. J Comput Sci 3(4):223–232

52. Wang C-M, Wu N-I, Tsai C-S, Hwang M-S (2007) A high quality steganographic method with pixel-value differencing and modulus function. J Syst Softw 1–8
53. Narayana S, Prasad G (2010) Two new approaches for secured image steganography using cryptographic techniques and type conversions. Signal Image Process Int J 1(2):60–73
54. Regalia PA (2008) Cryptographic secrecy of steganographic matrix embedding. IEEE Trans Inf Forensics Secur 3(4):786–791
55. Kharrazi M, Sencar HT, Memon N (2006) Improving steganalysis by fusion techniques: a case study with image steganography. Springer transactions on data hiding and multimedia security I, pp 123–137
56. Manjunatha Reddy HS, Raja KB (2010) High capacity and security steganography using discrete wavelet transform. Int J Comput Sci Secur 3(6):462–472
57. Westfeld A (2001) F5—a steganographic algorithm. In: Proceedings of the 4th international workshop on information hiding, pp 289–302
58. Hetzl S, Mutzel P (2005) A graph-theoretic approach to steganography. In: Lecture notes in computer science, pp 119–128
59. Dey N, Das P, Roy AB, Das A, Chaudhuri SS (2012) DWT-DCT-SVD based intravascular ultrasound video watermarking. In: 2012 World congress information and communication technologies (WICT), pp 224–229
60. Dey N, Mukhopadhyay S, Das A, Chaudhuri SS (2012) Analysis of P-QRS-T components modified by blind watermarking technique within the electrocardiogram signal for authentication in wireless telecardiology using DWT. Int J Image Graph Signal Process 4(7)
61. Dey N, Maji P, Das P, Biswas S, Das A, Chaudhuri SS (2013) Embedding of blink frequency in electrooculography signal using difference expansion based reversible watermarking technique. arXiv preprint arXiv:1304.2310
62. Al-Ataby A, Al-Naima F (2010) A modified high capacity image steganography technique based on wavelet transform. Int Arab J Inf Technol 7(4):358–364
63. Upham D (1999) JSteg steganographic algorithm. Available on the internet ftp://ftp.funet.fi/pub/crypt/steganography
64. Provos N, Honeyman P (2003) Hide and Seek: an introduction to steganography. In: IEEE Computer Society, pp 32–44
65. Marvel LM, Boncelet CG, Retter CT (1999) Spread spectrum image steganography. IEEE Trans Image Process 8(8):1075–1083
66. Dey N, Mishra G, Nandi B, Pal M, Das A, Chaudhuri SS (2012) Wavelet based watermarked normal and abnormal heart sound identification using spectrogram analysis. In: IEEE international conference on computational intelligence and computing research (ICCIC), pp 1–7
67. Dey N, Acharjee S, Biswas D, Das A, Chaudhuri SS (2013) Medical information embedding in compressed watermarked intravascular ultrasound video. arXiv preprint arXiv:1303.2211
68. Banerjee S, Chakraborty S, Dey N, Kumar Pal A, Ray R (2015) High payload watermarking using residue number system. Int J Image Graph Signal Process 7(3):1–8
69. Dey N, Samanta S, Yang XS, Das A, Chaudhuri SS (2013) Optimisation of scaling factors in electrocardiogram signal watermarking using cuckoo search. Int J Bio-Inspired Comput 5 (5):315–332
70. Provos N (2001) Defending against statistical steganalysis. In: Proceedings of the 10th conference on USENIX security symposium, vol 10, pp 24–24
71. Pal AK, Das P, Dey N (2013) Odd–even embedding scheme based modified reversible watermarking technique using Blueprint. arXiv preprint arXiv:1303.5972
72. Sallee P (2004) Model-based steganography. Springer, IWDW, digital watermarking, pp 154–167
73. Solanki K, Sullivan K, Madhow U, Manjunath BS, Chandrasekaran S (2005) Statistical restoration for robust and secure steganography. In: IEEE international conference on image processing, vol 2, pp II-1118–II-1121
74. Keanini T (2005) Protecting TCP IP. Elsevier J Netw Secur 2005:13–16

75. Murdoch SJ, Lewis S (2005) Embedding covert channel in TCP/IP. In: Lecture notes in computer science, vol 3727, pp 247–261
76. Bagade AM, Talba SN (2014) Secure transmission of morphed stego keys over Internet using IP steganography. Int J Inf Comput Secur 6(2):133–142
77. Ahsan K, Kundur D (2002) Covert channel analysis and data hiding in TCP/IP. M.A.Sc. thesis, Deptartment of Electrical and Computer Engineering, University of Toronto
78. Ahsan K, Kundur D (2006) Practical data hiding in TCP/IP. In: Proceedings of workshop on multimedia security at ACM multimedia
79. Panajotov B, Aleksanda M (2013) Covert channel in TCP/IP protocol stack. In: ICT innovations web proceedings, pp 190–199
80. Bellovin SM (2004) A look back at security problems in TCP/IP protocol suite. In: Proceedings of the 20th annual computer security applications, pp 229–249

Information Hiding: Steganography

V. Thanikaiselvan, Shounak Shastri and Shaik Ahmad

Abstract The sharing of digital media over the Internet has given rise to many issues concerning security and privacy of data. While encryption is used to secure the data, the appearance of encrypted images and video streams makes them vulnerable to attacks. This situation can be disastrous in cases where security and privacy are of utmost importance, such as in military, medical, or covert applications. This drawback of cryptography is one of the many factors which encourage research in the field of information hiding. Information hiding aims to protect the integrity of the data during transmission or while in storage. Steganography is a branch of information hiding which conceals the existence of the secret data hidden in a cover medium. This chapter discusses the idea behind information hiding with respect to reversible and non-reversible steganography. It starts with a general introduction about steganography and its classification into spatial and transform domain techniques. This chapter then moves on to give an explanation of the various spatial and transform domain techniques in use today. It then suggests the scope for future developments followed by the conclusion.

Keywords Steganography · Reversible · Non-reversible · Spatial domain · Transform domain · Steganalysis

V. Thanikaiselvan (✉) · S. Shastri · S. Ahmad
School of Electronics Engineering, VIT University, Vellore
Tamil Nadu 632014, India
e-mail: thanikaiselvan@vit.ac.in

S. Shastri
e-mail: shounak.mangesh@vit.ac.in

S. Ahmad
e-mail: shaik.ahmad@vit.ac.in

© Springer International Publishing Switzerland 2017
N. Dey and V. Santhi (eds.), *Intelligent Techniques in Signal Processing for Multimedia Security*, Studies in Computational Intelligence 660,
DOI 10.1007/978-3-319-44790-2_4

1 Introduction

Today, communication plays a vital role in our everyday lives because of the advances in Internet technologies. Digitization of media has made it simpler to duplicate and distribute data [1]. Digital information regarding business transactions, research, etc., is transferred overseas on a regular basis. Release of private information such as passwords and keys, staff information, and medical records, whether deliberate or accidental, can lead to huge losses. This raises concerns regarding security and privacy of the data. The effortlessness in making unauthorized copies on a large scale can result in huge losses to the companies which make music, movies, books, or software. Information security thus has a crucial role in maintaining the integrity of the information.

The most popular methods to provide security are cryptography and data hiding [2]. The classification is shown in Fig. 1. Cryptography transforms the data to an unintelligible form which cannot be reverted to the original form without a proper key. Information hiding conceals the existence of the data itself by embedding the data in some other files, usually images. Data hiding can further be classified into watermarking and steganography. Watermarking embeds a special pattern as watermark in the cover file. The primary objective of watermarking is to protect the copyright ownership of electronic data and guard it against unauthorized access. Sometimes, the watermarks are spread over the complete file and may appear as a segment of the file. These changes may not be visible, and an average person might not be able to detect the changes. Watermarking can be performed in both spatial [3–7] and transform domains [8–11], whereas in steganography, the primary objective is to provide security in the transfer of huge data.

Steganography can be applied in a variety of real-world applications [12], some good and some not so good. Steganography can primarily be used to carry out secret communication while maintaining confidentiality in order to avoid theft or unauthorized access of data. In most of the cases, images can be used as cover files to carry secret information because of their inherent redundancies and the inability

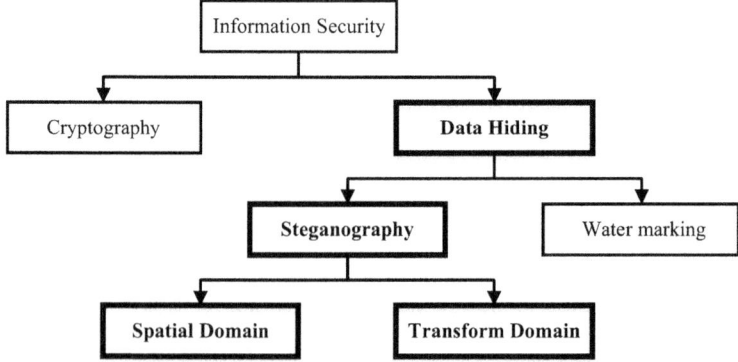

Fig. 1 Classification of information security techniques

of the human eye to detect small changes in the pixel intensity values. Besides secret communication, steganography can be used in personnel identification cards to store information about the individual, to store network information in TCP/IP packets, to embed patient information in medical images, etc.

Sadly, because of the nature of privacy it provides, steganography can also be used for immoral activities. For example, someone can leak information to an unauthorized party by hiding it in some innocent-looking document or file or it can be used for correspondence between terrorist organizations. So, steganography can be a useful tool if used for the right purposes.

This chapter deals with steganography and the various methods employed to hide secret data in images. Steganography is introduced first by giving a historical perspective and its applications in today's world. Then, this chapter moves on to explain about two basic forms of steganography which are reversible and irreversible. A brief explanation is then given about some popular spatial and transform domain steganography techniques. Finally, this chapter offers some explanation about the popular steganalytic techniques followed by conclusions with recommendations for future research.

1.1 History

The Greeks were one of the first ones to make use of steganography as a technique to secure the exchange of information. Ancient Greek emissaries tattooed messages on their shaved heads so that the message would be unnoticeable when the hair grew back [13]. One of the first usages of steganography can be found in *The Histories of Herodotus*. It indicates that a spartan named Demaratus scraped off the wax from one of the tablets to write a warning about an oncoming Persian invasion. He then again covered the tablet with wax to hide the message.

Information security techniques witnessed great progress during the World War II [14, 15]. Secrecy was given utmost importance. This resulted in innovations such as the Enigma machine, microdots [16], various null ciphers, and invisible inks made from milk, vinegar, and fruit juices. One example of null ciphers is 'Fishing freshwater bends and saltwater coasts rewards anyone feeling stressed. Resourceful anglers usually find masterful leapers fun and admit swordfish rank overwhelming anyday.' This can be decoded by taking the third letter from each word. The hidden message reads 'Send lawyers guns and money.' Such examples are key elements inspiring the research in this field.

2 Basic Procedure of Steganography

The basic procedure of steganography consists of two stages, namely embedding and extraction [17]. In the embedding stage, an appropriate embedding algorithm E $(\cdot)$ is chosen and the secret data 'd' is hidden in the cover image 'I_c'. Many times, a

Fig. 2 Basic steganography procedure

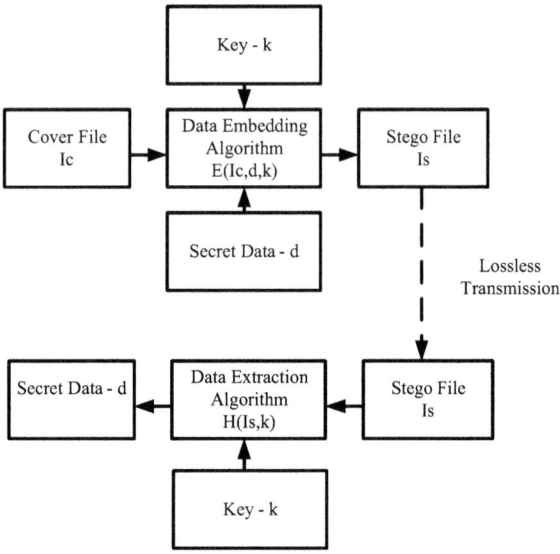

key 'k' needs to be used. The key contains the auxiliary information such as positions of embedded bits, thresholds, and user-defined constants and serves as a password for the extraction of the hidden secret data. Thus, the embedding algorithm can be modeled as a function of the cover image I_c, the secret data d, and the key k as shown in Eq. 1. This stage gives a stego image 'I_s' which is then transmitted over the communication channel.

$$I_s = E(I_c, d, k) \tag{1}$$

On reception of the stego image, the extraction algorithm $H(\cdot)$ is used to extract the hidden secret data. Similar to the embedding algorithm, the extraction algorithm can be modeled as a function of the stego image I_s and the key k as shown in Eq. 2. The whole procedure is summarized in Fig. 2.

$$d = H(I_s, k) = H(E(I_c, d, k), k) \tag{2}$$

2.1 Evaluation Metrics for Steganography

Steganography schemes can be evaluated by the following metrics:

- *Embedding Capacity* The size of the data that can be hidden in an image without a substantial degradation in the image quality is the embedding capacity. It is usually expressed in kilobits (kb) or kilobytes (kB) or bits per pixel (bpp).

- *Mean Square Error (MSE)* The deviation of the stego image from the original image is given by the mean square error. A higher MSE implies more distortions in the stego image. This increases the probability of detection of the hidden data. The MSE can be calculated as

$$\text{MSE} = \frac{1}{MN} \sum_{i=1}^{M} \sum_{j=1}^{N} (I_c(i,j) - I_s(i,j))^2 \tag{3}$$

where M and N are the dimensions of the cover image I_c and I_s represents the marked image.

- *Peak Signal-to-Noise Ratio (PSNR)* PSNR is a measure of the visual quality of a stego image. It is measured in decibels (dB) and is one of the most popular techniques to see how much the stego image resembles the original image. The PSNR is given as

$$\text{PSNR} = 10 \log_{10} \left(\frac{(\max(I_c(i,j)))^2}{MSE} \right) \tag{4}$$

$$\text{PSNR} = 10 \log_{10} \frac{255 \times 255 \times M \times N}{\sum_{i=1}^{M} \sum_{j=1}^{N} (I_c(i,j) - I_s(i,j))^2} \tag{5}$$

- *Structural Similarity (SSIM) Index* SSIM is another metric to measure the visual quality of stego images. Though it is not as popular as the PSNR, its usage is increasing among the research community as it is better in quantifying the subjective quality of the stego image. The SSIM can be calculated as

$$\text{SSIM}(x, y) = \frac{\left(2\mu_x\mu_y + c_1\right)\left(2\sigma_{xy} + c_2\right)}{\left(\mu_x^2 + \mu_y^2 + c_1\right)\left(\sigma_x^2 + \sigma_y^2 + c_2\right)} \tag{6}$$

where x and y are the window parameters, μ_x and μ_y are the average values of x and y, σ_x^2, σ_y^2 and σ_{xy} are the variances between x and y, and c_1 and c_2 are weights used for stabilization of the denominator.

2.2 Reversible and Non-reversible Steganography

Steganography can be reversible or non-reversible [18]. In the reversible case, the cover image can be recovered completely without any losses and it matches the original image, whereas in the non-reversible case, the cover image is destroyed after the extraction of the hidden data. Reversible steganography can be useful when the cover and the secret data are important. Consider a scenario in which a sender wants to send some files, but the combined size of all the data is too large for that channel. In such a case, the sender can embed some of the smaller files in a larger file using a reversible steganography algorithm so that the overall size of the data would be reduced. At the receiver end, on applying the extraction algorithm, the receiver will be able to recover the data along with the cover file without any losses. This can be useful in case of medical applications where the patient data, billing information, doctor's name, etc., can be embedded in a medical image like a CT scan or an MRI as watermarks [19–23]. Non-reversible steganography can be used when the lossless recovery of the cover is not a priority. Thus, it can be said that the expression given in Eq. 2 represents the extraction of non-reversible steganography. So, with a slight modification to the Eq. 7, reversible steganography can be represented as

$$[I_c, d] = H_{rev}(I_s, k) = H_{rev}(E_{rev}(I_c, d, k), k) \tag{7}$$

where $E_{rev}(\cdot)$ and $H_{rev}(\cdot)$ are the reversible embedding and extraction algorithms, respectively.

2.3 Spatial and Transform Domain Steganography

Steganography can further be classified on the basis of the domain in which the secret data is embedded. In spatial domain steganography, the data is embedded by directly modifying the pixel values of the cover image, while in transform domain steganography, the image is first converted into a transform domain image and the data is hidden in the transform coefficients. The following sections give an overview of some of the recent spatial and transform domain steganography techniques.

3 Spatial Domain Steganography

3.1 Least Significant Bit (LSB) Modification

LSB modification is the most basic form of steganography. In its simplest form, the LSBs of the cover image pixels are replaced by the secret data. This takes the advantage of the fact that the human visual system cannot detect small changes in gray-level intensities. One of the earliest schemes proposed [24] used LSB substitution, to embed the secret data. This scheme is capable of embedding data in multiple bit planes if the payload size is more. First, the data d is divided into k groups of n bits where n is the number of bit planes to be used for embedding. The stego image can be constructed as

$$I_s(i,j) = I_c(i,j) - I_c(i,j)\,2^n + d_k \tag{8}$$

where d_k is the kth n-bit group from the secret data in decimal form. The quality of the stego image depends heavily on the number of bit planes used for embedding. As the number of bit planes used for embedding increases, the changes in the cover image pixel values become more profound. This results in the degradation of the stego image quality.

The data extraction is simply a mod operation of the stego image pixels and 2^n. The extraction of the hidden data distorts the recovered cover image. In order to get reversibility, a location map indicating the locations of the pixels which contain hidden data needs to be transmitted. This adds to the payload and thus results in the deterioration of the stego image quality. To counter this, an optimal pixel adjustment process (OPAP) is used to enhance the visual quality of the stego image.

This technique gives a PSNR ranging between 31 dB for embedding in 4 LSB planes and 51 dB for embedding in 1 LSB plane. A scheme that gave a considerable rise in the visual quality of stego images was given in [25]. The scheme gave an average PSNR of 48–49 dB. The only problem with this scheme was that the average execution time for grayscale images was about 2.75 s. This is little slower than the times reported by some other techniques. A recent technique [26] modified sign digit to embed the secret bits. This scheme boasts of an average PSNR of about 52 dB while maintaining an embedding capacity of 1 bpp and is resistant to RS steganalytic attacks.

3.2 Lossless Compression-Based Steganography

This scheme was proposed in [27], and it used lossless compression techniques to compress bit planes and make room for the secret data. The cover image I_c is first quantized by using a quantization function 'Q', and the residual 'r' is obtained as given in Eq. 9.

$$r = I_c - Q(I_c) \tag{9}$$

The residual is compressed (r_c) and concatenated with the secret data 'd' to form the bit stream 'b' where $b = [r_c{:}d]$. This bit stream is then embedded in the cover image as given in Eq. 10

$$I_s = Q(I_c) + b \tag{10}$$

During extraction, the hidden data can be recovered by

$$b = I_s - Q(I_s) \tag{11}$$

Compression-based steganography techniques usually have the drawbacks of low PSNR and capacity and higher computation times. Recently, schemes using vector quantization (VQ) [28] and block truncation coding (BTC) [29] were proposed which improved the embedding capacity. An average visual quality of 30.91 dB was reported for [28]. The scheme in [29] embeds the data in iterations, thus increasing the capacity, but as the number of iterations increases, the visual quality of the stego image decreases. It also considerably improves the computation times.

3.3   Pixel Value Differencing (PVD)-Based Steganography

These techniques modify the difference between neighboring pixels for embedding data. They use the pixel differences to decide whether the pixels belong to a smooth region or an edgy region in the image. The data is embedded in the edges as the difference between the adjacent pixel values is more in edges as compared to those in smooth regions.

The basic scheme [30] divides the cover image into blocks of two pixels. The blocks are then classified into ranges R_i, where $i = 1, 2, 3, \ldots n$ according to their difference values 'δ'. The ranges decide the amount of data that can be embedded in a particular block. The capacity for each block is given by $c = \log_2(u_k - l_k + 1)$ where u_k and l_k are the upper and lower limits of the kth range, respectively. The difference δ is then modified to δ' as given in Eq. 12.

$$\delta' = \begin{cases} l_k + d & \text{if } \delta \geq 0 \\ -(l_k + d) & \text{if } \delta < 0 \end{cases} \tag{12}$$

where d is the integer form of c number of bits taken from the secret data. If 'p_i' denotes the ith pixel of the cover image and '$p_i{}'$' denotes the ith pixel of the stego image, then the modified pixel values are given by

$$\left(p'_i, p'_{i+1}\right) = \begin{cases} p_i - \frac{\delta' - \delta}{2}, p_{i+1} + \frac{\delta' - \delta}{2} & \text{if } \delta \text{ is odd} \\ p_i - \frac{\delta' - \delta}{2}, p_{i+1} + \frac{\delta' - \delta}{2} & \text{if } \delta \text{ is even} \end{cases} \tag{13}$$

The secret bits are extracted by following the same procedure as embedding, and the secret bits are extracted by $d = |\delta'| - l_k$.

A scheme was proposed in [31] which used 4 pixels instead of 2 to embed data. It gave an average PSNR of 38.68 dB with a large embedding capacity. As these techniques embed data according to the differences instead of the pixel values, they usually give a stego image with high embedding capacity and good visual quality. But they can be vulnerable to statistical steganalytic attacks involving comparison of difference histograms. Normal difference histograms follow a Laplacian-like curve. But due to the difference modification, the difference histograms show abnormal spikes. Because of this, most of the recent PVD-based schemes concentrate on increasing the security of the algorithms. Some schemes [32, 33] employ pseudo-random mechanisms to select blocks adaptively for embedding. This increases the security of the data. Another scheme that uses patched reference tables [34] was reported to better exploit the redundancies resulting in a better visual quality of stego images. This scheme is secure against some statistical steganalytic attacks. Hilbert curve-based embedding [35] was found to have high capacity and good security against RS steganalytic attacks.

3.4 Histogram Shifting (HS)-Based Steganography

HS-based steganography [36] shifts the histogram of the cover image to make space for the secret data. The histogram is scanned to find the bins of the peak (P) and zero (Z) points. If $P < Z$, then the histogram between $[P + 1, Z - 1]$ is shifted by one to the right. Similarly, if $P > Z$, then the histogram between $[P - 1, Z + 1]$ is shifted by one to the left. This creates an empty bin adjacent to the peak which can be used for embedding the data. The data is embedded by simply scanning the image once. Every time a pixel with value equal to P is encountered, the secret data is checked. If the secret bit to be embedded is 1, the value of the pixel is increased or decreased depending on the location of the empty bin. If the secret bit is 0, then the pixel value remains unchanged. At the end of the embedding, the histogram peak would be changed. The secret data can be recovered by simply scanning the image again and noting the occurrences of the pixels with values equal to P and $P + 1$ in case of $P < Z$ and values equal to P and $P - 1$ in case of $P > Z$. If the payload size is more than the number of pixels in the peak, then the second highest peak and a minimum point pair are found to embed the data. This method is simple and gives stego images of good quality for smaller payloads. But for larger payloads, the stego image quality reduces as more peaks are used for embedding.

Later, the HS technique was generalized [37] and two algorithms based on this were proposed. The first algorithm gave an average PSNR value of 39.38 dB for an embedding capacity of 0.5 bpp, and the second algorithm gave an average PSNR of 58.13 dB for a payload of 10 kb. The problem of low capacity was addressed by using histogram association mapping [38]. This scheme could embed data up to 2.88 bpp and did not need any preprocessing. But the visual quality of the stego image was low.

3.5 Difference Expansion (DE)-Based Steganography

The DE scheme [39] uses the pixel differences to embed the secret data, but unlike PVD, this scheme uses integer transform to expand the difference between pixel values. The embedding starts by calculating the mean (m) and the difference (δ) between each of the two pixels of the cover image. If the p_i and p_{i+1} represent two consecutive cover image pixels, then $m = \lfloor (p_i + p_{i+1})/2 \rfloor$ and $\delta = p_i - p_{i+1}$. Now, to embed the secret $b \in \{0,1\}$, the difference is expanded as $\hat{\delta} = 2\delta + b$. So the data can be embedded by modifying the pixel values as $p'_i = m + \lfloor (\hat{\delta}+1)/2 \rfloor$ and $p'_{i+1} = m - \lfloor \hat{\delta}/2 \rfloor$.

The secret bits can be extracted by simply extracting the LSBs of the difference δ' of the stego image pixels. The original cover image can be restored by using the inverse integer transform as $p_i = m' + \lfloor (\delta'+1)/2 \rfloor$ and $p_{i+1} = m' - \lfloor \delta'/2 \rfloor$ where $m' = \lfloor (p'_i + p'_{i+1})/2 \rfloor$

The DE method gives a better embedding capacity and visual quality than the compression-based and HS-based steganography techniques. But it suffers from capacity control and large location maps. A solution to this problem was provided in [40] by using wavelet lifting scheme and LSB prediction. The scheme in [41] improves the DE by using directional interpolation to get a better approximation of the embedding locations. Another scheme [42] used embedded the data in encrypted images to increase the overall security.

3.6 Prediction Error Expansion (PEE)-Based Steganography

PEE technique [43] is better in exploiting the image redundancies as compared to DE as it employs a larger neighborhood than the two adjacent neighborhoods in DE or PVD. This technique uses a predictor to generate a prediction image. The error between the original image and the predicted image is used to generate an error histogram. Secret data is then embedded in this error histogram using the HS technique. If pi is a cover image pixel and $\overline{p}_i$ is the predicted value of the pixel, then the prediction error can be given as $\varepsilon = p_i - \overline{p}_i$. If b is the secret data bit such that $b \in \{0,1\}$, and T is a user-defined threshold to control the embedding capacity, then the data can be embedded by using the equations given in Eq. 14.

$$\varepsilon'_i = \begin{array}{ll} 2\varepsilon_i + b & \text{if } \varepsilon_i \in [T, -T] \\ \varepsilon_i + T & \text{if } \varepsilon_i \in (T, +\infty) \\ \varepsilon_i - T & \text{if } \varepsilon_i \in (-\infty, -T) \end{array} \qquad (14)$$

The cover pixels can then be modified as $p'_i = \overline{p}_i + \varepsilon'_i$, and the stego image can be obtained.

During extraction, the original prediction errors are retrieved as given in Eq. 15. The hidden bits are the LSBs of the recovered prediction errors.

$$\varepsilon_i = \begin{cases} \lfloor \varepsilon'_i/2 \rfloor & \text{if } \varepsilon'_i \in (-2T, 2T) \\ \varepsilon'_i - T & \text{if } \varepsilon'_i \in (2T, +\infty) \\ \varepsilon'_i + T & \text{if } \varepsilon'_i \in (-\infty, -2T) \end{cases} \qquad (15)$$

PEE technique gives a good embedding capacity with low distortions in the stego image. Because of this, it has grown popular among the researchers as one of the most powerful techniques to hide data. The algorithm in [44] uses anisotropic diffusion to generate the prediction image. This scheme can give capacity of up to 0.2 bpp with a PSNR of about 58.56 dB. The use of diffusion can increase the computation times. This problem is avoided by using pixel value ordering [45, 46]. Pixel value ordering divides the image into blocks and arranges the pixels in each block in ascending or descending order. This generates a prediction image and reduces the computation times.

4 Transform Domain Steganography

Image transforms convert the spatial information into frequency information, and it classifies image content with respect to frequencies. In general, human visual system (HVS) is more sensitive to the changes in low frequencies (smooth areas) as compared to high frequencies (edge areas). This classification allows us to modify the high-frequency image content which is less sensitive to HVS. As the modifications are performed on the frequency coefficients rather spatial intensities, there is a minimum visible change in the cover image.

Transform domain steganography uses transforms such as Discrete Cosine Transform (DCT), Discrete Wavelet Transform (DWT) and Integer Wavelet Transform (IWT) to obtain frequency content of the image. Embedding algorithm modifies the cover image coefficients according to the secret data. In most cases, the embedding and extraction algorithms used in transform domain are similar to those used in the spatial domain. This section describes about the popular data hiding techniques available in the literature.

4.1 DCT-Based Steganography

DCT is one of the classic transforms in signal processing and has a wide range of applications in image processing. Data hiding in DCT coefficients follows the Joint Photographic Experts Group (JPEG) compression model; hence, it is also called as JPEG data hiding. Mathematical representation of two-dimensional DCT is shown in Eq. 16.

$$C(u, v) = \alpha(u)\alpha(v) \sum_{x=0}^{N-1} \sum_{y=0}^{N-1} f(x, y) \cos\left[\frac{(2x+1)u\pi}{2N}\right] \times \cos\left[\frac{(2x+1)v\pi}{2N}\right] \quad (16)$$

$$\text{where} \quad \alpha(u) = \begin{cases} \frac{1}{N}, & \text{for } u = 0; \\ \frac{\sqrt{2}}{N}, & \text{for } u = 1, 2, \ldots, N-1. \end{cases}$$

Here, $f(x, y)$ and $C(u, v)$ represent pixel intensity value and DCT coefficient value at the coordinates (x, y) and (u, v), respectively. $C(u, v)$ value at the origin ($u = 0$ and $v = 0$) is known as Direct Current (DC) component, which equals the average of all pixel intensities. The remaining coefficients of $C(u, v)$ are known as Alternating Current (AC) components. DC component and its nearby coefficients hold most of the spatial information; hence, any change in these coefficients creates a large distortion in the reconstructed image.

In DCT-based steganography, the cover image is divided into 8×8 pixel non-overlapping blocks. Then, each 8×8 block is converted into DCT coefficients using Eq. 16. DCT-transformed coefficients are usually floating-point numbers, and hence, quantization is performed using a quantization table. These quantized values are used for secret data embedding. Selecting the coefficients for data embedding is very important as altering a single coefficient would influence the complete 8×8 block in the stego image. After data embedding, inverse DCT generates the stego image pixel block [47]. Middle- and high-frequency coefficients are preferred for a stego image with good visual quality. Figure 3 describes the data embedding process in the DCT domain; the shaded portion indicates the data-embedded part.

DCT-based data hiding techniques are quite popular because of its easy implementation. Various DCT-based data hiding tools are available in the Internet, and one can use these tools for personal and commercial secret data sharing. JSteg, OutGuess, and F5 are some popular tools which use LSBs of the quantized coefficients for data hiding. But these techniques provide less embedding capacity, poor visual quality, and low security to the secret data. They are especially vulnerable to histogram-based stego attacks because the stego images have irregular histogram shapes. New schemes use quantization table modification [48, 49] techniques. The main objective of these methods is providing high capacity and good imperceptibility. Embedding capacity of 20 kb is achieved with a PSNR of 39.10 dB using scaled and modified 8×8 quantization table [48]. This method uses the middle-frequency coefficients for data hiding which improve the storage capacity of

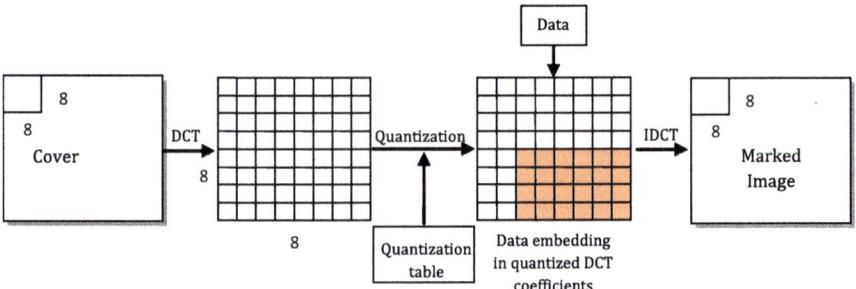

Fig. 3 DCT-based data hiding

each block while maintaining the imperceptibility, but this process is very time-consuming. Another algorithm is proposed to improve the speed [49] by using 16×16 blocks. Also this method achieves an embedding capacity of 24 kb with a PSNR value of 44.10 dB. Adaptive steganography approach with block discrete cosine transform (BDCT) and visual model [50] is proposed to enhance the security of these schemes. This approach uses sub-band coefficient adjustment process for data embedding. It is robust against common visual attacks, filtering, and compression operations. Another scheme uses piecewise linear chaotic map (PWLCM) [51]-based random LSB embedding and achieves a PSNR of 59.69 dB. It performs well against supervised universal steganalysis approach based on Fisher linear discriminator (FLD).

4.2 DWT-Based Steganography

DWT is a popular transform in image processing to separate high- and low-frequency components of an image. Wavelet transform provides both frequency and spatial information of an image. The general filter bank representation of one-dimensional DWT (1D-DWT) is shown in Eqs. 17 and 18.

$$a[p] = \sum_{n=-\infty}^{\infty} l[n - 2p] \times [n] \tag{17}$$

$$d[p] = \sum_{n=-\infty}^{\infty} h[n - 2p] \times [n] \tag{18}$$

In DWT-based data hiding, first cover image is decomposed using DWT [52]. Image decomposition normally can be performed using 1-D DWT transformations along the rows and coloumns. In row processing, each row in the cover image is applied to low-pass ($l[n]$) and high-pass ($h[n]$) filters followed by down sampler. These operations divide a row into low-frequency (L) and high-frequency (H) parts. The low-frequency band holds an approximation $a[p]$ of the image, and the high-frequency band holds the details $d[p]$ of the image. In a similar manner, the column processing divides the L and H bands into two sub bands. The low-frequency band (L) is divided into low–low (LL) and low–high (LH) sub-bands, and similarly, high-frequency band is divided into high–low (HL) and high–high (HH) sub-bands. So as a result, the 2D DWT divides the image into four equal-sized bands. Next-level decomposition can be carried out by considering the LL component as input signal. In these four components, approximation component (LL) holds the low-frequency or average information of the image. Just like the DC coefficient in DCT, the approximation coefficient in DWT is important for reconstruction. Any changes in approximation coefficients show serious changes in the reconstructed image.

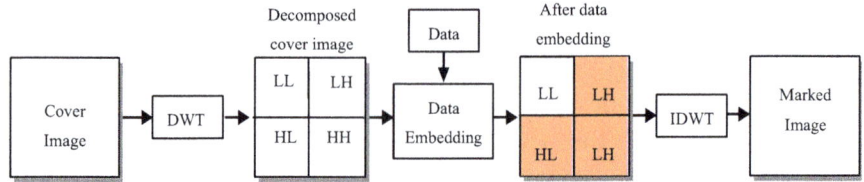

Fig. 4 DWT-based data hiding

In general, the decomposed coefficients are floating-point numbers; hence, quantization is performed to achieve integer coefficients which are used for data embedding. To attain good visual quality, most of the embedding techniques concentrate on the high-frequency sub-bands (LH, HL, HH). After that, data embedding inverse DWT achieves the stego image. Fig. 4 describes the DWT-based data hiding process.

DWT-based data hiding techniques provide high security against statistical steganalytic attacks. Embedding secret data in LSB positions of the DWT coefficients is one of the simplest data hiding techniques [53]. High-frequency sub-bands are the most commonly used embedding locations. But this technique is not reversible. In compression-based technique [54], Huffman coding is used to compress the high-frequency coefficients of Haar DWT. Then, secret data and residual data are stored in the high-frequency sub-bands. This method achieves an average PSNR of 44.78 dB, but it requires more auxiliary data to restore the cover image. To reduce the auxiliary information problem, a high payload frequency-based reversible image hiding (HPFRIH) [55] method was proposed. It uses adaptive arithmetic coding to compress high-frequency band coefficients.

In general, the conventional DWT produces coefficients with high fractional values; on the other hand, most of the data embedding algorithms work on integer values. Truncation of DWT coefficients is necessary to achieve integer coefficients for proper data hiding which results in an acceptable loss of information. In steganography, losing such small information will not show much difference in the stego image quality, but for reversible steganography applications, the fractional values need to be stored as an auxiliary information along with the secret data. But it would reduce the embedding capacity. The implementation of reversible integer-to-integer wavelet transform (IWT) gives a solution to this reversible problem in data hiding. This is explained in the next section.

4.3 IWT-Based Steganography

IWT transform converts the spatial domain pixels (integers) into transformed domain coefficients using lifting scheme. Unlike DWT where the coefficients are floating-point numbers, these coefficients are integers. Lifting scheme uses floor and ceiling functions to achieve integer values without quantization. And these values are reversible in nature. Prediction and update procedures using lifting scheme to construct 1D Haar wavelet transform are represented in Eqs. 19 and 20, respectively.

$$d_k = x_{2k+1} - x_{2k} \tag{19}$$

$$s_k = x_{2k} + \frac{d_k}{2} \tag{20}$$

where x is a one-dimensional input vector, whose length is equal to multiple powers of 2 and subscript k represents the position of sample. The predicted and updated coefficients in IWT are equal to the high-frequency and low-frequency filter coefficients in DWT.

LSB replacement of IWT coefficients is one of the simplest and earliest works in IWT-based steganography. To improve security in [56], the high-frequency sub-band coefficients are selected randomly using graph theory. Because of random embedding, it performed well against blind attacks. Histogram shifting and compression techniques are the two major techniques used on IWT coefficients to achieve reversibility. In compression-based techniques, the high-frequency sub-bands are replaced with the compressed version of cover data and the secret data. In the scheme [57], arithmetic coding is used to compress the high-frequency sub-bands of IWT and supports average embedding capacity of 139 kb with a good PSNR of 50.34 dB. But this technique requires very large auxiliary data for reversibility, whereas in IWT- and HS-based steganography techniques [58, 59], the data embedding is performed by modifying the transformed image histogram. In [58], histogram shifting is performed on the 3D IWT coefficients. It achieves robust stego image with an average embedding capacity of 49 kb and BER of 0.47 %. Another scheme [59] uses two-dimensional generalized histogram shifting (GHS) on middle sub-band coefficients.

There are some data hiding techniques which use combination of two transforms to enhance the security and visual quality further [60, 61]. Compared to the single transform techniques, this method provides better security to the secret data, but it requires more time to process.

Apart from DWT and IWT, several new transforms are also developed for steganography, e.g., Slantlet transform [62], Karhunen Loêve transform (KLT) [63], Fresnel transform (FT) [57], and mix column transform [64], each one of them offers specific feature over others. The Slantlet transform achieves better time localization and compression like wavelet transform. It outperforms the standard DCT and discrete Haar wavelet transforms. KLT is the optimal transform to exploit the correlation between multispectral bands. KLT followed by IWT is performed well on multispectral image data hiding. FT satisfies the unitary, duality translation, and dilation properties. FT and QR code-based image steganography is developed to enhance the security of transform domain steganography techniques.

Mixed column transform is based on irreducible polynomial which provides high capacity with reasonable imperceptibility. This method is computationally complex and time-consuming.

5 Steganalysis

Steganalysis is the science of detecting whether any data is hidden in a suspect file. Steganalysis poses as a challenge because usually the secret data is encrypted and coupled with noise which is present in almost all the images. So in order to know whether the steganalysis is successful, the analyst has to extract the data, possibly decrypt it, and then analyze it to see whether the suspect image has any data hidden it. This section gives an introduction to some basic steganalytic techniques.

- *Visual Inspection* The first and the most basic attack that a suspect image faces is the visual inspection. Changing the pixel data might lead to some drastic visual changes in the image. Abnormal patterns and sudden changes in smooth regions can lead to suspicions about the hidden data. It is shown in Fig. 5 that as the embedding rate increases from 2 to 5 bpp, the changes in the image become more and more visible.
- *Content and Histogram Analysis* This can be a problem when the cover is not carefully selected by the steganographer. In the case of the original image being taken from the Internet or some free open image databases, the original image can be easily acquired by the steganalyst. Thus, the hidden data can be exposed by simply converting the image into a hex code using a hex convertor or comparing the histograms of the original and the suspect image files. Figure 6 shows a comparison of original boat image and the stego boat image.
- *Chi-squared test* In statistics, this test is used to check whether there is any relationship between two sets of data. For images, if the stego image is found to have a significant relationship with the original image, the probability of detection of secret data increases. Figure 7 gives the results of chi-squared test for the boat images in Fig. 6.

(a)   **(b)**   **(c)**   **(d)**

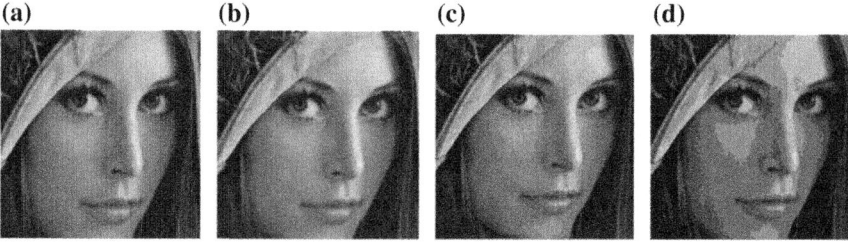

Fig. 5 Lena image: **a** original image, **b** 2 bpp embedding, **c** 4 bpp embedding, and **d** 5 bpp embedding

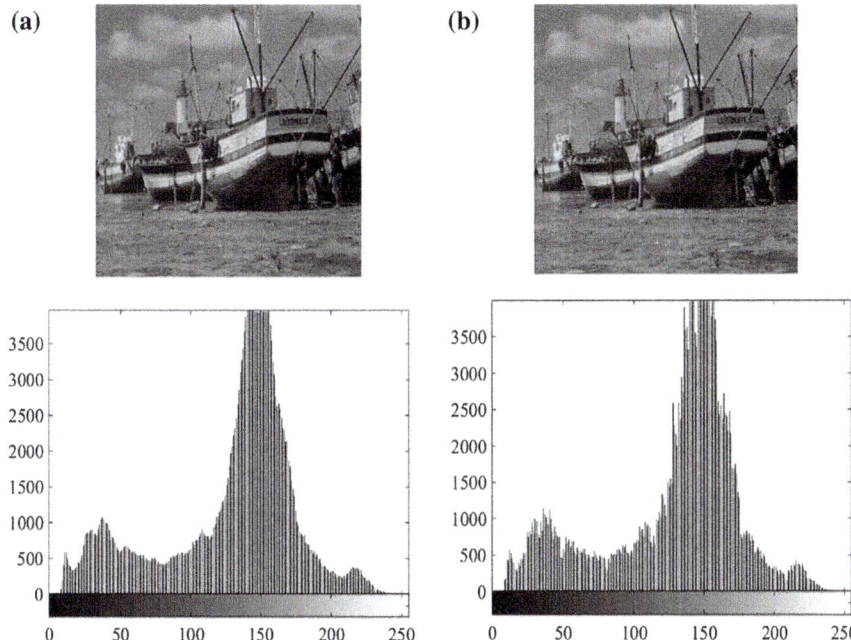

Fig. 6 Histograms of Boat.tiff. **a** Without any hidden data and **b** with hidden data

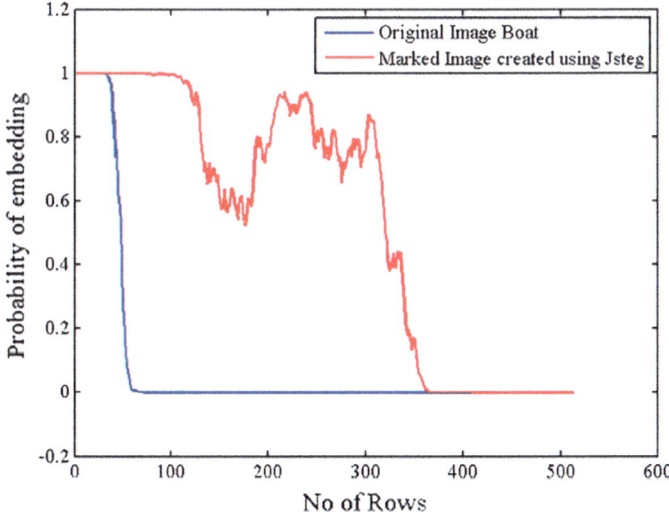

Fig. 7 Results of chi-squared test performed on the image Boat.tiff before and after data embedding

- *RS Steganalysis* [65] This is a tool specially developed for JPEG image steganalysis. The tool works by separating the image data into three parts, namely regular (*R*), singular (*S*), and unused (*U*). A mask M is generated along with its flipped version -M. These masks are then applied to the *R* and *S* groups, and the results are compared. If $R_{M} \cong R_{-M}$ and $S_M \cong S_{-M}$, then it can be said that no data is hidden in the image. However, if the condition is not satisfied, the analyst may use some other advanced techniques to extract the data.

6 Randomization in Steganography

The computational complexity of an algorithm plays an important role in enhancing the security of the algorithm. If the number of computations required to crack a stego system is small, the attacker would be able to extract the secret data with a simple brute-force attack. The computational complexity adds to the security by making it tougher for the attacker to extract data as the number of calculations involved is very large.

Randomization can add to the complexity of the system. It can be included in the embedding stage by using a pseudo-random number generator (PRNG) to choose random embedding locations. This makes the challenge of steganalysis more complicated because the data would appear as noise to the analyst.

Pseudo-random numbers can be generated using chaotic functions. Chaotic functions are iterative functions which use the output of one iteration as the input for the next. They are highly sensitive to the initial conditions, and even a slight change results in drastic changes in the output values. Figure 8a shows an embedding map constructed using the chaotic function given in Eq. 21 with the

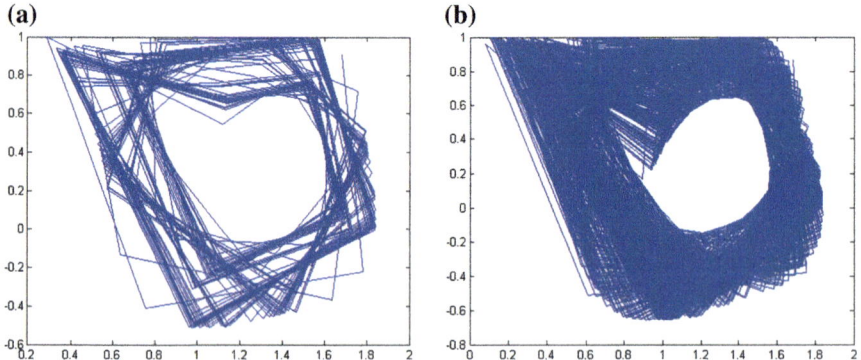

Fig. 8 Outputs of the chaotic function in Eq. 21. **a** Initial values: $x = y = z = 1$, $a = 2$, $b = c = 1$, and $d = 0$ and **b** initial values: $x = y = z = 1$, $a = 2.1$, $b = c = 1$, and $d = 0$

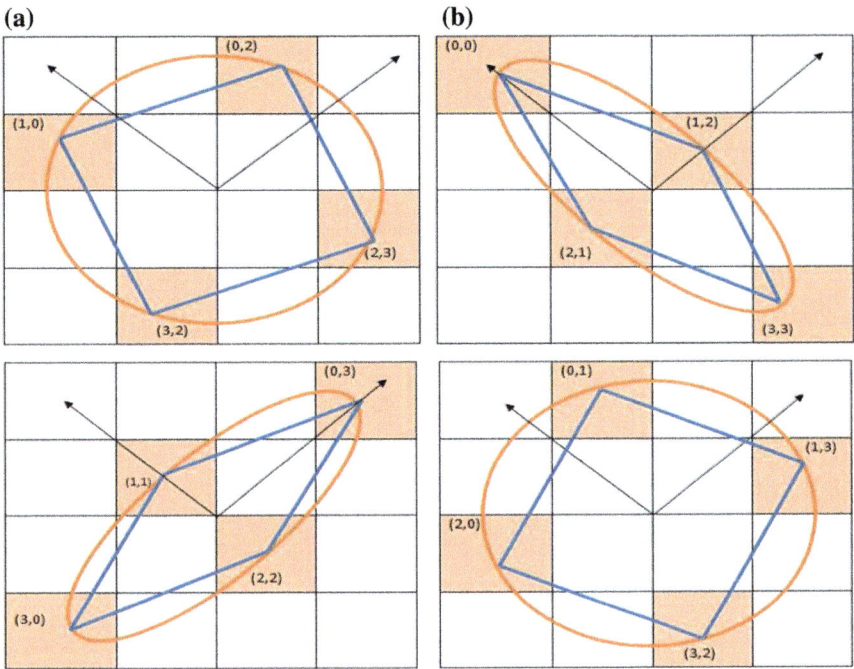

Fig. 9 Knight's tour: All the pixels are visited in one complete round. **a** Right circle, **b** left ellipse, **c** right ellipse, and **d** left circle

initial values $x = y = z = 1$, $a = 2$, $b = c = 1$, and $d = 0$. Figure 8b shows the output of the same chaotic function with a = 2.1.

$$x_{i+1} = \sin(y_i \times b) + c\,\sin(x_i \times b)$$
$$y_{i+1} = \sin(x_i \times a) + d\,\sin(y_i \times b) \tag{21}$$

Another randomization method is given in [66]. The method simulates the knight's movements in a chess on an 8×8 block of pixels as a map to embed data. The 8×8 block of pixels is divided into 4 blocks with 4×4 pixels. The knight's movement has 4 variations as shown in Fig. 9 which are applied to all 4 sub-blocks. A mathematical model is then created to embed the pixels in the complete 8×8 block. This can be repeated to embed data in the whole image.

7 Cryptography for Added Security

Security can further be increased by encrypting the secret data before embedding it [42, 67]. The effect of encryption on the visual quality will be almost negligible as long as the amount of data embedded remains the same. Encrypted data alone can

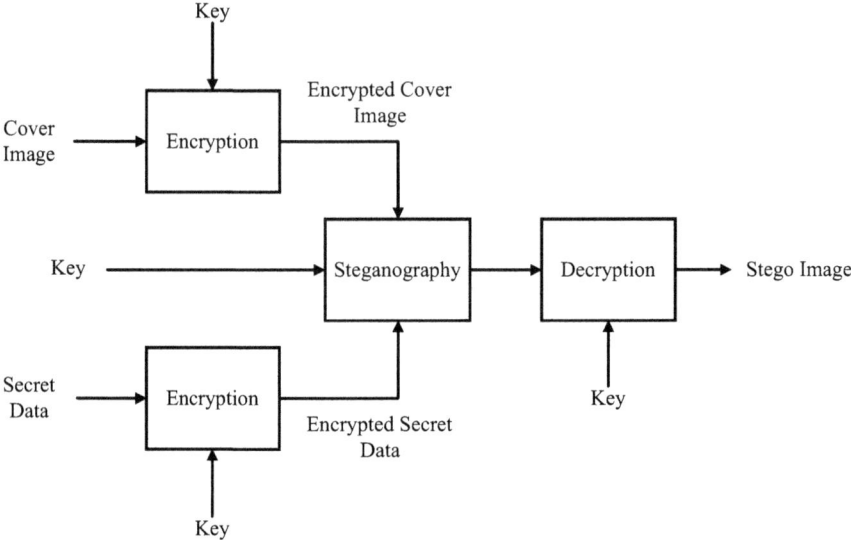

Fig. 10 Using cryptography for added security

be detected easily as it appears as gibberish. This can alert some attackers to intercept the ongoing transmission and steal the data or block the transmission completely. Using steganography, the encrypted data can be embedded so as to hide the evidence of any data being transferred. This adds another layer of security to the information exchange. Also, encryption often equalizes the probabilities of the occurrence of 0 and 1 in the secret data, thus improving the security against probabilistic attacks.

The system given in Fig. 10 uses a similar system for embedding secret data. The cover image is first encrypted, and the secret data is embedded in the encrypted cover image. For added security, the secret data is also encrypted. After the data embedding, the cover image is decrypted to get the stego image. The decryption of the image further jumbles the embedded data, thus adding another layer of security.

8 Applications

Data hiding in digital images is one of the advanced data security techniques for covert communication. Nowadays, medical, defense, and multimedia sectors need secret data communication as well as storage. The solution for this requirement is steganography because it conceals the existence of data. In the medical applications, steganography is used to hide the patient's information inside the medical images.

Secure transmission of sensitive data is critical for defense applications. Cryptography techniques fail to provide complete security because cipher data is unreadable in nature. But cryptography can be combined with steganography to provide higher security. Due to this, steganography took recognizable interest over cryptography where the secret data is hidden inside a cover file. Security organizations are using customized steganography tools for hiding encrypted secret data in the digital images for covert communications.

In the multimedia applications, the information hiding is treated as watermarking. Here, the cover is more valuable compared to the hidden data. It provides copyright protection to the digital content by embedding a predefined bit stream into it. Apart from these, steganography can be used for protection against data alteration, restricting access to digital files, and storing metadata of digital files.

9 Challenges

The main challenge in steganography is to devise a scheme that can give unconditional covertness just as the one-time pad gives unconditional secrecy. The basic metrics that define a steganography scheme include the amount of data that can be hidden using the scheme, i.e., the capacity, the imperceptibility, how close the stego image is to the original image, and the robustness of the scheme to various steganalytic attacks. It is a challenge to balance all three as favoring one will result in the degradation of other. Embedding data in an image is basically altering its original form. So the original and the stego images would have some differences. If the amount of data embedded in the image is increased, the changes in image would be more profound which would in turn reduce the visual quality of the stego image. Hence, it is a challenge to balance capacity, imperceptibility, and robustness. These three form a triangle as shown in Fig. 11 in which improving one parameter would result in the degradation of the other.

Complexity also plays an important role. If the algorithm is complicated to implement, it would consume more power and require more time to implement. This can be a problem when the transmitter or receiver is powered by batteries as in the case of wireless sensor networks. But, high computational complexity also means that the system is secure against brute-force attacks. Thus, the steganography

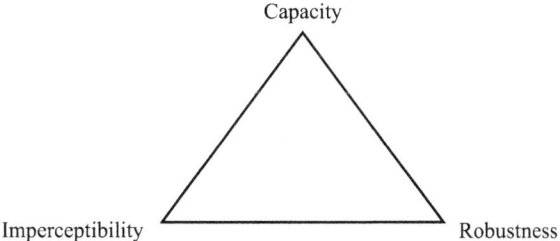

Fig. 11 The magic triangle showing the relation between capacity, imperceptibility, and robustness for steganographic algorithms

system should be such that it should be secure against brute-force attacks, and at the same time, it should not overwhelm the processing unit.

Also, a lot of researches are being carried out on steganalysis. Steganalysis is the science of detecting whether any data is hidden in a suspect image. Steganalysis is a challenging task because the suspect file may not contain any hidden data at all or the data might be encrypted or it might be disguised as noise. In order to confirm

Table 1 Advantages and disadvantages of the steganography techniques discussed in Sects. 3 and 4

Method	Advantages	Disadvantages
LSB	• Easy to implement and support higher embedding capacities with good visual quality	• Poor security and robustness against steganalytic attacks
Lossless compression-based steganography	• Provide decent security to the secret data	• Low PSNR, higher computation times, and low embedding capacity
PVD	• Store data in edge region where HVS is inefficient • Provide a stego images with high embedding capacity and good visual quality	• Vulnerable to statistical steganalytic attacks which involves the comparison of difference histograms
HS	• Simple RDH technique • Maintain good visual quality if the capacity is low • Hard to detect secret data in the stego image	• Limited embedding capacity and poor robustness against attacks
DE	• It provides better embedding capacity and visual quality than the compression- and HS-based techniques	• Require large location data to extract secret data • Lack of capacity control
PEE	• Achieve superior embedding capacity with good visual quality among all steganography techniques	• Require more number of calculations for predicting and data embedding
DCT	• Simple data hiding technique in transform domain steganography	• Provide less embedding capacity, poor visual quality, and low security to the secret data
DWT	• Highly secure and robust data hiding technique	• Offer moderate embedding capacity and require high auxiliary data to achieve reversibility
IWT	• Integer transform coefficients are quite useful for RDH • Provide high security to the secret data	• Embedding capacity is low as compared to spatial techniques

whether any secret data is hidden in the image, the analyst has to extract, probably decrypt, and then analyze the extracted data.

10 Discussion

The reversibility in data hiding techniques depends heavily on the way the features of the original image are compressed and modified. The data is generally hidden in the edge regions of the image as the changes in pixel intensities can be spotted easily in the smooth regions. This shows that there is a need for research in data hiding techniques for smooth areas. This would be useful in medical applications where the images have large smooth areas which can be used for embedding patient information.

In the spatial and transform domain techniques, it is evident that a compromise needs to be made between the embedding capacity and the security according to the application. Spatial domain techniques can hide huge amounts of data but are not as secure as the transform domain techniques. Whereas, the embedding capacity given by the transform domain techniques leaves much to be desired. The ongoing research uses randomization in embedding the data to increase the security of the spatial domain techniques. But this usually results in large location maps and thus increases the auxiliary information. This suggests the need for algorithms which are less dependent on location maps and can embed large amounts of data with minimal effect on the security. Table 1 gives the advantages and disadvantages of all the methods discussed in Sects. 3 and 4.

Cryptography can be used to enhance the security of the stego images while keeping the capacity same. The data can be embedded in an encrypted image to generate an encrypted stego image. This encrypted stego image can be decrypted to generate the final stego image. During extraction, the exact reverse process can be used to extract the hidden data.

11 Conclusions and Recommendations

In this chapter, the recent literature in the field of steganography is presented in the form of spatial and transform domain. It is clear that a lot of researches have been done on spatial domain techniques because of their ease of implementation. Initially, the basic functioning of the data hiding system is discussed. The features of spatial and transform domain techniques are discussed with respect to visual quality, capacity, PSNR, and security. A brief description about the steganalysis techniques is presented, and some basic attacks on the data hiding system are discussed.

The spatial domain techniques perform well with respect to image visual quality and capacity, but they fail to provide security. Transform domain techniques are secure and give a decent quality of marked image, but they offer a lower embedding capacity than the spatial domain techniques. From the discussion of transform domain techniques, it can be seen that there is a need for more research which will enable the user to increase the embedding capacity and the visual quality of the stego image. Due to the limitations of 2D transforms, the amount of research for transform domain techniques is less. For modern-day communication, new 2D transforms need to be developed for providing reversible steganography with high capacity, high security, and high robustness. Also, computational complexity of the transform should be balanced in such a way that it is not too heavy on the processor, but at the same time, the transform is not too easy to crack.

One needs to select the data hiding system according to the requirement of the application. For example, if the application requires large data to hide with good imperceptibility, then DE, PVD, and PEE are suitable choices. DWT and IWT methods are useful for covert communications where the secret data size is small, but security is essential. This chapter recommends, and non-LSB-based data embedding methods need to be developed for both spatial and transform domain steganography techniques. Also it recommends data hiding in encrypted cover files.

References

1. Petitcolas FAP, Anderson RJ, Kuhn MG (1999) Information hiding—a survey. Proc IEEE 87:1062–1078
2. Amirtharajan R, Archana P, Rayappan JBB (2013) Why image encryption for better steganography. Res J Inf Technol 5:341–351
3. Dey N, Biswas S, Roy AB, Das A, Chowdhuri SS (2013) Analysis of photoplethysmographic signals modified by reversible watermarking technique using prediction-error in wireless telecardiology. In: International Conference on Intelligent Infrastructure, 47th Annual National Convention at Computer Society of India, pp 1–5
4. Chakraborty S, Samanta S, Biswas D, Dey N, Chaudhuri SS (2013) Particle swarm optimization based parameter optimization technique in medical information hiding. In: 2013 IEEE international conference on computer and intelligent computing research, IEEE, pp 1–6
5. Dey N, Samanta S, Chakraborty S, Das A, Chaudhuri SS, Suri JS (2014) Firefly algorithm for optimization of scaling factors during embedding of manifold medical information: an application in ophthalmology imaging. J Med Imaging Heal Informatics 4:384–394
6. Chakraborty S, Maji P, Pal AK, Biswas D, Dey N (2014) Reversible color image watermarking using trigonometric functions. In: 2014 international conference on electronic systems, signal processing and computing technologies, IEEE, pp 105–110
7. Pal AK, Das P (2013) Odd-Even Embedding Scheme Based Modified Reversible Watermarking Technique using Blueprint. published in arXiv preprint arXiv: 1303.5972
8. Dey N, Mukhopadhyay S, Das A, Chaudhuri SS (2012) Analysis of P-QRS-T components modified by blind watermarking technique within the electrocardiogram signal for authentication in wireless telecardiology using DWT. Int J Image, Graph Signal Process 4:33–46
9. Dey N, Das P, Biswas S, Das A, Chaudhuri SS (2012) Lifting Wavelet Transformation based blind watermarking technique of Photoplethysmographic signals in wireless telecardiology. In: 2012 world congress on information and communication technologies, IEEE, pp 230–235

10. Dey N, Pal M, Das A (2012) A session based blind watermarking technique within the NROI of retinal fundus images for authentication using DWT, spread spectrum and Harris Corner. arXiv Prepr arXiv12090053 2:749–757

11. Dey N, Roy AB, Das A, Chaudhuri SS (2012) Stationary Wavelet Transformation Based Self-recovery of Blind-Watermark from Electrocardiogram Signal in Wireless Telecardiology. In: Thampi SM, Zomaya AY, Strufe T, Alcaraz Calero JM, Thomas T (eds) Proceedings on recent trends computing networks distributions on system security international conference on SNDS 2012, Trivandrum, India, Oct 11–12, 2012. Springer, Berlin, pp 347–357

12. Ker AD, Ox O, Bas P, Böhme R, Craver S, Fridrich J (2013) Moving steganography and steganalysis from the laboratory into the real world categories and subject descriptors. In: 1st IH&MMSec work

13. Cheddad A, Condell J, Curran K, Mc Kevitt P (2010) Digital image steganography: survey and analysis of current methods. Sig Process 90:727–752

14. Johnson NF, Duric Z, Jajodia S (2001) Information hiding: steganography and watermarking-attacks and countermeasures. Kluwer Academic Press, London

15. Johnson NF, Jajodia S (1998) Exploring steganography: seeing the unseen. IEEE Comput 31:26–34

16. Kahn D (1996) The Codebreakers: the comprehensive history of secret communication from ancient times to the internet. Simon and Schuster, New York

17. Subhedar MS, Mankar VH (2014) Current status and key issues in image steganography: a survey. Comput Sci Rev 13–14:95–113

18. Fridrich J, Goljan M, Du R (2002) Lossless data embedding—new paradigm in digital watermarking. EURASIP J Adv Signal Process 2002:185–196

19. Dey N, Samanta S, Yang XS, Das A, Chaudhuri SS (2013) Optimisation of scaling factors in electrocardiogram signal watermarking using cuckoo search. Int J Bio-Inspired Comput 5:315

20. Dey N, Maji P, Das P, Biswas S, Das A, Chaudhuri SS (2013) An edge based blind watermarking technique of medical images without devalorizing diagnostic parameters. In: 2013 international conference on advanced technologies engineering IEEE, pp 1–5

21. Dey N, Das P, Chaudhuri SS, Das A (2012) Feature analysis for the blind-watermarked electroencephalogram signal in wireless telemonitoring using Alattar's method. In: Proceedings of fifth international conference on security information networks - SIN '12. ACM Press, New York, pp 87–94

22. Dey N, Dey M, Mahata SK, Das A, Chaudhuri SS (2015) Tamper detection of electrocardiographic signal using watermarked bio-hash code in wireless cardiology. Int J Signal Imaging Syst Eng 8:46

23. Dey N, Maji P, Das P, Biswas S, Das A, Chaudhuri SS (2012) Embedding of blink frequency in electrooculography signal using difference expansion based reversible watermarking technique. Sci Bull Politeh Univ Timisoara Trans Electron Commun 57–71:7–12

24. Chan C-K, Cheng LM (2004) Hiding data in images by simple LSB substitution. Pattern Recogn 37:469–474

25. Lu T-C, Tseng C-Y, Wu J-H (2015) Dual imaging-based reversible hiding technique using LSB matching. Sig Process 108:77–89

26. Kuo W, Wang C, Hou H (2016) Signed digit data hiding scheme. Inf Process Lett 116:183–191

27. Celik MU, Sharma G, Tekalp AM, Saber E (2005) Lossless generalized-LSB data embedding. IEEE Trans Image Process 14:253–266

28. Ma X, Pan Z, Hu S, Wang L (2015) Reversible data hiding scheme for VQ indices based on modified locally adaptive coding and double-layer embedding strategy. J Vis Commun Image Represent 28:60–70

29. Chang I-C, Hu Y-C, Chen W-L, Lo C-C (2015) High capacity reversible data hiding scheme based on residual histogram shifting for block truncation coding. Sig Process 108:376–388

30. Wu D-C, Tsai W-H (2003) A steganographic method for images by pixel-value differencing. Pattern Recogn Lett 24:1613–1626

31. Liao X, Wen QY, Zhang J (2011) A steganographic method for digital images with four-pixel differencing and modified LSB substitution. J Vis Commun Image Represent 22:1–8
32. Thanikaiselvan V, Arulmozhivarman P, Amirtharajan R, Rayappan JBB (2012) Horse riding & hiding in image for data guarding. Proc Eng 30:36–44
33. Chen J (2014) A PVD-based data hiding method with histogram preserving using pixel pair matching. Signal Process Image Commun 29:375–384
34. Hong W (2013) Adaptive image data hiding in edges using patched reference table and pair-wise embedding technique. Inf Sci (Ny) 221:473–489
35. Shen S-Y, Huang L-H (2015) A data hiding scheme using pixel value differencing and improving exploiting modification directions. Comput Secur 48:131–141
36. Ni Z, Shi Y-Q, Ansari N, Su W (2006) Reversible data hiding. IEEE Trans Circuits Syst Video Technol 16:354–362
37. Li X, Li B, Yang B, Zeng T (2013) General framework to histogram-shifting-based reversible data hiding. IEEE Trans Image Process 22:2181–2191
38. Ong S, Wong K, Tanaka K (2014) A scalable reversible data embedding method with progressive quality degradation functionality. Signal Process Image Commun 29:135–149
39. Tian J (2003) Reversible data embedding using a difference expansion. IEEE Trans Circuits Syst 13:890–896
40. Kamstra L, Heijmans HJAM (2005) Reversible data embedding into images using wavelet techniques and sorting. IEEE Trans Image Process 14:2082–2090
41. Sabeen Govind PVS, Wilscy M (2015) A new reversible data hiding scheme with improved capacity based on directional interpolation and difference expansion. Proc Comput Sci 46:491–498
42. Shiu C-W, Chen Y-C, Hong W (2015) Encrypted image-based reversible data hiding with public key cryptography from difference expansion. Signal Process Image Commun 39: 226–233
43. Thodi DM, Rodríguez JJ (2007) Expansion embedding techniques for reversible watermarking. IEEE Trans Image Process 16:721–730
44. Qin C, Chang C-C, Huang Y-H, Liao L-T (2013) An inpainting-assisted reversible steganographic scheme using a histogram shifting mechanism. IEEE Trans Circuits Syst Video Technol 23:1109–1118
45. Peng F, Li X, Yang B (2014) Improved PVO-based reversible data hiding. Digit Signal Process 25:255–265
46. Qu X, Kim HJ (2015) Pixel-based pixel value ordering predictor for high-fidelity reversible data hiding. Sig Process 111:249–260
47. Fridrich J, Goljan M, Du R (2001) Invertible authentication watermark for JPEG images. In: Proceedings international conference on informational technology coding computers, 2001, pp 223–227
48. Chang C-C, Chen T-S, Chung L-Z (2002) A steganographic method based upon JPEG and quantization table modification. Inf Sci (Ny) 141:123–138
49. Almohammad A, Hierons RM, Ghinea G (2008) High capacity steganographic method based upon JPEG. In: Proceeding on ARES 2008—3rd international conference on availability, security reliability, pp 544–549
50. Sun Q, Qiu Y, Ma W, Yan W, Dai H (2010) Image steganography based on sub-band coefficient adjustment in BDCT domain. In: 2010 international conference on multimedia technology, ICMT 2010, p 3
51. Habib M, Bakhache B, Battikh D, El Assad S (2015) Enhancement using chaos of a steganography method in DCT domain. Fifth Int Conf Digit Inf Commun Technol Appl 2015:204–209
52. Tong LIU, Zheng-ding QIU (2002) A DWT-based color image steganography scheme. In: 6th international conference on signal processing, 2002, vol 2, pp 1568–1571
53. Kamila S, Roy R, Changder S (2015) A DWT based steganography scheme with image block partitioning. In: 2nd international conference on signal process integration networks, SPIN 2015, pp 471–476

54. Chan YK, Chen WT, Yu SS, Ho YA, Tsai CS, Chu YP (2009) A HDWT-based reversible data hiding method. J Syst Softw 82:411–421
55. Chang CC, Pai PY, Yeh CM, Chan YK (2010) A high payload frequency-based reversible image hiding method. Inf Sci (Ny) 180:2286–2298
56. Thanikaiselvan V, Arulmozhivarman P, Subashanthini S, Amirtharajan R (2013) A graph theory practice on transformed image: a random image steganography. Sci World J. doi:10.1155/2013/464107
57. Uma Maheswari S, Jude Hemanth D (2015) Frequency domain QR code based image steganography using Fresnelet transform. AEU—Int J Electron Commun 69:539–544
58. Fang H, Zhou Q, Li X (2014) Robust reversible data hiding for multispectral images. J Netw 9:1454–1463
59. Yamato K, Shinoda K, Hasegawa M, Kato S (2014) Reversible data hiding based on two-dimensional histogram and generalized histogram shifting. IEEE Int Conf Image Process 2014:4216–4220
60. Dey N, Biswas D, Roy AB, Das A, Chaudhuri SS (2012) DWT-DCT-SVD based blind watermarking technique of gray image in electrooculogram signal. In: 2012 12th international conference on intelligence systems design and applicaions, IEEE, pp 680–685
61. Dey N, Das P, Roy AB, Das A, Chaudhuri SS (2012) DWT-DCT-SVD based intravascular ultrasound video watermarking. In: 2012 world congress on informational communications technology, IEEE, pp 224–229
62. Thabit R, Khoo BE (2015) A new robust lossless data hiding scheme and its application to color medical images. Digit Signal Process 38:77–94
63. Fang H, Zhou Q (2013) Reversible data hiding for multispectral image with high radiometric resolution. In: 2013 seventh international conference on image graph, pp 125–129
64. Abduallah WM, Rahma AMS, Pathan A-SK (2014) Mix column transform based on irreducible polynomial mathematics for color image steganography: a novel approach. Comput Electr Eng 40:1390–1404
65. Fridrich J, Goljan M, Du RDR (2001) Detecting LSB steganography in color, and gray-scale images. IEEE Multimed 8:22–28
66. Thanikaiselvan V, Arulmozhivarman P (2015) RAND—STEG: an integer wavelet transform domain digital image random steganography using knight's tour. Secur Commun Networks 8:2374–2382
67. Amirtharajan R, Rayappan JBB (2012) Inverted pattern in inverted time domain for icon steganography. Inf Technol J 11:587–595

Digital Watermarking: A Potential Solution for Multimedia Authentication

Kaiser J. Giri and Rumaan Bashir

Abstract The digitization has resulted in knowledge explosion in the modern technology-driven world and has led to the encouragement and motivation for digitization of the intellectual artifact. The combining, replication, and distribution facility of the digital media such as text, images, audio, and video easier and faster has no doubt revolutionized the world. However, the unauthorized use and maldistribution of information by online pirates is the sole threat that refrains the information proprietors to share their digital property. It is therefore imperative to come up with standard means to protect the intellectual property rights (IRP) of the multimedia data, thereby developing the effective multimedia authentication techniques to discourage the illegitimate distribution of information content. Digital watermarking, which is believed to be the potential means among the various possible approaches, to encourage the content providers to secure their digital property while maintaining its availability, has been entreated as a potential mechanism to protect IRP of multimedia contents.

Keywords Multimedia authentication · Digital watermarking · Copyright protection · Intellectual property rights · Discrete wavelet transformation

1 Introduction

The realm of computer science and information technology has revolutionized the entire world. In particular, computer science is the scientific and practical approach to computation and information technology is the application of computers, the ever-evolving fields in human history. Both of these aspire on giving vast solutions

K.J. Giri (✉) · R. Bashir
Department of Computer Science, Islamic University of Science and Technology,
Awantipora, Pulwama, J&K 192122, India
e-mail: kaiser.giri@islamicuniversity.edu.in

R. Bashir
e-mail: rumaan.bashir@islamicuniversity.edu.in

© Springer International Publishing Switzerland 2017
N. Dey and V. Santhi (eds.), *Intelligent Techniques in Signal Processing for Multimedia Security*, Studies in Computational Intelligence 660,
DOI 10.1007/978-3-319-44790-2_5

to day-to-day problems with the focus to save time and effort. These have become an integral part of our lives, shaping virtually everything from the way we live to the way we work. The exponential growth in these high-speed computer networks and World Wide Web has converged the whole world into a very small place and have explored means of new scientific, economic, business, entertainment, and societal opportunities in the shape of data sharing, collaboration among computers, instant information delivery, electronic distribution and advertising, business transaction processing, product ordering, digital libraries and repositories, network video and audio, individual communication, and a lot more.

The panorama of economic feasibility together with recent advances in computing and communication technology has revolutionized the world. The cost-effectiveness of vending software by communication over World Wide Web in the form of digital images and video clips is significantly enhanced due to the advancement in technology. Of many technological advances, the digital media invasion in nearly every aspect of routine life was one of the biggest technological accomplishments. Digital media has several advantages over its analog counterpart. The key attribute of digital data is that it can be stored efficiently with very high quality, manipulated easily using computers by accessing its discrete locations, communicated in a fast and economical way through networks without losing quality [1]. Editing is simple because of access to the exact discrete locations that are required to be modified. The copying of digital data is easy without loss of reliability. The digital media duplicate is identical to the original; pursuant to which, a serious concern has emerged due to which proprietors refrain to share digital content in view of threats posed by the online pirates and to maintain intellectual property rights.

Consequently, as an interesting challenge, digital information protection schemes, multimedia authentication in the form of information security, have gained so much of attention. Since copyright enforcement and content verification are challenging issues for digital data, one solution would be to restrict access to the data by using encryption tools and techniques. However, encryption does not provide overall protection of the whole data as it can be freely distributed or manipulated once the encrypted data is decrypted. Unauthorized use of data creates several problems. The subject of multimedia authentication is becoming more and more important. The prevention of any illegal duplication to the data is of more concern to copyright owners [2]. Therefore, concrete measures are required in order to maintain the availability of multimedia information but, in the meantime, standard procedures/methods must be designed to protect intellectual property of creators, distributors, or owners. This is an interesting challenge and has led to the development of various multimedia authentication schemes over the past few decades.

Of the many possible approaches to protect visual data, digital watermarking is probably the one that has received most interest. Digital watermarking is the process of embedding information into a noise-tolerant digital signal such as image or audio data [3]. Information is embedded in such a manner that it is difficult to be removed due to which the relevant information is not easily identified, thus serving different purposes such as copyright ownership of the media, source tracking, and piracy deterrence.

2 Digital Watermarking

The concept of hiding or concealing some additional information (watermark) in the host data such as images, audio, video, text, or combination of these to establish the ownership rights is known as watermarking [4]. In digital watermarking, an invisible signal (message) generally known as watermark is embedded into digital media (host) such as in text, image, audio, or video. The watermark embedding is further integrating in an inseparable form from the host digital content [2]. While embedding the watermark information into the host media, the imperceptibility and robustness properties have to be taken into consideration so that the quality of watermarked data may not degrade and can replace the original unwatermarked data for all practical purposes. Besides, the embedding process has to be carried in such a manner so that the watermark data remains inseparable from watermarked data and can even be extracted later to make an assertion of the digital data. The authentication and copyright protection of the digital content are the two prime objective of the watermarking. Watermark embedding/extraction is primarily carried either in spatial domain or in frequency (transform) domain. In spatial domain, the host data pixels under consideration are directly modified with respect to the watermark pixels, whereas in transform domain, some reversible transformation is initially applied to host data and the resulting transformed coefficients are then modified with watermark pixels. Digital watermarking offers two principle advantages for authentication: firstly, the watermark is integrated as an inherent part of the host image avoiding the appended signature of cryptology, and secondly, the watermark will experience the similar transformations as that of the data due to the fact that it is concealed within the data [5, 6]. These transformations can be undone by observing the transformed watermark.

A generic watermarking model is presented in Fig. 1.

The idea of watermarking can be dated back to the Late Middle Ages. The earliest use has been to record the manufacturer trademark on the products so that authenticity could be easily established. The term "digital watermark" was first coined in 1992 by Andrew Tirkel and Charles Osborne. Actually, the term used by Andrew Tirkel and Charles Osborne was originally used in Japan—from the Japanese—"denshisukashi"—literally, an "electronic watermark" [7, 8]. The idea of watermarking has been used for several centuries, in the form of watermarks found initially on plain paper. However, the field of digital watermarking flourished during the last two decades, and it is now being used for many different applications. Therefore, a digital watermark is a message that is embedded in the digital media (audio, video, text, or image) which needs to be extracted later. These embedded messages carry the ownership information of the content. The procedure

Fig. 1 Generic watermarking model

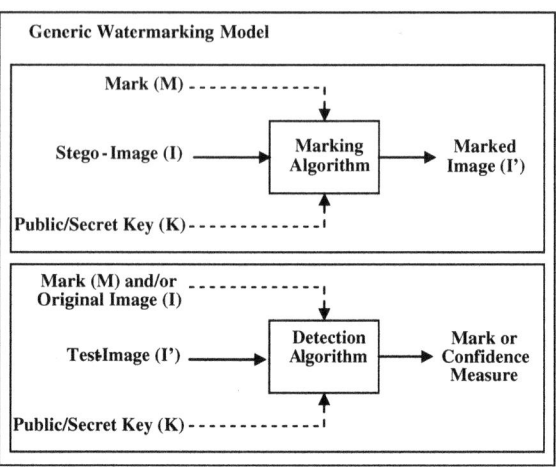

of embedding copyright information in the form of a digital watermark into digital media is called digital watermarking.

The idea of watermarking and steganography are closely related to each other in the sense that both of them hide a message inside the digital media. However, the difference lies in their objectives. In case of watermarking, the message that is embedded is related to the actual content of the digital signal, whereas in steganography the message being concealed has no relation to the digital signal. The technique of steganography embeds data inside a signal, either host or cover, in an unnoticeable manner. Various types of digital data that can be used as a cover medium for information hiding include text, images, audio, and video. The invisible inks that one could make use of to send a secret message to others was one of the simplest and oldest ways of steganography. Numerous types of invisible inks were available to conceal the messages such as lemon juice, onion juice, milk, and urine to name a few. For hiding a message, one would write it using these inks on a sheet of paper. Since the ink is invisible, nothing would appear on the paper. On reception, this paper was placed over flame in order to recover the message.

For securing the transmission of data, a technique known as cryptography, which converts information into an unintelligible form, is often used. Here a "key" is used for performing encryption, which disguises the original data. On reception, the encrypted message is decoded to retrieve the original message with the help of either same or different key. Hence, the objective of cryptography is to protect the contents of the message. However, once the data is decrypted, it is available to the intruder. Therefore, cryptography is a scrambling message so that it is not comprehensible to the unapproved user whereas in watermarking, neither the cover medium nor the copyright data is converted to an unintelligible form. Instead, the copyright data is concealed in order to provide the ownership information of the signal in which it is hidden.

Digital signatures are similar to written signatures and are mainly used to provide message authentication. It is an electronic signature which can authenticate the identity of the sender/signatory of a message during its transmission. They are transportable, non-imitable, and automatically time-stamped. It possesses the property of non-repudiation, which ensures that the original signed message has arrived. A digital signature is separately placed in the protected message and is vulnerable to distortions, whereas a digital watermark lies within the protected message and has to tolerate a certain level of distortion. Watermarking and digital signature both protect integrity and authenticity of a message.

Therefore, digital watermarking is the addition of "ownership" information in a signal in order to prove its authenticity [9]. This technique embeds data, an unperceivable digital value, the watermark, carrying information about the copyrights of the message being protected. Constant efforts are being made to devise effective and efficient watermarking techniques which are robust enough to all potential attacks [10]. An attacker tries to remove the watermark in order to violate copyrights. An attacker can cast the same watermark in the message after altering it in order to forge the proof of authenticity.

3 Characteristics of Watermarking

Watermarking provides an assortment of vast solutions to a variety of data. The technique of watermarking should possess a set of properties which make it a potential solution [2, 9–11]. An effective watermarking system should exhibit the following characteristics:

1. Imperceptibility
   The imperceptibility property signifies that the host medium should be minimally altered, once the watermark data is embedded within it. The cover medium being protected should not be affected by the presence of the watermark data. If a watermarking system does not uphold this property, post-insertion of a watermark data in a cover medium, the image quality may get reduced.
2. Robustness
   The robustness property of the watermark data indicates that the watermark data should not be destroyed if common operations and malicious attacks are performed on the message. Robustness is a primary requirement of watermarking, but it depends upon the application areas under consideration.
3. Fragility
   The fragility property indicates that the copyright data is altered or modified within a certain limit when on applying common manipulations and malicious attacks on the host media. Application areas such as tamper detection require a fragile watermark in order to detect any tampering with the host media. Other applications may require semi-fragility which indicates that the embedded watermark comprises of a fragile as well as robust components.

4. Resilient to Common Signal Processing
   The resilience property implies that the watermark should be retrievable even when common signal processing manipulations such as analog-to-digital and digital-to-analog, image contrast, brightness and color adjustment, conversion, audio bass and treble adjustment, high-pass and low-pass filtering, histogram equalization, format conversion, dithering and recompression, resampling, and requantization are applied to the host media.
5. Resilient to Common Geometric Distortions
   The watermarking scheme except for audio watermarking should embed watermarks which are resilient to geometric image operations. These operations include translation, rotation, cropping, and scaling.
6. Robust to Subterfuge Attacks (Collusion and Forgery)
   The watermark embedded should be robust to collusion attack which implies that multiple individuals possessing a copy of the watermarked data may destroy the watermark presence by colluding the watermark copies in order to create the duplicate of the original copy.
7. Unambiguousness
   After retrieving the watermark, the identity of the owner should be unambiguously identifiable. In addition, if the watermarked data is subjected to some attack, the owner identification should not degrade beyond an established threshold.

4 Digital Watermarks Techniques

Watermarking techniques can be classified on various parameters in view of the variety of digital media types and application requirements [2, 9] as shown in Fig. 2.

1. Embedding/Extraction Domain
   There are two types of watermarking schemes depending upon the embedding/extraction domain:

   - Spatial Domain-based Watermarking Schemes: The watermarking is carried out by directly modifying the host data pixels according to the watermark pixels. Number of schemes haven been developed by researches over the period of time in the spatial domain.
   - Transform Domain-based Watermarking Schemes: The watermarking embedding/extraction is performed by applying some reversible transformation to the host data and modifying transform domain coefficients with watermark data. The watermark data is later obtained by applying the inverse transformation. The transform-domain-based watermarking systems are more robust as compared to spatial domain watermarking systems. They are robust against simple image processing operations such as low-pass filtering,

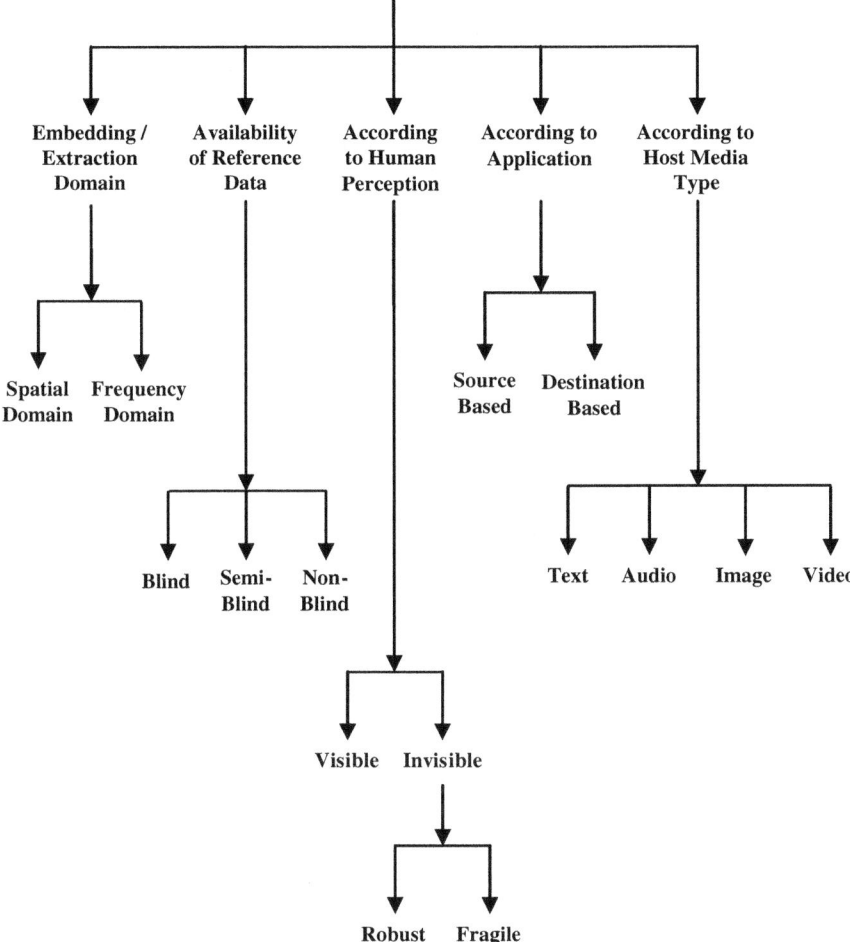

Fig. 2 Watermarking techniques

brightness and contrast adjustment, and blurring. However, they are difficult to implement and computationally more expensive. We can use either discrete cosine transform (DCT), discrete Fourier transform (DFT), or discrete wavelet transform (DWT) [12–15]; however, DCT is the most exploited one.

2. Availability of Reference Data
   As per the availability of reference data, the watermarking techniques can be of three types as follows:

   • Blind: The original unwatermarked data is not required for extraction of watermark [16–20]. Blind watermarking scheme is also known as public watermarking scheme. This is the most challenging type of watermarking system as it requires neither the cover (original) data, I, nor the embedded watermark, W. These systems extract n bits of the watermark data from the watermarked data (i.e., the watermarking image) and using key K recover/reconstruct the watermark. $I \times K \rightarrow W$, where K is the key.

- Non-Blind: The original unwatermarked data is required for extraction of watermark [21]. This scheme is also known as private watermarking scheme. This system requires at least the cover (original data) for detection.
- Semi-Blind: Some features derived from original unwatermarked data are required for extraction of watermark [22]. This scheme is also known as semi-private watermarking scheme. This system does not require the cover (original image) for detection.

3. According to Human Perception
   According to human perception, watermarking can be of the following two types:

   - Visible: The embedded watermark inlaid in the host data is transparent [23]. Visible watermarks can be seen by the user; logo and the owner details are identified by person. These technique changes the original signal.
   - Invisible: The embedded watermark inlaid in the host data is hidden and can be extracted only by an authorized user. Invisible watermarks cannot be seen by other party, and output signal does not change when compared to the original signal.

       - Fragile: The embedded watermark is destroyable by any kind of modification to the host data [24]. These techniques are more sensitive than other and can be easily destroyed with small modification.
       - Robust: The embedded watermark is resilient to image processing attacks. These methods are used for copyright protection because this type of watermark cannot be broken easily.

4. According to Application
   From application point of view, the watermark can be of two types, i.e. source based or destination based.

   - Source Based: To detect the host data tampering, source-based watermarks are desirable for ownership identification or authentication where a unique watermark identifying the owner is introduced to all the copies of a particular image being distributed. A source-based watermark could be used for authentication and to determine whether a received image or other electronic data has been tampered with.
   - Destination Based: To embed the watermark for copyright protection, the watermark could also be destination based where each distributed copy gets a unique watermark identifying the particular buyer. The destination-based watermark could be used to trace the buyer in the case of illegal reselling.

5. According to Host Media

According to the host media, the watermarking can be of four types:

- Text Watermarking: Most of the paper documents such as those from digital libraries, banks, journals, books contain more valuable information than other types of multimedia. In order to trace illegally copied, altered, forged, or distributed text documents, watermarking is being used as a tool to provide copyright information. Documents on which text watermarking can be applied have to be properly formatted unlike the text in raw form like source code. Unformatted text cannot be watermarked because of lack of "perceptual headroom" for the watermark to be placed.

- Image Watermarking: As there are multiple sources of digital images such as photographs, medical scans, satellite images, or computer-generated images, watermarking is therefore one of the commonly used watermarking schemes. Watermarks for images usually modify pixel values or transform coefficients. However, other features like edges or textures could also be modified to include the watermark. An image may be subjected to certain geometric transformations such as filtering, cropping, compression, and other hostile attacks; therefore, imperceptibility and robustness properties are usually the most desirous properties in case of image watermarking. In context of image compression such as JPEG, watermarking in the transform or wavelet domain is usually exploited.

- Video Watermarking: Video watermarking is the application of watermarking, wherein the sequence of moving still images are watermarked; hence, various image watermarking techniques have been used for the digital video [25, 26]. Contrary to simple images, digital video requires large bandwidth which implies that larger messages can be embedded within the video. As digital video is mostly stored and distributed in compressed form like MPEG, therefore, it is required that the watermarked video should not take more bandwidth as compared to its counterpart.

- Audio Watermarking: In case of audio files, digital audio watermarking is used to protect them from illegal copying. Keeping in view the ease with which the audio files can be downloaded and copied, audio watermarking is becoming necessary. Audio watermarks embedded into the digital audio are special signals. Audio watermarking techniques exploit the imperfection of the human auditory system. These techniques are difficult to design because of the inherent abilities of the complex human auditory system. Thus, good audio watermarking schemes are difficult to design.

6. Spatial Versus Transform Domain

Watermarking operations are mainly performed in either spatial domain or in the transform domain [9, 27]. In order to achieve improved performance in terms of robustness and perceptual transparency, a thorough understanding of the embedding/extraction domain is essential. Accordingly, a comparative analysis in terms of the computational cost, time, resources, complexity, robustness, capacity, and quality is given in Table 1.

Table 1 Comparative analysis of spatial domain and transform domain

S. No.	Factors	Spatial domain	Transform domain
1.	Technique	Simple technique to use by modifying pixel values	Complex to use by modifying transform coefficients
2.	Computation cost	Computation cost involved is low	Computation cost involved is high
3.	Robustness	Fragile, less robust, incompetent in dealing with various attacks	More robust against various attacks
4.	Perceptual quality	High control	Low control
5.	Computational complexity	Low	High
6.	Computational time	Less	More
7.	Capacity	Limited capacity to hold the watermark	High capacity to hold the watermark
8.	Example of application	Mainly authentication	Copy rights

5 Applications of Digital Watermarking

The field of watermarking has witnessed a great deal of research over the last few decades [9]. Due to the important applications of watermarking for copyright protection and management, the research is increasing day by day. A watermark can be applied for a variety of purposes. The following are a few noted applications of watermarking:

1. Copyright Protection
   Watermarking is chiefly used in an organization in order to assert its "owner-ship" of copyright with regards to the digital items [10, 28]. This watermarking application is the main focus of institutions which are vending objects of digital information (news/photograph) and to "big media" organizations. Here, a tiny amount of data (watermark) needs to be embedded which requires a high level of resistance to signal alteration. When digital data is broadcasted, it demands strong watermarking as any intruder can use it without paying the IPR charges to its owner.
2. Copy Protection
   In order to prevent illegal copying of digital data, watermarking is used as a highly potential tool. A system or software trying to copy an audio compact disc cannot do so if a watermark has been embedded in it. Additionally, if copying is done still the watermark will not get copied to the new duplicate compact disc. Hence, the replica compact disc can be recognized easily due to the absence of the watermark. Digital recording devices can be controlled by

the information stored in a watermark for copy protection purpose [29]. Here, the watermark detectors in the recording device decide whether the recording can take place or not. This is made possible by the presence of a copy-prohibited watermark which in the simplest form is a single bit that indicates whether the content is copyrighted or not. This copy protection bit cannot be removed easily as it is strongly tied to the content and when removed the digital content would get seriously affected.

3. Tamper Detection

   Tamper detection is used to ascertain the origin of a data object and to prove its integrity [30]. The classical example of tamper detection is presentation of photographic forensic information as evidence in the legal matters. Since digital media can be manipulated with great ease, there may be an obligation to prove the fact that a media (image/video) has not been altered. In such cases, a camera can be equipped with the capability for tamper detection, by a watermarking system which is embedded in digital cameras [31]. Here as an example, while proving charges of an over speeding driver in a legal court, the driver may claim that the video taken by the police department is tampered. Such a watermark would get destroyed when one would try to tamper the data.

4. Broadcast Monitoring

   A need to monitor the broadcasts of certain individual and institutions for their interests is an important application of watermarking. Here as an example, the exact airtimes purchased from the broadcasting firms received by the advertisers who want to advertise can be ensured. Similarly, celebrities like actors want to calculate their accurate royalty payments of their performances when broadcasted. In addition, the copyright owners do not want that their digital properties illegally rebroadcasted by pirate stations. Such incidents have been seen all over the globe. In Japan, advertisers were paying hefty amounts for commercials that were never broadcasted [32]. The said case had been undetected for nearly 20 years as there was no system to monitor the real broadcast of advertisements. This broadcast can be monitored by using a unique watermark in each video signal or audio clip prior to the broadcast. The broadcasts can be monitored by automated stations to check for the unique watermarks.

5. Fingerprinting

   If the same watermark is placed in all the copies of a single item, it might be a problem for monitoring and owner identification. If one of the legal users of the common digital item sells it illegally, it would be difficult to identify the culprit. Therefore, each copy of the digital data distributed is customized for each legal user. This method embeds a unique watermark to each individual copy. The owner can easily identify the user who is illegally vending the digital media by checking the watermarks. This application of watermarking is called fingerprinting [2]. The fingerprinting serves two purposes: It acts as prevention to illegal use and as an aid to technical investigation.

6. Annotation Applications

Here, the watermarks express the information specific to the digital item such as "feature tags" or "captions" or "titles" to users of the item. As an example, identification of the patient can be embedded into medical media-like images. There exists no need to protect such digital items against intentional tampering. They require comparatively large quantities of embedded data. Such digital items might be susceptible to transformations such as image cropping or scaling; therefore, the watermarking technique resistant to those types of modifications must be employed.

7. Image Authentication and Data Integrity

Image authentication is another application of watermarking. As an example, the authenticity of images/photographs used for surveillance in the military must be established. Digital images are exploited by image processing packages using seamless modifications. Digital images produced before courts as evidences may be modified or altered beforehand. Watermarking can enable the user to detect the imperative modifications of the images under consideration. The watermarks used for such verification purposes need to be fragile so that any alteration to the original image will obliterate or noticeably alter the watermark [33]. Fragile watermarks indicate whether the data has been changed. Further, they also provide information as to where the data was changed.

8. Indexing and Image Labeling

Comments and markers can be embedded in the video content for the purpose of indexing of video mail, news items, and movies that can be used by search engines. This feature is usually used by the digital video disc recordings in order to let the user select certain scenes or episodes without rewinding or forwarding which is otherwise necessary with a video cassette recording (VCR). When the information about the image content is embedded as a watermark, the application is called as image labeling. Such embedding methods are required for proving extra information to the viewer or for image retrieval from a database.

9. Medical Safety

A safety measure similar to fingerprinting could be used to embed the date and the name of the patient in medical images or a music recording with the owner/user information [34]. This application is becoming more important in view of tele-medicine. The watermarks could be used to authenticate the claims made by various bodies regarding the serious health condition of an important personality.

10. Data Hiding

The transmission of secret private messages can exploit watermarking techniques. Using tools such as invisible ink, microscopic writing, or hiding code words within sentences of a message, data hiding or steganography is a method of hiding the existence of a message [35]. Here, the communication using often enciphered messages without attracting the attention of a third party is performed. Here, the important and mandatory properties of the watermark are the invisibility and capacity while robustness requirement is low for steganography. Users may hide their messages due to the fact that various governments are restricting the use of encryption services, in other data.

6 Recent Advances: Wavelet Domain Image Watermarking

Watermarking techniques are usually applied either in the spatial domain or in the transform domain. Spatial domain techniques directly modify the pixels of an image or a subset of them with respect to the watermark pixels values. These methods when subjected to normal media operations are not more reliable. In the transform domain techniques, the image is subject to some kind of reversible transformation [36–39] such as discrete fourier transformation (DFT), discrete cosine transformation (DCT), discrete wavelet transformation (DWT) [13, 15] and then the watermarking is being carried on transform coefficients. Transform domain techniques provide higher image perceptibility and are more robust against various image processing attacks. However, the time/frequency decomposition characteristics of DWT which are very similar to human visual system (HVS) have made them more suitable for watermarking. Earlier, the work carried out in the field of watermarking was mainly focused on monochrome and grayscale media; however, due to widespread use of multimedia applications and keeping in view the fact that most of the organizations and business concerns nowadays mainly use color data especially images (labels, logs, etc.), the need for design and development watermarking schemes for color images has accordingly increased.

Discrete wavelet transformation (DWT) [40] is more appropriate for performing watermarking in transform domain, for it hierarchically decomposes an image [41]. The application of wavelet transform is more suitable so far as the processing of non-stationary signals is concerned. The whole concept of wavelet transform is based on small waves of limited duration called as wavelets having the property of multiresolution analysis [42]. Dissimilar to that of the conventional Fourier transform, the wavelet transform provides a multiresolution description of an image both in spatial as well as in frequency domain and further retains the temporal information also. The high- and low-frequency components are separated from a signal using DWT. The high-frequency components mainly contain the information related to edges, whereas the low-frequency components related to sharp details are further split into low- and high-frequency parts. The high-frequency components are mostly used for watermark embedding/extraction for they are less sensitive toward human visual system (HVS) in comparison with their low-frequency counterparts. The low-frequency components, however, are more robust to image distortions that have low-pass characteristics. These distortions include lossy compression, filtering, and geometric manipulations. So far as gamma correction, contrast/brightness adjustment, and cropping are concerned, low frequencies are

Table 2 Comparative analysis of important digital watermarking methods

S. No.	Algorithm	Advantages	Disadvantages
1.	LSB	It is easy to implement and understand. Image quality is not highly degraded and it possesses high perceptual transparency	It lacks basic robustness and is vulnerable to noise, scaling, and cropping
2.	Correlation	Gain factor can be increased which results in improved robustness	Very high increase in gain factor causes the image quality to deteriorate
3.	Patchwork	It displays high level of robustness against various attacks	It can conceal a diminutive amount of information
4.	Texture mapping	It conceals data within the continuous random texture patterns of a picture	It is only suitable for those areas with large number of arbitrary texture images
5.	DCT	The watermark is embedded into the coefficients of the center frequency, so the visibility of image is not affected. The watermark will not be removed by any kind of attack	Blockwise DCT destroys the invariance properties of the system. Some higher frequency components tend to be suppressed while quantizing
6.	DFT	It is rotation, scaling, and translation invariant therefore can recover from geometric distortions	It has complex implementation. The cost of computing may be higher
7.	DWT	It allows superior localization both in time and spatial frequency domain. It provides higher compression ratio which is pertinent to human perception	Cost of computing is somewhat higher. Time required for compression is more. Noise is present near edges of images or video frames

less robust. The watermark data inserted into middle- and high-frequency components on the other hand is typically less robust to lossy compression, low-pass filtering, and small geometric deformations of the image. The wavelet-based watermarking algorithms and schemes presented so far have been implemented using different wavelet filter banks and by using different decomposition levels. The type of wavelet to be used mainly depends on two important parameters, i.e. symmetry and perfect reconstruction. It has been observed that the appropriate selection of embedding subspace, wavelet filter bank, and decomposition level have sound bearing as far as the robustness and transparency properties are concerned.

A comparative analysis of some of the important digital watermarking methods in spatial domain as well as in transform domain is provided in the following table. From the table, it is evident that spatial domain techniques though simple in implementation are not so resistant to noise and other image processing attacks as compared to the transform domain counterparts (Table 2).

From the available literature, it has been observed that the work done by various researchers over the period of time for watermarking color images using discrete wavelet transformation has been carried in the following directions:

1. Types of Wavelet Filters and Level of Decomposition: Symmetry and perfect reconstruction are two important aspects to decide the type of wavelet being used. As the human visual system is less sensitive to symmetry, the wavelet used should be symmetric. Moreover, to preserve the image quality and imperceptibility, the decomposed image should be perfectly reconstructed. The different types of wavelet filters that have been used for watermarking are Haar wavelet, Daubechies wavelet, bi-orthogonal wavelets, complex wavelets, wavelet packets, balanced multiwavelets, stationary wavelets, morphological wavelets, non-tensor wavelets, and Berkeley wavelets.

2. Color Space Used: In the pervasive multimedia applications, color images are considered to be the basic and pivotal component of all the multimedia systems. The visual perception of these color images has a significant impact even if very small color changes are made. The representing models that can be used as candidates for color image watermarking are RGB, HSV, HSL, CMYK, CIE Lab, and CIE XYZ. RGB images to any other color spaces and then processing these images give good results.

3. Optimization Techniques Used: Optimizations being one of the important parameters to improve the efficiency and effectiveness of any algorithm, wavelet-based watermarking algorithms for color images have been optimized using number of techniques [43–45] such as singular value decomposition (SVD) [46], ant colony optimization (ACO), independent component analysis (ICA), differential evolution (DE), support vector machine (SVM), genetic algorithm (GA), fuzzy logic, cat swarm optimization (CSO), particle swarm optimization (PSO) [47], firefly algorithm (FA), bees algorithm (BA), and cuckoo search (CS).

7 Watermarking Attacks

The most common watermarking attacks [48–50] reported in the literature can be categorized into following four classes:

1. Removal Attacks:
   These attacks are being performed to completely remove the ownership information embedded as watermark data from the watermarked data without compromising on the security of the watermarking scheme under consideration. Some of the attacks included in this category are quantization, denoising, collusion, and remodulation. Though none of these methods significantly damage the watermark information; however, some of the methods come very close to complete removal of watermark information. In order to recover the embedded

watermark as realistically as possible, while maintaining the quality of data being attacked, some sophisticated removal attacks go for optimization of operations such as quantization or denoising.

2. Geometric Attacks:

These attacks intend to change watermark detector synchronization with the embedded watermark information instead of removing the embedded watermark itself [51]. On regaining the perfect synchronization, the embedded watermark information could be recovered by the detector. The amount of complexity involved in the required synchronization process, however, might be too high to be practical. In case of image watermarking, most of the commonly known benchmarking tools normally integrate a variety of geometric attacks. However, by using special synchronization techniques most of the recent watermarking methods withstand against these attacks.

3. Cryptographic Attacks:

These attacks are intended to crack the security of the watermarking schemes being used and thus finding a way to either remove the embedded watermark information or to embed misleading watermark information. The techniques included in this category are oracle attack, where a non-watermarked signal is being created once watermark detector device becomes available and brute-force search for the embedded secret information. However, the computational complexity involved restricts the application of these attacks.

4. Protocol Attacks:

These attacks are being performed with the objective of attacking the concept of the watermarking application as a whole. The concept of invertible watermarks forms the basis of one of the attacks in this category, wherein the attacker claims to be the owner of watermarked data after subtracting his own watermark information from the watermarked data. The idea behind this type of attack is to create ambiguity so far as the true owner of the data is concerned. Another form of protocol attack is the copy attack, wherein the objective is neither to destroy the watermark nor to impair its detection; instead, the watermark data is copied to some other data known as target data. This is usually done after analyzing the watermarked data and estimating the presence of watermark in it. To satisfy the imperceptibility property, the adoption of estimated watermark to the local features of the target data is considered (Fig. 3).

Fig. 3 Watermarking attacks

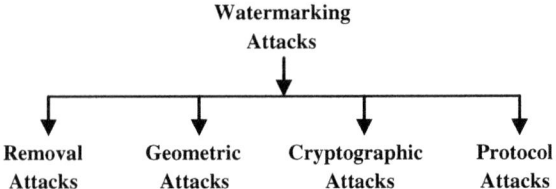

It has been observed that despite of having a clear separation between different attacks as presented in the above classification, an attacker most often applies these attacks in combination instead of using a particular attack in isolation.

8   Future Scope

The extensive creation and dissemination of multimedia content have attracted researchers from computer science and other allied fields to work in the area of digital watermarking to ensure the genuineness of the multimedia content, besides protection of the owner copyrights. Accordingly, lot of research work has been done in this area over the past few decades. Keeping in view the exponential growth of the multimedia data, there is a tremendous potential of further research in this field. Further studies may be conducted to evaluate the attack impacts on multimedia data and then watermarking schemes to be developed so that those impacts could be minimized before the start of watermarking so that a better recovery of the copyright data could be performed. Keeping in view the huge financial implications and aspects of the watermarking application areas, more characteristics against some attacks such as forgery attack or multiple watermarking can be embedded. More wavelet transforms should be examined for embedding of the watermark data and robustness against JPEG2000 format conversion. A watermarking scheme may have some relationship with the image on which it is going to apply. Performance of the watermarking scheme or selection of the watermarking scheme or at least few input parameters of the watermarking scheme must be related to image characteristics. The provision of embedding nested watermarks could be explored in future studies for improved security of the watermark under consideration from the pirates.

9   Conclusion

Digital watermarking is an evolving field of research in computer science and information technology besides many other fields which include signal processing, information security, cryptology, and communications. The diverse nature of this field with respect to multimedia has made research more exciting and challenging. Digital watermarking is still evolving and is an open problem for future researchers.

Acknowledgments We thank our parents for their unflagging belief and encouragement.

References

1. Singh N, Sharma D (2015) A review on watermarking & image encoding. Int J Comput Sci Mobile Comput 4(6):632–636
2. Giri KJ, Peer MA, Nagabhushan P (2014) A channel wise color image watermarking scheme based on discrete wavelet transformation. In: Proceedings of IEEE international conference on computing for sustainable global environment trans, pp 758–762
3. Nin J, Ricciardi S (2013) Digital watermarking techniques and security issues in the information and communication society. In: International conference on advanced information networking and applications, pp 1553–1558
4. Sequeira A, Kundur D (2001) Communications and information theory in watermarking: a survey. In: Proceedings of SPIE multimedia systems and application IV, vol 4518, pp 216–227
5. Cox IJ, Miller ML (2001) Electronic watermarking. In: IEEE fourth workshop on multimedia signal processing, pp 225–230
6. Cox IJ, Miller ML, Bloom JA, Friedrich J, Kalker T (2008) Digital watermarking and steganography, 2nd edn. Morgan Kaufman, San Francisco
7. Schyndel RGV, Tirkel AZ, Osborne CF (1994) A digital watermark. In: Proceedings of IEEE international conference on image processing, vol 2, pp 86–90
8. Tirkel AZ, Rankin GA, Van Schyndel RM, Ho WJ, Mee NRA, Osborne CF (1993) Electronic water mark. DICTA, Macquarie University, pp 666–673
9. Kaiser J, Giri M, Nagabhushan P (2015) A robust color image watermarking scheme using discrete wavelet transformation. Int J Image Graph Signal Process 1:47–52
10. Giri KJ, Peer MA, Nagabhushan P (2013) Copyirght protection of color images using novel wavelet based watermarking algorithm. In: Press: 2nd IEEE international conference on image information processing, JUIT, Shimla, India
11. Kutter M, Petitcolas F (1999) A fair benchmark for image watermarking systems. In: Electronic imaging 199: security and watermarking of multimedia content, vol 3657 of SPIE Proceedings, San Jose, California USA, 25–27 January 1999
12. Jithin VM, Gupta KK (2013) Robust invisible QR code image watermarking in DWT domain. Int J Electron Commun Eng Technol (IJECET) 4(7):190–195
13. Zhao M, Dang Y (2008) Color image copyright protection digital watermarking algorithm based on DWT & DCT. IEEE, pp 1–4
14. Anuradha RPS (2006) DWT based watermarking algorithm using haar wavelet. Int J Electron Comput Sci Eng 1:1–6
15. Zhang Y, Wang J, Chen X (2012) Watermarking technique based on wavelet transform for color images. In: 24th Chinese Control and Decision Conference (CCDC), pp 1909–1913
16. Dey N, Biswas S, Das P, Das A, Chaudhuri SS (2012) Lifting wavelet transformation based blind watermarking technique of photoplethysmographic signals in wireless telecardiology. In: 2012 World Congress on Information and Communication Technologies (WICT). IEEE, pp 230–235
17. Dey N, Mukhopadhyay S, Das A, Chaudhuri SS (2012) Analysis of P-QRS-T components modified by blind watermarking technique within the electrocardiogram signal for authentication in wireless telecardiology using DWT. Int J Image Graph Signal Process 4 (7):33
18. Dey N, Das P, Chaudhuri SS, Das A (2012) Feature analysis for the blind-watermarked electroencephalogram signal in wireless telemonitoring using Alattar's method. In: Proceedings of the fifth international conference on security of information and networks. ACM, pp 87–94
19. Dey N, Biswas D, Roy AB, Das A, Chaudhuri SS (2012) DWT-DCT-SVD based blind watermarking technique of gray image in electrooculogram signal. In: 2012 12th international conference on intelligent systems design and applications (ISDA). IEEE, pp 680–685
20. Dey N, Roy AB, Das A, Chaudhuri SS (2012) Stationary wavelet transformation based self-recovery of blind-watermark from electrocardiogram signal in wireless telecardiology. In:

Recent trends in computer networks and distributed systems security. Springer, Berlin, pp 347–357

21. Khalifa A, Hamad S (2012) A robust non-blind algorithm for watermarking color images using multi-resolution wavelet decomposition. Int J Comput Appl 37(8):0975–8887

22. Elbasi E, Eskicioglu AM (2006) A semi-blind watermarking scheme for color images. Int J Technol Eng Syst (IJTES) 2(3):276–281

23. Craver S, Memon N, Yeo B, Young M (1997) On the invertibility of invisible watermarking techniques. In: Proceedings of the IEEE International Conference on Image Processing, vol 1, pp 540–543

24. Hua Y, Wu B, Wu G (2010) A color image fragile watermarking algorithm based on DWT-DCT. In: Chinese control and decision conference, pp 2840–2845

25. Hartung F, Girod B (1997) Digital watermarking of raw and compressed video. In: Proceedings of SPIE 2952: digital compression technologies and systems for video communication, pp 205–213

26. Hartung F, Girod B (1998) Watermarking of uncompressed and compressed video. Signal Process 66:283–301

27. Chavan SK, Shah R, Poojary R, Jose J, George G (2010) A novel robust colour watermarking scheme for colour watermark images in frequency domain. In: International conference on advances in recent technologies in communication and computing, pp 96–100

28. Brassil J, Low S, Maxemchuk N, O'Gorman L (1994) Electronic marking and identification techniques to discourse document copying. Proc INFOCOM 13(8):1495–1504

29. Langelaar GC, Lagendijk RL, Biemond J (1998) Real-time labeling of MPEG-2 compressed video. J Vis Commun Image Represent 9(4):256–270

30. Dey N, Dey M, Mahata SK, Das A, Chaudhuri SS (2015) Tamper detection of electrocardiographic signal using watermarked bio–hash code in wireless cardiology. Int J Signal Imaging Syst Eng 8(1–2):46–58

31. Friedman G (1993) The trustworthy digital camera. IEEE Trans Consum Electron 39(4):93–103

32. Cox IJ, Miller M, Bloom J (2001) Digital watermarking: principles and practice. Morgan Kaufmann 10:1558607145

33. Wolfgang RB, Delp EJ (1999) Fragile watermarking using the VW2D watermark. Proc Electron Imaging 3657:204–213

34. Anderson RJ, Petitcolas FAP (1998) On the limits of Steganography. IEEE J Sel Areas Commun 16:474–481

35. Brassil J, Low S, Maxemchuk N, O'Gorman L (1995) Hiding information in document images. In: Proceedings of the 29th annual conference on information sciences and systems, pp 482–489

36. Dey N, Biswas S, Roy AB, Das A, Chowdhuri SS (2013) Analysis of photoplethysmographic signals modified by reversible watermarking technique using prediction-error in wireless telecardiology

37. Dey N, Biswas S, Das P, Das A, Chaudhuri SS (2012) Feature analysis for the reversible watermarked electrooculography signal using low distortion prediction-error expansion. In: 2012 International conference on communications, devices and intelligent systems (CODIS). IEEE, pp 624–627

38. Chakraborty S, Maji P, Pal AK, Biswas D, Dey N (2014) Reversible color image watermarking using trigonometric functions. In: 2014 International conference on electronic systems, signal processing and computing technologies (ICESC). IEEE, pp 105–110

39. Dey N, Maji P, Das P, Biswas S, Das A, Chaudhuri SS (2013) Embedding of blink frequency in electrooculography signal using difference expansion based reversible watermarking technique. arXiv preprint arXiv:1304.2310

40. Dey N, Pal M, Das A (2012) A session based watermarking technique within the NROI of retinal fundus images for authentication using DWT, spread spectrum and Harris corner detection. Int J Mod Eng Res 2(3):749–757

41. Liu K-C (2009) Human visual system based watermarking for color images. In: Fifth international conference on information assurance and security, vol 2, pp 623–626
42. Qiang S, Hongbin Z (2010) Color image self-embedding and watermarking based on DWT. In: International conference on measuring technology and mechatronics automation, vol 1, pp 796–799
43. Dey N, Samanta S, Chakraborty S, Das A, Chaudhuri SS, Suri JS (2014) Firefly algorithm for optimization of scaling factors during embedding of manifold medical information: an application in ophthalmology imaging. J Med Imaging Health Inform 4(3):384–394
44. Dey N, Samanta S, Yang XS, Das A, Chaudhuri SS (2013) Optimisation of scaling factors in electrocardiogram signal watermarking using cuckoo search. Int J Bio-Inspired Comput 5 (5):315–326
45. Acharjee S, Chakraborty S, Samanta S, Azar AT, Hassanien AE, Dey N (2014) Highly secured multilayered motion vector watermarking. In: advanced machine learning technologies and applications. Springer, Berlin, pp 121–134
46. Dey, N., Das, P., Roy, A. B., Das, A., &Chaudhuri, S. S. (2012, October). DWT-DCT-SVD based intravascular ultrasound video watermarking. In Information and Communication Technologies (WICT), 2012 World Congress on (pp. 224–229). IEEE
47. Chakraborty S, Samanta S, Biswas D, Dey N, Chaudhuri, SS (2013) Particle swarm optimization based parameter optimization technique in medical information hiding. In: 2013 IEEE international conference on computational intelligence and computing research (ICCIC). IEEE, pp 1–6
48. Voloshynovskiy S, Pereira S, Iquise V, Pun T (2001) Attack modelling: towards a second generation watermarking benchmark. In: Signal processing, Special Issue on Information Theoretic Issues in Digital Watermarking
49. Hartung F, Su JK, Girod B (1999) Spread spectrum watermarking: malicious attacks and counter-attacks. In: Proceedings of SPIE vol 3657: security and watermarking of multimedia contents, San Jose, CA, USA
50. Kutter M, Voloshynovskiy S, Herrigel A (2000) Watermark copy attack. In: Wong PW, Delp EJ (eds) IS&T/SPIE's 12th annual symposium, electronic imaging 2000: security and watermarking of multimedia content II, vol 3971 of SPIE Proceedings, San Jose, California USA, 23–28 January 2000
51. Liu L-M, Han G-Q, Wo Y, Wang C-S (2010) A wavelet-domain watermarking algorithm against geometric attacks based on SUSAN feature points. IEEE, pp 1–5
52. Pal AK, Das P, Dey N (2013) Odd-even embedding scheme based modified reversible watermarking technique using Blueprint. arXiv preprint arXiv:1303.5972

Real-time Implementation of Reversible Watermarking

H.R. Lakshmi, B. Surekha and S. Viswanadha Raju

Abstract In today's Internet-connected world, tampering of digital images by malicious users has become very common. This clearly is a direct violation of one's intellectual property rights, and hence, image protection by resolving rightful ownership is gaining utmost importance. Reversible watermarking (RW) techniques are a potential replacement for conventional watermarking systems, in case if the images to be protected are very sensitive. Studies have suggested that these methods offer greater balance among the requirements of watermarking such as invisibility, robustness, and capacity. This chapter gives an overview of the RW techniques, their vulnerabilities, and recommendations for future development. Performance issues related to the real-time implementation of watermarking systems through various algorithms are also featured. Finally, an extension of real-time implementation of RW approach based on asynchronous architectures is introduced.

Keywords Watermarking · Copyright protection · Intellectual property · FPGA · Asynchronous clock

1 Introduction

The wide outreach of Internet has resulted in a steady rise of transmission of digitized images, audio, and video. Due to this, the odds of an unauthorized person tampering and misusing the images (audio or video) also are on a steady rise [1–3].

H.R. Lakshmi (✉) · B. Surekha
K S Institute of Technology, Bangalore, India
e-mail: hrl.lakshmi@gmail.com

B. Surekha
e-mail: borrasurekha@gmail.com

S. Viswanadha Raju
JNTUHCEJ, Karimnagar, Telangana, India
e-mail: svraju.jntu@gmail.com

© Springer International Publishing Switzerland 2017
N. Dey and V. Santhi (eds.), *Intelligent Techniques in Signal Processing for Multimedia Security*, Studies in Computational Intelligence 660,
DOI 10.1007/978-3-319-44790-2_6

Thus, in order to protect one's privacy and intellectual property rights, protection of digital images is inevitable. One way of tamper detection and protecting digital images is through digital watermarking, where secret information called watermark is embedded in the owner's digital image. The embedded watermark can be text, bit patterns, or another image, which can later be extracted to attest ownership. Since watermarking involves modification of original image to include watermark, it may damage/distort some sensitive information present in original image [4]. In critical medical diagnosis and military imaging, this is highly unacceptable [5]. A general classification of digital image watermarking algorithms is shown in Fig. 1.

Based on the embedding domain, there are spatial and frequency categories, with frequency being more robust. In frequency domain techniques, the watermark is embedded after extracting features from image through transforms [6]. Usually, frequency domain techniques are more robust than spatial domain methods.

Based on the human perception, the watermarking schemes may be classified as visible techniques, where the watermark embedded is visible to the viewer, and invisible techniques, where the embedded watermark is inconspicuous to the viewer. While visible watermarking schemes are mainly used to declare ownership of the image, invisible techniques provide authentication of ownership apart from detection of tampering of image.

Based on the image recovery capability, the watermarking schemes are classified as non-reversible and reversible techniques [7]. While non-reversible techniques cannot recover the quality image after embedding; RW techniques can recover the original with no loss after extraction. Reversible watermarking techniques are particularly useful in critical applications such as medical imagining and defense, where even a slight distortion in the original image is unacceptable.

This chapter focuses on RW techniques and their real-time implementations. Section 2 briefs the concepts and applications of RW techniques. Section 3 presents the literature survey on RW techniques. This section also focuses on the real-time implementations of existing watermarking techniques, their performance issues, and their vulnerabilities. Section 4 presents a novel asynchronous architecture-based hardware implementation of reversible watermarking. Section 5 discusses about the results.

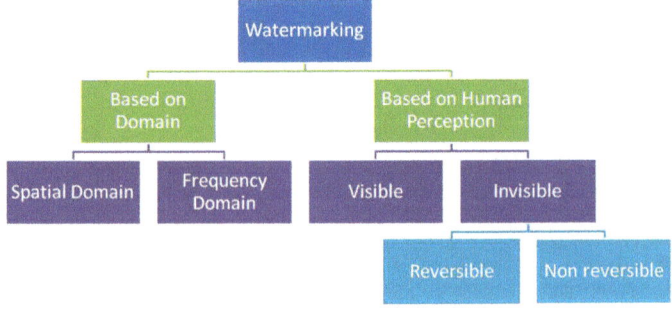

Fig. 1 Classification of digital image watermarking

2 Reversible Watermarking

2.1 Overview

In general, any reversible watermarking involves three stages: (i) watermark embedding, (ii) watermark extraction, and (iii) original image recovery [8]. All these stages need a secret key to ensure the security of the watermark. The block diagram of reversible watermarking is shown in Fig. 2. The key steps involved are as follows:

(i) Watermark embedding:
   Input: Original Image + Watermark, Output: Watermarked image
   In watermark embedding process, a watermark is chosen and is embedded into the image with the help of a suitable RW algorithm. A secret key/information is also generated which later aids in extraction process.
(ii) Watermark extraction:
   Input: Watermarked Image, Output: Watermark
   For the watermark extraction, a watermark extraction algorithm is needed which is usually the inverse of the watermark algorithm. This extracts the watermark.

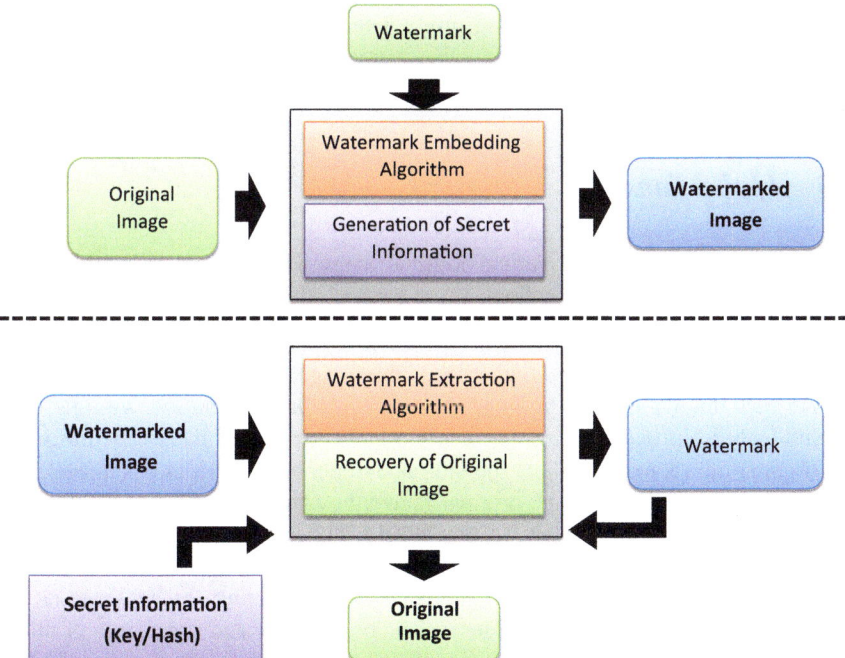

Fig. 2 General watermarking scheme

(iii) Original image recovery
       Input: Watermarked Image + Secret Key + Watermark,
       Output: Original Image
       From the watermark and secret key, original image is restored.
       If the watermarked image is subjected to any attacks/tampering, the extracted
       watermark is distorted indicating foul play [9].

2.2 Requirements

- *Robustness*: The watermark embedded must resist as much number of attacks as possible. The intentional or unintentional attacks should not distort the watermark.
- *Imperceptibility*: The watermark should remain invisible to the viewer. That is, the original image and watermarked image must be similar.
- *Security*: Unauthorized people should not be able to remove the watermark. Hence, the RW algorithm used must be secure.
- *Capacity*: The capacity of a watermark implies the amount of information that the watermark can convey. The capacity of the watermark should be high.
- *Tradeoffs*: There are various trade-offs in choosing a good RW scheme. The trade-offs may be among capacity, imperceptibility, and robustness [10].
- *Processing time*: The processing time for the watermark embedding or extraction should be less. That is, the RW scheme should be fast.

2.3 Applications

Due to its capability of recovering original image with no loss, reversible watermarking has following applications:

- Watermarks to authenticate ownership: The owner of the image will be able to assert her ownership. The owner A of the image generates a watermark using her own unique secret key. Now, if a malicious user B wants to wrongly claim his ownership over the image, or a modified version of it, A can furnish the original image and also show the presence of her watermark in B's image. As B does not have access to the original image, he cannot justify. For this to work, the watermark should resist various attacks, either intentional or unintentional. Also the watermark should be non-replicable.
- Watermarks for detection of fraudulence and tampering: In applications like news media, business transactions, medicine, and law and order, it is of utmost importance to guarantee that the content originated from legit source has not been tampered. The watermark embedded serves this purpose. The integrity of

watermark assures the integrity of the image. The watermark used should be invisible [10].

- Watermarks for ID card security: ID card frauds are one of the most persisting problems. This can be controlled through the use of watermarks. In the person's photograph that is in the ID card, the basic information regarding the person's identity can be embedded as the watermark. By extracting this information (watermark), person's identity can be verified. This controls the identity theft crime.
- Protection of library manuscript while digitization.
- Protection of radiological images in telemedicine.
- Protection of defense images.

The applications stated above are only few among the many applications of digital watermarking. For each application, the applied watermarking scheme should meet application-specific requirements [10].

3 Literature Survey

The image watermarking schemes can be either software or hardware implemented. For software implementation, an external processor and memory blocks are needed to run the watermarking algorithms. Since they run on an external processor, they cannot be optimized in terms of area, power, and performance. In contrast, for a real-time hardware implementation, full custom circuitry is used, and the above-mentioned parameters can be optimized. There are many software implementations available for RW in the literature, whereas the hardware implementations are limited in number.

3.1 Reversible Watermarking (RW) Algorithms

The very first reference to reversible data embedding is found in the patent filed by Barton [11] where the bits to be embedded are compressed and added to the cover data. Honsinger et al. [12] showed a lossless recovery of cover image by first reconstructing the embedded watermark and then subtracting it from the watermarked image. Hong [13] proposed a new method using a modified histogram shifting. Improvement in image quality and reduction in cost are the highlights of this scheme.

Luo et al. [14] proposed embedding watermark through interpolation technique. Interpolation succeeds well in estimating pixels but the capacity of watermark is low. Celik et al. [15] proposed a RW scheme where image pixels are L-level quantized. The watermark bits are added to these quantized pixels to get the watermarked image. This method exhibits low watermark capacity.

Rajendra et al. [16] proposed a modified histogram shifting technique which increases the capacity of watermark and also enhances image quality. In [17], Ni et al. used histogram zero and peak points to embed the watermark into the histogram bins. The drawback of this method was that additional information about the peak and zero points was needed at the decoder side. This drawback is overcome by Hwang et al. [18] where they use a binary tree structure.

Tian [19] proposed a difference expansion (DE)-based method which makes the watermark capacity to increase significantly as there is no need for compression to embed the watermark. In the DE methodology, the storage space is allocated by exploiting redundancy in images. Compared to compression-based techniques, this scheme exhibits better image quality with high embedding capacity. The watermark bit is embedded into an expandable difference value of two adjacent pixels. Restoring original data and getting back the watermark requires decoder knowledge of which difference values are embedded. This information is also embedded within the image in the form of a location map.

Coltuc et al. [20] proposed a RW scheme in spatial domain. This is based on an integer transform and reversible contrast mapping (RCM). High data rate and reduction in complexity are the highlights of this scheme. Even if the LSB's are lost, watermark can be extracted. The mathematical complexity of this technique is less when compared to [19] scheme.

Dey et al. [21] discussed a blind spread spectrum and discrete wavelet transform (DWT)-based watermarking scheme for authentication of medical images. A non-region of interest is chosen for the watermark to be embedded. Dey [22] further proposed an edge-based scheme where watermark is embedded at the edges, i.e., boundaries between region of interest (ROI) and the region of non-interest (RONI).

Pal et al. [23] proposed a reversible watermarking scheme where a blueprint of the cover image is generated using odd–even embedding technique. The watermarked image quality was found to be approximately 51 dB.

Dey et al. [24] proposed a reversible watermarking technique using difference expansion (DE) where computed parameters of electrooculography (EOG) signal is embedded as watermark. This avoids postcalculation of data and aids in communication of patient information only among the authorized parties. The same authors proposed [25] a pixel prediction error technique for reversible watermarking of EOG signals discussed. This method has improved SNR between embedded and extracted EOG signals.

Dey et al. [26] discussed a self recovery-based watermark retrieval scheme using stationary wavelet transformation (SWT), quantization, and spread spectrum. Dey et al. [27] discussed safe transmission of electrocardiogram (ECG) using a combination of reversible watermarking and bio-hash techniques. Any tampering in the ECG signal can be detected. Dey et al. [28] also presented a new robust watermarking scheme for embedding hospital logo or patient information into an ECG signal. DWT and Cuckoo Search (CS) algorithms are used.

Dey et al. [29] presented a robust biomedical authentication system using firefly algorithm (FA). The watermark is embedded within the retinal cover image using either DCT–DWT–SVD using the meta-heuristic algorithm.

Dey et al. [30] used a combination of DWT–DCT–SVD to embed watermark into medical images. Authors also [31] proposed a lifting wavelet transformation-based binary watermark for authentication of photoplethysmography (PPG) signals. These are tools to monitor the functions of the human heart.

Chakraborty et al. [32] discussed a particle swarm optimization (PSO) technique to optimize the scaling factors for watermark embedded. A DWT-based method is used to embed the watermark which is either hospital logo or patient information. Chakraborty et al. [33] proposed a color image reversible watermarking scheme using trigonometric functions. The watermark bits are embedded into gray planes of the color host image.

Alattar [34] discussed a DE scheme for colored images. A pair of bits are embedded into a triplets of pixels. One triplet may consist of three pixels from the same spectral component or three pixels from different spectral components. This scheme enables high imperceptibility and higher capacity for the watermark.

Aarthi et al. [35] discussed a LSB method. The watermark is embedded by performing operations on least significant bits. The complexity of this scheme is very low.

Dragoi et al. [36] presented an adaptive interpolation scheme. The interpolated values are calculated in one of the two ways—average of four neighboring pixels and average of horizontal or vertical pairs of pixels. To achieve minimum distortion, data are hidden in pixels with estimation error 'e' less than threshold T. If $e > T$, the pixels are grouped as 'not embedded.' Comparison of reversible watermarking algorithms is given in Table 1.

3.2 Real-time Implementation

Real-time implementation of watermarking refers to the algorithms implemented on a hardware platform, the choice of which can be field-programmable gate arrays (FPGA), digital signal processors (DSP), or system on chips (SoC's). DSP implementations [37] do not fully exploit the parallelism of the algorithms as multichannel processing is not straight forward in real-time task. The parallel processing capability of FPGA implementations allows multichannel processing and suits best for real-time image processing. Further, FPGA is better with respect to device cost, power consumption, and processing speed [38].

The desirable characteristics for a good real-time watermarking scheme are as follows:

- The scheme/device should utilize as little area as possible. Memory utilization in particular should be minimal.
- The power consumption should be very less.

Table 1 Comparison of reversible watermarking (RW) techniques

Author	Imperceptibility	Method	Capacity of watermark	Robust to attacks	Processing time
Tian et al. [19]	High	Difference Expansion	High	No	Fast
Yang et al. [37]	High	Histogram bin shifting	Less	Moderate	Slow
Celik et al. [15]	Moderate	Compression	Moderate	No	Fast
Dragoi et al. [36]	High	Prediction error	Moderate	Moderate	Fast
Alattar [34]	High	Difference expansion of triplets	High	No	Fast
Ni et al. [17]	High	Modified histogram bin shifting	Less	Moderate	Slow
Aarthi et al. [35]	Moderate	LSB	Less	No	Fast
Wien Hong [13]	High	Modified histogram shifting	Moderate	Moderate	Slow
Luo et al. [14]	High	Interpolation	Less	No	Fast
Rajendra et al. [16]	High	Modified histogram shifting	High	Moderate	Slow

- The device should have very high performance with very low delay times.
- The device has to be able to easily integrate with other devices [39].

Mohanty et al. [41] presented a hardware implementation of a JPEG encoding block. CMOS 0.35 μm technology is used. This watermarking chip has the capability to implement both fragile and robust watermarks. The algorithms discussed in [43–45] are used for this implementation. Garimella et al. [42] showed a watermarking scheme for color images and implemented using application specific integrated circuit (ASIC). The area and power are calculated using CMOS 0.13 μm technology. The algorithm used is difference encoding scheme [46]. ASIC is an IC that is specifically developed only for a particular function rather than for general purpose. The technology used is 0.13 μm CMOS.

Samanta et al. [47] aimed at reducing power consumption through the use of single-electron tunneling (SET) devices. These devices have ultra low-power consumption metrics. Lower bit plane modulation technique is used for embedding the watermark [47]. This is an example of ultra large-scale integration (ULSI).

Mathai et al. [48] discussed the various issues involved in hardware implementation of watermarking and proposed a video watermarking algorithm, namely JAWS focusing on area, power, and timing constraints. To reduce power, Mohanty et al. [49, 50] presented a VLSI implementation based on dual voltage, dual frequency, and clock gating. This method is defined for both visible and invisible watermarks using DCT. Karmani et al. [51] described a two-dimensional scan-based frequency domain watermarking for both video and image. The design is implemented in Altera Stratix—II FPGA. Results show that this method is robust against malicious attacks. Lad et al. [52] proposed a VLSI implementation of watermarking based on wavelets. The design is implemented on FPGA as well as through full custom circuitry and to optimize the area and power.

Morita [53] discussed implementation of a DCT visible watermarking scheme. A hardware–software co-design approach is used. Computationally intensive tasks are real-time implemented, whereas less complex processes are handled by software. Lim [54] proposed a FPGA implementation of watermarking scheme for a digital camera. This implementation was much faster than the existing software implementations available.

All the above-discussed schemes are synchronous and are clock based. The power consumption can further be optimized by choosing an asynchronous architecture. The next section discusses an asynchronous implementation of reversible watermarking.

4 Proposed Reversible Watermarking (RW) Implementation

4.1 Review of Dragoi [36] Algorithm

The reversible watermarking technique chosen for implementation is Dragoi et al. [36] scheme. It is defined in spatial domain and is based on enhanced rhombus interpolation and difference expansion. The algorithm consists of mainly three stages: watermark embedding, watermark extraction, and original image recovery. These stages are shown in Figs. 3, 4, and 5.

4.1.1 Watermark Embedding

Cover image P is first preprocessed by defining a rhombus neighborhood. The pixel of the cover image $p_{x,y}$ is modified as $\widehat{p_{x,y}}$ by taking the average of its four immediate neighbors $p_{x-1,y}, p_{x,y-1}, p_{x+1,y}, p_{x,y+1}$. If pixel region is uniform, estimation is done using Eq. 1.

$$\widehat{p_{x,y}} = [p_{x-1,y} + p_{x,y-1} + p_{x+1,y} + p_{x,y+1}] \tag{1}$$

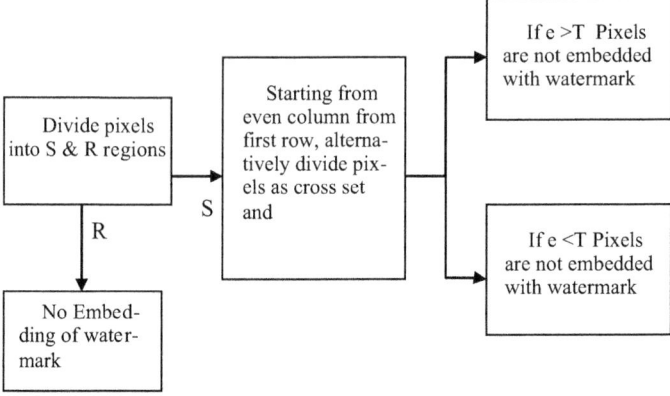

Fig. 3 Watermark embedding

Fig. 4 Watermark extraction

Fig. 5 Original imager restoration

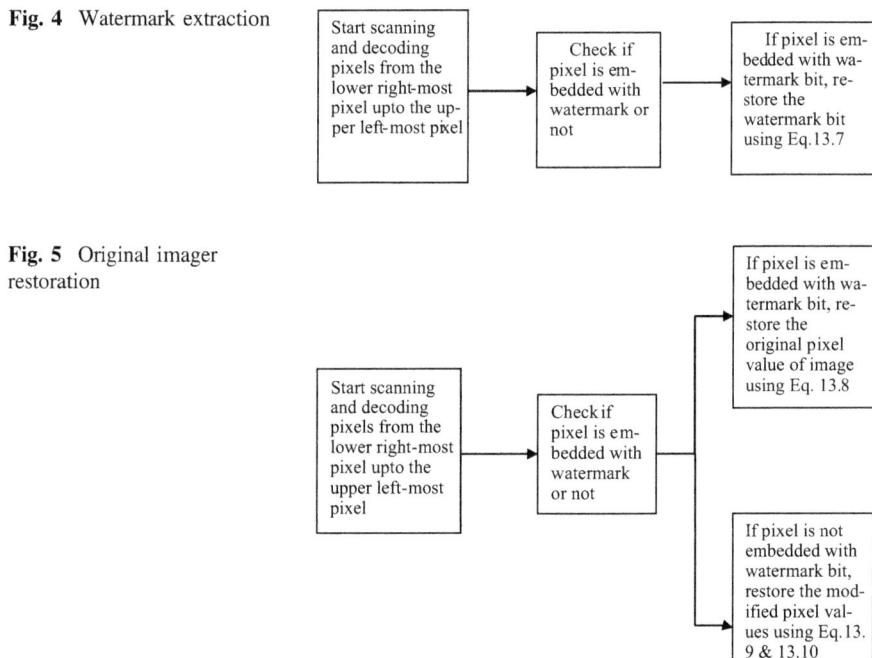

Or else, the pixel is considered to be native to the most homogeneous group. To know the homogeneity between the pixels, their distance is computed for both the groups where dst_h and dst_v are horizontal and vertical distances, respectively

$$\text{dst}_h = p_{x,y-1} - p_{x,y+1} \qquad (2)$$

$$\text{dst}_v = p_{x-1,y} - p_{x+1,y} \qquad (3)$$

Modulus values are considered in Eqs. 2 and 3. If the uniformity of the rhombus is within a fixed threshold value, then new value of the pixel is calculated through Eq. 1. If the vertical distance is more than horizontal distance, then the average of horizontal neighbors is considered. Or else, the average of vertical neighbors is considered. To embed the watermark, the estimation error parameter 'e' is used:

$$e = p_{x,y} - \widehat{p_{x,y}} \qquad (4)$$

The Watermark bit 'w' is embedded into the pixel $p_{x,y}$ of the original image, to form $p'_{x,y}$ as follows:

$$p'_{x,y} = e + w + p_{x,y} \qquad (5)$$

The limiting values for the pixel are $[0, L - 1]$. Here, for an 8-bit grayscale image, the value $L - 1 = 255$. After embedding, the new estimation error e' is computed as follows:

$$e' = 2e + w \qquad (6)$$

Steps for watermark embedding:

- The image is grouped into two divisions—S (selected) region and R (reserved) region. S consists of the pixels where watermark has to be embedded. R is used to send information needed at the decoding stage in the form of threshold values and flag bits.
- The watermark bits are embedded into the pixels with estimation error $e < T$, where T is the threshold. This condition is maintained to minimize distortions.
- Starting with $T = 1$, the embedding process is simulated. If the number of embeddable pixels does not suffice, T is incremented. This continues till the watermark is completely embedded.
- The pixels with $e > T$ form the 'not embedded group.' These pixels are either modified or retained without change so that they provide higher error in comparison with the embedded pixels.
- The pixels of the S region are further segregated as two sets, namely the dot set and the cross set.
- Starting from first row 1, the pixels of all even numbered columns are assigned to the cross set and the remaining pixels are assigned to the dot set. For row 2, the pixels of even columns are put into the dot set, and those of odd columns are put into cross set. The distribution of pixels is thus alternated in the same manner.
- While embedding the watermark bits, the cross set pixels are first checked whether they are embeddable or not, followed by dot set pixels.

4.1.2 Watermark Extraction

The extraction of watermark follows exact process of embedding in reverse. The pixels which are operated from top left to the bottom right, in case of embedding, are now operated in reverse order. The value of error and the value $p'_{x,y}$ of pixel decide if the watermark is embedded or not. The watermark bit w is then extracted from the error e as follows:

$$w = (e') \bmod 2 \qquad (7)$$

Since the encoding started from the cross set, decoding starts from dot set. Since the decoding is done in a direction opposite to encoding, in order to recover the embedded watermark, the bit stream has to be reversed.

4.1.3 Original Image Recovery

A pixel chosen for embedding is restored as follows:

$$p_{x,y} = p'_{x,y} - e' - w \qquad (8)$$

For $e > 0$, a pixel that is not embedded with watermark is restored as

$$p_{x,y} = p'_{x,y} - T \qquad (9)$$

For $e < 0$, a pixel that is not embedded with watermark is restored as

$$p_{x,y} = p'_{x,y} - 1 + T \qquad (10)$$

4.2 Architecture

This section describes a low-power implementation of reversible watermarking. Since the above-mentioned problems are due to clock, a lucrative solution would be to eliminate clock completely. Such clock-less design constitutes asynchronous designs. As asynchronous designs are relatively new, not many CAD tools are available to support this style. In contrast to the synchronous design styles where clock synchronizes all the operations, a separate synchronizing mechanism is needed, in case of asynchronous designs. Thus, asynchronous designs usually tend to have higher area requirements than their synchronous counterparts. In spite of these drawbacks, asynchronous design styles are emerging as they consume very little power.

In the absence of clocks, the blocks in asynchronous systems communicate via handshaking signals—request (req) and acknowledgment (ack). Only when the blocks are actively communicating, these req and ack signals are active. In contrast, in clocked designs, all the blocks switch on every clock cycle, thus increasing power consumption. Another advantage of asynchronous designs is that there is no need for clock routing, and hence, the design complexity reduces considerably [41].

Two basic protocols are used in asynchronous systems, viz. 2-phase and 4-phase handshaking. A 4-phase handshaking scheme needs four full transitions to complete one req–ack-based transaction between the send and the receive ends. Similarly, a 2-phase handshaking scheme needs two full transitions to complete one transaction. This is shown in Fig. 6.

The pipeline structure used is mousetrap pipelining as shown in Fig. 7. Mousetrap pipelining uses cells from the standard cell library. This style was chosen because the signaling is simpler, and the pipeline control circuitry is small and simple [57]. The pipeline includes a 2 I/P XNOR gate and a D latch that is level triggered. The initial state is same for all request (req) and acknowledge (ack) signals. A 'high' on req0 gets a 'high' on ack1 after a delay through one gate. The req1 of stage 1 is the delayed version of req0. This delay depends on the worst case delay for the combinational logic associated with that stage. The latch is made transparent or opaque by the latch controller based on the ack signal from the next stage. The ack2 does not become high till complete data are passed out of stage 1. Till this point, the latch of stage 1 is opaque, i.e., no new data are allowed to enter stage 1.

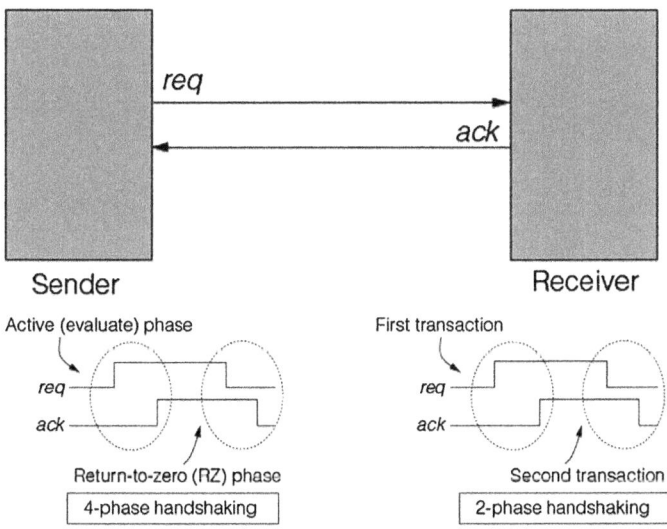

Fig. 6 Asynchronous protocols [57]

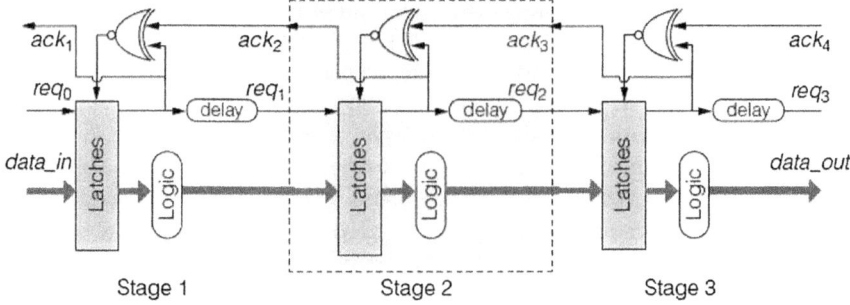

Fig. 7 Mousetrap pipeline [57, 58]

Once the ack2 is high, the XNOR gate output is high and the latch becomes re-enabled. Thus, the pipeline takes care that data are latched only when there is a request. New data are applied only when there is an acknowledgment from the next stage.

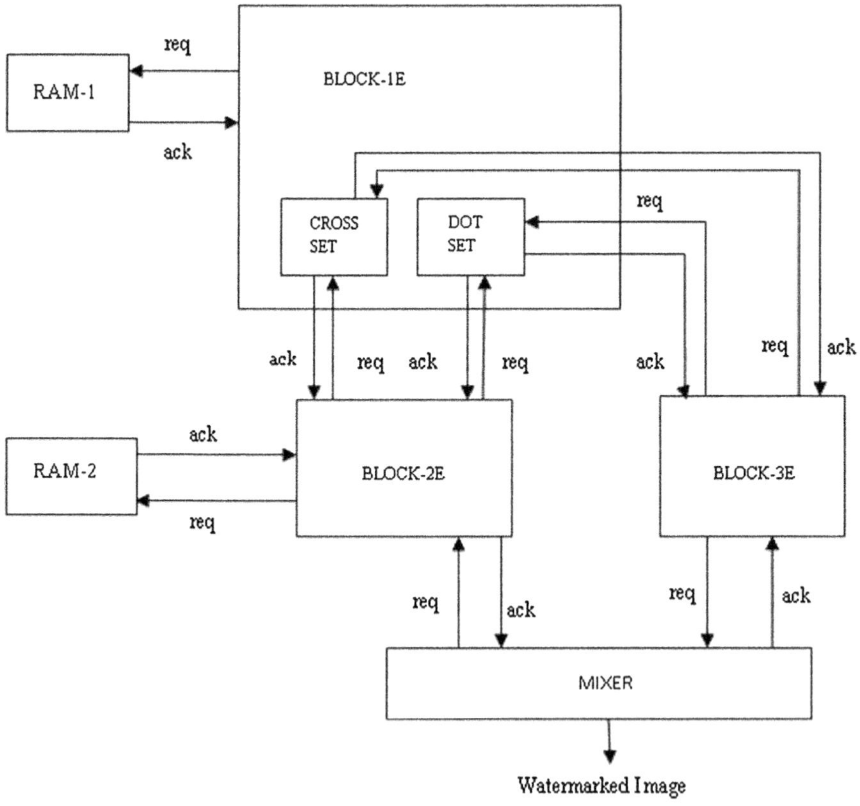

Fig. 8 Top level block for encoding

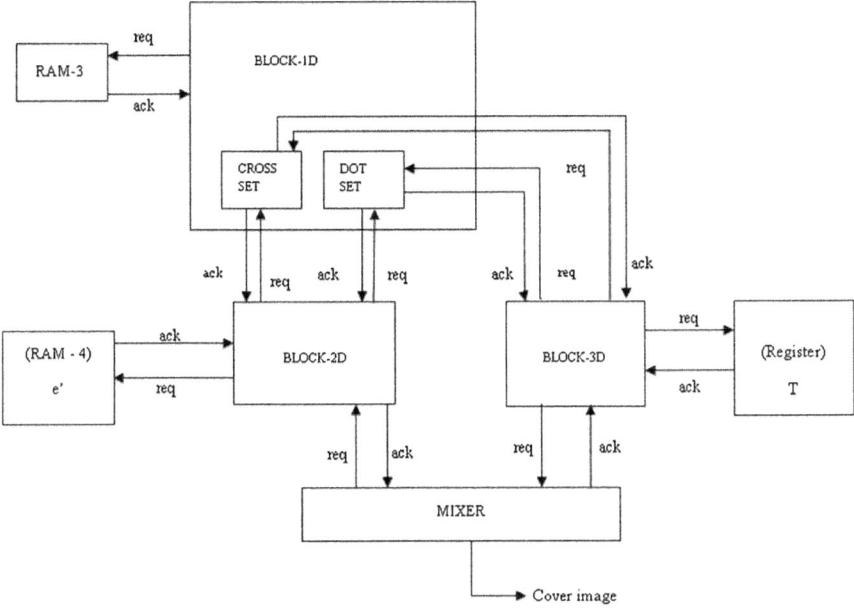

Fig. 9 Top level block for decoding

Figure 8 shows the top level block diagram for encoding operation using the above-mentioned scheme. The architecture mainly consists of three blocks. Block 1E for dividing the pixels of the original image into cross set and dot set regions. Block 2E embeds the watermark into the original image. Block 3E performs some modification operations on the not embedded pixels. Memory RAM-1 stores the original cover image. RAM-2 stores the watermark.

Figure 9 shows the block of decoding operation. Block 1D performs operation similar to Block 1E. Block 2D extracts the watermark from the watermarked image using the value of e'. This watermark is stored in RAM-3. Block 3D restores the pixel values using value of T. These are stored in RAM-4. As evident from the block diagram, each block communicates with the other using req and ack signals. The memory blocks used are also asynchronous memories.

5   Results and Discussion

In the proposed method, Verilog is used as the hardware description language. The synthesis tool Vivado is used to perform synthesis. Zynq-7000 Xilinx SoC is used for the implementation.

Table 2 Comparison of proposed asynchronous implementations with synchronous counterpart

Parameter	Synchronous	Asynchronous	% Change
Encoder			
I/O	0.92 %	4.19 %	3.27 % increase
LUT	70.56 %	75.26 %	4.7 % increase
POWER	4.2 W	1.3 W	30.95 % decrease
Decoder			
I/O	2.98 %	6.03 %	3.81 % increase
LUT	80.59 %	84.13 %	3.54 % increase
POWER	5.93 W	2.07 W	34.91 % decrease

The Dragoi et al. [36] RW algorithm is implemented for both synchronous and asynchronous cases. The results are presented in Table 2. A comparison in terms of area and power requirements is also listed in Table 2 for both asynchronous architecture and synchronous counterpart.

In comparison with the synchronous implementation, there is a 4 % increase in area. But this penalty is negligible compared to an average 33 % reduction in overall power consumption. A similar comparison is made between the synchronous implementation developed by Sudip et al. [59] and above-discussed asynchronous scheme and tabulated in Table 3. There is a 3 % increase in the area, whereas the average decrease in power consumption is 33 %.

The summary of some other real-time synchronous implementations available in the literature is shown in Table 4. Most of the hardware implementations listed in Table 4 are robust and are clock-based implementations. Further, there are many computer-aided design (CAD) tools to cater for the easy implementation of complex synchronous designs. But the drawback is that, the clock or its derivatives synchronize all operations of the IC. When compared with the proposed asynchronous architecture, all the above-mentioned synchronous schemes consume a very significant amount of power. This is because, in all the synchronous implementations, at each clock edge, all the blocks of the design switch irrespective of whether they have new data to process. As the designs are becoming more complex,

Table 3 Comparison of proposed asynchronous implementation with synchronous [59]

Parameter	Synchronous	Asynchronous	% Change
Encoder			
I/O	0.26 %	4.19 %	3.93 % increase
LUT	73.24 %	75.26 %	2.02 % increase
POWER	3.7 W	1.3 W	35.13 % decrease
Decoder			
I/O	0.26 %	6.03 %	2.77 % increase
LUT	73.24 %	84.13 %	1.31 % increase
POWER	3.7 W	2.07 W	30.89 % decrease

Table 4 Comparison of synchronous real-time implementations

Author	Perceptibility	Domain	Device	Hardware platform	Strength
Garimella et al. [46]	Invisible	Spatial	Chip	FPGA	Robust
Mohanty et al. [50]	Invisible	DCT	Chip	FPGA and ASIC	Robust/fragile
Samanta [47]	Invisible	Spatial	Chip	SET device	Robust
Garimella et al. [42]	Invisible	Spatial	Chip	ASIC	Fragile
Karmani et al. [51]	Invisible	Wavelets	Chip	ASIC	Fragile
Lad et al. [52]	–	Wavelets	Chip	FPGA and ASIC	Robust
Morita [53]	Visible	DCT	Chip	FPGA	Robust
Lim [54]	Visible	DCT	Camera	EP10K100ARC240-3	Robust

and the dimensions are shrinking, some of the issues that were insignificant are becoming consequential. Most of these issues are associated with clock, viz. clock slew, clock distribution, clock skew, and power dissipation due to clock.

In contrast to the synchronous design styles, the blocks in the proposed asynchronous systems communicate via handshaking signals—request (req) and acknowledgment (ack). Further, there is no need for clock routing, and hence, the design complexity reduces considerably. In addition, the design has lower power consumption compared to the existing synchronous architectures.

Reversible watermarking (RW) is a very useful tool in medical and critical military applications. Algorithms for RW of color images are scarce, and hence should be developed. Also, robustness, imperceptibly, and watermark embedding capacity trade-off should be minimized. Real-time implementation of RW schemes is the need of the hour. There is further scope to optimize the area, power, and performance metrics.

6 Conclusions

Reversible watermarking provides a solution to reconstruct original image with no loss besides authenticating ownership. Studies have suggested that these methods offer greater balance among the requirements of watermarking such as invisibility, robustness, and capacity. This chapter gives an overview of the RW techniques, their vulnerabilities, and recommendations for future development. Performance issues related to the real-time implementation of watermarking systems through various algorithms is also featured. Finally, an extension of real-time implementation of RW approach based on asynchronous architectures is introduced.

References

1. Surekha B, Swamy GN (2012) Digital image ownership verification based on spatial correlation of colors. In: IET conference on image processing, pp 1–5
2. Acharjee S, Chakraborty S, Samanta S, Azar AT, Hassanien AE, Dey N (2014) Highly secured multilayered motion vector watermarking. In: Advanced machine learning technologies and applications. Springer, Berlin, pp 121–134)
3. Dey N, Das P, Roy AB, Das A, Chaudhuri SS (2012) DWT–DCT–SVD based intravascular ultrasound video watermarking. In: 2012 World Congress on Information and Communication Technologies (WICT). IEEE, pp 224–229
4. Surekha B, Swamy GN (2014) Lossless watermarking technique for copyright protection of high resolution images. In: IEEE Region 10 Symposium, 2014, Malaysia, pp 73–78
5. Dey N, Mukhopadhyay S, Das A, Chaudhuri SS (2012) Analysis of P-QRS-T components modified by blind watermarking technique within the electrocardiogram signal for authentication in wireless telecardiology using DWT. Int J Image Graph Signal Process 4(7):33
6. Surekha B, Swamy GN (2012) A semi-blind image watermarking based on discrete wavelet transform and secret sharing. In: IEEE International conference on communication, information and computing technology, Mumbai, India, pp 1–5
7. Surekha B, Swamy GN (2012) Visual secret sharing based digital image watermarking. Int J Comput Sci Issues 9(3):312–317
8. Surekha B, Swamy GN, Rama Linga Reddy K (2012) A novel copyright protection scheme based on visual secret sharing. In: Third IEEE international conference on computing, communication and networking technologies, Coimbatore, India, pp 1–5
9. Lakshmi HR, Surekha B (2016) Asynchronous Implementation of Reversible Image Watermarking Using Mousetrap Pipelining, In: IEEE 6th International Conference on Advanced Computing, Bhimavaram, India, pp. 529–533
10. Pushpa Mala S, Jayadevappa D, Ezhilarasan K (2015) Digital image watermarking techniques —a review. Int J Comput Sci Secur 140–156
11. Barton JM (1997) Method and apparatus for embedding authentication information within digital data. U.S. Patent 5, 646 997
12. Honsinger CW, Jones PW, Rabbani M, Stoffel JC (2001) Lossless recovery of an original image containing embedded data. U.S. Patent 6278791
13. Hong W, Chen TS, Lin KY, Chiang WC (2010) A modified histogram shifting based reversible data hiding scheme for high quality images. Asian Netw Sci Inf Inf Technol J 179–183
14. Luo L, Chen Z, Chen M, Zeng X, Xiong Z (2009) Reversible image watermarking using interpolation technique. IEEE Trans Inf Forensics Secur 187–193
15. Celik MU, Sharma G, Tekalp AM, Saber E (2005) Lossless generalized-LSB data embedding. IEEE Trans Image Process 253–266
16. Kanphade RD, Narawade NS (2012) Forward modified histogram shifting based reversible watermarking with reduced pixel shifting and high embedding capacity. Int J Electr. Comput. Eng 185–191
17. Ni Z, Shi YQ, Ansari N, Su W (2006) Reversible data hiding. IEEE Trans Circuits Syst Video Technol 16(3):354–362
18. Hwang J, Kim JW, Choi JU (2006) A reversible watermarking based on histogram shifting. In: International workshop on digital watermarking, LNCS, vol 4283, pp 348–361
19. Tian J (2003) Reversible data embedding using a difference expansion. IEEE Trans Circuits Syst Video Technol 13(8):890–896
20. Coltuc D, Chassery JM (2007) Very fast watermarking by reversible contrast mapping. IEEE Signal Process Lett 14:255–258
21. Dey N, Pal M, Das A (2012) A session based watermarking technique within the NROI of retinal fundus images for authencation using DWT, spread spectrum and Harris corner detection. Int J Mod Eng Res 2(3):749–757

22. Dey N, Maji P, Das P, Biswas S, Das A, Chaudhuri SS (2013) An edge based blind watermarking technique of medical images without devalorizing diagnostic parameters. In: 2013 International conference on advances in technology and engineering (ICATE). IEEE, pp 1–5

23. Pal AK, Das P, Dey N (2013). Odd-even embedding scheme based modified reversible watermarking technique using Blueprint. arXiv preprint arXiv:1303.5972

24. Dey N, Maji P, Das P, Biswas S, Das A, Chaudhuri SS (2013) Embedding of blink frequency in electrooculography signal using difference expansion based reversible watermarking technique. arXiv preprint arXiv:1304.2310

25. Dey N, Biswas S, Das P, Das A, Chaudhuri SS (2012) Feature analysis for the reversible watermarked electrooculography signal using low distortion prediction-error expansion. In: 2012 International conference on communications, devices and intelligent systems (CODIS). IEEE, pp 624–627

26. Dey N, Roy AB, Das A, Chaudhuri SS (2012) Stationary wavelet transformation based self-recovery of blind-watermark from electrocardiogram signal in wireless telecardiology. In: Recent trends in computer networks and distributed systems security. Springer, Berlin, pp 347–357

27. Dey N, Dey M, Mahata SK, Das A, Chaudhuri SS (2015) Tamper detection of electrocardiographic signal using watermarked bio-hash code in wireless cardiology. Int J Signal Imaging Syst Eng 8(1–2):46–58

28. Dey N, Samanta S, Yang XS, Das A, Chaudhuri SS (2013) Optimisation of scaling factors in electrocardiogram signal watermarking using cuckoo search. Int J Bio-Inspired Comput 5 (5):315–326

29. Dey N, Samanta S, Chakraborty S, Das A, Chaudhuri SS, Suri JS (2014) Firefly algorithm for optimization of scaling factors during embedding of manifold medical information: an application in ophthalmology imaging. J Med Imaging Health Inform 4(3):384–394

30. Dey N, Biswas D, Roy AB, Das A, Chaudhuri SS (2012) DWT–DCT–SVD based blind watermarking technique of gray image in electrooculogram signal. In: 2012 12th international conference on intelligent systems design and applications (ISDA). IEEE, pp 680–685

31. Dey N, Biswas S, Das P, Das A, Chaudhuri SS (2012) Lifting wavelet transformation based blind watermarking technique of photoplethysmographic signals in wireless telecardiology. In: 2012 World Congress on Information and Communication Technologies (WICT). IEEE, pp 230–235

32. Chakraborty S, Samanta S, Biswas D, Dey N, Chaudhuri SS (2013) Particle swarm optimization based parameter optimization technique in medical information hiding. In: 2013 IEEE international conference on computational intelligence and computing research (ICCIC). IEEE, pp 1–6

33. Chakraborty S, Maji P, Pal AK, Biswas D, Dey N (2014) Reversible color image watermarking using trigonometric functions. In: 2014 International conference on electronic systems, signal processing and computing technologies (ICESC). IEEE, pp 105–110

34. Alattar AM (2003) Reversible watermark using difference expansion of triplets. Int Conf Image Process 1:501–504

35. Aarthi R, Jaganya V, Poonkuntran S (2012) Modified LSB watermarking for image authentication. Int J Comput Commun Technol 3(3):2231–2371

36. Dragoi C, Coltuc D (2012) Improved rhombus interpolation for reversible watermarking by difference expansion. IEEE Trans EUSIPCO 1688–1692

37. Yang B, Schmucker M, Niu X, Busch C, Sun S (2004) Reversible image watermarking by histogram modification for integer DCT coefficients. In: IEEE 6th workshop on multimedia signal processing, pp 143–146

38. Cumplido R, Feregrino-Uribe C, Garcia-Hernandez JJ (2011) Implementing digital data hiding algorithms in reconfigurable hardware—experiences on teaching and research. In: Reconfigurable communication-centric systems-on-chip, pp 1–6

39. Shirvaikar M, Bushnaq T (2009) A comparison between DSP and FPGA platforms for real-time imaging applications. SPIE 7244:724–806

40. Kougianos E, Mohanty SP, Mahapatra RN (2008) Hardware assisted watermarking for multimedia. Int J Comput Electr Eng 35(2):339–358
41. Mohanty SP, Ranganathan N, Namballa RK (2003) VLSI implementation of invisible digital watermarking algorithms towards the developement of a secure JPEG encoder. In: IEEE workshop on signal processing systems, pp 183–188
42. Garimella A, Satyanarayana MVV, Murugesh PS, Niranjan UC (2004) ASIC for digital color image watermarking. In: IEEE 11th digital signal processing workshop and IEEE signal processing education workshop, pp 292–296
43. Tefas A, Pitas I (2001) Robust spatial image watermarking using progressive detection. In: IEEE International Conference on Acoustics, Speech and Signal Processing, vol 3, pp 1973–1976
44. Barni M, Bartolini F, Tefas A, Pitas I (2001) Image authentication techniques for surveillance applications. IEEE 89(10):1403–1418
45. Mohanty SP, Ramakrishnan KR, Kankanhalli MS (1999) A dual watermarking technique for images. In: 7th ACM International Multimedia Conference (ACM-MM99), pp 49–51
46. Garimella A, Satyanarayana MVV, Kumar RS, Murugesh PS, Niranjan UC (2003) VLSI implementation of online digital watermarking technique with difference encoding for 8-bit gray scale images. In: IEEE 16th international conference on VLSI design, New Delhi, pp 283–288
47. Samanta D, Basu A, Das TS, Mankar VH, Ghosh A, Das M, Sarkar SK (2008) SET based logic realization of a robust spatial domain image watermarking. In: IEEE 5th international conference on electrical and computer engineering, Dhaka, pp 986–993
48. Mathai NJ, Kundur D, Sheikholeslami A (2003) Hardware implementation perspectives of digital video watermarking algorithms. IEEE Trans Signal Process 51(4):925–938
49. Mohanty SP, Ranganathan N, Balakrishnan K (2006) A dual voltage frequency VLSI chip for image watermarking in DCT domain. IEEE Trans Circuits Syst II Express Briefs 53(5): 394–398
50. Mohanty SP, Kougianos E, Ranganathan N (2007) VLSI architecture and chip for combined invisible robust and fragile watermarking. Comput Digit Tech 1:600–611
51. Karmani S, Djemal R, Tourki R (2013) Efficient hardware architecture of 20 scan based wavelet water-marking for image and video. Comput Stand Interfaces 31(4):801–811
52. Lad TC, Darji AD, Merchant SN, Chandorkar AN (2011) VLSI implementation of wavelet based robust image watermarking chip. In: International symposium on electronic system design, pp 56–61
53. Morita Y, Ayeh E, Adamo OB, Guturu P (2009) Hardware/software co-design approach for a DCT based watermarking algorithm. IEEE, pp 683–686
54. Lim H, Park S-Y, Kang S-J, Cho W-H (2003) FPGA implementation of image watermarking algorithm for a digital camera. IEEE, pp 1000–1003
55. Ali MO, Rao R (2011) An overview of hardware implementation for digital image watermarking. In: International conference on signal, image processing and applications, vol 21
56. Werner Tony, Akella Venkatesh (1997) Asynchronous processor survey. IEEE Comput 30 (11):67–76
57. Nowick Steven M, Singh Montek (2011) High-performance asynchronous pipelines: an overview. IEEE Des Test Comput 28:8–22
58. Singh M, Nowick SM (2007) MOUSETRAP: High-speed transition-signaling asynchronous pipelines. IEEE Trans Very Large Scale Integration (VLSI) Syst 15:684–698
59. Ghosh S, Das N, Das S, Maity SP, Rahaman H (2014) FPGA and SoC based VLSI architecture of reversible watermarking using rhombus interpolation by difference expansion. In: Annual IEEE India Conference, pp 1–6

Comparative Approach Between Singular Value Decomposition and Randomized Singular Value Decomposition-based Watermarking

Sayan Chakraborty, Souvik Chatterjee, Nilanjan Dey,
Amira S. Ashour and Aboul Ella Hassanien

Abstract Threats and attacks on networks have increased significantly with the recent advancement of technologies. In order to avoid such problems, watermarking techniques was introduced. However, several popular watermarking techniques consume much more time to embed and extract the hidden message. In this current work, a randomized singular value decomposition (rSVD)-based watermarking technique has been proposed in order to avoid such problems. In this work, firstly, the watermark is embedded into the cover image using singular value decomposition, and then the rSVD is employed. Finally, the obtained results were compared to check the time complexity of both methods and to compare the quality of watermarked image and recovered watermarks.

Keywords Singular value decomposition · Randomized singular value decomposition · Correlation · Watermarking · Peak signal-to-noise ratio

S. Chakraborty · S. Chatterjee
Department of CSE, Bengal College of Engineering and Technology,
Durgapur, India
e-mail: sayan.cb@gmail.com

S. Chatterjee
e-mail: souvikcha2008@gmail.com

N. Dey (✉)
Department of Information Technology, Techno India College of Technology,
Kolkata, India
e-mail: neelanajn.dey@gmail.com

A.S. Ashour
Department of Electronics and Electrical Communications Engineering,
Faculty of Engineering, Tanta University, Tanta, Egypt
e-mail: amirasashour@yahoo.com

A.E. Hassanien
Faculty of Computers and Information, Cairo University, Cairo, Egypt
e-mail: aboitcairo@gmail.com

© Springer International Publishing Switzerland 2017
N. Dey and V. Santhi (eds.), *Intelligent Techniques in Signal Processing for Multimedia Security*, Studies in Computational Intelligence 660,
DOI 10.1007/978-3-319-44790-2_7

133

1 Introduction

The expeditious growth of the Internet increases the access to multimedia data [1, 2]. Digital multimedia development becomes a must as a grave need for protecting multimedia data in the Internet. In order to achieve the required protection, digital watermarking techniques can be employed to provide copyright management and protection for the digital data. Digital watermarking in image forensics has been widely studied. Forensic watermark enhances the content owner's capability to detect any misuse of its resources, including documents, images, and videos. Forensic watermarking provides irrefutable and positive evidences of misuse for the content. In the context of forensic, a digital image can provide information about the used image acquisition device and accordingly can offer a way to know the person who probably took the contents/resources. Typically, digital image forensics domain can be generally divided into image source identification that is based on precise characteristics of the image acquisition device, and the determination of whether a certain digital image has suffered from tampering or malicious post-processing. Thus, forensic algorithms are designed to disclose either the characteristic traces of image processing operations or to authenticate the integrity of specific features presented in the image acquisition process.

Watermarking can be considered to be the act of veiling a message related to a digital signal within the signal itself. A digital watermark is a slice of information that is embedded in the digital media and hidden in the digital content in such a way that it is indissoluble from its data. However, later it can be detected or extracted to create an assertion about the object. The object may be an image, text, audio, or video. The expanding amount of research on watermarking over the past decade has been largely driven by its important applications such as (a) broadcast monitoring, i.e., make the entire monitoring process easier, (b) owner identification that helps to identify the owner of a specific digital work, (c) transaction tracking, where watermarking can be used to record the recipient of every legal copy of a digital work by implanting a different watermark in each copy, and (d) secret communication, where techniques can be used to transmit any secret information by a hidden way.

Digital watermarking has several uses for forensics in several applications as it (i) creates a powerful prevention from leaking controlled content, (ii) identifies quickly and precisely the leaked content source, (iii) provides evidence of content misuse, and (iv) gains visibility. Till date a number of methods have been reported regarding watermarking. Among these methods, the SVD-based watermarking [3–5] is considered to be a well-known and useful approach. It is a powerful numerical technique for matrices, which achieves minimum least square truncation error. The SVD theorem [6–8] crumbles a digital image A of size $M * N$ as follows:

$$A = USV^{\mathrm{T}} \tag{1}$$

where U and V matrices are of size $M \times M$ and $N \times N$, respectively. S is a diagonal matrix containing the singular values. In watermarking, SVD is applied to the image matrix and then watermark resides by altering the singular values.

In the present work, SVD-based and randomized SVD-based image water-marking scheme have been proposed and the results have been compared.

The structure of the remaining sections is as follows. Section 2 described the related works in this domain. Methodology is discussed in Sect. 3, while detailed explanation of the proposed method is introduced in Sect. 4. Simulation results are shown and discussed in Sect. 5. Finally, the conclusion is presented in Sect. 6.

2 Related Work

Watermarking techniques has attracted several researchers. In 1997, Cox et al. [1] proposed a frequency-domain watermark scheme based on the spread-spectrum approach. Through this approach, the watermark was spread over broadband so that it can be imperceptible in any frequency beam. However, selecting the best strength value of the watermark for hiding was a complex topic. In 1998, Huang and Shi [2] proposed an adaptive spread-spectrum watermark scheme that was based on human visual mask: brightness sensitivity and texture sensitivity. The brightness sensitivity was estimated by the DC components in the discrete cosine transform (DCT) domain. Conversely, the texture sensitivity was estimated by quantizing the DCT coefficients using the JPEG quantization table and calculating the numbers of nonzero coefficients. Thereafter, all the blocks were clustered into three classes: (a) dark and weak texture, (b) bright and strong texture, and (c) remaining situation. Subsequently, three numbers of the watermark sequences were embedded in low-frequency coefficients of each block. However, the image visual model was not satisfied exactly and smoothly.

Another image watermarking scheme based on multi-band wavelet transfor-mation was proposed by Mohammed and Sidqi [3]. Initially, the scheme was tried out on the spatial domain in order to juxtapose its results with a frequency domain. In the frequency domain, an adaptive scheme was enacted based on the bands selection criteria to root the watermark. These criteria hinge on the number of wavelet passes. However, the entanglement of a large number of wavelet bands in the rooting process was the main disadvantage of this scheme. In 2012, a semi-fragile watermarking technique based on block-based SVD was presented by Arathi [4]. Semi-fragile watermark is basically fragile to malicious modifications while robust to incidental manipulations. The major advantage of this scheme is its ability to extract the watermark without the original image. The SVD transforma-tion preserves both one-way and non-symmetric properties that are not obtainable in the DCT and discrete Fourier transform (DFT) . It can also detect tamper made on the image. Ganic and Eskicioglu [5] presented a SVD-based digital image watermarking process in discrete wavelet transform (DWT) domain. Embedding was done by modifying the singular values of the wavelet transformed sub-bands with the singular values of the watermark image. This method was robust because of using all bands in the embedding process. However, it has the disadvantages of being a non-blind method and the transparency of the watermarked image was not so effective. In 2010, Bhagyashri and Joshi [9] proposed a method that is similar to the previous one. Moreover, they showed that modifications in all frequencies made the watermarking schemes robust to a wide range of attacks. However, it was more robust to geometric attack during embedding data in high-frequency band.

In 2012, Zhang et al. [6] proposed a randomized singular value decomposition (rSVD) method to achieve lossless compression, reformation, classification, and target detection with hyperspectral (HSI) data. Approximation errors for the rSVD were evaluated on HSI, and comparisons were made to settled techniques as well as to different randomized low-rank matrix approximation methods including compressive principal component analysis. Another rSVD algorithm was presented by Drinea et al. [7] which samples a fixed number of rows or columns of the matrix and scales them suitably to form a small matrix, and then computed the SVD of that matrix. The exactness and speed of this algorithm for image matrices using assorted probability distributions to perform the sampling were experimentally evaluated in the work. A modular framework was proposed by Halko et al. [8] for constructing randomized algorithms that compute partial matrix decompositions. These methods used random sampling to spot a subspace that captures most of the activity of a matrix. In many cases, this approach beats its previous combatants in terms of exactness, speed, and robustness.

3 Methodology

Recently, some novel watermarking techniques based on SVD are being developed. All these SVD-based watermarking techniques have high robustness and good embedding quality, but they consume long processing time. Thus, the proposed approach is interested with the improvement of the SVD-based watermarking technique for superior time-consuming.

3.1 Singular Value Decomposition (SVD)

In the SVD, for a matrix A of size $m \times n$ and has a rank of k admits a factorization $Z = XSY^*$, where X is an $m \times k$ orthonormal matrix, Y is an $n \times k$ orthonormal matrix [4], and S is a $k \times k$ nonnegative diagonal matrix, which given by:

$$S = \begin{bmatrix} \sigma 1 & & & \\ & \sigma 2 & & \\ & & \cdots & \\ & & & \sigma k \end{bmatrix} \tag{2}$$

The numbers σ_j are called the singular values of Z. They are organized in weakly decreasing order: $\sigma 1 \geq \sigma 2 \geq \cdots \geq \sigma_k \geq 0$. The columns of X and Y are called left singular vectors and right singular vectors, respectively. The singular values are connected with the approximation of the matrices. For each j, the number $\sigma_j + 1$ equals the spectral norm [5] discrepancy between Z and an optimal rank-j approximation, where

$$\sigma_j + 1 = \min\{kZ - Bk : B \text{ has rank } j\} \tag{3}$$

3.2 Randomized Singular Value Decomposition (rSVD)

The SVD of a matrix $Z \in R^{m \times n}$ is defined as follows:

$$Z = X \sum Y^T \tag{4}$$

where X and Y are orthonormal, and $\sum$ is a rectangular diagonal matrix whose diagonal entries are the singular values signified as σ_i. The column vectors of X and Y are left and right singular vectors, respectively, symbolized as x_i and y_i. Define the truncated SVD (TSVD) approximation [6] of Z as a matrix Z_k such that

$$Z_k = \sum_{i=1}^{k} \sigma_i x_i y_i^T \tag{5}$$

The randomized SVD (rSVD) of A can be defined as follows:

$$\hat{Z}_k = \hat{X}\hat{\Sigma}\hat{Y}^T \tag{6}$$

where $\hat{X}$ and $\hat{Y}$ are both orthonormal and $\hat{\Sigma}$ is diagonal with diagonal entries symbolized as $\hat{\sigma}_i$. Designate the column vectors [7] of $\hat{X}$ and $\hat{Y}$ as $\hat{x}$ and $\hat{y}$ and $\hat{\sigma}_i$ correspondingly. Elucidate the residual matrix of a TSVD approximation as follows:

$$R_k = Z - Z_k \tag{7}$$

The residual matrix of a rSVD approximation is as follows:

$$\hat{R}_k = Z - \hat{Z}_k \tag{8}$$

Elucidate the random projection of a matrix as follows:

$$Y = \Omega^{\mathrm{T}} Z \text{ or } Y = Z\Omega \tag{9}$$

where Ω is a random matrix with independent and identically distributed [8] entries.

The rSVD algorithm [6, 7] explores approximate matrix factorization using random projections to split the process into two stages. In the first stage, random sampling is operated to acquire a reduced matrix whose range approaches the range of Z. Thereafter, in the second stage the reduced matrix is factorized. Using the first stage on matrix Z to find the matrix Q with orthonormal columns for $\xi > 0$, thus it is obtained that

$$\left\| Z - QQ^{\mathrm{T}}Z \right\|_F^2 \leq \xi \tag{10}$$

In the second stage of the rSVD method, SVD of the reduced matrix $Q^{\mathrm{T}}Z \in R^{l \times m}$ is obtained, where $l << n$. By using $\hat{X}\hat{\Sigma}\hat{Y}^{\mathrm{T}}$ to denote the SVD of $Q^{\mathrm{T}}Z$, the following expression is obtained:

$$Z \approx (Q\tilde{X})\hat{\Sigma}\hat{Y}^{\mathrm{T}} = \hat{X}\hat{\Sigma}\hat{Y}^{\mathrm{T}} \tag{11}$$

where $\hat{X} = Q\tilde{X}$ and $\hat{Y}$ are orthogonal matrices.

4 Proposed Method

In this work, the cover image was watermarked using singular value decomposition (SVD). A methodology for secure biomedical image authentication and information hiding is proposed by combining the image embedding with the optimal scaling factors. The DWT–DCT–SVD-based proposed watermarking [10–12] is as follows.

4.1 Watermark Embedding

The steps for the watermark embedding are as follows:

Step 1. Original gray cover image size is calculated.

Step 2. Three gray scale watermark images of the same size as that of the cover image are taken.

Step 3. Cover image is decomposed into four sub-bands LL_1, LH_1, HL_1, and HH_1 by applying DWT.

Step 4. The DCT is applied on the HH_1, HL_1, and LH_1 sub-bands individually.

Step 5.	The SVD and rSVD are applied on all the resultant sub-bands.

Step 6.	Each watermark image is decomposed into four sub-bands (LL_1, LH_1, HL_1, and HH_1) by applying DWT.

Step 7.	The DCT is applied on all sub-bands of the decomposed watermark images.

Step 8.	After applying the DCT on the four sub-bands of the decomposed watermarks, the SVD and rSVD are applied on each resultant sub-band.

Step 9.	Singular values of the entire resultant sub-bands obtained from the cover image are modified using singular value of the corresponding HH_1 sub-bands of the watermark images after applying DWT and DCT on it.

Step 10.	The IDCT is applied followed by IDWT in order to produce the final watermarked image.

4.2 Watermark Extraction

The watermark extraction has the following steps:

Step 1.	Watermarked image is decomposed into four sub-bands by applying DWT. (LL_1, LH_1, HL_1, and HH_1)

Step 2.	The DCT is applied on the HH_1, LH_1, HL_1 sub-bands.

Step 3.	The SVD and rSVD are applied on the resultant sub-bands.

Step 4.	Singular values of the watermarked image are modified.

Step 5.	The IDCT is applied followed by IDWT on the resulting image in order to extract the watermark image from the watermarked one.

Based on the preceding algorithm, the cover image is decomposed into four sub-bands, namely LL_1, LH_1, HL_1, and HH_1 by applying the DWT. The DCT is applied on the HH_1, HL_1, and LH_1 sub-bands individually. The SVD and rSVD are applied accordingly on all the resultant sub-bands. Each watermark image [13–15] is decomposed into the four sub-bands by applying DWT. The DCT is applied on all sub-bands of the decomposed watermark images. After applying the DCT on the four sub-bands of the decomposed watermarks, SVD and rSVD are applied on each resultant sub-band. Singular values of all the resultant sub-bands obtained from the cover image are modified using singular value of the corresponding HH_1 sub-bands of the watermark images after applying DWT and DCT on it. The IDCT is applied followed by IDWT which leads to generating the watermarked image. In this method, scaling factor was set to an average value of 0.5. Figure 1 shows the embedding procedure.

During the watermark extraction process as illustrated in Fig. 2, the watermarked image [16–18] is decomposed into the four sub-bands by applying DWT. The DCT is applied on the HH_1, LH_1, HL_1 sub-bands. The SVD and rSVD are again applied on the resultant sub-bands. Singular values of the watermarked image are modified. To extract the watermark [19, 20] image from the watermarked image

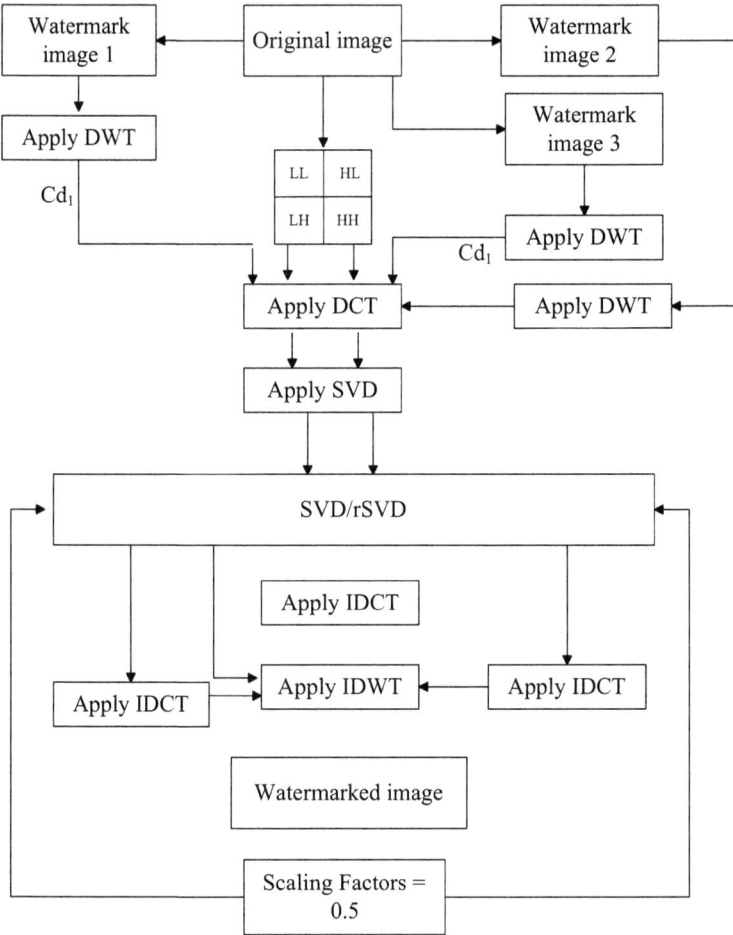

Fig. 1 Watermark embedding using SVD and rSVD

[21, 22], the IDCT is applied followed by IDWT on the resultant image and also the scaling factors (=0.5) were used.

The proposed method has been implemented using MATLAB R2012a on a 3.60 GHz Pentium IV computer with Microsoft Windows XP operating system.

5   Results and Discussion

In order to evaluate the proposed approach, the original image is illustrated in Fig. 3a. Figure 3b, c demonstrate the watermarked images [23, 24] using the randomized singular value decomposition and singular value decomposition, respectively.

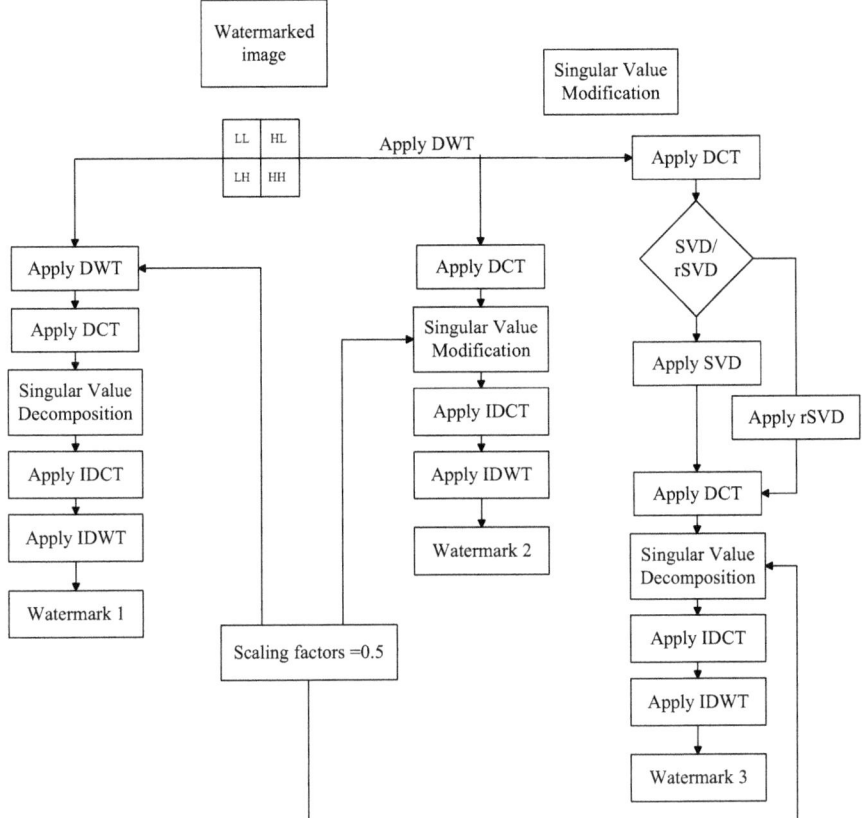

Fig. 2 Watermark extraction using SVD and rSVD

Originally, the watermarking has been done using MATLAB R2012a by using the watermarks shown in Fig. 4 that embedded into the cover image (Fig. 3a).

Figure 4a, d, g show the original watermark [25–27]. As previously noted, these watermarks are embedded in the cover image [28, 29] (shown in Fig. 3a) using rSVD and SVD, respectively. Following the extraction process [30, 31], the recovered watermark [32, 33] from rSVD [7, 8] watermarking framework is shown in Fig. 4b, e, h. Figure 4c, f, i show the recovered watermarks [34, 35] from the SVD framework. In order to observe both of the technique's efficiency and robustness, the proposed method used various sizes of single image. Figure 5 illustrates the processing time versus the various sized images for both the SVD and the rSVD.

Figure 5 depicts that the proposed rSVD requires less computational time compared to the SVD approach. These results are tabulated in Table 1.

Table 1 shows the result obtained from the proposed framework. The time taken for embedding [36–38] three watermarks (in Fig. 5) in the various sized cover

Fig. 3 **a** Original image. **b** Watermarked image (using rSVD). **c** Watermarked image (using SVD)

Fig. 4 **a**, **d**, **g** Original
watermark. **b**, **e**, **h** Recovered
watermark using rSVD. **c**,
f i Recoveredwatermark
usingSVD

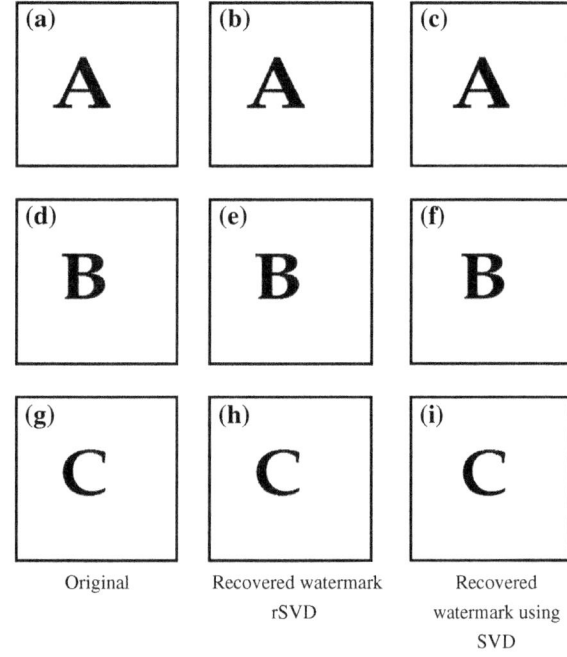

images (in Fig. 3) is shown in Table 1. Table 1 and Fig. 5 depict that the
rSVD-based watermarking [39, 40] took much lesser time than the SVD-based
framework [41]. Hence, the rSVD-based watermarking technique is faster than the
SVD-based watermarking technique [41].

In order to study the proposed approach performance, the peak signal-to-noise
ratio (PSNR) is calculated. The PSNR [41] is defined as the ratio between the
maximum achievable power of a signal and the corrupting noise power that affects
the reliability of the signal representation. This ratio can be used as an evaluation of

Fig. 5 Time comparison
graph

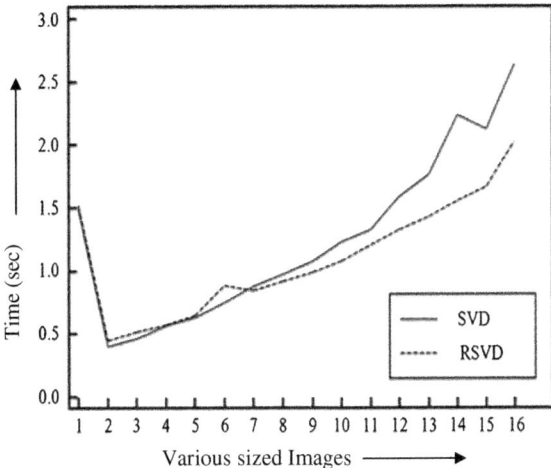

Table 1 Time comparison

Image size (pixel^2)	Time taken in rSVD (s)	Time taken in SVD (s)
32 × 32	1.527855	1.490118
64 × 64	0.448777	0.402552
96 × 96	0.517844	0.459246
128 × 128	0.572309	0.564563
160 × 160	0.647941	0.626511
192 × 192	0.888074	0.748964
224 × 224	0.842116	0.875024
256 × 256	0.917742	0.975061
288 × 288	0.988023	1.078293
320 × 320	1.075146	1.226406
352 × 352	1.206069	1.327856
384 × 384	1.323861	1.590968
416 × 416	1.424275	1.75936
448 × 448	1.556513	2.230931
480 × 480	1.668832	2.125843
512 × 512	2.028288	2.640091

the quality of a watermarked image. It can be used as a performance metric to find
out the perceptual transparency of the watermarked image with respect to the cover
image. Therefore, this ratio also measures the invisibility of the embedded watermark. The high PSNR value implies better invisibility of the watermark. The PSNR
ratio can be defined as follows:

$$\text{PSN\_R} = \frac{AB \max_{a,b} D_{a,b}^2}{\sum_{a,b} \left(\left(D_{a,b} - \bar{D}_{a,b} \right)^2 \right)} \tag{12}$$

where A and B are the numbers of rows and columns, respectively in the input image, $D_{a,b}$ is the original signal, and $\bar{D}_{a,b}$ is the watermarked image. Hence, the PSNR can be a good method to measure any image's quality. The obtained PSNR from the watermarked images using rSVD and SVD method are reported in Table 2 for various sized cover images.

The obtained results in Table 2 are used to construct the graph in Fig. 6.

The obtained results were further processed to construct a graph that shown in Fig. 6. Figure 6 and Table 2 show that the obtained results by applying the SVD method are better than that of rSVD.

Moreover, since the similarity of the recovered logo (a') and the original logo (a) can be determined by using the standard correlation coefficient [42] (Corr), once the watermark embedding process is complete, the correlation is given by:

$$\text{Corr} = \frac{\sum_i^j \sum_i^j \left(a_{ij} - a' \right) \left(b_{ij} - b' \right)}{\sqrt{\left(\sum_i^j \sum_i^j \left(a_{ij} - a' \right)^2 \right) \left(\sum_i^j \sum_i^j \left(b_{ij} - b' \right)^2 \right)}} \tag{13}$$

where y and y' are the transforms of x and x', respectively.

In the current work, correlation has been used to verify the robustness of recovered watermarks. Table 3 shows the obtained correlation values for three watermarks [40, 41].

Table 2 The PSNR comparison

Image size (pixel^2)	rSVD	SVD
32 × 32	20.68365	32.12484
64 × 64	22.54088	37.5255
96 × 96	23.6611	40.0231
128 × 128	24.84681	41.31281
160 × 160	25.58887	43.24419
192 × 192	26.32708	44.49747
224 × 224	26.84416	45.33338
256 × 256	27.49373	46.54702
288 × 288	28.06805	47.60425
320 × 320	28.75351	47.18845
352 × 352	29.07014	48.66764
384 × 384	29.65768	50.51544
416 × 416	30.08054	50.58948
448 × 448	30.37014	51.22097
480 × 480	30.79012	51.04328
512 × 512	30.72017	51.20362

Fig. 6 PSNR comparison graph

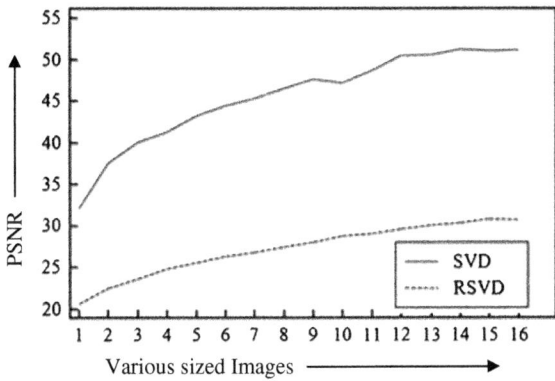

Table 3 is used further to construct the graph which is shown in Fig. 7.

Figure 7 establishes that the obtained correlation values using randomized singular value decomposition are superior to that of obtained values by the singular value decomposition.

Consequently, from the preceding results it can be depicted that the rSVD-based watermarking is faster than the SVD-based watermarking. Although, the obtained PSNR values are superior (higher) in case of the SVD-based watermarking compared to the rSVD approach. The obtained values of PSNR using the rSVD

Table 3 Comparison of recovered watermark's correlation

Image size (pixel^2)	rSVD			SVD		
	Recovered watermark 1	Recovered watermark 2	Recovered watermark 3	Recovered watermark 1	Recovered watermark 2	Recovered watermark 3
32 × 32	0.9902	0.9550	0.9400	0.8766	0.9001	0.8445
64 × 64	0.9964	0.9726	0.9649	0.9347	0.9560	0.8976
96 × 96	0.9986	0.9778	0.9680	0.9557	0.9657	0.9318
128 × 128	0.9989	0.9842	0.9757	0.9662	0.9776	0.9516
160 × 160	0.9993	0.9888	0.9859	0.9718	0.9835	0.9639
192 × 192	0.9995	0.9937	0.9877	0.9770	0.9905	0.9730
224 × 224	0.9996	0.9951	0.9889	0.9807	0.9923	0.9770
256 × 256	0.9997	0.9969	0.9905	0.9829	0.9946	0.9781
288 × 288	0.9998	0.9980	0.9920	0.9845	0.9966	0.9814
320 × 320	0.9998	0.9967	0.9917	0.9854	0.9947	0.9824
352 × 352	0.9998	0.9962	0.9954	0.9871	0.9947	0.9879
384 × 384	0.9999	0.9963	0.9946	0.9883	0.9948	0.9881
416 × 416	0.9999	0.9967	0.9948	0.9896	0.9957	0.9898
448 × 448	0.9999	0.9971	0.9951	0.9895	0.9959	0.9896
480 × 480	0.9999	0.9973	0.9960	0.9902	0.9961	0.9910
512 × 512	0.9999	0.9983	0.9960	0.9912	0.9976	0.9914

Fig. 7 Comparison graph of recovered watermark's correlation

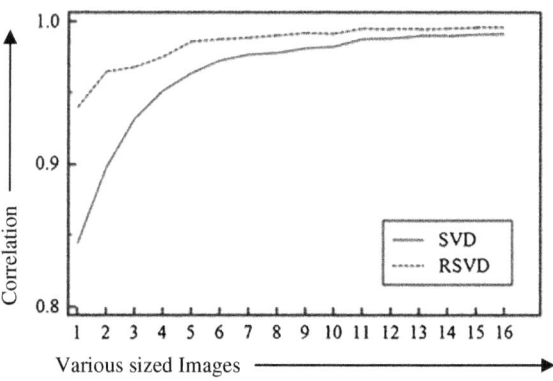

framework were mostly near or above 25, hence this method is acceptable. On the contrary, during the watermark extraction, the correlation coefficient values of watermarks using rSVD technique were better than that of SVD technique. Therefore, it is recommended to enhance the randomized singular value decomposition-based watermarking to obtain better results as a future work. In addition, optimization algorithms [41–44] can be employed to optimize the scaling factors for both of the watermarking techniques and can also be included as a future scope of this work.

6   Conclusion

Digital forensic enables the determination and investigation of the digital evidence reliability for the ownership. In addition, digital watermarking schemes have been broadly applied to protect rightful ownership of digital images. Thus, it becomes a widely research domain to hide messages by embedding and extracting the secret message in less amount of time. Hence, this paper shows the comparison of the SVD-based watermarking technique and the rSVD-based watermarking technique. The experimental results proved that the rSVD was faster than the normal SVD-based watermarking method. However, the results obtained using the SVD-based watermarking was better than that of rSVD-based watermarking.

References

1. Cox IJ, Kilian J, Leighton FT, Shamoon T (1997) Secure spread spectrum watermarking for multimedia. IEEE Trans Image Process 6(12):1673–1687
2. Huang J, Shi YQ (1998) Adaptive image watermarking scheme based on visual masking. IEEE Electron Lett 34(8):748–750

3. Mohammed AA, Sidqi H-M (2011) Robust image watermarking scheme based on wavelet technique. Int J Comput Sci Secur 5(4):394–404
4. Arathi C (2012) A semi fragile image watermarking technique using block based SVD. Int J Comput Sci Inf Technol 3(2):3644–3647
5. Ganicand E, Eskicioglu AM (2005) Robust embedding of visual watermarks usig DWT-SVD. J Electron Imaging 14(4):1–9
6. Zhang J, Erway J, Hu X, Zhang Q, Plemmons R (2012) Randomized SVD methods in hyperspectral imaging. J Electr Comput Eng 2012:1–15
7. Drinea E, Drineas P, Huggins P (2001) A randomized singular value decomposition algorithm for image processing applications. In: Panhellenic Conference on Informatics (PCI)
8. Halko N, Martinsson PG, Tropp JA (2011) Finding structure with randomness: probabilistic algorithms for constructing approximate matrix decompositions. SIAM Rev 53(2):217–288
9. Bhagyashri K, Joshi MY (2010) Robust image watermarking based on singular value decomposition and discrete wavelet transform. In: 3rd IEEE International Conference on Computer Science and Information Technology (ICCSIT) 5:337–341
10. Dey N, Bose S, Chakraborty S, Mukherjee A, Madhulika, Samanta S (2013) Parallel image segmentation using multi-threading and k-means algorithm. In: IEEE international conference on computational intelligence and computing research (ICCIC), pp 26–28. doi:10.11.9/ICCIC.2013.6724171
11. Zhanh L, Li A (2009) Robust watermarking scheme based on singular value of decomposition in DWT domain. In: Information processing, Asia Pacific Confrence (APCIP), vol 2, pp 19–22. doi:10.1109/APCIP.2009.141
12. Kamble S, Maheshkar V, Agarwal S, Srivastava VK (2011) DWT-based multiple watermarking for privacy and security of digital images in e-commerce. In: International conference on multimedia, signal processing and communication technologies (IMPACT), pp 224–227. doi:10.1109/MSPCT.2011.615048
13. Singh SP, Rawat P, Agarwal S (2012) A robust watermarking approach using DCT-DWT. Int J Emerg Technol Adv Eng 2(8)
14. Dey N, Das P, Das A, Chaudhuri SS (2012) DWT–DCT–SVD based intravascular ultrasound video watermarking. In: Second World Congress on information and communication technologies (WICT 2012), Trivandrum, India: October 30–November 02, 2012 (IEEE Xplore)
15. Dey N, Das P, Das A, Chaudhuri SS (2012) DWT–DCT–SVD based blind watermarking technique of gray scale image in electrooculogram signal. In: international conference on intelligent systems design and applications (ISDA-2012), Kochi, 27–29 Nov 2012 (IEEE Xplore)
16. Wang A, Zhang C, Hao P (2010) A blind video watermark detection method based on 3d-DWT transform. In: 17th IEEE international conference on image processing (ICIP)
17. Bhattacharya T, Dey N, Chaudhuri SRB (2012) A session based multiple image hiding technique using DWT and DCT. Int J Comput Appl 38
18. Dey N, Bardhan Roy A, Dey S (2011) A novel approach of color image hiding using RGB color planes and DWT. Int J Comput Appl 36(5). ISSN: 0975–8887
19. Pal AK, Das P, Dey N (2013) Odd-even embedding scheme based modified reversible watermarking technique using Blueprint. arXiv preprint arXiv:1303.5972
20. Dey N, Chakraborty S, Samanta S. Optimization of watermarking in biomedical signal. Lambert Publication, Heinrich-Böcking-Straße 6, 66121 Saarbrücken, Germany. ISBN-13: 978-3-659-46460-7
21. Dey N, Nandi B, Das P, Das A, Chaudhuri SS (2013) Retention of electrocardiogram features insignificantly devalorized as an effect of watermarking for a multi-modal biometric authentication system. In: Advances in biometrics for secure human authentication and recognition, pp 450. ISBN: 9781466582422
22. Dey N, Das P, Das A, Chaudhuri SS (2012) Feature analysis for the blind–watermarked electroencephalogram signal in wireless telemonitoring using Alattar's method. In: 5th International conference on security of information and networks (SIN 2012). In Technical

Cooperation with ACM Special Interest Group on Security, Audit and Control (SIGSAC), 2012, Jaipur, India

23. Das P, Munshi R, Dey N (2012) Alattar's method based reversible watermarking technique of EPR within heart sound in wireless telemonitoring. Intellectual Property Rights And Patent Laws, IPRPL-2012, Jadavpur University, August 25

24. Dey N, Dey M, Biswas D, Das P, Das A, Chaudhuri SS. Tamper detection of electrocardiographic signal using watermarked bio-hash code in wireless cardiology. Spec Issue Int J Signal Imaging Syst Eng Inderscience (in press)

25. Deb K, Al-Seraj MS, Hoque MM, Sarkar MIH (2012) Combined DWT–DCT based digital image watermarking technique for copyright protection. In: Electrical & Computer Engineering (ICECE), 7th International conference

26. Bhattacharya T, Dey N, Chaudhuri SRB (2012) A session based multiple image hiding technique using DWT and DCT. Int J Comput Appl 38

27. Dey N, Dey G, Chakraborty S, Chaudhuri SS (2014) Feature analysis of blind watermarked electromyogram signal in wireless telemonitoring. Concepts and Trends in Healthcare Information System of the Annals of Information Systems Series (In press)

28. Dey N, Samanta S, Yang XS, Chaudhuri SS, Das A (2013) Optimization of scaling factors in electrocardiogram signal watermarking using Cuckoo Search. Int J Bio-Inspired Comput 5 (5):315–326

29. Dey N, Mukhopadhyay S, Das A, Chaudhuri SS (2012) Using DWT analysis of P, QRS and T components and cardiac output modified by blind watermarking technique within the electrocardiogram signal for authentication in the wireless telecardiology. I.J. Image, Graphics and Signal Processing (IJIGSP). ISSN: 2074-9074

30. Chakraborty S, Maji P, Pal AK, Biswas D, Dey N (2014) Reversible color image watermarking using trigonometric functions. In: International conference on electronic systems, signal processing and computing technologies, Nagpur, 09–11 January 2014, pp 105–110

31. Dey N, Biswas S, Das P, Das A, Chaudhuri SS (2012) Lifting wavelet transformation based blind watermarking technique of photoplethysmographic signals in wireless telecardiology. In: Second World Congress on information and communication technologies (WICT 2012), pp 230–235

32. Dey N, Biswas S, Das P, Das A, Chaudhuri SS (2012) Feature analysis for the reversible watermarked electrooculography signal using low distortion prediction-error expansion. In: 2012 International conference on communications, devices and intelligent systems (CODIS), pp 624–627

33. Dey N, Das P, Das A, Chaudhuri SS (2012) DWT–DCT–SVD based intravascular ultrasound video watermarking. In: Second World Congress on information and communication technologies (WICT 2012), pp 224–229

34. Dey N, Das P, Das A, Chaudhuri SS (2012) DWT–DCT–SVD based blind watermarking technique of gray scale image in electrooculogram signal. In: International conference on intelligent systems design and applications (ISDA-2012), pp 680–685

35. Chakraborty S, Samanta S, Mukherjee A, Dey N, Chaudhuri SS (2013) Particle swarm optimization based parameter optimization technique in medical information hiding. In: 2013 IEEE international conference on computational intelligence and computing research (ICCIC), pp 1–6

36. Dey N, Mishra G, Nandi B, Pal M, Das A, Chaudhuri SS (2012) Wavelet based watermarked normal and abnormal heart sound identification using spectrogram analysis. In: 2012 IEEE international conference on computational intelligence and computing research (ICCIC), pp 1–7

37. Dey N, Maji P, Das P, Das A, Chaudhuri SS (2013) An edge based watermarking technique of medical images without devalorizing diagnostic parameters. In: International conference on advances in technology and engineering, pp 1–5

38. Dey N, Das P, Biswas D, Maji P, Das A, Chaudhuri SS (2013) Visible watermarking within the region of non-interest of medical images based on fuzzy C-means and Harris corner

detection. In: The fourth international workshop communications security & information assurance (CSIA-2013) [Springer], pp 161–168

39. Dey N, Biswas S, Roy AB, Das A, Chaudhuri SS (2012) Analysis of photoplethysmographic signals modified by reversible watermarking technique using prediction-error in wireless telecardiology. In: International conference of intelligent Infrastructure, 47th annual national convention of CSI, McGraw-Hill Proceeding

40. Réthoré J, Gravouil A, Morestin F, Combescure A (2005) Estimation of mixed-mode stress intensity factors using digital image correlation and an interaction integral. Int J Fract 132 (1):65–79

41. Dey N, Samanta S, Chakraborty S, Das A, Chaudhuri SS, Suri JS (2014) Firefly algorithm in optimization of scaling factors for manifold medical information embedding. J Med Imaging Health Inf 4(3):384–394

42. Samanta S, Dey N, Das P, Acharjee S, Chaudhuri SS (2012) Multilevel threshold based gray scale image segmentation using cuckoo search. In: International conference on emerging trends in electrical, communication and information technologies-ICECIT, Dec. 12–23

43. Tyagi S, Bharadwaj K (2014) A particle swarm optimization approach to fuzzy case-based reasoning in the framework of collaborative filtering. Int J Rough Sets Data Anal 1(1)

44. Ashour AS, Samanta S, Dey N, Kausar N, Abdessalemkaraa WBK, Hassanien A. Computed tomography image enhancement using cuckoo search: a log transform based approach. J Signal Inf Process 6(4)

Biometric-Based Security System: Issues and Challenges

Ujwalla Gawande, Yogesh Golhar and Kamal Hajari

Abstract Though biometric systems have been successfully engaged in a number of real-world applications, they are error prone. The challenge is to develop a biometric system that is highly accurate and secure, convenient to use, and easily scalable to a large population. Systems that have the ability to authenticate persons accurately, rapidly, reliably, cost-effectively, friendly to use, and without drastic changes to the existing infrastructures are desired. Despite rapid growth in biometric systems, in the past few decades, a number of core research issues have not yet been fully addressed. Multibiometrics system design is certainly a challenging task since it is very difficult to choose the best possible sources of biometric information and fusion strategy for a particular application. In this chapter, different issues and challenges in designing biometric-based security systems and methodologies involved in overcoming these issues are discussed. Many researchers developed biometric-based security system despite that many challenges remain unaddressed. Only few challenges are addressed completely; still, there is a requirement of robust algorithms. In this chapter, we propose two approaches of feature extraction, i.e., block sum for iris and modified minutiae method for fingerprint. The main aim of this chapter was to provide an informative analysis and solutions of key issues and challenges of biometric-based security system.

Keywords Noisy artifacts · Iris · Occlusion · Classifiers

U. Gawande (✉) · Y. Golhar · K. Hajari
Department of Information Technology, Yeshwantrao Chavan College
of Engineering, Hingna Road, Wanadongri, Nagpur 441110, India
e-mail: ujwallgawande@yahoo.co.in

Y. Golhar
e-mail: yj999@ymail.com

K. Hajari
e-mail: kamalhajari123@gmail.com

© Springer International Publishing Switzerland 2017   151
N. Dey and V. Santhi (eds.), *Intelligent Techniques in Signal Processing for Multimedia Security*, Studies in Computational Intelligence 660,
DOI 10.1007/978-3-319-44790-2_8

1 Introduction

There are numerous reasons that motivate researcher's interest in enhancing the performance of biometric recognition approaches. First of all, unimodal biometric features are not exactly the same every time while they are gathered. For instance, no two fingerprints are ever exactly the same. The quality of fingerprint images may be degraded as a result of physical problems such as dry skin, oily skin, dirt in finger, dirt on sensor surface, scars, and other factors or simply because the user has positioned his or her finger on the fingerprint sensor in a different orientation. Moreover, limited iris texture information is available due to occlusion of eyelids and eyelashes in iris image, thereby making iris recognition, a difficult task. Again, there are several issues regarding other biometric traits such as face, ear, DNA, and speech. So fusion of the complementary information in multimodal biometric data has been a research area of considerable interest in recent years, as it plays a critical role in overcoming certain important limitations of unimodal systems. It mainly relies on fusing separate information from different modalities to provide complementary information for achieving more reliable recognition of individuals. Information fusion may be done at four different levels, viz. sensor, feature, matching score, and decision levels.

There is a considerable amount of literature that details different approaches for multimodal biometric systems, e.g., fusion of fingerprint, palm print, and hand geometry at the score level [1]; fusion of the iris and palm print at the score level [2]; feature-level fusion framework for fingerprint, iris, and face [3]; decision-level fusion of iris and fingerprint [4]; and audiovisual person authentication [5]. After going through the literature, the fusion methods in present scenario were understood. Efforts in multimodal area are mainly focused on fusing the information obtained from a variety of independent modalities. Sensor-level fusion is expected to enhance the biometric recognition accuracy but it cannot be applied to incompatible data, gathered from different modalities [6]. Scanty references are available on feature-level fusion. Since the feature set contains rich information about the raw biometric data, integration at feature level is expected to provide better recognition results. Fusion at feature level is difficult to accomplish [3] due to the reasons like the feature sets from different modalities may be incompatible and converting them into a homogeneous form without any loss of information is difficult task. Sequential or parallel concatenation of feature sets may result in high-dimensional feature vector. This requires more memory space and more time for recognition. Fusion at the match score and decision levels has been extensively studied in the literature and is the most common approach in multimodal biometric systems [1, 2]. From the literature, it is noted that the match scores generated by the individual matchers may not be homogeneous and in same numerical range. Also individual biometrics often shows significantly different performance at different times [1]. Combining them directly by score fusion may degrade the performance of the fused system compared to the performance of the best individual modality. Decision-level fusion delimits the basis for enhancing the system accuracy through the fusion process [4, 5]. Due to these reasons, some different methods of fusion, resulting in higher performance, were felt necessary. We propose two approaches for two different unimodal traits, i.e., block sum feature extraction for iris and modified minutiae method for extracting reliable fingerprint for feature extraction.

2 Issues and Challenges in Unimodal Systems with Methods to Address Challenges

In this section, issues and challenges in biometric-based security systems and approaches to overcome these issues are discussed.

2.1 Iris-based Recognition

Among all biometric traits, iris-based recognition systems are used everywhere due to its discriminable pattern. Iris patterns consist of several important distinct and tedious patterns of collarets, pigment spots, furrows, crypts, arching, etc. [7]. These factors result in discriminating textural patterns that are distinct to each eye of a human being, and even twins eyes are distinct [8]. It is very precise and most stable [9–11] personal identification biometric. Despite the aforementioned advantages and claim of highly accurate performance by existing iris recognition systems, the quality iris image acquisition framework designing is yet a challenging problem [8, 12–14]. This is due to various artifacts present at the time of capturing iris images. The effects of different artifacts are demonstrated here with the help of sample images from UBIRIS database [12] in Fig. 1. For example, there is an occlusion

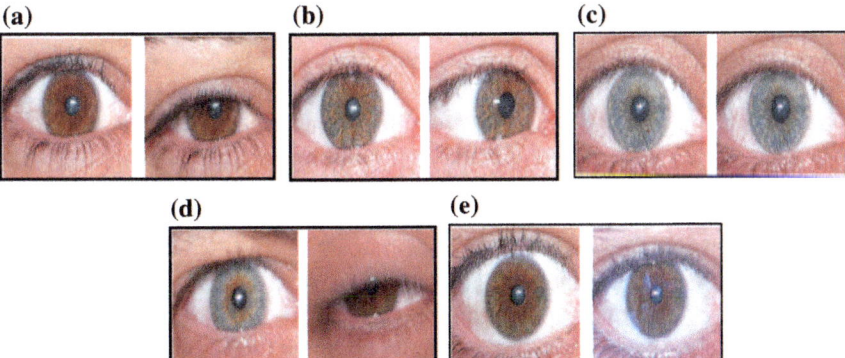

Fig. 1 Iris images from UBIRIS database showing artifacts. **a** Texture occluded by eyelids and eyelashes. **b** Different gaze direction. **c** Effects of contraction and dilation. **d** Motion blurriness. **e** Effect of the natural luminosity factor

(a)   (b)   (c)   (d)   (e)

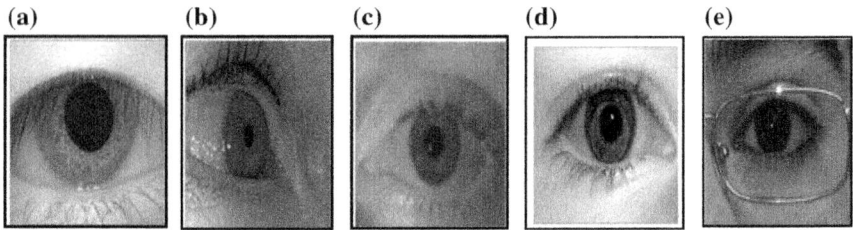

Fig. 2 **a** Specular reflections. **b** Off-angle. **c** Close-up of an iris image acquired at a large standoff distance. **d** Contact lenses. **e** Person with spec

due to the presence of eyelids and eyelashes of the iris as shown in Fig. 1a. There may be a natural position shift of left and right iris, also known as different gaze direction which is depicted in Fig. 1b. Iris images consist of motion blur artifacts as shown in Fig. 1d. Similarly, there may be artifacts due to contraction and dilation of pupils or excessive pupils or excessive luminosity due to natural light. These effects are presented in Fig. 1c, e respectively.

Similarly, the effects of other artifacts are demonstrated here with the help of sample images from other iris databases [12] as shown in Fig. 2. Figure 2a shows the specular reflection that degrades the quality of iris image. The off-angled iris images are caused, when angle of orientation of camera sensor used for acquiring iris is improper as shown in Fig. 2b. The off-angle iris image localization process is most critical. The standoff distance term refers to the distance between the camera location and subject. The pixel resolution depends on distance between object and camera at the time of image acquisition. The total pixel is less in iris image, if the distance is large. In such situations, the boundaries may not be detected accurately and vice versa. Figure 2c shows an acquired iris image at a large standoff distance. The more noisy artifacts are added, when the acquired iris images consist of lenses as shown in Fig. 2d.

There are very less algorithms for detection of iris with eye glasses. Here, somehow localization accuracy is very less because pupil is occluded by the contact lenses and structural features of contact lenses and pupil region are similar, which results in false identification of circular boundaries and poor recognition rate. The acquired iris image with spec causes problems on localizing the iris boundaries which causes detection of non-circular iris boundaries. Spec consists of glasses because of which reflection noise can also be added in the acquired iris image, and this affects the preprocessing algorithms accuracy. Figure 2e shows the iris image with spec.

There are some other factors like head orientation, camera diffusion, camera view angle, reflection component, and contrast which may cause improper segmentation of iris and ultimately degrade the performance of recognition. For any biometric recognition system, the rejection rate is a critical parameter. The above discussed artifacts result in an increased FRR of the recognition system. From the study of standard iris challenge evaluation [13–16], it is revealed that most of the iris recognition algorithms have high FRR. So the main focus of research is to device different techniques to reduce FRR. Many algorithms have been developed to reduce FRR [17, 18]. Different multiresolution-based methods like multichannel Gabor filter [17], multiscale 2D Gabor wavelets [18], four-level Laplacian pyramid [19], and quadratic spline wavelets [18] have been reported in the literature. Further, any effort to reduce FRR increases the FAR also.

From the above literature, it is also noticed that to reduce FRR, researcher used high-dimension feature vector that increases the overall execution time of recognition system which makes the system not viable for real-time application. So there is a need to propose an algorithm for iris-based recognition which has less computational complexity with acceptable FRR and FAR. We developed a novel iris-based recognition algorithm to tackle the above issues. We present a novel feature set for iris which is compact in size and hence reduces the computational complexity. The new feature allows us to present the texture information uniquely, and it results in an increase in genuine recognition rate. As part of the novel feature set, we introduce block sum. Based on these new features, we present iris recognition algorithm [20–22] called as a block sum method. This algorithm provides efficient recognition with reduced computational complexity and compact dimension of feature sets retaining adequate discriminating information for recognition which is the major contribution of our proposed algorithm.

2.2 Fingerprint-based Recognition

Among all biometric traits, the fingerprint is highly accepted by society and extensively used by forensic experts in criminal investigations [23]. Fingerprints are unique across individuals and across different fingers of the same individual. Even twins having same DNA have distinct fingerprints [24]. All these factors have led to the popularity of automatic fingerprint-based recognition systems in civilian, commercial, government and law enforcement applications. A fingerprint image is a combination of various ridge patterns flowing in different direction. The ridge flow exhibits anomalies in local regions of the fingertip as shown in Fig. 3. The ridge pattern in a fingerprint may be viewed as an oriented texture pattern, having a fixed spatial frequency and orientation in local neighborhood. The frequency occurs due to ridge spacing in the fingerprint, and the angle occurs due to pattern flows present in ridges of fingerprint. Frequency and orientation of non-overlapping ridges are responsible for distinct representation of fingerprint [25]. However, to match two fingerprints suitable ridge structures with proper alignment is essential.

Though most of the problems of fingerprint recognition have been extensively studied, there are a variety of unresolved issues that need to be addressed effectively. Some of the challenges are described below [13, 15, 26]. The fingerprint matching performance is affected by the nonlinear distortions present in the fingerprint image. These distortions have to be accounted prior to the matching stage for performance improvement. The fingerprint of a person does not change over

Fig. 3 Fingerprint image
with marked core and four
minutiae points

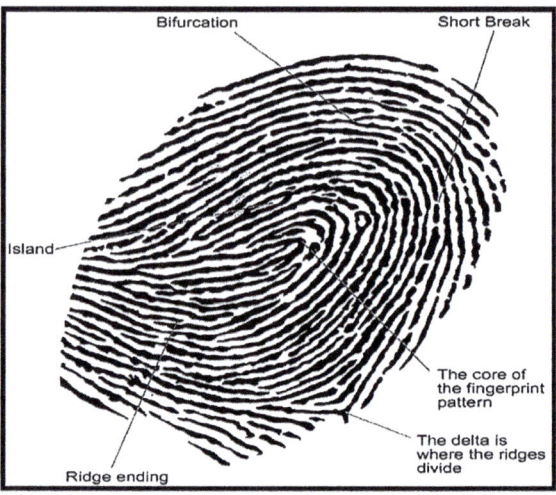

time, but it is possible that a person may have minor cuts or bruises which may alter
the ridge structure of a fingerprint. Moreover, the moisture content of the fingertip
may change over time which affects the quality of the fingerprint image being
acquired from a user. Due to this, template obtained at enrollment time and verification time may vary.

It is difficult to extract features from poor-quality images. Users having such
noisy fingerprint data may find it difficult to enroll in and interact with a biometric
system that uses only a fingerprint.

Due to advancement in technology, day by day, different types of sensors are
developed that are compact in nature. Compact sensors can be easily incorporated
into existing computer peripherals like the keyboard, as shown in Fig. 4a, mouse, as
shown in Fig. 4b. Similarly, latest cell phones are having embedded fingerprint
sensor, as shown in Fig. 4c. Compact sensors have increased the demand of fully
automated, highly accurate, real-time systems, but at the same time, it has increased
the challenges in fingerprint recognition.

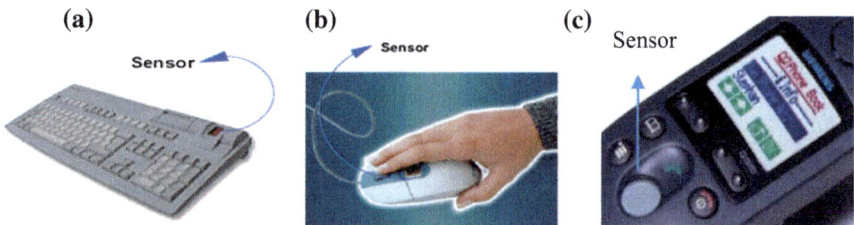

Fig. 4 Fingerprint sensors installed on keyboard, mouse, and cell phone [27]. **a** A keyboard (the
cherry biometric keyboard). **b** A mouse (the ID mouse manufactured by Siemens). **c** A cell phone
(the cell phone manufactured by Siemens)

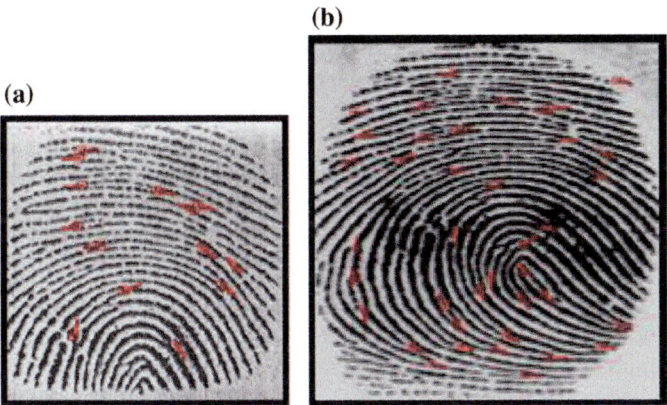

Fig. 5 Detected minutiae points in the fingerprint images. **a** 11 minutiae found in compact sensor. **b** 39 minutiae found in optical sensor

The above sensors when attached in compact systems like desktop, laptop, mouse, and cell phones provide a small contact area for the fingerprint and, therefore, sense only a small part of the fingerprint. Generally, the compact sensors provide the small contact area of $0.6'' \times 0.6''$, whereas standard optical sensors provide contact area of $1'' \times 1''$. This difference in contact area results in varying size of the fingertip images. With compact sensors, the fingertip size is of about 200×200 pixels and that of with optical sensors is about 380×408 pixels [15, 25, 27, 28]. Hence, the total number of minutiae features is more in optical sensor. Figure 5 illustrates this difference, where Fig. 5a depicts the fingerprint image obtained using solid-state sensor and Fig. 6b shows the fingerprint image obtained using optical sensor. As can be observed from Fig. 5, 11 minutiae are detected from the fingerprint image obtained using solid-state sensor and 39 minutiae are detected from the fingerprint image obtained using optical sensor. Also, multiple impressions of the same finger acquired at different instances may result in different images due to the rotation and translation [29–31], as shown in Fig. 6. Such images

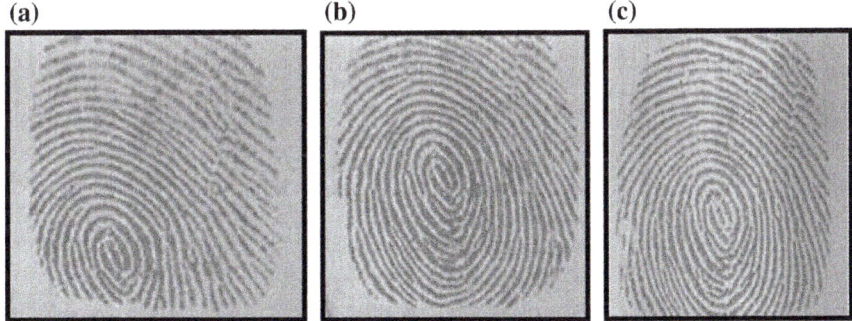

Fig. 6 Effect of orientations on a fingerprint. **a** Impression 1. **b** Impression 2. **c** Impression 3

need an additional alignment before the matching process of fingerprints. This increases the computational complexity of the recognition process.

Though there have been a large number of methods reported in the literature for extracting the features from fingerprint, for example, minutiae points also known as local pattern [32, 33], correlation-based, known as a global pattern [23–34], still the reduced feature vector with more discriminating power and reduced computational complexity is the need of an hour. Fingerprint image consists of minutiae's features. There are a large number of minutiae-based methods reported in the literature [32, 35, 36]. We also use the minutiae points-based feature for our proposed algorithm. In case of solid-state sensors, minutiae information may not be very discriminating. So an alternate representation, to supplement minutiae information, was felt necessary to improve the recognition performance in solid-state sensors. We propose novel minutiae points, which handle the variations in the positions of fingerprints. We choose minutiae-based features because no correlation approaches proposed in the literature gain performance comparable to this method [37]. The novelty of our algorithm [38] is that we first detect the core point, and using this core point, we follow the ridges to extract the minutiae points. Combining this information provides much discriminating feature set and increases the recognition rate for solid-state sensors.

2.3 Face-Based Recognition

Face of a human being changes day by day. Hence, face recognition system is difficult to develop. The accuracy of face recognition system gets affected by many artifacts reported in the literature. These artifacts are non-adaptive to illumination variations, makeups, aging effects, accessories, etc. Our face changes according to variation in illumination, as shown in Fig. 7b and pose (which causes due to small rotational angles variation) as shown in Fig. 7d [39]. For example, the FERET dataset sample face images appear different due to change in illumination that results in misclassification and wrong identification of person based on face images.

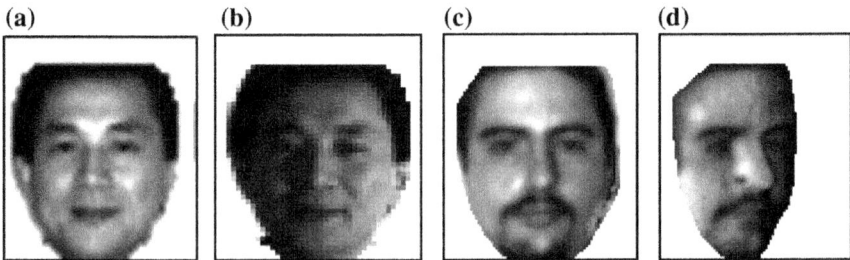

Fig. 7 Sample face images having illumination problem. **a** Original image. **b** Illumination problem. **c** Original image. **d** Pose problem

In the literature, there are many algorithms for face recognition. Zhao et al. [40] discussed several methods such as holistic template-based systems, geometric shape-based feature scheme, and hybrid approaches. Each method has its own advantage and disadvantage. The illumination problem has been solved using domain knowledge-based approach. In this, face has been divided into homogeneous classes. Under this, there are several different methods such as (1) image comparison and representation based on distance metric like Euclidean, Mahalanobis, and Hamming distance. (2) Heuristic methods include the principal component analysis of face features. (3) Class-based methods and multiple faces of same individuals are acquired in fixed position but different illumination and (4) 3D modeling of face images. In pose problem, the rotation problems are mainly divided into three categories: (1) availability of multiple images per individuals, (2) applicability of hybrid approach when multiple images have been used for training and testing, and (3) use of shape features for recognition without training.

Most of the work in face recognition is based on single face detection from complex background, skin color based, neural network based, deformable feature based, etc. [40]. The most popular approach is Eigen faces, which is the statistical approach. In this method, the face image has been represented as the vector of weights. During testing, the obtained test image vector has been compared with the database vector using distance metric. Other face recognition issue is overfitting problem. Mostly, this problem occurs in large face image database where training images have been very less. Zhao [41] discussed in his thesis about subspace LDA. This method has been divided into two steps: (1) the face images have been projected into small subspace using principal component Analysis (PCA). Then, (2) PCA projection has been transformed to linear discriminant analysis (LDA) for linear classification. The subspace size has been constant for each image in the database. Subspace dimensions have been mainly based on the eigenvector's and not on eigenvalues. Due to these types of flexibility for choosing, the subspace has been available in LDA, and hence, performance of classification gets improved. To resolve the overfitting issue, weighted distance metric has been added for improvement in performance of LDA method.

Again, the most challenging problem in face recognition is recognition of face of human being from video. Due to facial expression and behavior detection, face recognition is still difficult from video. Here, the performance of face detection system is also affected due to low-quality video, face mages size variation, motion estimation of moving object, background noise, and processing time. A video-based face recognition system using radial basis function-based neural network has been proposed by [42]. In this approach, features have been extracted using Gaussian filters and Gabor wavelet transform. Here, radial basis kernel has been used, due to its fast learning ability and motion-based segmentation. Motion pixels have been grouped together, and group object has been tracked using Kalman filters, and finally, RBF network has been used for object detection.

2.4 Palm Print-based Recognition

Though the most widely used biometric features are iris, fingerprint, and face [2], each one of them is having its own limitations. For example, it is challenging task to extract minutiae points from noisy fingerprint sample. Comparatively, palm print contains more rich features than fingerprints, so they are more discriminative. Also, in terms of cost, palm print is less costly than other biometric traits. It consists of several important features such as palm geometry, ridge, valley, principal lines and wrinkles, and termination and bifurcation points. These features are useful for developing highly reliable and accurate system. Apart from these structural features, palm print also has textural and geometrical features such as minimum axis, maximum axis, width, length, and area of palm. Many researchers have developed offline palm print recognition system [12], but it cannot work in dynamic environment.

Some researchers make attempts to solve problems, by making the online palm print recognition system. The main problem here was that fake palm print can be created and misused. Hence, point features are used nowadays. The main advantage of this system is that (1) palm print verification becomes more accurate, (2) palm print can be identified in the presence of noise and from low-quality images, and (3) the datum point-based features are more accurate and stable. Figure 8 shows the process of palm line feature extraction steps. Wu et al. [43] proposed a novel approach of palm line feature extraction and matching for authenticating individuals. Different directional line detectors and chain code representation have been used for extracting palm line features without loss of palm line structure details. At the end, matching score has been computed between two palm prints according to

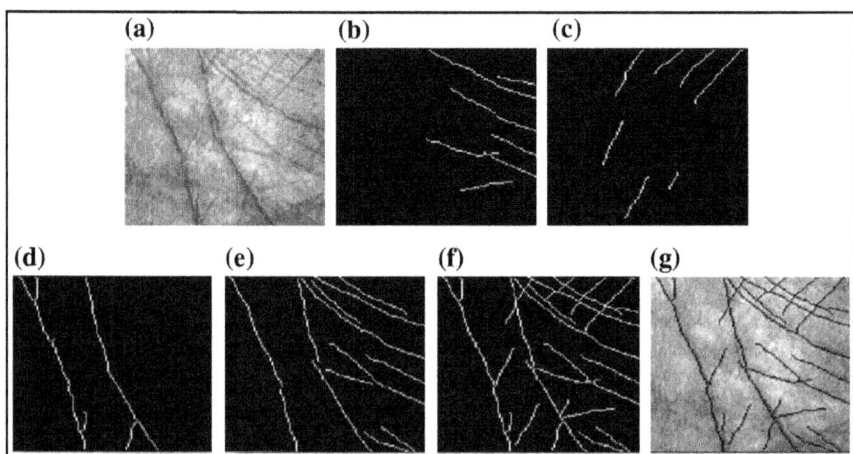

Fig. 8 Palm line feature extraction. **a** Palm print sample image, **b–e** the directional line images (0°, 45°, 90°, and 135°), **f** output palm lines feature identified image, **g** palm print image overlapped with lines features

the points of their palm lines using distance metric. Similarly, Zhang [44] proposed combined approach of datum point invariance and line feature-based palm print verification techniques. These techniques have been rotational invariant. Pradeep [45] proposed novel algorithm which has been robust to noise and distortions in the palm print and also subjected to rotation and translation invariant.

Kanchana and Balakrishnan [46] described a novel quadtree decomposition-based palm print recognition system. In this type of system, palm print image has been divided into different resolution, and sometimes, it is known as multiresolution decomposition. The main advantage of this method is that segmentation and recognition accuracy get increased. Similarly, Han et al. [47] used combined approach of gradient edge detection Sobel operator and morphological operations to extract line features. Wang [48] proposed a palm print identification approach using local binary pattern (LBP) histogram bins to extract and represent the local features of a palm print image.

2.5 Speech Recognition

Nowadays, human and machine interaction is possible using automatic speech recognition system. Most of the deployed speech recognition system performs well in real-world environment. Speech recognition is the process of recognition of words and transforms it into machine-understandable language. It plays vital role in many applications like the following:

1. **Language Translation**: Converting spoken words into required language which user can understand.
2. **Air Force**: Establishing long-distance communication between fighter aircraft and helicopters in battle.
3. **Academia**: Sharing knowledge among students while teaching Foreign language with correct language semantics, pronunciation, and vocabulary.
4. **Health Care**: Advanced speech recognition systems used in critical surgery and operations.
5. **Security system**: Speech-based security system has been deployed at high-security zones.

Currently, no speech recognition system is accurate and there are several issues and challenges as follows, which affect the performance:

1. **Environment**: In day-to-day life, voice of a human being has been affected by bad weather conditions. Also, it is very difficult to identify human voice if he/she suffers from cough or a fever.
2. **Noise generated by transducer**: Various frequency range changes between caller and receiver. Again in microphone, telephone frequency encoding and decoding process affects the performance.

3. **Medium/Channel noise**: Due to band amplitude modulation, distortion echoes additive noise need to be added in the speech.
4. **Vocabulary**: Due to training dataset of available metadata, specific or generic vocabulary.
5. **Speech styles**: Due to voice accents, tone like quiet, normal, shouted, and speed of voice like slow, normal, and fast, rough voice.

Therefore, the main objective of researchers is to develop accurate voice recognition system that can handle above issues. In the literature, many features have been investigated. The most important of them is Mel frequency cepstral coefficients (MFCC) feature [49]. It identifies the linguistic terms and removes the other noise such as emotions and environmental background noise from the audio signal. Abushariah et al. [50] described speech recognition, in which they used combined approach of MFCC and Hidden Markov Model (HMM). MFCC feature extraction technique has been used for English digits database, and classification has been performed using HMM. Jiang et al. [51] proposed an approach that uses a principal of max and min multiclass separator for estimating density of HMM.

2.6 DNA Recognition

DNA is the most complex and reliable form of biometric trait. DNA never changes during whole human's life and even after death. There are lot of issues involved while designing DNA-based identification system such as privacy of human being, accessibility of DNA information, health issues, social issues, storage, and processing time. Other challenges are like the following: Real-time DNA matching is not possible; a DNA sample does not tolerate the uncertainty; and automated system is difficult to develop. DNA features required more time for matching as compared to other biometric traits. On the contrary, DNA-based identification system is more accurate [52].

2.7 Biomedical Signal Processing

Now days, telemedicine is in widespread use and is gaining attention of a lot of people all over the world. Telemedicine is the combined approach of information technology and telecommunication to provide medical services everywhere. The main use of telemedicine is to provide clinical health care at a long distance. Many e-consultants are available online at different healthcare sites. Obviously, certain secure measure is required to transfer medical content via public medium. Dey et al. [53] proposed a secure and accurate approach to biomedical content authentication by embedding logo of the hospital within the electrocardiogram signal by using combined approach of discrete wavelet transformation (DWT) and cuckoo search

(CS). Again, Dey et al. [54] proposed a secure blind watermarking technique using DWT–DCT (discrete cosine transformation)–SVD (singular value decomposition) for grayscale biomedical images in electrooculogram signal. He further extended his work using wireless blind watermarking method with secret key by embedding black and white watermark image into ECG signal [55]. Also, he embeds the watermark in edges of source image by using canny edge gradient operator [56], low-distortion prediction-error expansion watermark insertion and extraction technique in EOG signal without affecting its diagnostic measures [57], and reversible black and white watermark embedding into the PPG signal and a watermark extraction mechanism using prediction error-based algorithm [58]. Chakraborty et al. [59] proposed a DWT-based method for embedding a Hospital Logo or Electronic Patient Record (EPR), where the embedding parameters are optimized by particle swarm optimization (PSO). Dey et al. [60] proposed Electrocardiogram Feature based Inter-human Biometric Authentication System. The system performance is evaluated by each other's ECG features stored in database and found more improved accuracy. Acharjee et al. [61] presented the comparative performance analysis of different cost functions and most suitable cost function for motion vector estimation. Dey et al. [62] proposed DWT-based heart disease identification system. Due to the presence of cumulative frequency components in spectrogram analysis, DWT is applied on spectrogram to find normal and abnormal heart sound identification. It has been useful for tele-diagnosis of heart diseases. Dey et al. [63] proposed to apply DWT–DCT–SVD on IVUS frames for embedding the watermark. The inverse DWT–DCT–SVD has been applied on the watermark frames for extracting watermark image. He further extended his work in [64] by proposing a novel approach of extracting features using Harris corner detection and wavelet and then fusing the image. In this section, different issues and challenges in designing biometric-based security systems and methodologies involved in overcoming these issues are discussed for each trait such as iris, fingerprint, face, palm print, speech, DNA, and biomedical signal processing. Also, the accuracy of recognition can be improved further by making use of multimodal systems.

3   Issues and Challenges in Multimodal Systems

Multibiometric system design is certainly a challenging task since it is very difficult to choose the best possible sources of biometric information and fusion strategy for a particular application. This difficulty is related to many issues such as follows:

(a) **Benchmark multimodal datasets**: The multimodal databases, those are publicly accessible, are quite limited. Due to this reason, some researchers have assumed that different biometric traits of the same person are statistically independent [26]. Most of the experiments in multimodal biometrics have been conducted on virtual databases [65, 66]. The virtual database is generated by combining different biometric traits belonging to different persons to form one subject [67–69].

(b) **Incompatibility of the information resources**: The integration of biometric information in the early stages is thought to be more valuable since the amount of information available in the raw data is more [6, 10, 66]. Nevertheless, fusion at early stages such as sensor and feature levels is not always possible due to the incompatibility of the gathered information. For example, in a multimodal biometric system based on fusing fingerprint and voiceprint, it is not possible to fuse the raw images of the fingerprint with the voice signal.

(c) **Social acceptance and privacy issues**: Privacy concerns are related to data collection, unauthorized use of recorded information, and improper access to biometric records. As such, for security concerns and privacy issues, data protection laws and standards should be developed.

(d) **Optimum Design Issues**: A number of factors need careful consideration while designing a multimodal biometric system [6]. These include the following:

   (a) The choice and number of biometric traits involved in recognition. It depends upon the type of application, the required level of security concerns, and overhead introduced by multiple traits.

   (b) Multiple traits information should be integrated at feature level. The choice of the fusion level is the most important design issue in a multi-biometric system, and it has a substantial impact on the performance of the system.

   (c) Given the type of information to be fused, a number of techniques are available for fusion of information provided by the multiple sources. Many of these fusion schemes may be admissible in an application, and the challenge is to find the optimal one.

   (d) The cost involved in developing and deploying a real-time system [1, 7] versus matching performance trade-off. The cost could be a function of the number of sensors deployed, the time required for acquisition and processing, storage, and computational requirements.

As a result, many 'diversity' measures exist while no measure is a satisfactory predictor of the higher performance. This work addresses five critical issues in the design of a multibiometric system, namely efficient feature extraction, fusion methodology, better accuracy, reduced error rates, and fast response time.

4 Proposed Work

In this section, proposed algorithms for addressing issues and challenges of iris and fingerprint are discussed in detail.

4.1   Block Sum Feature Sets Algorithm for Superior Representation of Texture of Iris

In this section, we propose iris recognition using a novel feature set, namely block sum method. Proposed algorithm consists of three main steps, as shown in Fig. 9: (1) preprocessing, (2) extracting the feature sets, and (3) recognition using support vector machine [21]-based classifier. The preprocessing step further consists of iris localization and normalization steps. In our algorithm, we first localize the iris area between the inner and outer boundaries of the iris. This step also detects and removes any specular reflection and eyelash or eyelids noise from the image. For this, we use a very popular technique, namely Hough transform [8]. Hough transform is robust with respect to breaks in the shape boundary. For iris normalization, we use very well-known standard technique, the Daugmans rubber sheet model [8]. It takes into account pupil dilation and size problem present in normalized image representation with constant dimension. The output of the preprocessing step is the region of interest (ROI).

The next step is to extract distinctive features of the iris texture pattern from ROI. Here, we extract the features of the iris using the proposed block sum method. For our proposed technique [21], we extract features based on the blocks, instead of individual pixels, to reduce computational complexity. For final recognition of these feature sets, we use radial basis function-based SVM classifier as it provides better generalization performance than conventional techniques [21].

4.1.1   Feature Extraction

This proposed algorithm is based on blocks. A block represents texture pattern in a better manner as compared to pixel-by-pixel representation. For this, given normalized iris strip is divided into blocks of size m × n for optimal performance, as shown in Fig. 10. The information from each block is extracted based on its entropy

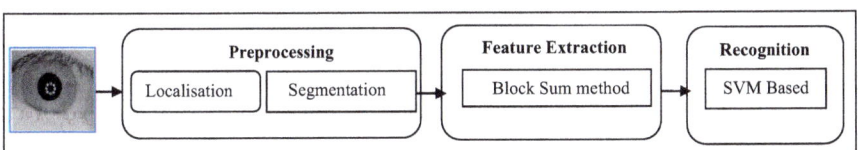

Fig. 9 Proposed block sum-based iris recognition

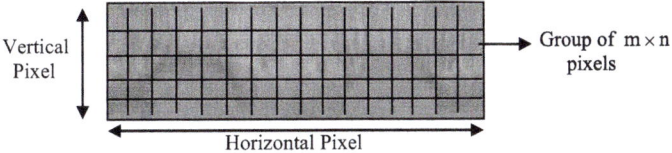

Fig. 10 Division of normalized image into cells

value and gray level value of each pixel value within the block. A representative value of each block is computed by first evaluating entropy 'E' for each block.

The gray level value for each pixel is represented as Ci. Now, for every pixel in a block, the feature Si is found by the mean of Si, which is the representative value for block under consideration. We obtain the feature vector of size $1 \times L$ using the block sum method, where 'L' is the number of elements present in the feature vector [20]. This algorithm is simple and does not need much processing, but at the same time, it helps in representing feature in a compact manner. The block sum feature vector is further used to train RBFSVM classifier.

4.2 Proposed Modified Minutiae-based Method

In this section, we propose fingerprint recognition using novel feature set extraction technique. The proposed fingerprint recognition algorithm consists of four steps, namely fingerprint image acquisition, fingerprint preprocessing, feature extraction, and recognition or classification as shown in Fig. 11.

The preprocessing step further consists of normalization, segmentation, orientation, image enhancement, image binarization, and thinning steps for obtaining the ROI for feature extraction. First, the proposed algorithm estimates the ridge orientation and frequency from the original fingerprint image [67–69]. Then, it performs filtering to remove the noise and enhance the fingerprint image. This enhanced image is then binarized and thinned to obtain the ridge skeleton, as shown in Fig. 12. The binarized ridge skeleton of the fingerprint is then used for feature

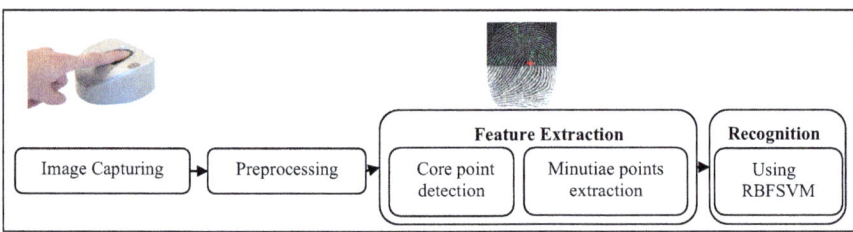

Fig. 11 Proposed modified minutiae-based fingerprint recognition

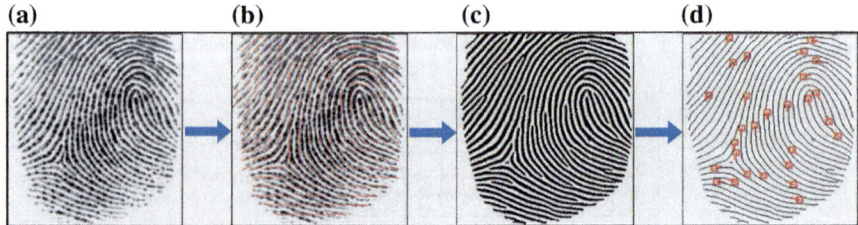

Fig. 12 Preprocessing steps of minutiae feature extraction algorithm. **a** Normalized image. **b** Enhanced image. **c** Binarized image. **d** Marked minutiae

extraction. Here, we propose novel minutiae-based algorithm for extracting features from the fingerprint sample. Minutia features represent locations where ridges end abruptly or where a ridge branches into different ridges. A quality fingerprint sample image contains about 30–70 minutiae points [31]; the actual feature size depends on the sensor surface and fingerprint acquisition framework. Fingerprint ridge skeleton consists of several important features such as core point, termination, and bifurcation points. To improve the performance of recognition, the proposed algorithm then applies logic to detect and remove spurious minutiae. The novel fingerprint feature set is based on the concept of distance between core point and minutiae points along the angular directions. The RBFSVM classifier is trained using these novel feature sets. The main advantage of this method is that it is invariant to rotation and translation. In most cases, image registration is a necessary step to take care of rotation and translation.

The registration steps for estimating rotation and translation are avoided in our proposed approach, which results in reduction in computational complexity. Further reduction in computational complexity is achieved by reducing the number of extracted minutiae points.

4.2.1 Feature Extraction

In this section, we propose a novel feature set extraction technique from a fingerprint. The output of preprocessing steps is the binarized thinned fingerprint image. While a filtered image itself can be used as a representation of fingerprint, the resultant feature vector may not be very relevant. With the assumption that the combination of two feature representations can improve the recognition performance, we propose a novel feature set that captures global point, i.e., core and local points, i.e., minutiae features of a fingerprint. The distance between these points is considered as a novel feature set. This provides a compact fixed length vector with better representation of the feature set. It also takes into consideration the variations in the positions of fingerprint images, by following the minutiae points from the core point along the radial direction. For proposed feature extraction, first we describe core point detection. For this, we used the method of [70] due to its more accurate results in detecting the core point. At each pixel, the directional component 'θ' is available based on the gradient. The input image is divided into blocks of size $k \times 1$, i.e., 3×3 pixels. In every block, the difference of direction components is computed using Eqs. 1 and 2.

$$\text{Diff } Y = \sum_{k=1}^{3} \sin 2\theta \, (k, 3) - \sum_{k=1}^{3} \sin 2\theta \, (k, 1) \tag{1}$$

$$\text{Diff } X = \sum_{k=1}^{3} \sin 2\theta \, (k, 3) - \sum_{k=1}^{3} \sin 2\theta \, (k, 1) \tag{2}$$

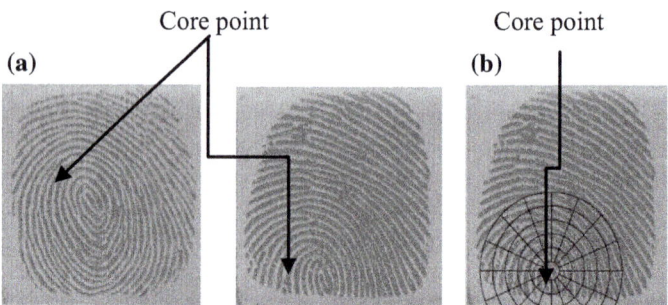

Fig. 13 Samples of core point detection. **a** Sample 1. **b** Sample 2

The core point (X) is located at the pixel (i, j) where Diff X and Diff Y are negative. The result of this core point detection is shown in Fig. 13.

The preprocessed binarized image is divided into blocks of size 3×3. We perform a morphological operation that connects the ridges for non-breakable ridge structure. At each pixel, 3×3 window sliding is performed. The minutiae points are marked based on number of ones in these 3×3 blocks. These blocks are categorized as terminations, if a central pixel is 1 and count of one valued neighbor is also one, as shown in Fig. 14a. Similarly, blocks are categorized as bifurcations, if a central pixel is 1 and counts of one valued neighbor are three, as shown in Fig. 14b. As a result of this step, all the minutiae in the fingerprint image are marked.

All the earlier stages may introduce some artifacts which later lead to spurious minutia. If such false minutiae are considered as genuine minutiae points, then it will eventually lead to poor recognition rate. So it is essential to remove such false minutiae to achieve a higher recognition rate [71]. After removing the false minutiae, the fingerprint image consists of genuine marked minutiae points, as shown in Fig. 15.

Fig. 14 Minutiae marking in the fingerprint. **a** Termination. **b** Bifurcation

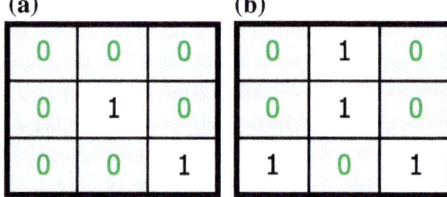

Fig. 15 Marked minutiae points around the core point

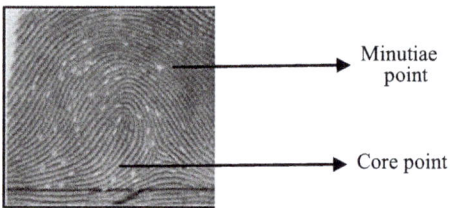

Now, using the particular points marked as terminations and bifurcations, we follow the ridges, starting from core point, in increasing direction of radial distance. In our case, we extract only first 60 minutiae points, which are encountered along the ridges, starting from core point. The value 60 is derived based on an observation that maximum fingerprint information will be captured by first countable number of minutiae points, i.e., 60.

The distance from the core point to each of initial 60 minutiae points is considered as new feature set. RBFSVM classifier is trained by these novel feature sets. In the following section, we discuss the simulation results of the proposed approaches, i.e., feature extraction for iris and fingerprint.

5 Experimental Results

The localized iris, in CASIA database, has iris radius from 80 to 150 pixels and pupil radius from 30 to 75 pixels. In the proposed block sum algorithm, the normalized iris strip is of size 240×20. This normalized image is divided into 60 blocks of size 16×5. The representative values of these 60 blocks are considered as a feature set using block sum method, which is of size $1 \times L$, where 'L' is nothing but 60 elements, for iris under consideration. We have obtained the accuracy of 89 % with 2 % FAR and 11 % FRR by proposed algorithm using novel block sum feature set. The average response time required for block sum-based iris recognition is 5.09 s. We have observed that RBFSVM classifier turn out to be superior. The above results are summarized in terms of recognition rates, error rates, and response time in Table 1. The comparison plot of recognition rates for each classifier and distance measure is shown in Fig. 16. The recognition rates of RBFSVM classifier using a block sum algorithm and approaches of [72–74] are also shown in Fig. 16, where it is observed that block sum algorithm performs better than the feature sets of [72–74]. Similarly, the recognition rates using PNN, RBFNN, and PolySVM classifiers are depicted. Here, also our feature sets perform well as compared to the approaches of [72–74]. The highest recognition rate is obtained using the RBFSVM classifier for our feature sets. Similarly, from the experimental results, we have obtained 86 % of recognition rate with 6 % of FAR and 14 % of FRR using the proposed algorithm, where RBFSVM classifier is used for recognition. The average response time required for recognition of one subject is 4.48 s. We trained PNN classifier using our novel feature sets. From the simulation results, it is found that 80 % recognition rate is obtained, using PNN classifier. We further tested this classifier for 50 imposter cases. Experiments show FAR of about 14 % and FRR of 20 %, using the PNN classifier for the proposed algorithm. The mean response time is 4.86 s. We further trained RBFNN classifier using novel minutiae feature sets. The number of training and testing samples is same for all classifiers. From the simulation results, it is found that for novel minutiae points, the recognition rate of 85 %, FAR of 10 %, and FRR of 15 % are obtained, using RBFNN classifier. The response time required using RBFNN classifier is 4.82 s.

Table 1 Comparison of recognition rates, error rates, and response time of proposed algorithms on CASIA iris database

Recognition using		Accuracy rate (%)	Error rates		Execution time (s)
Feature set extraction technique	Classifier/distance measure		FAR (%)	FRR (%)	
Our proposed block sum	RBFSVM	89	2	11	5.09
	PNN	84	4	16	5.88
	RBFNN	88	2	12	5.37
	PolySVM	85	2	15	5.39
	Hamming distance	88	4	12	5.95
Cumulative-sum-based [73]	RBFSVM	81	4	21	8.02
	PNN	76	8	24	8.92
	RBFNN	78	6	22	8.83
	PolySVM	80	6	20	8.09
	Hamming distance	80	8	20	8.99
2D Gabor filter [74]	RBFSVM	80	6	20	7.07
	PNN	77	8	23	7.87
	RBFNN	79	6	21	7.32
	PolySVM	79	6	21	7.39
	Hamming distance	82	8	18	7.90
	Euclidian distance	76	5	24	7.95
Discrete wavelet transform [72]	RBFSVM	82	4	18	5.32
	PNN	79	6	21	6.25
	RBFNN	80	4	20	6.12
	PolySVM	82	4	22	6.09
	Hamming distance	83	6	17	6.55
	K-NN	75	9	25	6.38

Fig. 16 Iris-based recognition rate comparison

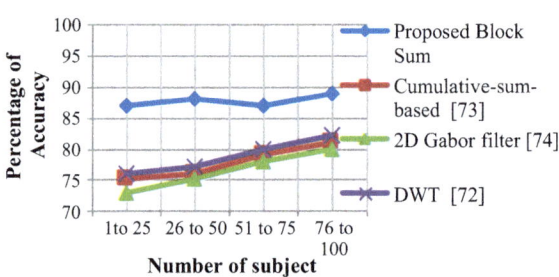

The number of training and testing samples is same for all classifiers. From the simulation results, it is found that for novel minutiae points, the recognition rate of 85 %, FAR of 10 %, and FRR of 15 % are obtained, using RBFNN classifier.

The response time required using RBFNN classifier is 4.82 s. These results clearly indicate the improvement in performance of recognition rates and response

Table 2 Comparison of recognition rates, error rates, and response time of proposed algorithm on YCCE fingerprint database

Recognition using		Accuracy rate (%)	Error rates		Execution time(s)
Feature set extraction technique	Classifier/distance measure		FAR (%)	FRR (%)	
Proposed novel minutiae points	RBFSVM	86	6	14	4.48
	PNN	80	14	20	4.86
	RBFNN	85	10	15	4.82
	PolySVM	82	10	18	4.62
	Hamming distance	80	18	20	4.92
Standard minutiae points [75]	RBFSVM	80	18	20	6.23
	PNN	76	20	24	6.72
	RBFNN	78	22	22	6.67
	PolySVM	78	18	22	6.30
	Hamming distance	80	20	20	6.88
DWT [76]	RBFSVM	81	16	19	5.43
	PNN	78	18	22	5.51
	RBFNN	80	20	20	5.63
	PolySVM	80	16	20	5.50
	Hamming distance	82	22	18	5.99
	K-NN	79	17	11	5.45

time using RBFNN classifier, as compared to PNN classifier. Table 2 represents the above results in terms of recognition rates, error rates, and response time, with proposed algorithm and comparison of the approaches of standard minutiae points [75] and DWT [76].

The comparison plot of recognition rates and error rates of each classifier and distance measure is shown graphically in Fig. 17. From the simulation results, we have observed that the highest accuracy of 80 %, with FAR of 20 % and FRR of 20 %, is obtained using Hamming distance, for the approach of Sardana et al. [75]. This accuracy rate of [76] is same as that of PNN classifier and Hamming distance of our approach. But the size of the standard minutiae feature vector [75] is of size 1 × 120. This ultimately results in high response time and FAR. We have obtained

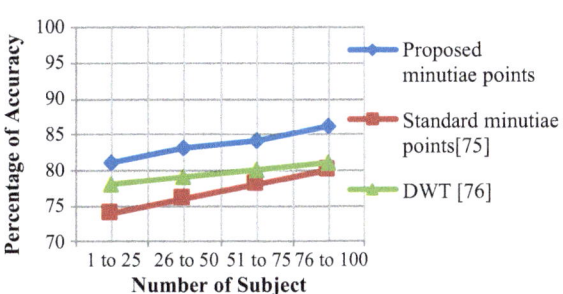

Fig. 17 Fingerprint-based recognition rate comparison

better results with reduced size feature vector of size 1 × 60. For the approach [76], the highest accuracy of 82 % is obtained using Hamming distance, with FAR 22 % and FRR of 18 %. Though this accuracy is same as that of our feature set using PolySVM classifier, its FAR and response time are much higher as compared to our proposed approach. The DWT feature sets by [76] show better results as compared to the standard minutiae points by [75], in terms of all assessment parameters.

6 Conclusions

In this chapter, various issues and challenges in designing biometric-based security system have been discussed. Again each aspect of biometric traits, i.e., for unimodal as well as multimodal has been depicted in brief here. The existence algorithms for handling these issues have been also described. We have proposed two feature extraction techniques. One approach is iris recognition using block sum, and another approach is fingerprint recognition using modified minutiae. Our proposed algorithm using novel feature sets and RBFSVM classifier perform better than other algorithms in the literature, in terms of all the assessment parameters, i.e., recognition rate, error rates, and response time. Block sum feature sets give a better representation of texture as compared to pixel-by-pixel representation. This also proved to be the better feature set for recognition. This is a simple algorithm with reduced computational complexity and fast response time. Our proposed technique generates compact, unique, and invariant feature sets. To overcome the limitations of existing feature sets, like a large computational burden, high memory requirement, high processing time, and high FRR, proposed block sum retains most of its desired properties, with the small size of feature set. They retain most of its desired properties, with the small size of feature set. For fingerprint, a novel fingerprint feature set extraction algorithm, combining core and minutiae points, has been presented. Using our proposed algorithm, it is possible to extract the desired features even with the smaller size of the input image; hence, such algorithm can be utilized with the compact sensors. The novel method reduces the computational complexity, as there is no need of translation and rotation steps, still taking care of variations in the positions of fingerprint images. Simulation results indicate that the proposed technique performs much better than a standard minutiae-based method and reduced feature set using DWT. The combination of proposed technique and RBFSVM classifier achieves a higher recognition rate, low false acceptance rate, and fast response time for fingerprint recognition as well. The main aim of this chapter was to highlight the key issues and challenges of each biometric-based system.

References

1. Yang F, Ma B (2007) A new mixed-mode biometrics information fusion based on fingerprint, hand-geometry and palmprint. In: International conference on image and graphics, China, August 22nd–24th, pp 689–693
2. Subbarayudu VC, Prasad MVNK (2008) Multimodal biometric system. In: IEEE international conference on emerging trends in engineering and technology (ICETET), Maharashtra, July 16th–18th, pp 635–640
3. Nagar A, Nandakumar K, Jain AK (2012) Multibiometric cryptosystems based on feature-level fusion. IEEE Trans Inf Forensics Secur 7(1):255–268
4. Besbes F, Trichili H, Solaiman B (2008) Multimodal biometric system based on fingerprint and iris recognition. In: IEEE international conference on information and communication technologies: from theory to applications, Damascus, Syria, April 7th–11th, pp 1–5
5. Sanderson C (2003) Automatic person verification using speech and face information. Ph.D. Thesis, Griffith University, Queensland, Australia
6. Ross A, Nandakumar K, Jain AK (2006) Handbook of multibiometric. International series on biometrics, vol 6. Springer, New York
7. Daugman J (2001) Statistical richness of visual phase information: update on recognizing persons by iris pattern. Int J Comput Vis 45(1):25–38
8. Daugman J (2004) How iris recognition works. IEEE Trans Circuits Syst Video Technol 14 (1):21–30
9. Jain AK, Ross A, Prabhakar S (2004) An introduction to biometric recognition. IEEE Trans Circuits Syst Video Technol Spec Issue Image Video Based Biom 14(1):4–20
10. Jain AK, Ross A, Pankanti S (2006) Biometrics: a tool for information security. IEEE Trans Inf Forensics Secur 1(2):125–143
11. Flom L, Safir A (1987) Iris recognition system. US Patent 4641394
12. Proenca H, Alexandre LA (2005) UBIRIS: a noisy iris image database. In: Proceeding of the 13th international conference on image analysis and processing (ICIAP), Cagliari, Italy, vol 1, September 6th–8th, pp 970–977
13. Bowyer K, Hollingsworth K, Flynn P (2008) Image understanding for iris biometrics. J Comput Vis Image Underst 110(2):281–307
14. Iris Challenge Evaluation. http://www.nist.gov. Accessed 25 Dec 2015
15. LG. Home Page. http://www.lgIris.com/ps. Accessed 30 Sept 2011
16. Liu X, Bowyer KW, Flynn PJ (2005) Experiments with an improved iris segmentation algorithm. In: 4th IEEE workshop on automatic identification advanced technologies, Buffalo, NY, USA, October 17th, pp 118–123
17. Wildes RP (1997) Iris recognition: an emerging biometric technology. IEEE J Circuit Syst Video Technol 85(9):1348–1363
18. Ma L, Tan T, Wang Y, Zhang D (2004) Local intensity variation analysis for iris recognition. Pattern Recogn Lett 37(6):1287–1298
19. Park C, Lee J, Smith M, Park K (2003) Iris-based personal authentication using a normalized directional energy feature. In: 4th International conference on audio-and video-based biometric person authentication (AVBPA), Berlin, Heidelberg, UK, June 9th–11th, pp 224–232
20. Gawande U, Zaveri M, Kapur A (2011) Improving iris recognition using haar, multiresolution and new block sum method: novel multialgorithmic approach. Biom Technol Today J 2011 (4):8–10
21. Gawande U, Zaveri M, Kapur A (2011) A novel multialgorithmic approaches forimproving iris recognition using Haar, multiresolution and new block sum method. In: International conference and workshop on emerging trends in technology 2011, ACM, Mumbai, vol 2, February 25th–26th, pp 576–584
22. Gawande U, Zaveri M, Kapur A (2011) An effective iris recognition system based on efficient multialgorithmic fusion technique. Int J Comput Appl 5(13):24–31

23. Ross A, Nandakumar K, Jain AK (2006) Handbook of multibiometric. International series on Biometrics, vol 6. Springer, New York
24. Pankanti S, Prabhakar S, Jain AK (2002) On the individuality of fingerprints. IEEE Int J Pattern Anal Mach Intell 1(8):805–812
25. Senior A (2001) A combination fingerprint classifier. IEEE Trans Pattern Anal Mach Intell 23 (10):1165–1174
26. Biometrics Market Intelligence. http://www.biometricsmi.com. Accessed 30 Jan 2010
27. http://www.9to5mac.com/. Accessed 21 Oct 2012
28. Daugman JG (2003) The importance of being random: statistical principles of iris recognition. Pattern Recogn Lett 27(2):279–291
29. Jain AK, Feng J, Nandakumar K (2008) On matching latent fingerprint. In: IEEE workshop of computer vision and pattern recognition (CVPRW'08), Computer society, June 23rd–28th, pp 36–44
30. Hong L, Wan Y, Jain AK (1998) Fingerprint image enhancement: algorithm and performance evaluation. IEEE Trans Pattern Anal Mach Intell 20(8):777–789
31. Sha L, Tang X (2004) Orientation improved minutiae for fingerprint matching. In: 17th International conference on pattern recognition, Cambridge, UK, vol 4, August 23rd–26th, pp 432–435
32. Kovacs-Vajna ZM (2000) A fingerprint verification system based on triangular matching and dynamic time warping. IEEE Trans Pattern Anal Mach Intell 22(11):1266–1276
33. Cappelli R, Maio D, Maltoni D, Wayman JL, Jain AK (2006) Performance evaluation of fingerprint verification systems. IEEE Trans Pattern Anal Mach Intell 28(1):3–18
34. Pankanti S, Prabhakar S, Jain AK (2002) On the individuality of fingerprints. EEE Int J Pattern Anal Mach Intell 1(8):805–812
35. Gu J, Zhou J, Yang C (2006) Fingerprint recognition by combining global structure and local cues. IEEE Trans Image Process 15(7):1952–1964
36. Liu L, Jiang T, Yang J, Zhu C (2006) Fingerprint registration by maximization of mutual information. IEEE Trans Image Process 15(5):1100–1110
37. FVC 2002—Fingerprint Verification Competition. http://bias.csr.unibo.it/fvc2002/databases. asp. Accessed 1 July 2010
38. Gawande U, Zaveri M, Kapur A (2013) Bimodal biometric system efficient feature level fusion of iris and fingerprint. Int J Biom Technol Today 2013(3):7–9
39. Zhao WY, Chellappa R (2002) Image-based face recognition: issues and methods, image recognition and classification, pp 375–402
40. Zhao W, Chellappa R, Rosenfeld A, Phillips PJ (2003) Face recognition: a literature survey. ACM Computing Surveys, pp 399–458
41. Zhao W (1999) Robust image based 3D Face recognition, Ph.D. Thesis, University of Maryland
42. Howell AJ, Buxton H (1996) Towards unconstrained face recognition from image sequences. In: International conference on automatic face and gesture recognition, pp 224–229
43. Wu X, Zhang D, Wang K (2006) Palm line extraction and matching for personal authentication. IEEE Transactions Syst Man Cybern Part A Syst Hum 36(5)
44. Zhang D, Shu W (1999) Two novel characteristics in palmprint verification: datum point invariance and line feature matching. Pattern Recogn Lett 32(4):691–702
45. Pradeep N, Jain M, Prakash C, Raman B (2006) Palmprint recognition: two level structure matching. In: International joint conference on neural networks sheraton Vancouver Wall Centre Hotel, Vancouver, BC, Canada, July 16th–21th
46. Kanchanal S, Balakrishnan G (2012) Quadtree decomposition for palm print feature representation in palmprint recognition system. In: IEEE International conference on advanced communication control and computing technologies (ICACCCT)
47. Han C, Chen H, Lin C, Fan K (2003) Personal authentication using palmprint features. Pattern Recogn 36(2):371–381

48. Wang X, Gong H, Zhang H (2006) Palmprint identification using boosting local binary pattern. In: 18th International conference on pattern recognition (ICPR'06), Hong Kong, vol 3, August, pp 503–506
49. Han W, Chan C-F, Choy C-S, Pun K-P (2006) An efficient MFCC extraction method in speech recognition. In: IEEE International symposium on circuits and systems, (ISCAS), 4th, May
50. Abushariah AAM (2010) English digits speech recognition system based on hidden Markov models. In: IEEE International Conference on Computer and Communication Engineering, Kuala Lumpur, Malaysia, 11th–13th, May
51. Jiang H, Li X, Liu C (2006) Large margin Hidden Markov models for speech recognition. IEEE Trans Audio Speech Lang Process 14(5):1584–1595
52. Foister S (2004) Introduction to DNA recognition by minor groove-binding polyamides. Ph. D. Thesis, California Institute of Technology, CaltechETD: etd-05112004-101833, 14th May
53. Dey N, Samanta S, Yang XS, Das A, Chaudhuri SS (2013) Optimisation of scaling factors in electrocardiogram signal watermarking using cuckoo search. Int J Bio-Inspired Comput 5 (5):315–326
54. Dey N, Biswas D, Roy AB, Das A, Chaudhuri SS (2012) DWT–DCT–SVD based blind watermarking technique of gray image in electrooculogram signal. In: International Conference on Intelligent Systems Design and Applications (ISDA), pp 680–685
55. Dey N, Mukhopadhyay S, Das A, Chaudhuri S (2012) Analysis of P-QRS-T components modified by blind watermarking technique within the electrocardiogram signal for authentication in wireless telecardiology using DWT. Int J Image Graph Signal Process 4(7)
56. Dey N, Maji P, Das P, Biswas S, Das A, Chaudhuri SS (2013) An edge based blind watermarking technique of medical images without devalorizing diagnostic parameters. In: International conference on advances in technology and engineering (ICATE), January, pp 1–5
57. Dey N, Biswas S, Das P, Das A, Chaudhuri SS (2012) Feature analysis for the reversible watermarked electrooculography signal using low distortion prediction-error expansion. In: International conference on communications, devices and intelligent systems (CODIS), December, pp 624–627
58. Dey N, Biswas S, Roy AB, Das A, Chowdhuri SS (2013) Analysis of photoplethy smographic signals modified by reversible watermarking technique using prediction-error in wireless telecardiology
59. Chakraborty S, Samanta S, Biswas D, Dey N, Chaudhuri SS (2013) Particle swarm optimization based parameter optimization technique in medical information hiding. In: IEEE international conference on computational intelligence and computing research (ICCIC), December, pp 1–6
60. Dey M, Dey N, Mahata SK, Chakraborty S, Acharjee S, Das A (2014) Electrocardiogram feature based inter-human biometric authentication system. In: IEEE international conference on electronic systems, signal processing and computing technologies (ICESC), January, pp 300–304
61. Acharjee S, Chakraborty S, Karaa WBA, Azar AT, Dey N (2013) Performance evaluation of different cost functions in motion vector estimation. Int J Serv Sci Manag Eng Technol 5 (1):45–65
62. Dey N, Das A, Chaudhuri SS (2012) Wavelet based normal and abnormal heart sound identification using spectrogram analysis. arXiv: 1209.1224
63. Dey N, Das P, Roy AB, Das A, Chaudhuri SS (2012) DWT–DCT–SVD based intravascular ultrasound video watermarking. In: IEEE World congress on information and communication technologies, October, pp 224–229
64. Dey N, Das S, Rakshit P (2011) A novel approach of obtaining features using wavelet based image fusion and Harris corner detection. Int J Mod Eng Res 1(2):396–399
65. Poh N, Bengio S (2006) Database, protocols and tools for evaluating score-level fusion algorithms in biometric authentication. J Pattern Recogn 39(2):223–233
66. Nandakumar K (2008) Multibiometric systems: fusion strategies and template security. Ph.D. Thesis, Department of Computer Science and Engineering, Michigan State University

67. Gawande U, Zaveri M, Kapur A (2013) A novel algorithm for feature level fusion using SVM classifier for multibiometrics-based person identification. Applied Computational Intelligence and Soft Computing, Hindawi, 2013(6), 1st–11th, June
68. Gawande U, Hajari K, Golhar Y (2014) Novel cryptographic algorithm based fusion of multimodal biometrics authentication system. In: International conference on computing and communication technology (ICCCT), December, 11th–13th, Hyderabad, pp 1–6
69. Gawande U, Zaveri M, Kapur A (2013) Fingerprint and iris fusion based recognition using RBF Neural Network. Int J Signal Image Process 4(1):142–148
70. Ross A, Jain A (2003) Information fusion in biometrics. Pattern Recogn Lett Spec Issue Audio Video Based Biom Pers Authentication 24(13):2115–2125
71. Feng J, Ouyang Z, Cai A (2006) Fingerprint representation and matching in ridge coordinate system. In: IEEE International conference on pattern recognition, Hong Kong, vol 39(11), August 20th–24th, pp 2131–2140
72. Rao P, Devi TU, Kaladhar D, Sridhar G, Rao AA (2009) A probabilistic neural network approach for protein superfamily classification. J Theor Appl Inf Technol 6(1):101–105
73. Chikkerur S, Cartwright AN, Govindaraju V (2007) Fingerprint enhancement using STFT analysis. J Pattern Recogn 40(1):198–211
74. Thai R (2003) Fingerprint image enhancement and minutiae extraction. B.E. Thesis, School of Computer Science and Software Engineering, The University of Western Australia
75. Monwar Md, Gavrilova M, Wang Y (2011) Novel fuzzy multimodal information fusion technology for human biometric traits identification. In: IEEE International conference on cognitive informatics and cognitive computing, Banff, Alberta Canada, August 18–20, pp 112–119
76. Garje PD, Agrawal SS (2012) Multibiometric identification system based on score level fusion. J Electron Commun Eng 2(6):7–11

Parametric Evaluation of Different Cryptographic Techniques for Enhancement of Energy Efficiency in Wireless Communication Network

Alka P. Sawlikar, Z.J. Khan and S.G. Akojwar

Abstract Security in the wireless network is a critical issue, so by most wireless and wired communication standards, orthogonal frequency division multiplexing (OFDM) is the most reliable modulation technique which has been adopted. To reduce the redundancy, storage requirements, and communication costs and to protect our data from eavesdropping, data compression algorithms are used. With the increasing demand of secure multimedia, there is a need to develop a new secured compression and encryption technique whose data contain graphic, video, images, and text files. Data compression offers an approach for reducing communication costs using effective bandwidth and at the same time considers the security aspect of the data being transmitted which is vulnerable to attacks. In this chapter, we have introduced new compression algorithm known as KSA, which is based on bit quantization and is the best technique which requires less encoding and decoding delay. The proposed algorithm is found most effective in terms of energy, delay and throughput when data travels wirelessly through OFDM. Here, the objective was to carry out an efficient implementation of the OFDM system using different combinations of encryption and compression algorithm for the energy optimization on data transmission. Lots of text encryption algorithms based on rounds and keys have been proposed earlier. Some of them are time consuming and complex, some have little key space. While working, the best combination for the encryption and compression of the data transmission and energy optimization is found out and implemented in NS2 for finding different parameters such as delay, energy, and throughput. The latest trend in text encryption is chaos based for some

A.P. Sawlikar (✉) · S.G. Akojwar
Department of Electronics Engineering, R.C.E.R.T, Chandrapur, India
e-mail: alkaprasad.sawlikar@gmail.com

S.G. Akojwar
e-mail: sudhirakojwar@gmail.com

Z.J. Khan
Department Electronics and Power Engineering, R.C.E.R.T, Chandrapur, India
e-mail: zjawedkhan@gmail.com

© Springer International Publishing Switzerland 2017
N. Dey and V. Santhi (eds.), *Intelligent Techniques in Signal Processing for Multimedia Security*, Studies in Computational Intelligence 660,
DOI 10.1007/978-3-319-44790-2_9

unique characteristics and parameters. In this chapter, we further proposed a unique chaos-based encryption technique which when combines with LZW compression gives best results. This encryption differs from earlier encryption technique such as RSA, ECC, Interleaving, Hill-climber, AES, and DES and is suitable for practical applications having large data capacity. The combination of compression, cryptography, and chaotic theory forms an important field of information security in text. Tabulation of compression time, encryption time, decompression time, decryption time, compressed data size, total time taken for compression, entropy, and total time required for decompression and compression ratio is found.

Keywords Orthogonal frequency division multiplexing (OFDM) · Lempel Ziv Welch · Run length encoder (RLE) · Huffman · Discrete cosine transform (DCT) · Discrete wavelet transform (DWT) · Network simulator 2 (NS2) · Chaotic map · Logistic map · Entropy · Socket secure layer (SSL)

1 Introduction

Today, information is more vulnerable to abuse because of the immense divergence of the network communication which has induced digitalized system dependency. The Web is going toward multimedia due to the development of network and multimedia technology. Multimedia data consist of image, audio, video, text, etc. The digital data become one of the most promised data carrier systems which are thoughtful for biometric authentication, medical science, military, etc.

When there is a high percentage of unreliability of transmission of any information, then at that time, some special types of modes of transmission should be selected [1]. So to overcome the effects of collision or interference, multichannel signaling is introduced in wireless communication systems. By doing so, one can transmit same message or information which contains signal over channels so that the information can be recovered. Security of information is always in demand since last few years, and lot of instances highlight the significance of the security of text data. The most important issue in world is the vast amount of valuable information that is spread in various networks demands the swap of information with more reduction and security that means the time for data transmission and data storage space. This can be achieved by compression and encryption, such type of system is called as compression—crypto system. That means ciphering/encryption and data compression is a coding technique, whose purpose is to reduce data storage space and the time for data transmission, and it is a secure coding technique. Hence, compression ratio becomes an most important parameter which we have to always keep in mind. Protection of data is increasing on a rapid rate and can be handled if we can remove the redundancy, and for this, cryptosystem technique is applied and made them work with integrity, so that our information will be of compressed and ciphered form and manageable which extends many advantages.

Imposing security everywhere promises that information is organized, stored, or transferred with integrity, reliability, and authenticity and available to all authorized entities [2]. Those encoding techniques are the best which offer moderate compression ratio so a module has been developed for combining compression and encryption on the different files of data and compare these outputs with the existing techniques or algorithms. To reduce storage cost by eliminating redundancies which occur in number of files, data compression techniques are used [3]. Over a period of last consecutive years, we found an unrivalled explosion in the amount of particulars, so to reduce the traffic, there is a need of compression, so that wide amount of information can be transferred.

Two types of data compression techniques are as follows: One is lossy and another is lossless [4]. Lossy compression technique is used for video, audio, and text compression. Well-known data compression algorithms are available, and this chapter deals with new compression and encryption algorithms. The mathematical analysis of these algorithms can be used for finding various parameters which is applied in this proposed system. As their classification is done on the basis of transformation, compression algorithm should be rearranged or modified to optimize input. Without the identical crypto-compression key, the decompression process will be disabling and the cryptanalysis cannot be completed. That means ciphering/ encryption and data compression is indeed a secure coding technique where the data which need to be protected are increasing at a rapid rate and can be handled a bit if we can remove the redundancy. There are two types of text encryption process called symmetric and asymmetric. Symmetric encryption [5] allows the cipher text expansion, how much long the encrypted text as compared to the plaintext.

In this chapter, comparison of the various compression algorithms along with new compression technique is clearly discussed and is combined with various cryptographic techniques and then transferred over channels using OFDM transmitter and receiver. At the OFDM transmitter, text file is randomly selected and compression is performed and then encryption is done. At the OFDM receiver, encrypted data are first decrypted and then decompressed. MATLAB simulation for the best crypto system is shown, and the results were observed. This crypto system is then applied to wireless communication network using NS2 software. Observation shows that the new proposed compression technique when combined with cryptographic technique gives the best results in terms of various parameters such as delay, throughput, and energy and when new compression algorithm is combined with cryptographic algorithm through protocol performance is improved as compared with common algorithms.

2 Fundamental Concepts of OFDM

To achieve the large data rate which is necessary for data intensive applications, OFDM is the standard being used that must now become routine [6]. Different coding schemes have been studied and compared earlier, and their implementation and simulation have been done with combination of the coding schemes. But in order to establish the context and need for the work undertaken, it is necessary to discuss the fundamental concepts behind the work [7]. The various compression algorithms are used to compress and transfer the data from OFDM in the network and encryption is used for increasing security as well as efficiency.

This chapter brings out the need for channel coding in OFDM system and discusses the various compression, decompression, encryption, and decryption algorithms used to transfer the data and shows the various aspects of the data used for in the OFDM transmitter and receivers.

Orthogonal means when the signal's dot product is zero and in OFDM the spacing should be correctly chosen so that each sub-bearer (carrier) is orthogonal to the other sub-bearer (carriers), and the range of each carrier has a zero at the center frequency. These consequences with no intervention, between the carriers, permit them to be spaced as close as theoretically possible.

Consider that there is a set of signals ψ; then, the signals are said to orthogonal [8]

$$\int_a^b \Psi_q^*(t)\mathrm{d}t = k \rightarrow \text{for} \rightarrow p = q \tag{1}$$

$$\int_a^b \Psi_q^*(t)\mathrm{d}t = 0 \rightarrow \text{for} \rightarrow p \neq q \tag{2}$$

if the integral value is zero where interval $[a, b]$ is a symbol period.

where ψ_p and ψ_q are pth and qth elements in the set.

As per the definition of orthogonalism the bearer (carriers), the nulls of one bearer (carrier) coincide with the top of another sub-bearer (carrier), and because of this, it is feasible to extract the sub-carrier of concern (interest). OFDM transmits a huge amount of narrow range of frequencies of sub-channels. The band between carriers is carefully selected in order to make their dot product zero on one another. In matter of fact, the carriers are split by an interval of $1/T$, where T represents the time of an OFDM sign (symbol). It is conventional that at the receiver, orthogonal signals can be separated by correlation techniques and the receiver behaves as a collection (bank) of demodulators, relocating each bearer (carrier) down to the band of frequencies from close to 0 Hz up to a higher cutoff frequency or maximum bandwidth. The outcoming signal then is integrated over a symbol period to reconstruct the data. If the another carriers all hit down to frequencies which, in the time domain, means an integer number of states per symbol period (T), then the integration process results in a null contribution from all these carriers.

2.1 Orthogonal Frequency Division Multiplexing (OFDM) Technique

In a standard serial data system, the bits (symbols) are transmitted in regular succession (sequentially), one by one, with the range of frequency (spectrum) of each data symbol allowed to engage (occupy) the entire obtainable bandwidth. A high rate data transmission assumes very short symbol duration, running at a large range of frequency (spectrum) of the modulation symbol.

There is a good probability that the frequency selective channel response affects in a very prominent way to the different spectral elements of the data symbol, hence introducing the inter symbol interference (ISI) [9]. The same event regarded in the time domain consists in diverging of message symbols such that the energy from one symbol interferes with the power of the ones, in such a way that the received signal has a good chance of being incorrectly illuminated.

Allegedly, one can supposes that the frequency selectivity of the channel can be diminished if, instead of transmitting a single bit at high data rate stream, we transmit the complete data for a given overall data stream, increasing the number of carriers which lessens the data rate that each particular carrier must transport; therefore, it is preferred to lengthen the symbol duration on each sub-bearer (carrier). To reduce effects of ISI, slow data rate (and long symbol duration) on each sub-channel is preferred. This is the basic idea that lies behind OFDM.

2.2 Block Diagram of an OFDM System

The block diagram of OFDM transmitter and receiver system is shown in Fig. 1. In OFDM:

- Some operations are done on the source data such as interleaving, mapping of bits onto symbols, and coding for correcting errors.
- With IFFT, modulation of symbols onto orthogonal sub-carriers takes place.
- This orthogonality can be achieved by adding a cyclic prefix to the OFDM frame which is to be sent and maintained during channel transmission.
- Correlation should be achieved as cyclic prefix is used to detect the start of each frame, and this is done by using the certainty that the first and last samples are the same, and thus synchronized.
- By using FFT, demodulation of the received signal is done.
- To maintain timing and equalize the channel or sending modulation symbols at predefined sub-carriers, the channel can be estimated which is channel equalization.
- Then decoding and de-interleaving.

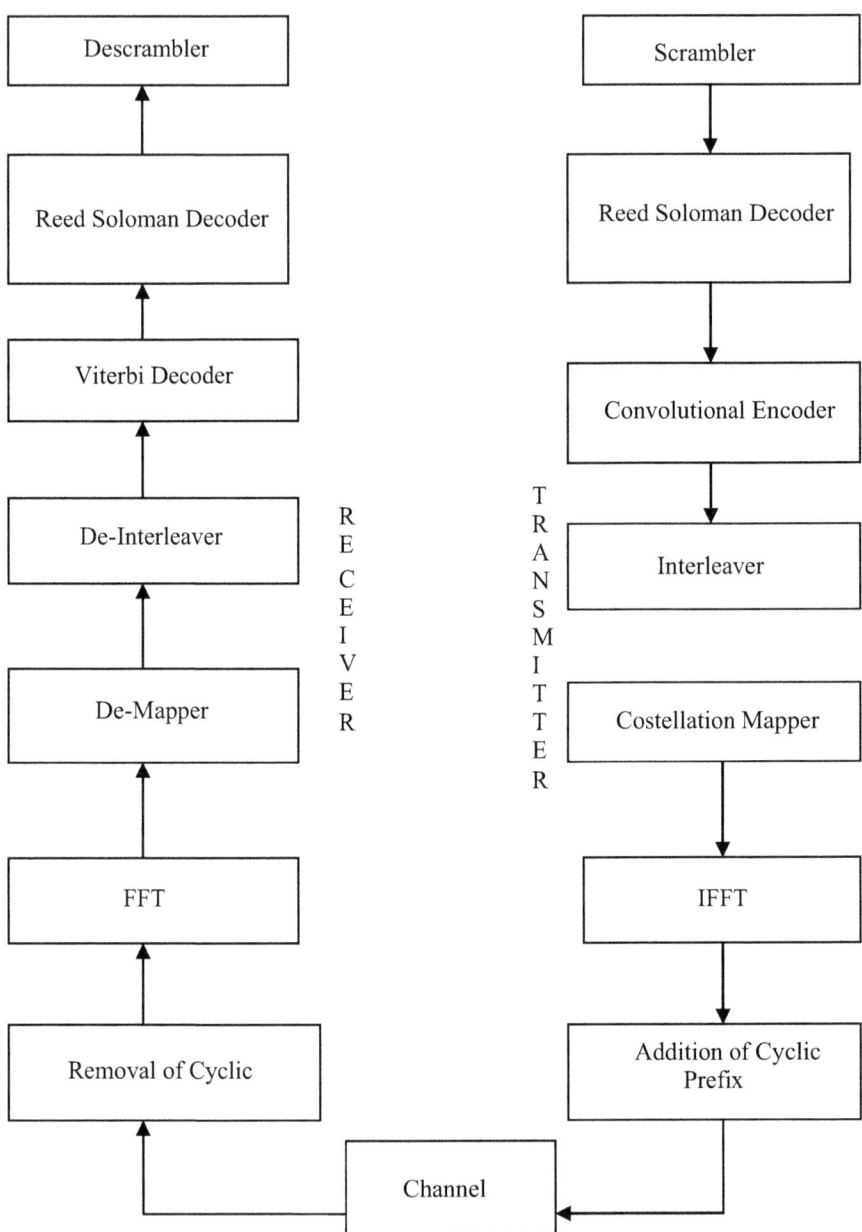

Fig. 1 Complete OFDM system

2.2.1 Scramble/Descramble

Data bits are inputs given to the transmitter. The input signal's power spectrum on the transmitted data can be vanished if these bits passes through a scrambler that randomizes t and makes the input sequence more disperse.

At the receiver end, descrambling is done. To recover original data bits from the scrambled bits, descrambler is used [10].

2.2.2 Reed–Solomon Encoder/Decoder

The bits after scrambling are fed to the encoder which is a part of error correction as Reed–Solomon coding is technique for error correction. The in-essential (redundant) bits are added to the actual message which provides strength against most channel conditions [11]. RS code is abbreviated in the form RS (n, k), where

$$n = 2^m - 1 \tag{3}$$

$$k = 2^m - 1 - 2t \tag{4}$$

where m is the number of bits/symbol, k is the number of input numbers (to be converted), n is the entire number of symbols (data + parity) in the codeword, and t is the maximum number of symbols that can be corrected. At the receiver, Reed–Solomon coded symbols are decoded by removing parity symbols

2.2.3 Convolution Encoder/Decoder

RS error-encoded bits are again coded by convolution encoder which adds unessential bits. Here, each m bit symbol is modified into an n bit symbol; m/n is known as the code rate. Depends upon the last k data symbols, the transformation of m bit symbol into n bit symbol is possible; thus, k is known as the constraint length of the code [12].

2.2.4 Interleaver/De-Interleaver

To protect the data from errors that occur in many consecutive bits, interleaver is added where the stream bit is repositioned in such a way that adjacent bits are no more adjoined to one other. The data are burst into blocks, and the bits within a block are rearranged. In terms of OFDM, the bits within an OFDM symbol are reconstructed in such a way so that adjoining bits are placed on non-adjacent sub-carriers. During reception, it again rearranges the bits into original form and with de-interleaving [13].

2.2.5 Constellation Mapper/De-Mapper

Various modulation techniques can be employed for different sub-carriers. The incoming (interleaved) bits onto different sub-carriers maps by constellation mapper. The bits are extracted from the modulated symbols at the receivers end with the help of de-mapper.

2.2.6 Inverse Fast Fourier Transform/Fast Fourier Transform

The most important in the OFDM communication system is inverse fast Fourier transform/fast Fourier transform. The orthogonality of the OFDM is given by IFFT. Each spectrum is transformed into a time domain signal by IFFT. FFT performs the reverse task at the receiver's end [14].

2.2.7 Addition/Removal of Cyclic Prefix

A cyclic guard interval is introduced in order to preserve the sub-carrier orthogonality and the independence of subsequent symbols. To remove exaggeration due to severe channel conditions, cyclic prefix is done at transmitter end, and at the receiver end, removal of cyclic prefix is done.

2.3 OFDM Model

Following figure shows OFDM system used for simulation with which various parameters of the system can be varied and tested. This model is used to measure the performance of OFDM under different channel conditions. Four main criteria were used to measure the performance of the OFDM system, and they are tolerance to multipath delay spread, peak power clipping, and channel noise and time synchronization errors [15] Fig. 2.

2.4 Advantages of OFDM

Over single carrier modulation systems, OFDM has various advantages and these make it a feasible alternative for CDMA in future wireless communication networks. Some of the advantages are as follows:

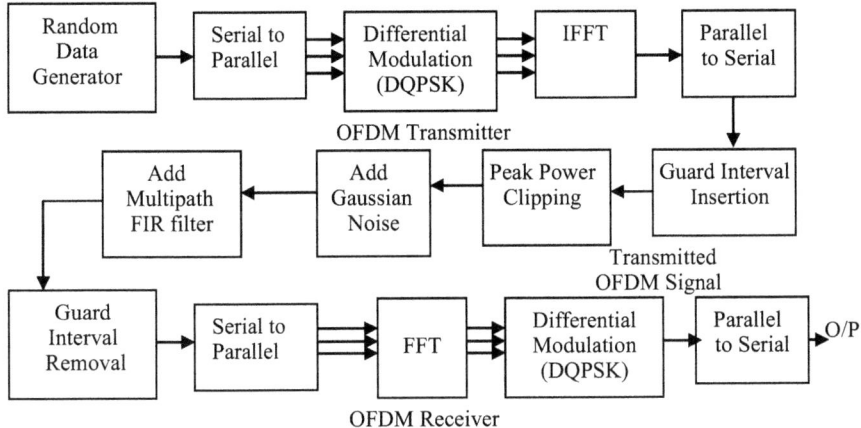

Fig. 2 OFDM model used for simulations

- Multipath delay spread tolerance:
  OFDM is highly resistant to multipath delay spread that causes ISI in wireless channels by converting a high data rate signal into low rate signals. Since duration of the symbol is made larger, the effect of delay spread is reduced by the same factor. Also by including the concepts of cyclic extension and guard time, the effects of ISI and ICI can be completely removed.
- Immunity to frequency selective fading channels:
  If the channel undergoes selective fading, then compound equalization techniques are required at the receiver for single carrier modulation, but with OFDM, the available bandwidth is divided among many orthogonal barely spaced sub-carriers and converted into many narrow flat-fading sub-channels [16].
- Spectrum efficiency:
  A significant OFDM advantage is that it makes efficient use of the available spectrum using close-spaced overlapping sub-carriers.

3 Simulation Analysis of a New Lossy Compression Technique Using Decimated Bit Level Quantization

The new data compression technique is proposed which is based on bit quantization level.

Reducing size of data and reducing the sampling rate of a signal or data rate are known as decimation, and the decimation factor is usually a rational fraction greater than one or an integer and this factor multiplies or equivalently divides the sampling rate. For example, 2048 bit text file is decimated to 1024 bits. For decimation, we can choose any image file, audio file and here-used text file, and that text file can be of random length or specified [17].

The different steps of the proposed algorithm are as follows:

- Take input data string of either random in length or specified.
- Decimate the data length by 2
- Convert decimal data string to binary
- Find the polarity of consecutive input sample
  if next_sample > current_sample
  Make LSB bit of current output as '1'
  Else make LSB bit of current output as '0'
- Convert binary data string to decimal
- We will get compressed output data string

Decompression:

- Copy first compressed output bit as it is
- Consider two consecutive bits and check current _sample and next_sample
- Convert decimal data string to binary
- If current_sample has 1 in LSB position, then current _sample is always equal to new_output
- Else add two consecutive samples and divide it by 2 that will be new_sample
- Convert binary data string to decimal.

By repeating the same process for complete data length, we will get decompressed output bits. The length of input bits will always be equal to the length of decompressed data bits. This algorithm is lossy compression algorithm and can be applied to any number of data bits and is named as KSA algorithm.

3.1   Comparison Results

Various compression algorithms are there but for comparison Huffman and LZW is considered. When Huffman and LZW are used, we get the following results for different text files having different data size as follows (Table 1, 2, 3).

From above, it has been observed that when we have added proposed KSA algorithm, its CR is 50 % while with other compression algorithm such as Huffman,

Table 1 Huffman encoding and decoding time

File size (KB)	Time for encoding (s)	Compression ratio (%)	Time for decoding (s)
1.05	0.32794	58.59	0.1958
3.69	0.46043	78	0.36201
4.24	0.49342	73	0.49456
15.7	1.503	78	1.2593

Table 2 LZW encoding and decoding time

File size (KB)	Time for encoding (s)	Compression ratio (%)	Time for decoding (s)
1.05	0.095469	27.62	0.089781
3.69	0.27039	74.05	0.18968
4.24	0.29809	71.8	0.20893
15.7	0.95395	85	0.53367

Table 3 New proposed KSA algorithm's encoding and decoding time

File size (KB)	Time for encoding (s)	Compression ratio (%)	Time for Decoding (s)
1.05	0.0509	50	0.0141
3.69	0.1016	50	0.0298
4.24	0.182	50	0.0576
15.7	0.546	50	0.1728

LZW compression ratio is more than proposed. The reason behind is they have some inherent drawbacks, some needs only sequential input data like 11110000 and spaces are considered as one individual character or bit, and if we give data like 10110010111000, then instead of compression, it is getting expanded because for every alternate 1's and 0's, it will encode separately, and thus, it is time consuming and energy required is also more.

Moreover, this algorithm verifies each sample on byte level and then performs operation on bit level for polarity which requires time which is encoding time, while in decoding, there is no need to compare with next sample.

Following are outputs showing input data, compressed data, de-compressed data, compression time, de-compression time, and compression ratio which we get from MATLAB. Tables 4 and 5 show that data can be numerals or alphabets or alphanumeric, the only thing is alphabets are converted into ASCII and then algorithm can be applied. In the same way, we can find outputs for random numbers or say file.

Table 4 Output of KSA for number of input bits = 12 showing compression time, compressed, and de-compressed data bits

Input data 1 2 3 4 5 6 7 78 12 15 22 25	
Compressed data 1 3 5 7 13 23	
De-compressed data 1 1 3 3 5 5 7 7 13 13 23 23	
Compression time = 0.0448 s	
De-compression time = 0.0131 s	
Compression ratio = 50 %	

Table 5 Output of KSA(new compression algorithm) for alphabets of input bits = 10 showing compression time, compressed, and de-compressed data bits

Input data
a d g e t y u o l h
Compressed data
a f u t l
De-compressed data
97 97 102 109.5 117 117 116 112 108 108
Compression time = 0.1537 s
De-compression time = 0.0149 s
Compression ratio = 50 %

4 KSA Algorithm Used for Energy Saving in Wireless Network

Combination of compression and encryption of large text data gives an efficient way of handling huge amount of data as this reduces the size first and then makes the reduced size secured, which is a less time-consuming process. Such methods can be helpful in saving memory, cost, and data transfer. To prove cryptanalysis efficiently, we have taken the sample of text, i.e., say 100, 200 bits, compressed it with common algorithms such as RLE, DCT, Huffman, DWT, and LZW. Also we have combined this technique with different encryption algorithm such as AES, Interleaver, Hill-Climbing, RSA, and ECC. Then, the parameters such as normal data size, compressed data size, time required for compression, time required for encryption, total time required for complete cryptanalysis, and compression ratio are found [18]. We observed that the combination with KSA technique gives better results as compared to the above techniques.

The parametric evaluation of proposed method provides good results when combining with encryption algorithms and is found to be the best and efficient in all terms. It has been observed that for 200 bits, when we perform simulation of AES, Interleaver, Hill-Climbing, RSA, and ECC with new compression algorithm KSA, compression time and compressed size of KSA is better.

In this algorithm, the data are converted to binary and a flag is added, and at the time of decompression, it is removed. It is a lossy compression but the most efficient. The aim of the system was to design an efficient compression technique in OFDM environment. The technique designed is the KSA Algorithm. This technique compresses the data by nearly 50 % of saving a lot of resources, and while sending the data, the only disadvantage of the technique is the lossy decompression but as it is highly efficient, the loss can be adjusted (Table 6).

Table 6 Parametric evaluation of compression and encryption techniques

Compression tech	Encryption tech	Time (s)	Normal data size	Compressed data size	Compression ration (%)
RLE	AES	6.85	1719	208	87.91
	INTERLEAVING	8.9556	208	104	50
	HILL	7.85	208	208	0
	RSA	5.77	208	208	0
	ECC	5.51	208	208	0
DWT	AES	4.82	1299	208	83.99
	INTERLEAVING	2.55	208	104	50
	HILL	3.28	208	208	0
	RSA	3.52	208	208	0
	ECC	2.56	208	190	8.65
DCT	AES	3.10	1219	208	82.94
	INTERLEAVING	2.61	208	104	50
	HILL	2.05	208	208	0
	RSA	3.38	208	208	0
	ECC	2.991	208	187	10.10
HUFFMAN	AES	2.11	1191	208	82.54
	INTERLEAVING	2.34	208	104	50
	HILL	2.15	208	208	0
	RSA	2.20	208	208	0
	ECC	3.60	208	191	8.17
LZW	AES	2.97	1283	208	83.79
	INTERLEAVING	2.51	208	104	50
	HILL	2.44	208	208	0
	RSA	2.03	208	208	0
	ECC	2.73	208	187	10.10
KSA	AES	7.28	208	104	50
	INTERLEAVING	2.27	208	104	50
	HILL	3.10	208	104	50
	RSA	2.92	208	104	50
	ECC	2.86	208	104	50

4.1 Simulation Results

Figures 3 and 4 show the graphical representation output when random values out of 200 and 1000 bits are combined with new compression KSA and encryption RSA.

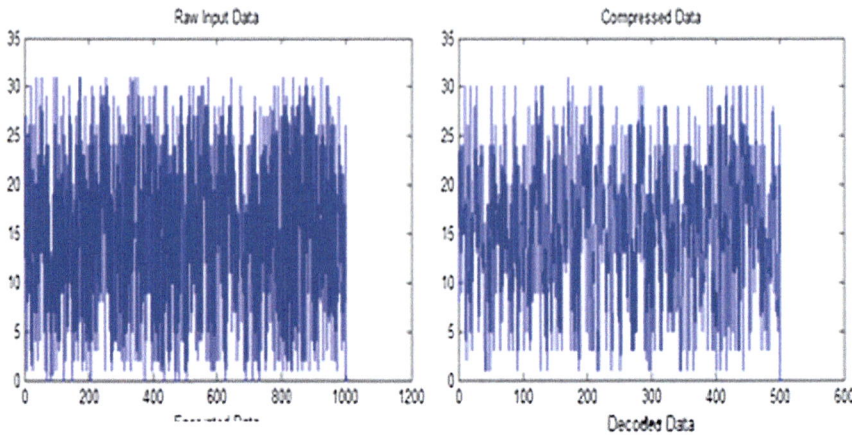

Fig. 3 MATLAB simulation output when KSA is combined with RSA for number of input bits = 1000

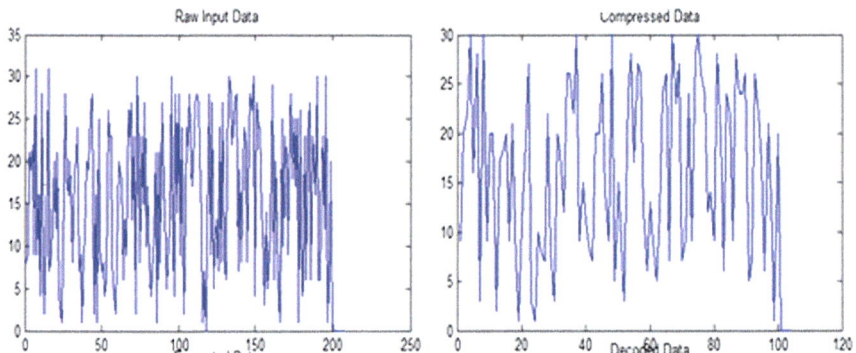

Fig. 4 MATLAB simulation output when KSA is combined with RSA for number of input bits = 200

4.2 Experimental Analysis and Results

After the combination of compression and encryption techniques, the best combination has been found out and energy optimization is carried out over channels using OFDM transmitter and receiver using security protocol Socket Secure Layer (SSL) in wireless communication network.

The number of sensor nodes can be varied as 30, 40, 50, and accordingly the parameters such as energy, delay, and throughput are observed. KSA_energy, KSA_delay, and KSA_throughput are energy, delay, and throughput, respectively, when KSA combination is to be best found.

Table 7 Energy comparison table when protocol is not applied and using KSA for nodes = 30

Simulation time	O-energy	Energy	KSA_energy
0	110	82	50
5	200	142	92
10	210	170	99
15	220	178	99
20	230	180	100
25	220	182	100
30	210	180	100

Table 8 Delay comparison table when protocol is not applied and using KSA for nodes = 30

Simulation time (s)	O-delay (s)	Delay	KSA_delay
1.02	3	0.1	0.19
1.03	0	0	0.2
1.04	0.8	0.2	0.00
1.05	0.6	0.6	0.8
1.06	0.00	0	0.00
1.07	0.3	0.8	0.82
1.08	0.6	0.7	0.3
1.09	0	0	0

Table 9 Throughput comparison table when protocol is not applied and using KSA for nodes = 30

Simulation time (s)	O-throughput	Throughput	Ksa_throughput
1	100	0	100
1.005	29	49	49
1.010	0	0	49
1.015	50–90	100	100
1.02	0	0	49

Comparison of energy, delay and throughput, and graphs for 30 nodes and stop time with 0.07 s is shown below using NS 2 software (Tables 7, 8, 9).

Above is the comparative analysis of parameters such as energy, delay, and throughput, respectively, and is calculated from Network Simulator 2. Similarly, we can change the number of nodes and stop time by changing the values in program and can find further results of all these parameters.

4.3 Limitations

This technique compresses the data by nearly 50 % saving a lot of resources while sending the data, the only disadvantage of the technique is the lossy decompression, but as it is highly efficient, the loss can be adjusted. As the proposed method is lossy compression algorithm, the inserted data will not exactly receive at the receiver side, but the characters and the numbers (numerals) are equivalent, the only thing is after decompression content is not same. Moreover, in alphanumeric file, there is a need to convert all alphabets in ASCII characters only.

5 A New Approach of Text Encryption Using Chaotic Map for Security Enhancement

During the last decade, chaos-based cryptography has received little bit attention due to signal such as noise for unauthorized person, randomness, mixing, and reactivity to initial conditions because of good ciphers. We have generated chaos sequence and for every cipher text expansion, chaos aims at providing the best-possible authenticity, integrity, and confidentiality [19]. To scrutinize whether this is actually achieved or not, we had first compressed text file by LZW and then guaranteed that symmetric cryptography can provide security for any cipher text expansion. Our characterization discloses not only that this encryption technique reaches the claimed objective, but also achieves full confidentiality with compression and cipher text expansion which provides new perception into the limits of symmetric cryptography. XOR operation is one of the most used value techniques as it is the two-way encryption. In order to improve the security performance of the text, reduction of data in the plain text and then changing the encrypted values of the shuffled text is used. In this work, we proposed encryption technique by using rotation and XOR-based encryption technique using chaos for secured and enhanced communication. The simulation result presents performance of our method against different types of attack. The comparative analysis shows that proposed technique is best and is ready for practical applications. One common attribute to all chaos-based encryption algorithms is that their security is not analyzed in terms of the techniques developed in cryptanalysis [20].

5.1 Encryption Algorithm

There are four stages to complete the overall encryption process. They are as follows:

- Chaos generation.
- Key generation
- Row rotation.
- XOR operation.

Steps of algorithm:

(1) Input text files have been prepared in notepad.
(2) All the files have been saved individually so that they can be read and compared.
(3) For chaos generation, equations used are

$$x_{n+1} = \mu x_n (1 - x_n) \qquad (5)$$

The condition to make this equation chaotic is $0 < xn < 1$ and $\mu = 4$.
For enhanced security, the following formulas are given:

$$x_{n+1} = \gamma x_n (1 - x_n) + \beta y_n^2 x_n + \alpha z_n^3 \qquad (6)$$

$$y_{n+1} = \gamma y_n (1 - y_n) + \beta z_n^2 y_n + \alpha x_n^3 \qquad (7)$$

$$z_{n+1} = \gamma z_n (1 - z_n) + \beta x_n^2 z_n + \alpha y_n^3 \qquad (8)$$

The value of α, β, and γ has to be assigned as 0.0125, 0.0157, and 3.7700, and for chaos generation
$0 < $ alpha $ < 0.015$, $0 < $ beta $ < 0.022$, and $3.53 < $ gamma $ < 3.81$
(4) Then, enter input of sequence to generate, here we had used 100000+ value
(5) Initialize $x(1) = 0.2350$; $y(2) = 0.3500$; $z(3) = 0.7350$;
6) Assume Keys N1 = 500; N2 = 1000; N3 = 600; N4 = 1000; N5 = 700;
(7) Row rotation: For the purpose of security, this rotation technique is used and is same as like as a combination lock of a briefcase. For the rotation of row, select chaos sequence say M number and then generate N1, which is a large number(random); then, from index N1, select M number of chaos and according to the value of chaos rotate row.
(8) As implementation is done in MATLAB round off keys with formula given

$$x = \text{mod}(floor(x * N_2), 256) \qquad (9)$$

$$y = \text{mod}(floor(y * N_4), 256) \qquad (10)$$

(9) XOR operation: The last step of this encryption process is XOR operation. At first, we generate a large random number N5. After that we XOR the chaos (starting from index N5) and row shifted text and finally we get encrypted output.
Following Fig. 5 represents the overall encryption process where N1, N2, N3, N4, and N5 are the keys as shown in Fig. 6.

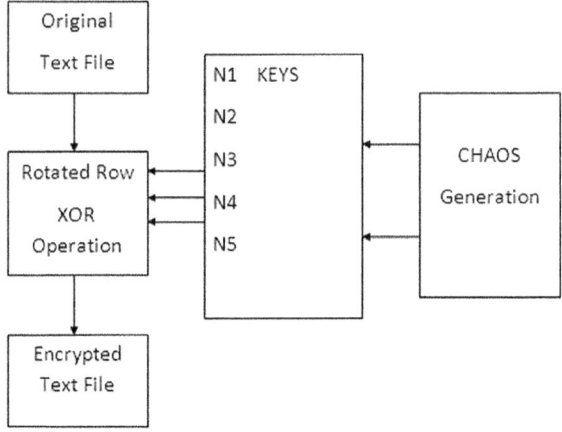

Fig. 5 Encryption technique using chaos

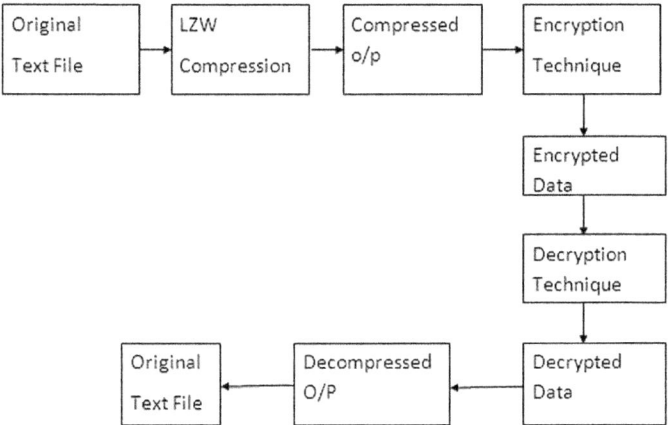

Fig. 6 Complete block diagram of methodology

5.2 Simulation Results and Comparison

The simulation results for different file sizes and outputs such as compression time, decompression time, encryption time, decryption time, total time for encryption and decryption, and decompression time are shown in Tables 10, 11, and 12.

Table 10 Results of proposed encryption and LZW compression technique

Size of file (KB)	Com pressed size (KB)	Compression time (s)	Encryption time (s)	Total time taken by LZW + new encryption technique (s)
1.05	0.82275	0.092725	0.083267	0.17599
3.69	2.120	0.25758	0.19996	0.45754
4.24	2.4676	0.29839	0.2692	0.56759
15.7	8.477	0.941	1.3211	2.2805
16.6	8.526	1.0082	1.3825	2.3907

Table 11 Results of decryption and LZW technique

Compressed file (KB)	Decompressed size (KB)	Decompression time (s)	Decryption time (s)	Total time taken by methodology (s)
0.82275	1.05	0.072345	0.12591	0.19825
2.120	3.69	0.16525	0.24916	0.41442
2.4676	4.24	0.19767	0.29689	0.49456
8.477	15.7	0.49919	1.4395	1.9387
8.526	16.6	0.50283	1.433	1.9359

Table 12 Simulation time comparison

Text file (kB)	DES (K4) (s)	RC4 (K4, K5) (s)	Double encryption (K1, K2, K3) (s)	Proposed methodology
0.013	0.3	0.19	0.03125	0.012823
0.636	4.8	0.2	0.1565	0.037765
6.17	245	2.13	1.48437	0.38522
8.21	438	3.57	2.14062	0.56521

6 Conclusions

Reduction of large data file with decimation method and ciphering gives an efficient way of data handling. Combination of compression and encryption techniques reduces the size first and then makes the reduced size secured, which is a less time-consuming process. Such methods offer many advantages such as saves space, manageable, easily transferrable, practical, and feasible.

The new compression algorithm, KSA, which is a lossy compression, is introduced with less compression time. To strengthen the security of the communication network, this technique has been suggested, and with the experimental results, it has been proved that it leads to increase not only security but is also energy efficient. The complete suggested design makes the cryptanalysis difficult for the intruder which is achieved effectively through a module which is a perfect blend of new compression technique with cryptography principles. This design is very influential in providing big challenge to the intruders who attempt to break the algorithms by any means. Thus, concluded that the secret data can be transmitted securely using this module and are viewed best in terms of speed, cost, throughput, and transfer rate.

Further a chaos-based text encryption technique with compression technique is experimented successfully. Though XOR operation for value transformation is not a new concept for text encryption, but to our knowledge, it is the first time that chaos has been used for large number of text files. People can use this algorithm for security purpose. A detailed statistical analysis on both compression system and the encryption scheme is given. However, experimental results show that this algorithm is sensitive to initial conditions and strong against the attacks and outperforms existing schemes in term of security. Having a very less delay, the proposed system is ready to be applied in real-time encryption applications by using security protocols and use in the secure data transmission.

Acknowledgments This chapter would not have been written without the support and help of many people. It is impossible for us to list all people who contributed along the way. We would like to thank the cryptographers who contributed to developing the theory on the design of symmetric ciphers, and from whom we learned much of what we know today. We would like to mention explicitly the people who gave us feedback in the early stages of the design process.

References

1. Sagheer AM, Al-Ani MS, Mahdi OA (2013) Ensure security of compressed data transmission. In: IEEE sixth international conference on developments in eSystems Engineering, pp 270–275
2. Bisht N, Singh S (2015) A comparative study of some symmetric and asymmetric key cryptography algorithms. Int J Innov Res Sci 4(3):1028–1031. ISSN:2347-6710
3. Sangwan N (2013) Combining Huffman text compression with new double encryption algorithm. In: IEEE international conference on emerging trends in communication, control, signal processing and computing applications (C2SPCA), pp 1–6
4. Jain A, Lakhtaria KI, Srivastav P (2013) A comparative study of lossless compression algorithm on text data. In: International conference on advances in computer science, AETACS, ElsevierDigital Library, pp 536–543
5. Joshi MR, Karkade RA (2015) Network security with cryptography. IJCSMC 4(1):201–204. ISSN:2320-088X
6. Mev N, Khaire BRM (2013) Implementation of OFDM transmitter and receiver using FPGA . Int J Soft Comput Eng (IJSCE) 3(3):199–202. ISSN:2231-2307
7. Bhardwaj M, Gangwar A, Soni D (2012) A review on OFDM: concept, scope and its applications. IOSR J Mech Civil Eng (IOSRJMCE) 1(1):7–11. ISSN:2278-1684
8. Parab SD, Limkar MB, Jadhav MS (2015). Improving the performance of smartphone application traffic using 4G/LTE technology. Int J Tech Res Appl (31):90–98. ISSN:2320-8163
9. Pathak N (2012) OFDM (orthogonal frequency division multiplexing) Simulation using Matlab. Int J Eng Res Technol (IJERT) 1(6):1–6. ISSN:2278-0181
10. Sawlikar AP, Khan ZJ, Akojwar SG (2015) Wireless network data transfer energy optimization algorithm. Int J Adv Res Sci Eng (IJARSE) 4(01):412–418. ISSN:2319-8354

11. Clarke CKP (2002) Reed–Solomon error correction. BBC Research & Development White Paper, WHP 031, pp 1–47
12. Sandesh Y, Rambabu K (2013) Implementation of convolution encoder and Viterbi decoder for constraint length 7 and bit rate ½. Int J Eng Res Appl 3(6):42–46. ISSN:2248-9622
13. Upadhyaya BK, Sanyal SK (2009) VHDL Modeling of Convolutional InterleaverDeinterleaver for Efficient FPGA Implementation. Int J Recent Trends Eng 2 (6):66–68
14. Proakis J (2007) Digital communication, 5th edn
15. Kansal1 R, Gupta S (2014) Survey paper on PAPR reduction techniques in WiMax OFDM MODEL. Int J Adv Res Comput Sci Softw Eng 4(7):790–793. ISSN:2277-128X
16. Ballal BR, Chadha A, Satam N (2013) Orthogonal frequency division multiplexing and its applications. Int J Sci Res (IJSR) 2(1):325–328. ISSN:2319-7064
17. Fowler ML (2000) Decimation vs. quantization for data compression in TDOA systems. In: Conference on mathematics and applications of data/image coding, compression, and encryption III SPIE's international symposium on optical science and technology, San Diego, CA, pp 56–67
18. Stallings W (2005) Cryptography and network security: principles and practices, 5th edn. Prentice Hall, Englewood Cliffs
19. Shukla PK, Khare A, Rizvi MA, Stalin S, Kumar S (2015) Applied cryptography using chaos function for fast digital logic-based systems in ubiquitous computing. Entropy 17:1387–1410. ISSN:1099-4300
20. Jakimoski G, Kocarev L (2001) Chaos and cryptography: block encryption ciphers based on chaotic maps. IEEE Trans Circuits Syst I Fundam Theory Appl 48(2):163–169

Part II
Personal Authentication and Recognition Systems

Hand Image Biometric Based Personal Authentication System

Ravinder Kumar

Abstract Hand geometry is widely accepted biometric modality for identification of human beings. This is considered as safest biometric indicator due to its strong resistance against the unauthorized access and easy to use modality from the user point of view. This chapter presents an approach for the personal authentication using geometrical structure of hand images. The proposed approach consists of many phases like acquisition of hand images of the user to the system, normalization of images, normalized contour and palm region extraction etc. The contour of the hand region from Region of Interest (ROI) is computed and is used to extract structural information, which describe the shape of the hand. The features of the test and the trainee images are matched using machine learning based classifier at the verification stage.

Keywords Hand geometry · Finger width · Support vector machine · Feature extraction

1 Introduction to Biometrics

The major objective of biometrics recognition is to provide automatic discrimination between subjects in a reliable way based on one or more physiological and behavioral traits. All these personal traits are commonly called as biometrics.

Biometrics is thus defined as the process of individual identification based on his/her distinguished characteristics. More precisely the biometrics is the science of identifying or verifying an individual identity based on his/her physiological or behavioral characteristics. Some of physiological characteristics are based on fingerprints, facial features, hand geometry, iris, and finger geometry; and some of behavioral characteristic of an individual are keystroke style, signature and voice print. A good quality biometric should have:

R. Kumar (✉)
Department of CSE, HMR ITM, GGSIPU, Delhi, India
e-mail: ravinder_y@yahoo.com

© Springer International Publishing Switzerland 2017
N. Dey and V. Santhi (eds.), *Intelligent Techniques in Signal Processing for Multimedia Security*, Studies in Computational Intelligence 660,
DOI 10.1007/978-3-319-44790-2_10

- **Uniqueness:** The features should be as unique as possible, Which means that the features of individual must be different
- **Universality:** The related biometric features should be present in all the persons enrolled.
- **Permanence:** They should not change with time.
- **Measurability:** The features should be measurable by relatively simple methods.

The authentication of a person can be done in two modules Identification and Verification. In the identification process the individual presents his/her biometric feature and the system associates an identity to that Individual. However, in verification mode the biometrics features of an individual are matched against the claimed identity to verify the individual. The system senses the biometrics measurements, extracts features, compare the input features to the features enrolled in the system database under the subject's ID. The system then either determines that the subject is who he claims to be or rejects the claim.

1.1 Performance of Biometric System

The features of the persons enrolled in the system are used for the verification. These features are related to the face, a hand print, a fingerprint or a voice print. These features of an unknown person need not be same as those computed during the registration process. The changes in the features are due to the environment and internal effect. This will result in mismatch of the test features (unknown person) and training features (enrolled or registered person). A suitable matching algorithm is required to eliminate this mismatch.

So the algorithm of matching needs to be designed to return the closest match. This may cause the problem in case person is not registered with the system and the system may give some closest match. To avoid this problem, a threshold is used. Only the matches who are above a certain threshold are valid and the others are rejected.

Biometric system's performance is measured in terms of two errors. These are:
False acceptance rate (FAR): It is the measure of the likelihood that the biometric security system incorrectly accepts an access attempt by an unauthorized

$$\text{FAR} = \frac{\text{Number of accepted imposter claims}}{\text{Total number of imposter claims}} \tag{1}$$

False rejection rate (FRR): It is the measure of the likelihood that the biometric system will incorrectly reject an access attempt by an authorized person.

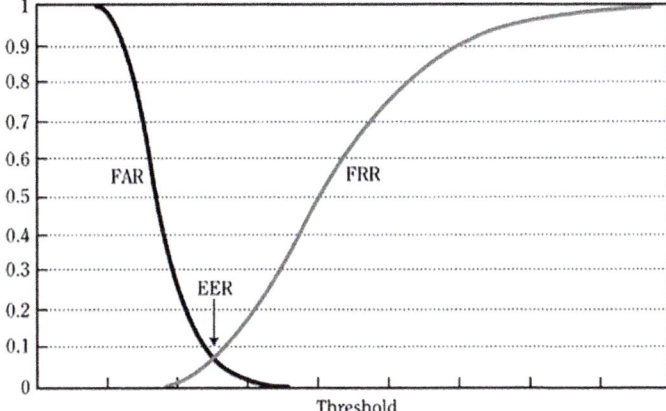

Fig. 1 Equal error rate

$$FRR = \frac{\text{Number of rejected genuine claims}}{\text{Total number of genuine claims}} \qquad (2)$$

Equal error rate (EER): It is also called crossover error rate (CER). EER is the value where FAR and FRR are equal. False acceptance causes a more serious problem than false rejection. That's why the biometric systems should have minimal FAR. It can be achieved by taking a high threshold value so that only closest the matches are accepted and all others are rejected.

FRR also depends on the threshold. FRR increases as the threshold increases. Owing to a high threshold, matches which are correct but less than the threshold due to noise or any other factors will not be accepted. For this reason EER serves our purpose. The receiver operating characteristic (ROC) is used to show the overall performance of a matcher/classifier and a varying threshold.

Figures 1 and 2 depict the EER and ROC respectively.

Fig. 2 The receiver operating characteristic (ROC)

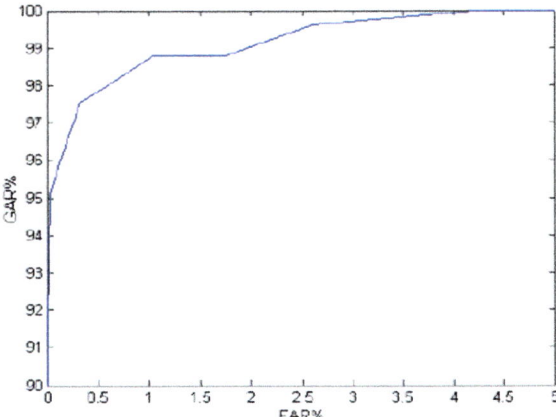

1.2   Biometric Technologies

There are different types of biometric systems based on human body's physical and behavioral characteristics.

1.2.1   Face Recognition

It is an important biometric modality. Owing to the increased terrorist threats a number of airports around the globe employ face based biometric systems to check the people entering and leaving the country. Though it is a difficult biometric as the changes in the face due to age or makeup and facial expressions may result in wrong authentication. There are two broad types of a face recognition systems acted under the uncontrolled and the controlled environment.

The recognition of passengers on an airport without their knowledge or consent is an example of trying to identify an individual from a group of individuals in an uncontrolled environment.

In the other type is a controlled environment where the distance between the person and the sensor is fixed. The output of systems in a controlled environment is obviously enhanced than in an uncontrolled environment.

1.2.2   Fingerprints

Fingerprint already gained an acceptance as biometrics all over the world. Nearly all law enforcement agencies in the world use fingerprints as an exact and helpful means of identification. Principally fingerprints are utilized as a method for confirmation. However now with law enforcement agencies building up electronic fingerprint databases fingerprints are also serving as an identification tool. The majority of fingerprint based biometric systems [1] utilize either minutiae of the ridges and bifurcation of the fingerprints. Image-based fingerprint recognition systems have also been proposed in the literature [2–8].

1.2.3   Signature Recognition

Signature Recognition is a behavioral biometric. Signatures have been used for authentication since a long time in legal, government and commercial transactions. Automated systems are available in both offline and online. The Offline systems [9] work only using the distinctiveness of the signature. While the online systems [10] can utilize features obtained from the sample. The online signature feature vector include total time for signature, speed of writing, the pen pressure, number of pen up and pen downs inclination.

1.2.4 Handwriting Recognition

It is also a behavioral biometric. It is applied in forensic document validation for user detection. A lot of research has been conducted to determine the uniqueness of a person's handwriting [11]. Several Handwriting recognition systems both online [12] and offline [10] have been proposed. In addition to the shape and size of letters; pen strokes, crossed lines and loops can be used as features from a handwritten document.

1.2.5 Ear Recognition

This is a comparatively new biometric feature. Even though it may seem that human ear does not have a fully random structure, it is distinctive enough for people to be identified using the features collected from ear. Besides it does not suffer from makeup effects and expression changes like face do. However hair present on the ear might cause problems for the biometric system. Also a variation in the brightness of the surrounding environment will unfavorably affect the system. When the ear is sheltered by hair or by clothing then a thermogram of the ear which gives the temperature profile of the ear can be used for authentication as discussed in literature [13, 14].

1.2.6 Iris Recognition

The iris, the colored area nearby the pupil holds substantial details and is found to be a very accurate biometric trait. Among all the biometric traits it is perhaps the most effective one. The probability of two individuals having the same iris is very rare. It is generally used government agencies where a very high security biometric system is required. The biometric system requires user co-operation and controlled environment for operation. The user is required to position his/her iris at a prede-termined position from the focal plane of the camera so that a high resolution image of the iris can be obtained. Also the brightness of the environment has to be controlled to obtain the consistent results.

1.2.7 Retina Recognition

The retina is the layer of blood vessels in the white portion of the eye. The chances of the two individuals having the same retina blood vessels patterns are found to be infinitesimal. Even eyes of alike twins are found to have dissimilar retina patterns. It is therefore applied in situations where a high security biometric system is com-pulsory. The system needs a controlled environment and user co-operation for operation. The user must point the eye at a fixed distance from the camera, look directly into the lens and remain absolutely still while the retina is being scanned.

The retina patters remain constant for a large number of human beings during their lifetimes. Few symptoms like diabetes and high blood pressure may affect the biometric.

1.2.8   Keystroke Dynamics

This is a behavioral biometric that can be used in grouping with passwords. The patterns of the keystroke are gathered as the user types including the total speed, the time required to find the keys, etc. Besides being significantly cheaper to implement than other biometrics it is much more inconspicuous as typing on a keyboard is much easier than the data gathering method of most other biometrics. This, however, is not very strong as it is vulnerable to the user moods and tiredness. The keyboard being used also plays a major factor as a user will have different dynamics on keyboards with different layouts. Also the actual dynamics may vary over time. The system in [15] uses a text length of 683 characters to achieve an FRR of 4 % and an FAR of less than 0.01 %. Though the results are very heartening, the text length is excessive and cannot be used easily.

1.2.9   Voice Recognition

This is one of the biometrics that is the most suitable to the users. It works by analyzing the air pressure patters and the waveform patterns [12] created by an individual's speech. Voice recognition systems deals with a succession of words and numbers or may need some haphazard input till the system can decide whether to validate or reject the user. The systems developed so far experience pitiable precision as an individual's voice can differ due to the circumstance. These systems have not been very successful against mimicry.

1.2.10   Gait Recognition

This is the latest recognized biometric trait. It works by analyzing the gait, i.e., the walking style of an individual. The style of walking is helpful in identifying a person at a distance hence it is a good behavioral biometric [16]. The variation in gait due to the person's mood and injury are a few of the key concerns for this biometric. Additionally biometrics like fingerprint and handprint don't change according to the individual's mood but gait is one of the biometrics which can be easily affected. In [16] promising recognition rates of up to 95 % are achieved using gait as the biometric.

Other biometric modalities have also been evolved in the recent time. A cancelable biometric approach called Bio-Hashing based authentication methods using electrocardiograph (ECG) features have also proposed in the literature recently [17–27].

1.3 Motivation

Hand based biometric authentication system is attractive in many security applications for a number of reasons. One of the most valid reason might be that almost all of the working population possess hands except the people with disabilities. During data capturing of hand geometry, data for palm prints and fingerprints can also be collected simultaneously. There is no hassle to the user as the matching accuracy of the system may greatly improve due to the addition of other biometric features. The latest biometric systems for most secure domains use fingerprints of all the fingers. They do not use the features of the hand geometry or the hand palm-print. Hand geometry based biometrics features are very useful in multi-modal systems. Most of the hand geometry based biometrics systems use pegs for setting up the placement of the palm on the scanner. In this proposed approach we aim to get rid of this requirement by allowing the user to vary the positioning of the palm on the scanner. Combining the other modality like fingerprints and palm prints to the current biometric systems is very easy because the acquisition of hand images are done in the peg-free environment.

Hand-based biometric modality is easy to acquire and it is very less vulnerable to disturbances and insensitive to the environmental settings and to individual anomalies. In contrast, face based recognition is quite susceptible to the problems like, facial accessories, pose, variation in lighting and facial expression; Iris or retinal-based recognition requires special illumination and is much less responsive; fingerprint imaging requires high-quality frictional skin, etc., and up to 5 % of the population may not be able to get enrolled. Therefore, verification based on hand shape can be an attractive substitute due to its unobtrusiveness, easy to interact, less costly and less data storage requirements.

Hand geometry based biometric systems are most among other biometric systems due to their universality, uniqueness and most prominently resistance against the fraud attacks. As it is a concealed inside the human body, it is not susceptible to stealing or loss like passwords and it is nearly impossible to repudiate the precise geometry. Hand geometry based systems are not affected by human body skin parameters i.e., color, moles, hair, etc.,

In this chapter, the shape of the hand based biometric system is proposed for individual identification/verification. Although the commercialization of hand-based biometrics product are growing rapidly, still actual implementation and documentation in the literature is limited as compared to that on other biometrics like voice or face.

1.4 Chapter Overview

The organization of the chapter is as follows:

Section 2 gives a literature review on the use of hand geometry as a biometric. Different approaches, different feature sets and result obtained using these approaches are given. It also presents the hand geometry biometric system. Section 3 is about the preprocessing step. In preprocessing the major task is to normalize the hand image the normalization of the hand images includes image segmentation, ring effect removal, finding extremities etc. Section 4 includes all the details about the feature extraction. It lists out all the special features and their extraction from the hand image. Extraction of the features includes the width of fingers, length of the fingers and contour of fingers from a particular point. Section 5 discusses about the matching process and also the SVM classifier. Section 6 gives the conclusions and suggestions for future work.

2   Literature Review

A set of diverse approaches exists for hand geometry recognition. The approaches vary mostly in the manner of mining and manipulation the features of hand.

In [11] a 4 B-spline curve symbolizes fingers. This facilitates the elimination of fixed pegs utilized in most systems. The classification is done by utilizing the differences between the curves created by a range of hand geometries and utilizing the curves as a signature for the person. Only the fingers get represented by the curves, the thumb is not a component of the signature. In a collection of 6 images from each of 20 persons a recognition rate of 97 % has been achieved in [28]. The error rate in the verification for the same database is 5 %.

Because of their interpolation property implicit polynomial 2D curves and 3D surfaces have been used for analyzing the handprint in [15]. This method proceeds by bringing an implicit polynomial function to fit the hand print. To keep the variation in the coefficients of the polynomial function to a bare minimum due to slight change in the data new methods like 3L [29] fitting [29] and Fourier fitting [30] are applied instead of the traditional least square fitting. In [31] the success rate is found to be 95 % and the verification rate 99 %.

In [29] Authentication is achieved by analyzing the vein patterns of the hand image. By using heat and conduction law a number of characteristic is extracted from each unique point of the vein pattern. By this method FAR is 3.5 % and FRR is 1.5 %.

The palm-print recognition based on the Eigen palm is discussed in [32]. The method utilizes the original palm prints and then palm prints are transformed to a set of characteristics. These characteristics are the Eigen vectors of the training set. For this, the Euclidian distance classifier is used. System works on a set of 200 people yielding FAR of 0.03 % at FRR of 1 %.

Bimodal systems, which are a grouping of two biometric modalities like hand geometry and palm prints are recognized as an effective tool for authentication. In [13] a bimodal biometric system using fusion of shape and texture of the hand is proposed. Palm-print validation is done using distinct cosine transform. New hand shape features are also proposed in [13]. The score level fusion of hand shape features and palm prints is done by the product rule. Making use of either of the hand shape or palmprints alone produces high FRR and FAR but when the two modalities are combined to construct a bimodal system, both FAR and FRR are significantly reduced. It is especially efficient when the hand shapes of two different individuals are alike and in such cases the palmprints improve the performance. On a database of 100 users the FRR is found to be 0.6 % and the FAR is found to be 0.43 %.

In [33] geometric classifiers are applied for hand recognition. The authentication system in [32] uses a document scanner to gather hand data. Also a few limitations are imposed on the positioning of the hand on the scanner. Overall a total of 30 diverse features is obtained from a hand. For each person 3 to 5 images are taken as the training set. A bounding box is found in the 30 dimensional feature spaces for each of these training sets. The distance of these bounding boxes to the query image is utilized as the measure of resemblance. The threshold is determined by experimentation on the database.

In [16] 3-D hand modeling and gesture recognition techniques are discussed. This planned system utilizes 2-D hand shape modeling. A hand shape model of a geometrical form can be approximated using splines as they are complicated geometrical shapes. The model can be made very precise but this will increase the difficulty as the parameters and the control points will swell. Simpler and less complex models can also be utilized but the precision of the modeling will suffer in return. Besides parameters, a set of 3-D points can also be utilized to model the hand. The polygon meshes represented by these set of points in 3-D space, approximate the hand shape. Yet another way is to construct the hand model from a set of images of the hand from different views.

3   Overview of Proposed Biometric System

A block diagram of hand-based biometric system is shown in Fig. 3. The proposed hand-based biometric system consists of following two phases; enrollment and verification phase. The major steps in hand-based biometric system are preprocessing of input image, feature extraction and feature matching. The enrollment phase includes; preprocessing of input image, feature extraction and generating template from extracted features and to store in a database. Preprocessing module is the most important part of a biometric system as it helps in efficient feature extraction. If an image is not preprocessed accurately the results might not be up to the mark. Next step of the enrollment phase of a biometric system is the feature extraction. In this phase, we measured length of five finger contours from pivot point to tip, finger widths are measured from three positions and finger lengths. Third module of the enrollment phase generate template and stores all the extracted features in a database. Here we have stored all the features of each image of every user in the database.

At the verification phase, a user is verified using input of present hand image. The steps at verification phase are very similar to that of enrollment phase except the last step is matching the current template against the stored template in the database.

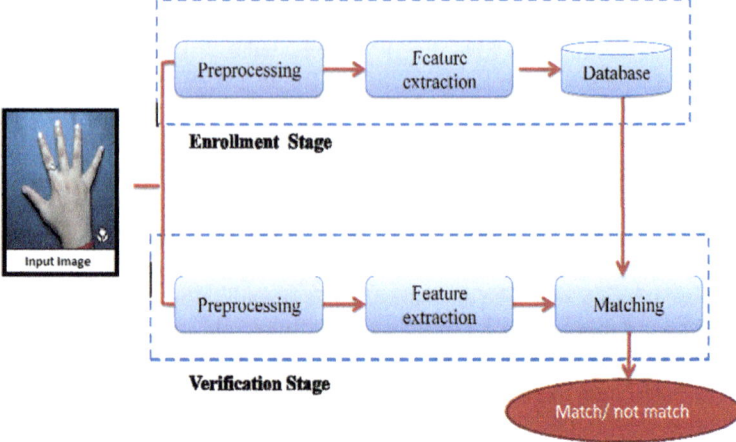

Fig. 3 A block diagram of a biometric system

3.1 Preprocessing

Preprocessing is an important step of any biometric/image processing system. Preprocessing enables the system to extract features/key points more accurately as compared to raw images. In this system we used a peg free environment during image acquisition step. In peg-based acquisition, users are restricted to put their hand in a fixed position. In peg-free environment a user is free to put their hand in any direction and in any position. Figure 4 shows the user's image in different positions.

The major steps of preprocessing is to normalize the image which includes the following steps:

- Image Segmentation
- Detection of Important Points
- Detection of Contour
- Contour Normalization

3.1.1 Image Segmentation

During segmentation of hand images, we separate the hand region from the background region. The extraction of hand region is relatively very important and easy task. This step basically includes separation of the hand contour from the background region and ring effect removal.

Image Separation

For the image segmentation, the main steps are:

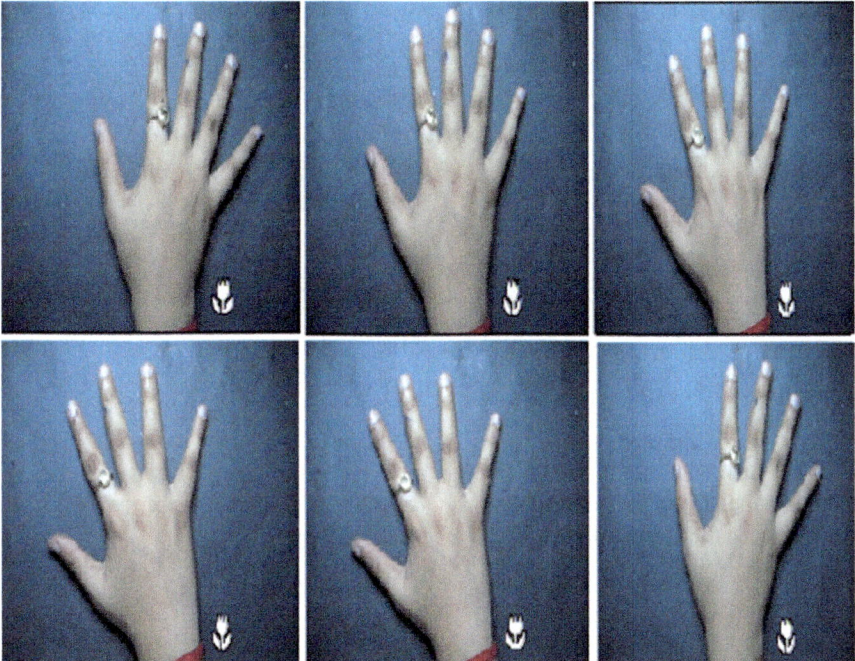

Fig. 4 Different hand positions

- Color to gray transformation
- Gray- to binary transformation
- Noise removal of binary image

At each pixel red, green and blue values are extracted. Using the thresholding operations we detect the skin color and separation of the hand image region from the background region. If Ir, Ig and Ib denote the red, green and blue values respectively. Following conditions are used for skin detection.

$$if\ (I_r < I_g\ and\ Ig < I_b\ and\ I_r < 150)$$
$$Image\ (Channel\ 1) = 0,\ Image\ (Channel\ 2) = 0,$$
$$Image\ (Channel\ 3) = 0$$
$$else$$
$$Image\ (Channel\ 1) = 255,\ Image\ (Channel\ 2) = 255,$$
$$Image\ (Channel\ 3) = 255$$

The acquired image is converted into the gray level image. Binarization is used get a binarized image using some threshold value. The original hand image and the segmented hand image are shown in Fig. 5 which shows the contours if we don't perform normalization.

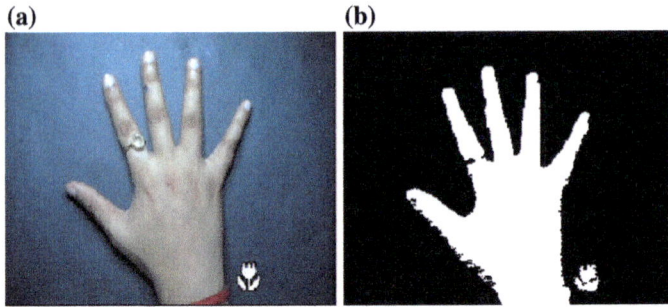

Fig. 5 **a** Original image, **b** separated image

Ring Effect Removal

If a ring is present on the finger of an individual then it may create a break between palm and finger. This finger can be identified or located easily. For this, we calculate the size of the connected components and identify the finger as the smaller part of the hand. Now to remove ring effect steps are:

Step 1: Find the hand Contour.
Step 2: Find Hand extremities.
Step 3: Finger's information Matrix.

Contour Extraction

For contour extraction, we first find out the major axis of a hand in the direction of the highest Eigen-value of inertia matrix. Inertia matrix can be calculated from Eq. (7). We compute the different moments of binary image using

$$m_{i,j} = \sum \sum_{(x,y) \in \text{object}} x^i y^i \qquad (3)$$

Here summation runs over the all the pixels of an object. The centroid is computed as:

$$\bar{x} = \frac{m_{1,0}}{m_{0,0}} \qquad (4)$$

$$\bar{y} = \frac{m_{0,1}}{m_{0,0}} \qquad (5)$$

Thus the central moments can be calculated as:

$$\mu_{ij} = \sum \sum (x - x_i)^i (x - x_j)^j \qquad (6)$$

The inertia matrix I is given by

$$\begin{bmatrix} \mu_{2,0} & \mu_{1,1} \\ \mu_{1,1} & \mu_{0,2} \end{bmatrix} \tag{7}$$

Orientation of the object is given by the angle

$$\theta = \frac{1}{2}\arctan\left(\frac{2\mu_{1,1}}{\mu_{2,0} - \mu_{0,2}}\right) \tag{8}$$

The reference point on wrist is determined as the point of intersection of hand's boundary with the ray, this is aligned in the direction of hand and originates from the hand's centroid. Once we found the first intersection then the intersection point is considered as the starting point of the contour, subsequently successive contour points are located by following the nonzero neighbors to it in the clockwise direction. By following this approach, we obtain the hand contours.

Hand Extremities

Hand extremities are basically the finger tips and valleys of a hand image and their extraction is very important step in the feature extraction process. Valley between any two fingers is the lowest point between them. In similar way the extremities are detected using highest curvature on the tips of the fingers in hand contour. But there is always a possibilities of detecting false extremities as these are sensitive to irregularities in the contour such as fake cavities and kinks. We use other approach that sketch a plot of radial distance from a detected reference point on the wrist. This reference point is seen as the initial point of the major axis with the wrist line. The resultant sequence includes radial distances having maxima and minima. These distances are not affected by irregularities of contour and kinks and also these maxima and minima are gives the location of nine extremities as shown in Fig. 6.

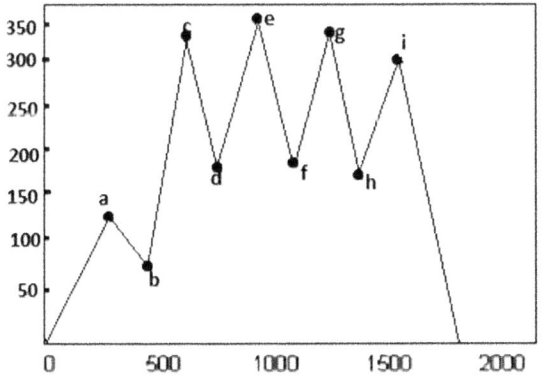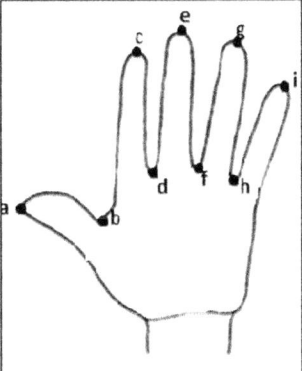

Fig. 6 Determination of extremities

The first row presents extremities for thumb and others rows are ordered in clockwise for other fingers of the hand. This gives enough information about the hand geometry to remove ring cavity from hand contour using this finger's information matrix. Finger profiles (i.e., every finger has profiles on both sides of the finger, left side valley to tip, and tip to right side valley) so that 1D distance of contour curve segments w. r. to the finger's central major axis that helps to detect cavities. These cavities are correspond to the locations where the profile drops below 75 % of its median and they are morphologically bridged by fitting a line segment on it. So in this way we remove the ring cavity.

We used the distance from contour to major axis of each fingers to identify the presence of thumb. When this distance either on left and/or right side of the finger is above a threshold, it is assumed to be a cavity and it is removed by bridging over the cavity as shown in Fig. 7.

3.1.2   Image Normalization

Image normalization is a pixel based operation which decreases the pixel grey value variation and normalizes the pixel intensity distribution. At this step binary hand image obtained from segmentation step is normalized using adaptive threshold values. This normalization is required because, we used a peg-free system where users are allowed to put their hand in any direction, and thus we may observe the condition as shown in Fig. 8. Normalization of images is used for registration of hand images, which in turn involves rotation and translation as well as re-orientation of the fingers along the standard directions.

The registration of images is done by following these steps:

- Translation of the centroid.
- Rotation in the direction of highest eigenvector of the inertia matrix.

Extraction of Fingers

From the tip of any finger we move along the boundaries of all the fingers to reach up to the adjacent valley points. We chose the shortest of these valleys and then swung like a pendulum towards the other valley point. This mark is be used to determine the finger length and also the shape of the finger.

Pivots of Finger

The length of each finger is extended to locate a point known as Pivots of finger. Using these finger pivots of each finger, we plot the hand pivotal axis as shown in Fig. 6 and defined in the below step. Similar steps are used to find the pivot of thumb. In total we locate five such 5 pivots with respect to each finger.

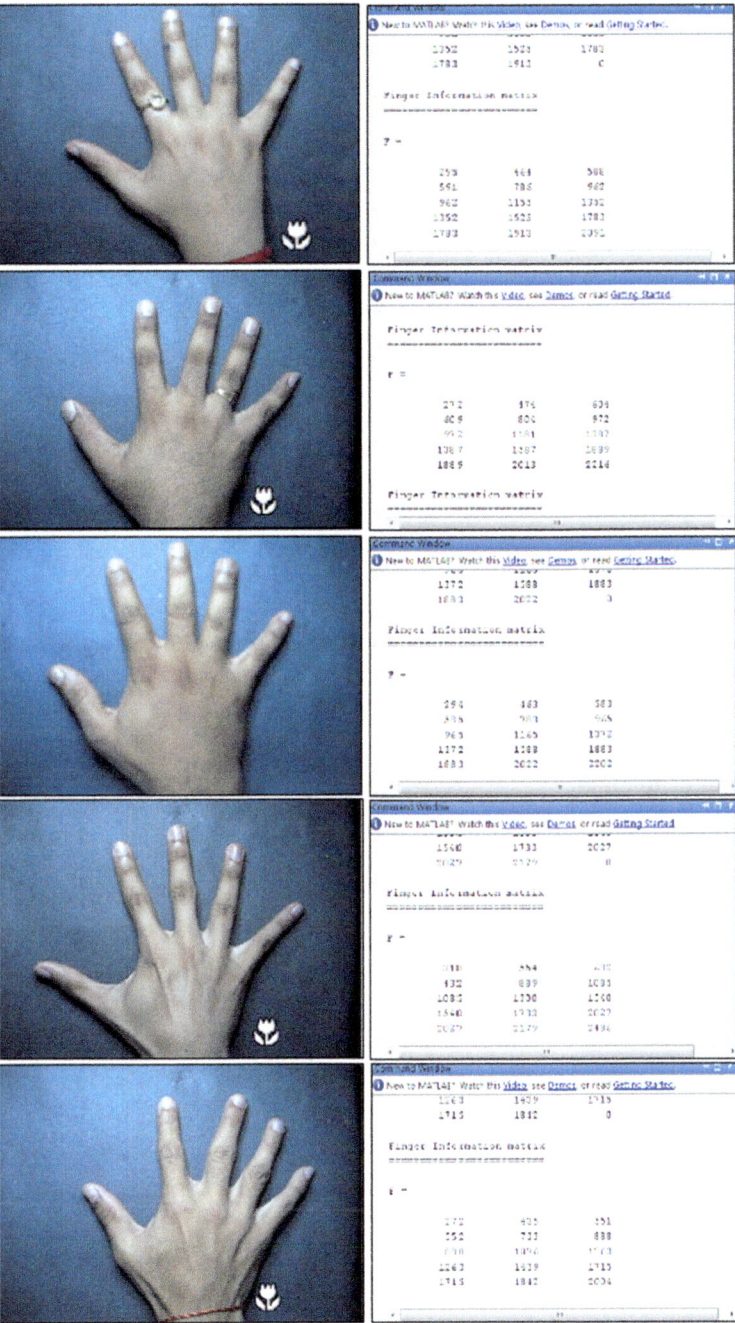

Fig. 7 Located extremities and valley

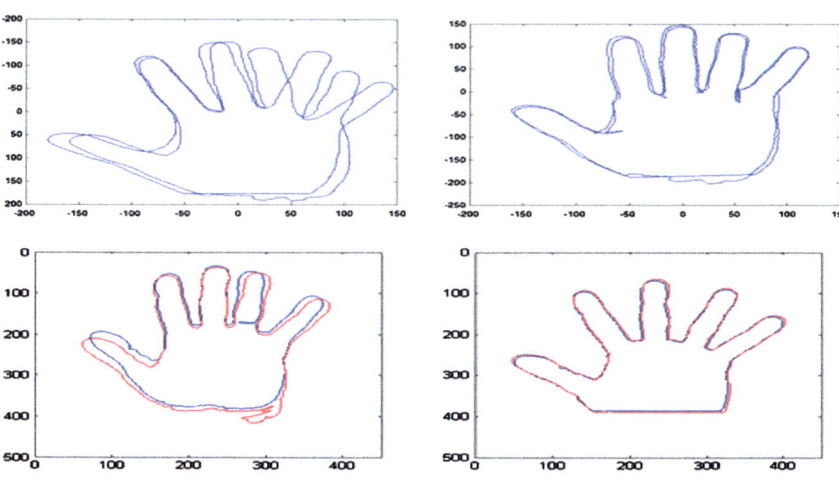

Fig. 8 Different hand contour positions

Hand Pivotal Axis

The four pivot points with respect to each finger are then joined together by drawing a line between last two pivot points using least square method or simply joining the last two pivot points to get the hand pivotal axis as shown in Fig. 9. Hand pivotal axis is very critical handle the rotation of the hand geometry.

Rotation of the finger

To handle the rotation of hand geometry we calculate the angle of rotation ψ_i in respect to major axis of each hand using the procedure discussed above. Each finger i is then rotated by the angle $\Delta\theta_i = \psi_i - \theta_i$. Here θ_i is the desired orientation of the finger. The finger rotation is done by multiplying the position vector of finger pixels by the rotation matrix $R_{\Delta\theta}$ given by

$$R_{\Delta\theta} = \begin{bmatrix} \cos\,\Delta\theta & -\sin\,\Delta\theta \\ \sin\,\Delta\theta & \cos\,\Delta\theta \end{bmatrix} \tag{9}$$

Fig. 9 a Extracted finger,
b pivotal axis

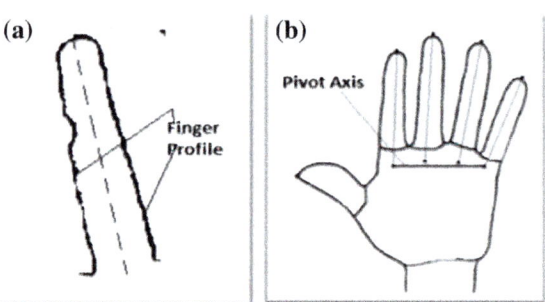

Processing for the thumb

The orientation of the thumb is rather complicated process in comparison to the fingers as thumb involves the rotations with respect to two diverse joints. Actually, both the metacarpal-phalanx joint as well as the trapezium-metacarpal joint of the thumb participate in the thumb orientation. We address this complication of orientation by a rotation followed by a translation. These issues arise from the fact that the stretched skin between the index finger and the thumb confuses the determination of valley points and extraction of thumb becomes more difficult. This is the reason, why we rely on the basic hand anatomy. A line from the major axis of the thumb is drawn and the point on the line from the tip of the thumb to 120 % to the length of the little finger marks the thumb pivot. After that the thumb is translated so that its pivot coincides with the tip of the hand pivot line, when it is swings by 90° in clockwise direction. Thumb is then rotated to its final angle of rotation and brought back to the original position.

Normalized Image

The normalized hand is translated so that its centroid, which is the mean of four finger pivot points (index, middle, ring, little) is taken as the reference point on the image surface. The hand image is then rotated such that its pivot line exactly matches with chosen orientation. Otherwise, the hands would be aligned to the major inertial axis and its center would be defined with respect to the hand contours (and not with respect to the pivotal centroid).

Contour of Normalized Hand

After finding the normalized hand image as shown in Fig. 10, we find again the contour of the normalized image as in Fig. 11.

Fig. 10 Normalized fingers and hand

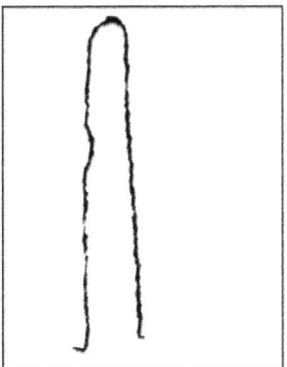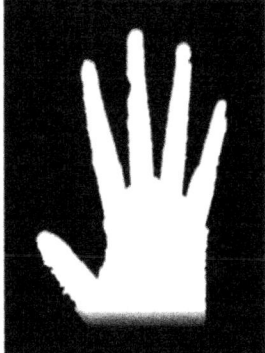

Fig. 11 Contour of
normalized hand

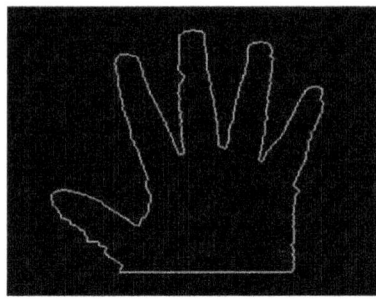

4 Feature Extraction

Set of features are extracted to form the feature vector. The steps described in Sect. 3 are used preprocess the image for feature extraction. Extraction of reliable features is a very important module for any biometrics recognition system. Feature extraction module extract and store features from the input image to the template. The feature extraction module gives out the measure of features like finger widths, lengths, and palm width etc.

The hand geometry-based biometric authentication system heavily depends on the geometric invariants of a human hand. Typically these features includes length and width of the fingers, the aspect ratio of the palm and fingers, the thickness of the hand, etc. Presently, the existing commercial systems do not take into account of any non-geometric attributes of the hand, e.g., color of the skin.

There are several other features those can be extracted from the palm-geometry. We are only interested in the features which are stable and consistent, i.e., features which are robust to the hand pose variation. In this work, we extracted the features like length of all fingers including thumb, width of all fingers at 3 different locations and four the distances from a fixed point on the palm to the tip of each finger. The additional features like width of the finger at three positions make this approach more efficient and robust. The length of the feature vector thus formed is 24 now.

In the Sect. 3, we have discussed about the hand extremities (five tips and four valley point), information matrix of finger, contour of finger and hand. Hand extremities are determined using radial distances from the reference point around the wrist. Fingers information matrix is a 5×3 matrix which contains all the information about the fingers of a hand. In finger information matrix 2nd column corresponds to the contour indices of finger tips and the 1st and 3rd columns correspond to indices of the contour indices of surrounding valleys of the particular tip.

4.1 Finger Length

In Sect. 3.1.2 we have discussed the normalization of hand image. Determination of finger length is done at the stage of normalization of hand image. In the present hand based biometric system, hand images are normalized by normalizing individual fingers and the palm region.

The process of detecting finger length start by cutting each finger from the palm. Two end point of the of finger's contour segment are joined through a straight line is used to defined the cut. The two adjacent valley points of finger are obtained from 5 × 3 finger information matrix. A binary line of zeros is drawn between two adjacent valley points differentiates the connected components. Now the fingers are cut by using the connected components algorithm, the larger label is used to be the palm and the other one to be the finger being cut. Distance from the fingertip to the mean of start and end point gives the length of the finger. Same method is applied for each finger to determine its length.

Algorithm for finger length extraction

Input: finger information matrix, binary image, hand's boundary contour
Output: finger length

1. Find the start point and the end point
2. Draw-line between the start point and the end point
3. Find the mid-point of line.
4. Find the distance between the fingertip and the mid-point
5. Repeat Step 1 for each finger.

4.2   Finger Width

As discussed in the above normalization Section, we extract finger length from the fixed point of the palm. Accuracy of individual finger length is important in hand based biometric system. For finger width extraction, we used the finger contour and fingertip to separate it from others fingers. Using the fingertip points obtained from the 2nd column of finger information matrix, we obtained the fingertip position on the contour. From this pivot point, we move in both directions 35 pixels with line of contour of each finger and calculate the distance from these two end points, which is the width of the finger 35 pixels far from the fingertip. At each 35 pixels away we calculated it three times. Generally, the length of finger is appx. 140 pixels. Moving by 35 pixels from finger-tip uniformly compute the width at three positions and these additional features make the proposed method more robust and efficient Fig. 12 describes the above approach.

Algorithm for finger width extraction
Input: Finger image contour, finger information matrix
Output: Finger width at three positions.

1. Fix fingertip as starting point
   Move away 35 pixels in both (left and right) directions from fingertips.

Fig. 12 Calculating finger
width

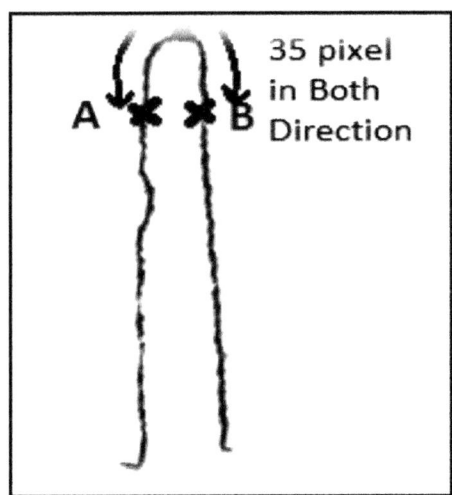

Obtain two points, say A and B
Find the distance between these two points.
Calculate two next distances in the similar fashion.
2. Repeat Step 1 for each finger.

4.3 Distance from a Fixed Point

We located the pivots by extending the length of each finger towards palm as
discussed in 3.1.2. A fixed point on palm at 50 pixels below the middle finger is
considered as pivot point. This point is considered as a reference point for computing all the features of hand image. This point is used to estimate the length of
each fingers from tip of each finger to the reference.

Algorithm for Special distances
Input: Normalized hand image, Normalized image contour, information matrix
Output: Four distances in spatial domain.

1. Fix middle finger's pivot as the starting point
2. Calculate reference point just 50 pixel below the middle finger's pivot position.
3. Calculate the distance from fingertip to this reference point for each finger.

We have calculated the four distances from the finger tips except thumb.
So now we have a total of 24 features: 5 finger lengths, 15 finger widths (three
for each) and 4 special distances. All these 24 features are shown in Fig. 13.

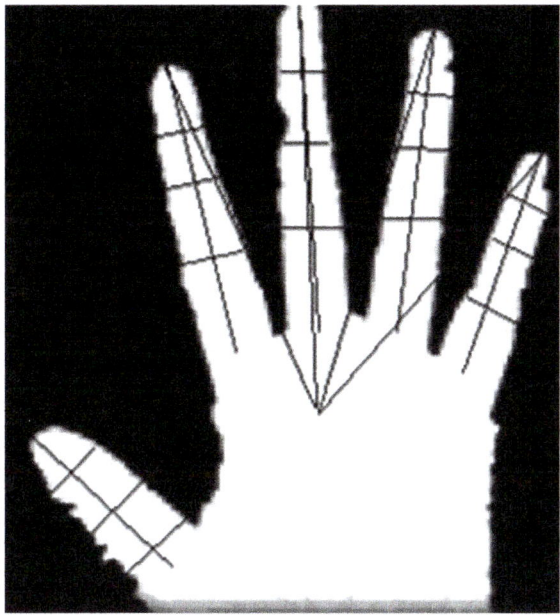

Fig. 13 All 24 features

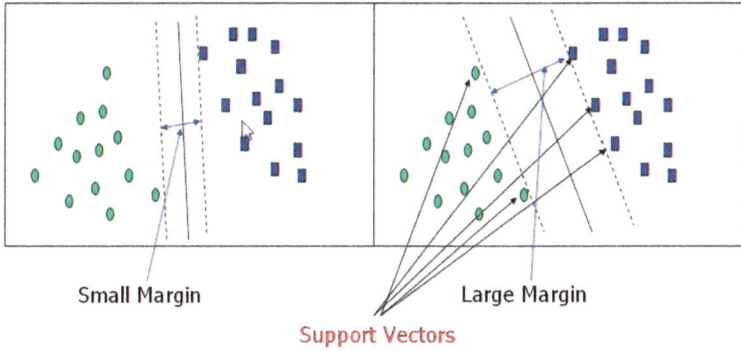

Fig. 14 Linear SVM; Curtsey [31]

5 Matching and Experimental Results

In this section, user authentication is carried out using the feature vector obtained from the normalized hand, which includes 24 distances as in Fig. 13.

In enrollment phase, features are extracted from preprocessed hand image and a template is generated and stored in the database. For matching feature template extracted from test image is matched with the templates stored in database. Two class SVM is used for matching.

Algorithm for matching:

Step 1. Input the Color image.
Step 2. Preprocessing of the input image

- Image Segmentation
- Ring Effect removal.
- hand image Normalization
- Finding Extremities

Step 3. Feature Extraction.
Step 4. Classification using SVM.

5.1 Training and Testing

A binary SVM classifier is used for matching the feature vector of test and trainee images. The system is first trained during enrollment phase by learning from the input data sets using SVM [10] classifier. Since SVM is only two-class classifier, a multi class classifier can be built by combining two class SVMs. For a given set of training data, each data sample belongs to one of two classes i.e.,i.e., match or non-match. These non-match samples are further classified using SVM and finally the samples are being into all classes as discussed in details in next paragraph. Figure 14 shows support vectors built by SVM [31].

5.2 Multiclass SVM

Multiclass SVM allocates labels to instances by using the support vectors. Here the labels are taken from a limited set of elements. Here the multiclass problem is converted into multiple binary classification problems. This can be accomplished in two ways:

1. One-versus-all
2. One-versus-one

Categorization of new samples in the case of one-versus-all is achieved by a winner-takes-all approach, in which the classifier with the maximum output function assigns the class. In the one-versus-one case, categorization is achieved by a max-win voting approach, in which every classifier assigns the sample to one of the two classes, then vote count of the assigned class is increased by one, and lastly the class with the majority votes determines the sample classification.

In the proposed system we have used one-versus-all approach to classify between each class and all the remaining classes.

5.3 Identification Results

The database used in this study consists of ten right hand colored images of one hundred people captured at the Biometric research lab at IIT-Delhi. Each image is of size 768×576 pixel in PNG format.

The database used in this study consists of 480 colored images of right hand for 48 users captured at the Biometric lab at IIT-Delhi. The users are of different age, sex, and colors group. Also these images are captured at different time, place, and at the different hand positions. We used peg free environment without posing any restriction on a user. In this scenario users are free to put their hand in any direction and in any position also there is no any restriction on Individuals that they should remove their hand or finger accessories like bracelet, ring and wrist watch. At the time of collection of hand images all kind of possible hands positions were involved.

We begin by considering classification problems with only two classes. Each instance i is classified as one element of the set {p, n} of true and false class labels. A classification is a mapping of instances to the predicted classes. Some classification algorithms produce a continuous output (e.g., an estimate of an instance's class membership probability). Different thresholds may be applied to predict class membership. Other algorithms produce a discrete class label indicating only the predicted class of the instance. To distinguish between the actual class and the predicted class we use the labels {T, F} for the class predictions produced by the model.

Given a classifier and an instance, there are four possible outcomes. If the instance is P and it is classified as positive, then it is counted as a true positive; if it is classified as negative, then it is counted as a false negative. If the instance is negative and it is classified as negative, then it is counted as a true negative; if it is classified as positive, it is counted as a false positive.

The database used to test the proposed system is consists of 480 images. For experiments 7 images are randomly selected for training and the rest 3 are used for testing. The matching accuracy of 95.84 is obtained using proposed methods.

The FRR obtained from the proposed system is given as

$$\text{FRR} = \frac{\text{Number of rejected genuine claims}}{\text{Total number of genuine accesses}} \times 100 \text{ \%}$$
$$\text{FRR} = \frac{6}{144} \times 100 = 4.16 \text{ \%}$$

(10)

Efficiency = 95.84 %
FRR   = 4.16 %

The chapter has explored the geometrical features only. Issues which will be taken further is the use of the contours of fingers by some polynomial function like spline functions. If we take the contours of all the fingers together the problem arises with respect to ever varying gaps between the fingers. This makes the applicability of polynomial function more difficult because of variation in contours. One of the easy method is to take each finger contour separately by fitting the polynomial curves using spline functions.

6 Conclusion

The rapid growth in the use of e-commerce applications requirement of reliable and secure method of biometric authentication is in great demand. Biometrics is being used all over the globe and is undergoing rapid development. Recently hand geometry based authentication system has proven its reliability. The proposed work exploited the shape of the hand to extract reliable features using very simple method.

The proposed biometric modality is not only user-friendly but also provides good results. The only disadvantage of the system is that this is affected by illumination. Peg free environment gives freedom to an individual to put their hand in any direction and position. This approach is subjected to the rotation and normalization such that the fingers are erect having substantial gaps among them.

We extract the invariant features from the hand geometry consisting of 5 finger lengths, 15 finger widths (3 for each) and 4 spatial distances (which are the distances from the finger tips to a fixed point of each finger) with a total of 24 features.

The proposed biometric system in tested on a database of 480 images collected from 48 users i.e., 10 images per user. Three sample images from each user are used for verification purpose and samples for the training. The results of verification are obtained using the SVM classifier which gives the accuracy around 95.84 % on the database.

References

1. Sanchez-Reillo R, Sanchez-Avila Gonzalez-Marcos A (2000) Biometric identification through hand geometry measurements. IEEE Trans Pattern Anal Mach Intell 22(10):1168–1171
2. Kumar R, Chandra P, Hanmandlu M (2011) Fingerprint matching based on orientation feature. In: Advanced materials research, vol 403. Trans Tech Publications, pp 888–894
3. Kumar R, Chandra P, Hanmandlu M (2013) Fingerprint matching based on texture feature. In: Mobile communication and power engineering. Springer, Berlin, p 86–91
4. Kumar R, Chandra P, Hanmandlu M (2013) Local directional pattern (LDP) based fingerprint matching using SLFNN. In: 2013 IEEE second international conference on image information processing (ICIIP). IEEE, pp 493–498
5. Kumar R, Chandra P, Hanmandlu M (2013) Fingerprint matching using rotational invariant image based descriptor and machine learning techniques. In: 2013 6th International conference on emerging trends in engineering and technology (ICETET). IEEE, pp 13–18
6. Kumar R, Chandra P, Hanmandlu M (2014) Rotational invariant fingerprint matching using local directional descriptors. Int J Comput Intell Stud 3(4):292–319

7. Kumar R, Chandra P, Hanmandlu M (2012) Statistical descriptors for fingerprint matching. Int J Comput Appl 59(16)
8. Kumar R, Hanmandlu M, Chandra P (2014) An empirical evaluation of rotation invariance of LDP feature for fingerprint matching using neural networks. Int J Comput Vis Robot 4 (4):330–348
9. Kumar A, Zhang D (2006) Combining fingerprint, palmprint and hand-shape for user authentication. In: 18th international conference on pattern recognition, 2006 (ICPR 2006), vol 4. IEEE, pp 549–552
10. Moore AW (2001) support vector machines, tutorial slides. http://www.autonlab.org/tutorials/svm.html
11. Ma Y, Pollick F, Hewitt WT (2004) Using b-spline curves for hand recognition. In: Proceedings of the 17th international conference on pattern recognition, 2004 (ICPR 2004), vol 3. IEEE, pp 274–277
12. Han CC, Cheng HL, Lin CL, Fan KC (2003) Personal authentication using palm-print features. Pattern Recogn 36(2):371–381
13. Kumar A, Zhang D (2006) Integrating shape and texture for hand verification. Int J Image Graph 6(01):101–113
14. Dey N, Nandi B, Das P, Das A Chaudhary SS (2013) Retention of electrocardioGram features insiGnificantly devalorized as an effect of watermarkinG for. In: Advances in biometrics for secure human authentication and recognition, p 175
15. Erçil A, Yöldöz VT, Körmözötas H, Büke B (2001) Hand recognition using implicit polynomials and geometric features. In: Audio-and video-based biometric person authentication. Springer, Berlin, pp 336–341
16. Wu Y, Huang TS (1999) Human hand modeling, analysis and animation in the context of HCI. In: Proceedings of the international conference on image processing, 1999 (ICIP 99), vol 3. IEEE, pp 6–10
17. Nandi S, Roy S, Dansana J, Karaa WBA, Ray R, Chowdhury SR, Chakraborty S, Dey N (2014) Cellular automata based encrypted ECG-hash code generation: an application in inter human biometric authentication system. Int J Comput Netw Inf Secur 6(11):1
18. Biswas S, Roy AB, Ghosh K, Dey N (2012) A biometric authentication based secured ATM banking system. Int J Adv Res Comput Sci Softw Eng. ISSN:2277
19. Dey N, Nandi B, Dey M, Biswas D, Das A, Chaudhuri SS (2013) Biohash code generation from electrocardiogram features. In: IEEE 3rd international advance computing conference (IACC). IEEE, pp 732–735
20. Dey M, Dey N, Mahata SK, Chakraborty S, Acharjee S, Das A (2014) Electrocardiogram feature based inter-human biometric authentication system. In: International conference on electronic systems, signal processing and computing technologies (ICESC). IEEE, pp 300–304
21. Acharjee S, Chakraborty S, Karaa WBA, Azar AT, Dey N (2014) Performance evaluation of different cost functions in motion vector estimation. Int J Serv Sci Manag Eng Technol (IJSSMET) 5(1):45–65
22. Dey N, Das A, Chaudhuri SS (2012) Wavelet based normal and abnormal heart sound identification using spectrogram analysis. arXiv:1209.1224
23. Dey N, Das S, Rakshit P (2011) A novel approach of obtaining features using wavelet based image fusion and Harris corner detection. Int J Mod Eng Res 1(2):396–399
24. Kaliannan J, Baskaran A, Dey N (2015) Automatic generation control of thermal-thermal-hydro power systems with PID controller using ant colony optimization. Int J Serv Sci Manag Eng Technol (IJSSMET) 6(2):18–34
25. Bose S, Chowdhury SR, Sen C, Chakraborty S, Redha T, Dey N (2014) Multi-thread video watermarking: a biomedical application. In: 2014 International conference on circuits, communication, control and computing (I4C). IEEE, pp 242–246
26. Bose S, Chowdhury SR, Chakraborty S, Acharjee S, Dey N (2014) Effect of watermarking in vector quantization based image compression. In: 2014 international conference on control, instrumentation, communication and computational technologies (ICCICCT). IEEE, pp 503–508

27. Chakraborty S, Samanta S, Biswas D, Dey N, Chaudhuri SS (2013) Particle swarm optimization based parameter optimization technique in medical information hiding. In: 2013 IEEE international conference on computational intelligence and computing research (ICCIC). IEEE, pp 1–6

28. Hanmandlu M, Kumar A, Madasu VK, Yarlagadda P (2008) Fusion of hand based biometrics using particle swarm optimization. In: Fifth international conference on information technology: new generations, 2008 (ITNG 2008). IEEE, pp 783–788

29. Lin CL, Fan KC (2004) Biometric verification using thermal images of palm-dorsa vein patterns. IEEE Trans Circuits Syst Video Technol 14(2):199–213

30. Kumar A, Hanmandlu M, Gupta HM (2009) Online biometric authentication using hand vein patterns. In: IEEE symposium on computational intelligence for security and defense applications, 2009 (CISDA 2009). IEEE, pp 1–7.

31. Linear SVM image, http://areshopencv.blogspot.in/2011/07/artificial-intelligencesupport-vector.html

32. Lu G, Zhang D, Wang K (2003) Palmprint recognition using eigenpalms features. Pattern Recogn Lett 24(9):1463–1467

33. Bulatov Y, Jambawalikar S, Kumar P, Sethia S (2004) Hand recognition using geometric classifiers. In: Beusl N (ed) Biometric authentication. Springer, Berlin, pp 753–759

A Study on Security and Surveillance System Using Gait Recognition

M. Sivarathinabala, S. Abirami and R. Baskaran

Abstract Security is an important aspect and international attention in smart environments. The surveillance cameras are deployed in all commercial and public places in order to improve the security against terrorism activities. Nowadays, more and more government and industry resources are involved in the researches of security systems, especially in multimedia security, i.e., to enforce security measures, from the images and videos taken from suspicious environment. Therefore, there exists a need to ensure the originality and authenticity of multimedia data as well as to extract intelligent information from enormous images/video streams taken from suspicious environments to build stronger security systems. In this scenario, person identification plays a major role in security systems from the footages of suspicious environments. Without any human assistance, video analyst can identify the person from a large number of videos.

Keywords Person identification · Gait recognition · Security · Surveillance videos

1 Introduction

The identification of the people based on their physical and behavioral characteristics is called as biometrics. Biometrics had reached its highest level in terms of security in various situations. The physical biometrics include the physical char-

M. Sivarathinabala (✉) · S. Abirami
Department of Information Science and Technology, Anna University,
Chennai, Chennai, India
e-mail: sivarathinabala@gmail.com

S. Abirami
e-mail: abirami_mr@yahoo.com

R. Baskaran
Department of Computer Science and Engineering, Anna University,
Chennai, Chennai, India
e-mail: baaski@annauniv.edu

© Springer International Publishing Switzerland 2017
N. Dey and V. Santhi (eds.), *Intelligent Techniques in Signal Processing for Multimedia Security*, Studies in Computational Intelligence 660,
DOI 10.1007/978-3-319-44790-2_11

acteristics of the people such as fingerprint, palm print, hand geometry, iris, and retinal recognition. These physical characteristics are enrolled in the database so that the system can identify the person based on the person's characteristics. The biometrics can be obtained with the help of that person's assistance. Furthermore, every security systems are based on the biometrics. The physical biometrics are mostly used for identifying the small group of people. While accessing the security system, the system scans the biometrics and checks with the database entries. If the biometric matches with the system, then the person's identity has been verified. On the other hand, behavioral biometric includes the characteristics such as gait, voice, and rhythm. In this behavioral biometrics, gait recognition plays a major role. The walking characteristics of each person are unique identity, and gait characteristics are extracted without the knowledge of the particular person. Thus, the gait biometrics is used to watch the abnormal or suspicious behavior of the person from the long distance.

1.1 Behavioral Biometric: Gait Recognition

In video surveillance, first and foremost step is to provide authenticated user access. The authenticated person is identified from the ongoing videos. A video analyst analyzes the video to identify the person present in the scene and his/her activities. Nowadays, a new technology has been emerged to extract the person's features without the knowledge of that person. The person can be identified by the way he/she walks; i.e., the gait of the person is extracted. In the videos, gait recognition plays an important part to identify the person in order to tighten the security. From the gait sequence, a system is able to recognize the familiar person or even unfamiliar persons recognize the gender, estimate the mental state of the person, estimate weight, and provide medical assessment and diagnosis/treatment of gait-related disorders. Depending upon the context, gait recognition can be operated in verification (authentication) mode and identification mode. The verification mode has 1:1 match between the stored template (gallery) and the probe template to confirm or to reject the claimed Id. The identification mode has 1:N match between the stored template (gallery) and the probe template to confirm or to reject the claimed Id.

The complete gait recognition system involves the following stages: video acquisition, preprocessing, tracking, feature extraction, classification, and recognition. Preprocessing as the first stage is to extract the silhouette from the video frames and to represent the silhouette conveniently. The background subtraction process has been used here to detect the moving objects in every frame. Tracking is the process to track the moving objects, and the blob tracker has been used to track the objects. Feature extraction is an important phase in recognition. Features are

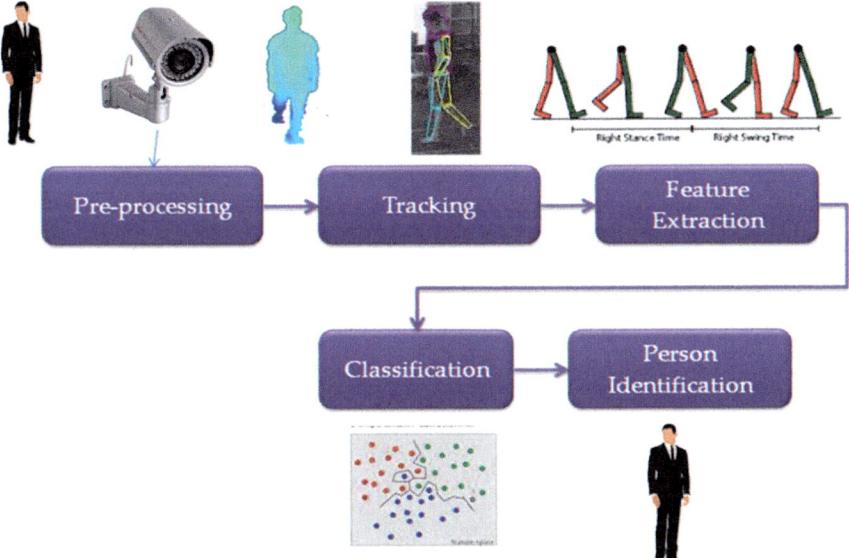

Fig. 1 Gait recognition system

considered from model-free methods or model-based methods. Model-free methods focus on the spatial and temporal information contained in the silhouette. Model-based methods serve as prior knowledge to interpret the human dynamics and to predict the motion parameters. Figure 1 shows the gait recognition system.

This chapter is intended to provide an advanced multimedia application in gait recognition with its focus on security. Two major areas will be covered in this chapter to elicit the secure uses of multimedia data and how to use multimedia data for security applications. This chapter deals with the following: (1) introduction of gait recognition methodologies, (2) feature extraction algorithms in research of gait recognition, (3) learning methodologies that are involved in person identification, and (4) research issues and applications of gait recognition systems in video surveillance.

2 Gait Recognition Methodologies

Gait recognition has been carried out in two ways: vision-based and sensor-based approaches. The vision-based approaches [1] are classified into model-based and model-free methods. The silhouette measurements based on the movement of the subject are done using model-free approaches, and in the model-based approach, the subject is modeled as a skeleton. From the skeleton model, the features are extracted and are tend to be difficult to implement. It is classified using SVM,

Fig. 2 Five-link biped model
—model-based approach and
GEI–model-free approach

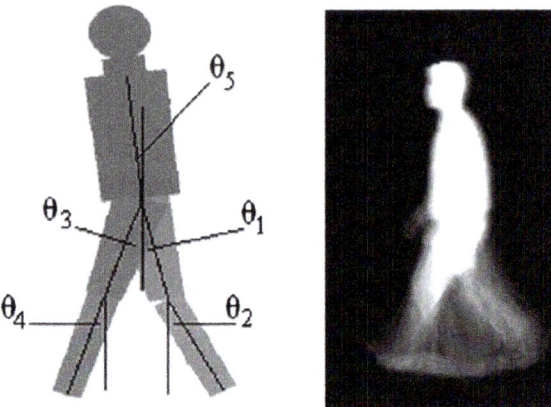

HMM, nearest neighbor, neural networks, etc., to recognize the subject. The main advantage of the model-free approach is less computational complexity. The sensor-based approaches are classified into floor sensor and wearable sensor approaches. The floor sensor is installed in the particular place or floor to collect the pressure signal generated while walking. The wearable sensors are placed on the different body parts to extract the gait features such as the speed, acceleration, and other information. The example of the skeleton model and the gait energy image is shown in Fig. 2. Gait recognition methodologies have been depicted as a tree diagram in Fig. 3.

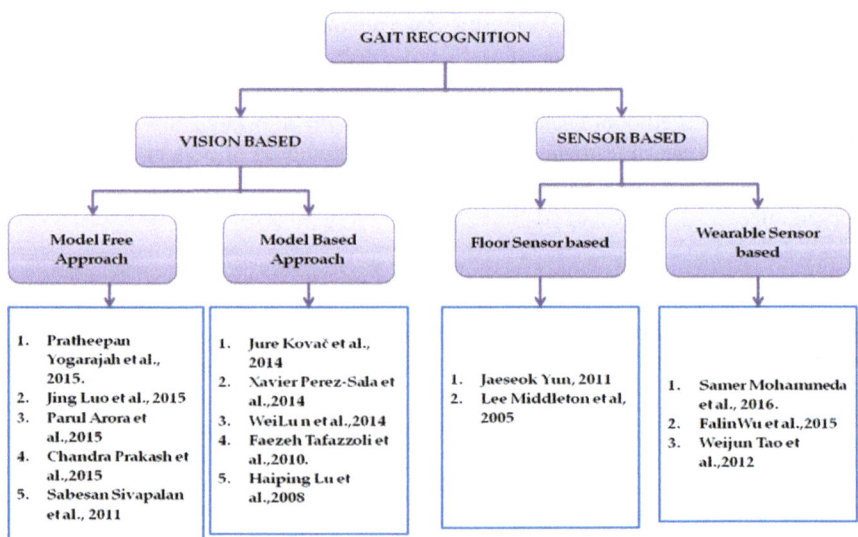

Fig. 3 Gait recognition methodologies [2]

Fig. 4 Example silhouettes from NLPR database [3]

2.1 Model-Free Approaches

Model-free approach is otherwise called as an appearance-based approach where the features are considered from the silhouette of the subject. Gait signatures are obtained using the shape or silhouette of the subject to estimate the human motion. Gait recognition is the more suitable biometric cue for person identification because it is usable even with low-resolution images. Figure 4 shows the examples of silhouettes from NLPR database.

The authors [4] proposed gait Gaussian image (GGI) as a spatiotemporal feature for human gait recognition. Gait Gaussian image is a feature of the gait image over a gait cycle. This work explains about the gait period calculation, where gait cycle is defined as one complete cycle, which is the repetitive pattern of the steps and strides. Gait cycle [5] starts from the flat position to right foot forward and flat position to left foot forward and again to the flat position. The gait cycle period has been calculated only from the lower half of the image as it has dynamic information. The number of foreground pixels has been counted to measure the gait cycle. If the count of the pixels is high, then it is considered as the legs are far apart, and otherwise, the legs are overlapped. So the count of each pixel in the frame gives a periodical signal. The GGI features are extracted, and the classification is done using nearest neighbor method. In order to deal with nonlinearities, the authors introduced fuzzy logic concepts. The authors carried out experiments using CASIA-B and Soton datasets.

In this work [6], authors considered gait energy image (GEI) as a feature to solve the problem of person identification. The gait energy image (GEI) feature can produce better results for gait recognition rate under normal walking conditions. The authors showed that GEI feature is not quite enough to solve the person identification problem. Apart from GEI, there are many appearance-based gait features to identify the covariate conditions. The authors have added dynamic parts to reduce the effect of covariate factors. The joint sparsity model (JSM) is applied to GEI features. The extracted gait features are called as GEIJSM, and the dimensions of GEIJSM are reduced using random projection (RP) technique. The sparse representation is used to classify gait features. From this work, it is known that

RP-based approach and sparse representation-based classification achieve good results than other identification approaches.

The view change is an important issue in the person identification model. Many researches [7–12] were carried out in order to solve the view-invariant challenge. The authors [13] proposed a view transformation model (VTM). The system works well when the training and testing views are similar. However, the gait recognition accuracy decreases whenever the testing views and training views are not similar. The authors proposed an arbitrary VTM (AVTM), in which the gait features are observed from the arbitrary views. In this AVTM model, 3D gait volume sequences are constructed for the training subjects, and the AVTM is trained using gait sequences. Part-dependent view selection scheme (PDVS) is incorporated in the AVTM model. The AVTM_PDVS model divides the gait features into many parts to define the destination view for transformation. The part-dependent view selection can reduce the transformation errors and increases the recognition rate. The datasets were collected, and the experiments were performed in different environments and show that AVTM is efficient for cross-view matching.

2.2   Model-Based Approaches

An increased interest in gait biometric has been used in many applications such as automated recognition systems for surveillance applications and forensic analysis. From many literatures [14–19], it is well known that person can be identified through their walking style, i.e., gait. Yet, many researchers were undergoing the study of person identification using the motion-related gait features. In this model-based approach, initial experiments [20–23] had proven that gait angular measurements that are derived from the joint motions mainly the hip, knee, and ankle angles are the important cues for gait identification. Gait features include the spatial displacement and angular measurements of the human body. Also, the gait features can be analyzed from the video data that are captured at a larger distance (Fig. 5).

A model-based approach provides in-depth details of motion dynamics of human gait in the feature extraction phase. The joint positions are marked, and from the joint positions, the joint angles are measured. The joint angles such as hip, knee, and ankle joint angles are called as gait dynamics. However, model construction is the main thing by applying complex features. This may lead to complex models with many parameters. The five-link biped model [23] is constructed as the subject model using gait features. The lower limbs of the silhouette are modeled as trapezoids, and the upper body is modeled without considering arms. It is very difficult to identify the position of the arm while walking at a long distance. In this model, authors have reduced the complexity by neglecting the arm dynamics. The sagittal plane elevation angles are measured from the certain body parts and the main axis of the body. The biped model M has been constructed with seven degrees of freedom, and it is given as $M = (C = x, y)$ and $\theta = (\theta 1, \theta 2, \theta 3, \theta 4, \theta 5)$, where C is the position of the body center of the image and θ is the orientation vector which

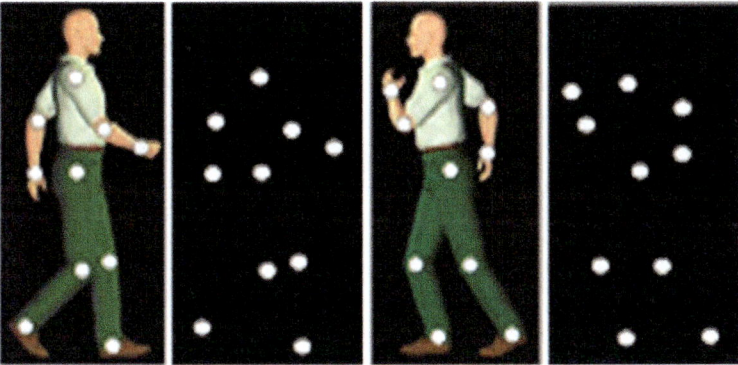

Fig. 5 Marker-based approach—figure reproduced from [Martin A. Giese & Tomaso Poggio, "Neural mechanisms for the recognition of biological movements", Nature Reviews Neuroscience 4, 179–192 (March 2003)]

includes sagittal plane elevation angles (SEAs) for the five body parts. The size of each body part in the images may vary with persons to persons or even the same person's size may vary at different distances from the camera.

The model fitting for each person is complex, and it is manually done using the joint positions on the image. The authors proposed a model fitting method, in which the size of each body part is fitted with the human shape model. The authors [24] proposed a model-based method using elliptic Fourier descriptor. The spatial model templates are described in a parameterized form that is invariant to scale and rotation. A feature selection is applied based on the proximity distance of the neighbors.

2.3 Floor Sensor-Based Approaches

The author [25] designed a new gait recognition system based on the floor sensor. This biometric system uses floor sensor consisting 1536 individual sensors arranged in a rectangular strip. The floor sensor operates at a sample frequency rate of 22 Hz. The gait features extracted are stride length and stride cadence. This information from the floor sensor has been considered as a feature vector. This system performance has been tested on a small database which consists of 15 subjects, and these features proved to be sufficient to achieve an 80 % recognition rate. The authors proposed a new gait feature called as the heel-to-toe, ratio and this feature illustrates the periodical property of the gait (Fig. 6).

The authors [3] proposed a methodology based on the measurements using pressure floor sensor. The footstep information has been extracted from the floor sensors. The fusion of gait information and footstep information has been considered as a feature vector and used for performance comparison. Regarding gait

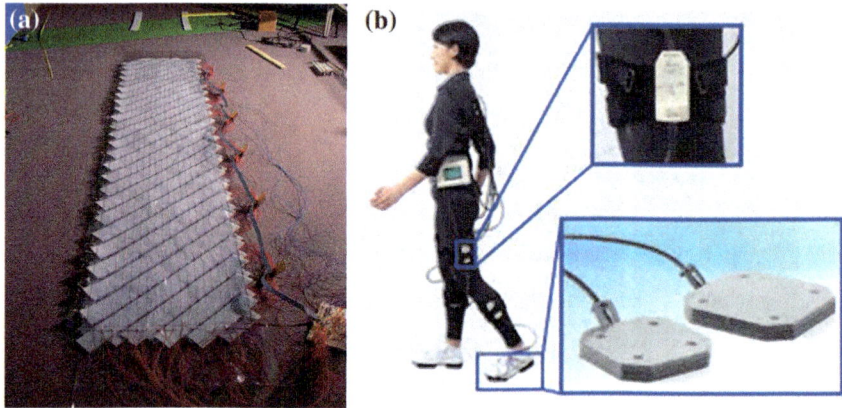

Fig. 6 **a** Floor sensor [25] and **b** wearable sensor [26]

mode, there are two approaches have been followed: enhanced GEI (EGEI) and MPCA for the fusion. The result obtained for gait database is 8.43 % of EER (equal error rate). Always EER value is expected to be lower for the better performance. On the other hand, spatiotemporal information is fused in the footstep mode. The result obtained is 10.7 % of EER. It is known that the performance is not good when the EER value is high. Thus, the performance of the gait system is not good as per the EER value.

A final fusion of the two modes has been performed at the score level and obtained the result of 42.7 %. The fusion result shows better performance compared with the individual performance, with an EER of 4.83 %.

2.4   Wearable Sensor-Based Approaches

At present, many motions' capture system is available for both 2D and 3D analyses. A real-time kinematic gait analysis can be measured in two ways: direct measurement- or contact-based techniques; optical- or non-contact-based techniques. The direct measurements are done using accelerometers and goniometers sensors. The optical-based techniques use active or passive markers for the gait analysis [27]. The marker that emits light by itself is called active marker. It uses light-emitting diode (LED), and the position of the marker is specified by this signal. The advantage of the active marker is identifying its predefined frequencies. Passive markers are the markers that reflect the incident light.

The author [28] presents the largest inertial sensor-based gait database and its application to perform the evaluation of person authentication. They have constructed several datasets for both accelerometer and gyroscope. The inertial measurement units are done using a smartphone which is placed around the waist of a subject. The database has 744 subjects (389 males and 355 females) from 2 to 78 years of age. In order to analyze the gait authentication performance, the number of factors such as gender, age-group, sensor type, ground condition, and sensor location is considered. In this work, the author [29] proposed a new technique with the Kinect Xbox device. It minimizes the segmentation errors, and the automatic background subtraction technique has been followed. The human skeleton model is generated from the background-subtracted images, altered by covariate conditions, such as the change in walking speed and variations in clothing type. The gait signatures are derived from joint angle trajectories of left and right hip and knee of the subject's skeleton model. The Kinect gait data have been compared with the sensor-based biometric setup called as an Intelligent Gait Oscillation Detector (IGOD). The Fisher discriminant analysis has been applied for training, gait signature, and Naïve Bayesian classifier that gives promising classification results on the dataset captured by Kinect sensor.

Most of the gait capture systems use direct measurement techniques to capture the gait parameters. From this information, the healthcare professionals can able to distinguish between normal and abnormal gaits. In the wearable sensor-based method, that subject has to carry sensors and cables. In order to avoid this, the author [15] uses passive marker-based gait analysis system. In the home setting arrangement, there are 5 passive markers and a personal computer. The marker coordinates have been obtained using this arrangement. The gait features such as stride length, gait cycle, cadence, and stride rate are derived using gait algorithms. The time and complexity have been reduced using this marker placement. The effect of the marker is based on the subject's movement. The system provides spatiotemporal and kinematics gait parameters that can help the healthcare professionals. From the error-free skeleton detection and tracking methodologies, the Kinect sensor can be used as a medical assistive tool for the detection of disorder gait.

The future work includes the study of Kinect sensor [30] by increasing the training subjects. It is desired to develop a feature model for selecting the best possible combination of features which are correlated with each other. This prototype may also release difficulties in gait acquisition for physically challenged persons by altering the wearable sensor-based biometric suit with Kinect sensor.

3 Feature Extraction Methods

Feature selection and feature extraction are the two important phases in gait recognition. The features are extracted from the silhouette of the human or the skeleton model of the human. The feature selection techniques have been presented in many works of the literatures [31–34]. The first model-based experiments are done by Johansson [35]. The human motion has been captured based on the movement of the light bulbs that are attached to the body parts of the human. The contour of the human body in each frame has been computed. From the body contour, a stick model is created. The static features such as silhouette height, the

Table 1 Model-based features [37]

Basic features	Kinematic features	Kinetic features
Absolute stance, swing, and double support time	The ankle angle at heel contact and toe off	Max vertical heel contact force
Normalized stance, swing, and double support time	The knee angle is measured at heel contact and toe off	Min vertical mid-stance force
Stride length (m)	Ankle ROM (range of motion) stance, swing, and stance-swing phases	Max vertical push-off force
Walking speed (m/s)	Knee ROM (range of motion) stance, swing and stance-swing phases	Max horizontal heel contact
Cadence (steps/min)		Max horizontal push-off force

distance between left and right foot, and all other distance measurements are calculated from the model-based approach.

Lee and Grimson [36] proposed a model-based gait recognition process. In their work, they have extracted the features from the silhouette images. The feature vector has been composed based on the features extracted such as the aspect ratio and the centroid of the silhouette for recognition. As humans, skeleton structures are proposed in the model-based gait recognition process. Cunado et al. [22] modeled the lower limb of the human body in which the thighs are modeled as pendulums and angular movements are extracted using Fourier series. Yam et al. (2004) designed a model based on the anatomical events that describe the two legs in walking and running. Begg and Kamruzzaman [37] proposed gait recognition by fusing the kinematic and kinetic features. Table 1 shows the features that are involved in model-based approaches. Kinematic features describe the movement of the human body without involving the forces in it. Kinetic features involve the forces acting on the body during movement with respect to the time and forces. Kinematic and kinetic features involve the motion dynamics.

Based on the tracking results from the bipedal link model [23], the authors proposed the space-domain features such as knee stride width, ankle stride width, knee elevation, and ankle elevation. For each of these four features, Discrete Fourier Transform (DFT) is computed. The authors have chosen S_i over the window size of 32 frames close to a typical gait cycle. The DFTs for the feature data reveals the periodic properties and the relative strength of the periodic components. It is known that the zeroth-order frequency component does not provide any information about the periodicity of the signal. Therefore, the authors sampled the magnitude and phase of the second to fifth lowest frequency components. This gives a feature vector consists of 4 magnitudes and 4 phase measures for each of the four space-domain-based features $(S_1,...,S_4)$ leading to an overall dimension of 32 (Fig. 7).

In model-free approaches [40–42], the features considered from silhouette extraction measure the human locomotion. From the literature, it is analyzed that the gait energy image (GEI) is an important feature considered in the model-free

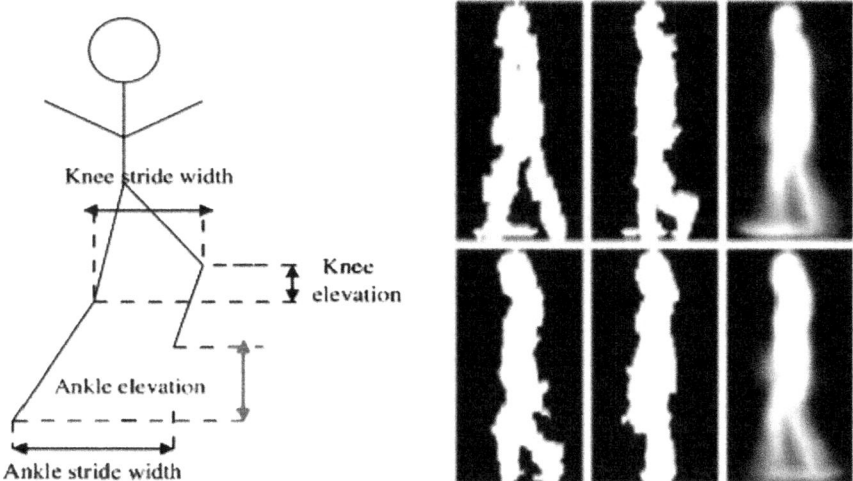

Fig. 7 Model-based features [38] and model-free features—GEI [39]

methods. GEI is a simple energy image used to reflect the gait sequences, and it is calculated using the weighted average method. The gait sequences in a gait cycle are processed to align the binary silhouette. If the gait cycle image sequence is $B(x, y, t)$, gait energy image can be calculated using Eq. (1).

$$G(x, y) = \frac{1}{N} \sum_{t=1}^{N} B(x, y, t) \tag{1}$$

where (x, y, t) is the gait cycle image sequence, N is the number of frames in a gait sequence of a cycle, and t is the number of gait frames. The luminance of the pixels in the GEI figure can indicate the size of the body parts when the person is walking. The white pixel points and the gray pixel points represent the moving parts (slightly and significantly). So, the gait energy image retains the static and dynamic characteristics of human walking, and thus, the computational complexity is reduced. To improve the accuracy of gait recognition, the author [39] proposed the accumulated frame difference energy image (AFDEI). Frame difference energy image is calculated by fusing the forward frame difference image and the backward frame difference image.

Always the lower body part represents the walking gait. Appearance will not vary in the lower limb, while carrying bag and other different movements. Therefore, gait energy-based features are extracted and it has high dimension as they use image pixels or voxels. In order to obtain the complete volume reconstruction, gait energy volume (GEV) [43] is proposed. This GEV requires multiple camera views of a subject. GEV has an advantage over GEI, that is, GEV provides much information than GEI. In this work, the authors implemented GEV by

Table 2 Summary of feature extraction methods from recent researches

References	Features extracted	Challenges' focused
Kovač and Peer [44]	(i) Body part angles: head, neck, torso, left and right thighs, and left and right shins; (ii) joint locations: left and right knees and ankles; (iii) masses and centers of mass (COM) for the whole silhouette and individual body parts; (iv) silhouette width and height	Gait recognition with varying walking speeds
Tafazzoli and Safabakhsh [17]	Joint position movement of leg and arm	Gait recognition in normal walking condition
Lu et al. [45]	Layered deformable model—model-based approach	Gait recognition—various clothing condition and person carrying some objects
Luo et al. [39]	GEI and AFDEI (accumulated frame difference energy image)	Gait recognition—normal waking, wearing, and packaging
Arora et al. [4]	Gait Gaussian image	Gait recognition in normal walking condition
Kusakunniran [46]	Histogram of STIP detectors—gait feature	Normal walking and walking with bag and different clothes
Middleton et al. [47]	Stride lengths, gait period, and heel-to-toe ratio	Gait recognition in the designed sensor
Vera-Rodriguez et al. [25]	EGEI—enhanced gait energy image and MPCA—multilinear principal component analysis	Gait recognition in normal walking

applying GEI for every single view to construct the voxel volumes. The individual views of feature vector are calculated and concatenated to obtain the final feature vector (Table 2).

3.1 Feature Dimensionality Reduction

Principal component analysis (PCA) is one of the familiar feature extraction and dimensionality reduction techniques. The eigenvectors and the covariance matrix of the original inputs are calculated, and then, PCA transforms uncorrelated components in a high-dimensional input vector into a low-dimensional one in the linear way. The authors [48] proposed kernel principal component analysis (KPCA) that generalizes the kernel method into the PCA. The original inputs are mapped in the high-dimensional space using the kernel method, and then, PCA is calculated. Recently, similar to PCA, transformation method called independent component analysis (ICA) is developed [49]. Instead of transforming uncorrelated components, ICA transforms independent components in the transformed vectors. ICA has been generalized for feature extraction.

The authors [7] compared the performance of PCA, KPCA, and ICA for feature extraction in the context of SVM. The different gait features are extracted in PCA, KPCA, and ICA using different algorithms. The new features are then given as the inputs of SVM, and this will solve the time series forecasting problems. The authors [50] proposed a fuzzy-based PCA (FPCA) as gait recognition algorithm. In this work, gait image is preprocessed and gait energy image is generated using a feature extraction process. The fuzzy components such as eigen values and eigen vectors are obtained using FPCA. The classification is done using nearest neighbor classifier. The experiments have been performed in the CASIA database and obtained 89.7 % correct recognition rate (CRR).

The authors [51] introduced a new technique based on the two stages of principal component analysis (PCA) on motion capture data. The first stage of PCA provides a gait feature in low dimension. The components of this representation correspond to spatiotemporal features of gait. A second stage of PCA captures the trajectory shape within the low-dimensional space during a given gait cycle across different individuals. The projection space of the second stage of PCA has separate clusters with respect to the individual identity and individual type of gait.

4 Learning Methodologies

The selection of classifier or the learning pattern makes more important in the gait recognition system. The previous research [7, 52, 53, 54] mainly focused on learning models in order to recognize the gait patterns. Here, in this work, some of the learning patterns are summarized.

4.1 Discriminative Classifier

From many literatures, several classifiers have been proposed to solve the classification problems. The discriminative classifiers provide simple classification techniques. The authors [55] proposed a method of automatic gait recognition system using discriminative classifier. In this work, they have considered three types of gait features: basic temporal/spatial, kinetic, and kinematic. They also analyzed that support vector machine (SVM) will be the suitable machine learning approach to recognizing gaits. Gaits will vary due to aging. Thus, the gait features are recorded for 12 young and 12 elderly subjects. Twenty-four gait features are extracted based on the three types of gaits. The PEAK motion analysis system and a force platform were used to analyze the gait features during normal walking. The recognition rate was obtained as 91.7 % by using SVM. The classification ability was unaffected in all kernel functions of the SVM. In SVM, the kernel functions are linear, polynomial, radial basis, exponential radial basis, multilayer perceptron, and spline. The authors [55] use the simple K-nearest neighbor (KNN) classifier to show the strength of the proposed feature vector. In the experiments, the leave-one-out rule is followed because of the nature of the data. $K = 1$ is used which means that for every query data point, they considered its nearest neighbor in the training set.

4.2   Hidden Markov Model

The author [56] proposed the distributed classifier which is based on the hierarchical weighted decision. The outputs of Hidden Markov Models (HMM) are applied to angular velocities of foot, shank, and thigh for the recognition of gait phases. Here, 10 healthy subjects are considered and their angular velocities were measured using three uniaxial gyroscope. As already discussed in the literature, after validation of the classifiers with cross–validation, classifiers are compared with the performance measure. The performance metrics such as sensitivity, specificity, and load have been measured for the different combinations of the anatomical segments. Moreover, the performance of the distributed classifier has been measured in terms of mean time and coefficient of variation. When the angular velocity of the foot was processed, then the specificity and sensitivity were obtained as high (>0.98). When the angular velocity of shank and thigh was analyzed, the distributed classifiers reach the acceptable values (>0.95). The proposed distributed classifier can be implemented in the real-time application because of the better performance and less computational load in gait phase recognition. Thus, gait variability in patients is evaluated for the recovery of movement in lower limb joints.

The authors [57] designed a system to recognize the abnormality of gait in the elderly people in the indoor environment. Gait of the person is captured using a single conventional camera for gait analysis. Next to the feature extraction phase, the gait data had to be analyzed for abnormality. In order to achieve this, the authors employed HMM. Intuitively, walking action exhibits quasiperiodicity characteristics, where the motion from one cycle to the next is not perfectly repeated and the number of cycles is not a predefined value. So as to model the walking actions, cyclic HMM (CHMM) [58] is proposed in order to find a variation of left-to-right structure with a return transition from the ending state to the beginning state. For this implementation, the discrete HMM [59] version is used to avoid the quantization distortion created by the vector quantization process.

4.3   Deterministic Learning

Performance of gait recognition is reduced due to the main factor such as view invariant. In this work, the authors [60] presented a new method for view-invariant gait recognition using deterministic learning. The gait feature is the width of the silhouette. This gait feature can extract spatiotemporal characteristics and models the periodic deformation of human gait shape. The gait recognition approach consists of a training phase and a testing phase. The gait features of individuals from various view angles are approximated in the training phase. The approximation has been done using radial basis function (RBF) neural networks. The gait dynamics are stored in constant RBF networks. In order to solve the view-invariant problem, the training phase contains all different kinds of gait dynamics of each individual observed across various views with the prior knowledge of human gait dynamics. The test gait pattern is compared with the set of estimators, and the recognition errors are generated. The similarity between the training gait and the test gait pattern is measured using average L1 norm error. According to the smallest error principle, the similarity between test and training pattern is recognized. The experiments are carried out on the multiview gait databases such as CASIA-B and CMU MoBo to prove the effectiveness of the approach.

The authors [61] presented a new method to recognize cross-speed gait using manifold approach. In this method, the authors considered the walking action and they employed kernel-based radial basis function (RBF) interpolation to fit the manifold. The kernel-based RBF interpolation separates the learned coefficients into an affine and a non-affine component, respectively, which encodes the dynamic and static characteristics of the gait manifold. The authors used non-affine component, and it is represented as cross-speed gait. This is also called as speed invariant gait template (SIGT). They also proposed an enhanced algorithm for reducing the dimension of SIGT called as globality LPP (GLPP). In GLPP, the Laplacian graph of intrasubject and intersubject parts is separately constructed and then combined as a new Laplacian graph.

4.4 Distance-Based Metric Learning

In the literatures, the distance-based metric learning is widely used for recognition purposes. The authors [33] extracted the three different features, namely stride length, stride cadence, and heel-to-toe ratio. However, the heel-to-toe ratio is the new feature and a good measure of identity. This proves that the dynamic features of the foot are an important cue. The three features were combined together to form a feature vector and plotted in 3D space. The individuals are recognized with 80 % accuracy. A feature vector, g, can be formed:

g = (LS, TS, RHT), where $g \ \varepsilon \ \mathbb{R}^3$. LS represents the stride length, TS represents the stride cadence, and RHT represents the heel-to-toe ratio. The analysis has been carried out for all 12 trials. It is noticed that each subject is separated in space. The similarity of the subjects has been identified using a standard distance metric, S_{ij}, of subjects i and j. Euclidean distance has been measured, and the similarity between subject $g1$ and $g2$ is denoted as follows:

$$S_{12} = \|g_1 - g_2\| \tag{2}$$

The authors [39] presented the gait energy-based features which represent the gait cycle. The distance is measured between probe and gallery cycle. Here, in this work, the distance d_{ij} between ith probe cycle and jth gallery cycle is determined by computing the Euclidean distance between the gait cycles' feature vectors (F) as follows:

$$d_{ij} = \sqrt{\sum \|F_i - F_j\|^2} \tag{3}$$

Boulgouris and Chi [62] proposed an algorithm to find the distance between a particular probe and the gallery subject, in which each subject has multiple gait sequences. The distance to the closest gallery cycle from each probe cycle and the distance to the closest probe cycle from each gallery cycle have been measured using the following Eqs. (4) and (5).

$$d^P_{\min i} = \min_j(d_{ij}) \tag{4}$$

$$d^G_{\min j} = \min_i(d_{ij}) \tag{5}$$

The final distance, D, between the probe and gallery sequence is as follows:

$$D = \frac{1}{2}\left((\mathrm{median}(d_{\min}P)) + (\mathrm{median}(d_{\min}G))\right) \tag{6}$$

5 Advantages and Disadvantages of Gait Recognition Techniques

Every system has its own advantages and disadvantages. From the previous works, the following comparison has been done and it is shown in Table 3.

6 Performance Evaluation Metrics

6.1 Frame-Based Metrics

Every system has been evaluated using statistical metrics to measure the performance of the system. The study of these metrics gives more details about the strength and weakness of the system. The performance evaluation metrics [63] are described as follows:

Table 3 Comparison of gait recognition techniques

Techniques	Features	Advantages	Disadvantages
Model-free approaches	Contour, width, projection, silhouette, angle, and speed	Lower computational complexity, easy implementation, and high robustness	The features extracted are noisy, and the structure is often unclear
Model-based approaches	Joint angle and combination of shape and parameters	Lossless representations can handle view angle variations and occlusions	Hard to capture, and accuracy depends on the features considered. High complexity
Wearable sensor-based gait recognition	stride length, step length, and gait phase duration	Transparent analysis—using sensor attached to the body part and monitoring of gait in daily activities	Cost of instrument is high and high complexity
Floor sensor-based gait recognition	Ground reaction force and heel-to-toe ratio	This technique can acquire accurate gait measurements	Floor mat is expensive and indoor measurement

True positive (TP), true negative (TN), false positive (FP), and false negative (FN) are called as frame-based metrics, where TP is for correct identification; TN for correct rejection; FP for incorrect identification; FN for incorrect rejection.

Precision [Positive Predictive value (PPV)]

Precision is defined as the number of true positives relative to the sum of the true positives and the false positives. That is, precision is the fraction of correctly detected items. Precision = TP/(TP + FP)

Recall (Sensitivity or True Positive Rate)

Recall is defined as the number of true positives relative to the sum of the true positives and the false negatives. Recall = TP/(TP + FN).

Specificity (True Negative Rate)

Specificity is defined as the true negatives relative to the sum of the true negatives and the false positive.

$$TNR = TN/(TN + FP).$$

$$\text{False Positive Rate (FPR)} = 1 - \text{specificity} = FP/(FP + TN)$$

$$\text{False Negative Rate} = 1 - \text{sensitivity} = FN/(TP + FN)$$

$$\text{Negative Predictive Value (NPV)} = TN/(TN + FN)$$

Precision and Recall Curve

The system evaluation can be done using precision and recall measures based on the correct identities generated. The precision and recall curve identifies the system that maximizes recall for a given precision.

F-Score

The F-score can be used as a single measure of performance test for the positive class. The F-score is the harmonic mean of precision and recall and is given as follows:

$$F - Score = 2^*(Recall^*Precision)/(Recall + Precision)$$

Accuracy

Accuracy is defined as the sum of the true positives and the true negatives relative to the total number of ground truth objects. This is a measure of the actual performance of the system with regard to both correctly detecting and correctly rejecting target identity.

$$Accuracy = (TP + TN)/(TP + TN + FP + FN)$$

6.2 Gait Recognition Performance Measures

Some of the metrics play a major role in measuring performance of the gait recognition system. The extracted gait features has been used in the experiments. The signal sequence was divided into the probe and gallery data equally for the personal authentication scenario. ROC, CMC, DET, CCR, FAR, FRR, EER, and occlusion success rate are some of the important metrics [63] that are defined below:

ROC (Receiver Operator Characteristic) Curve

Gait recognition performance is evaluated using receiver operating characteristic (ROC) curve. The ROC curve denotes a trade-off curve between a false acceptance rate (FAR) and false rejection rate (FRR) in x- and y-axes, respectively, when the acceptance threshold is varied by a receiver.

CMC (Cumulative Match Curve)

The cumulative match curve (CMC) is used to measure the performance of the identification system. CMC curve ranks the person identities that are enrolled in the database with respect to testing image. The CMC curve is dependent on the training size.

Detection Error Trade-off Curve (DET curve)

The DET curve is the modified ROC curve that plots the error rates across the range of operating points. False negative rate (FNR) is in x-axis and false positive rate (FPR) is in y-axis.

Correct Classification Rate (CCR)

CCR is defined as the rate at which the correct identity is done by the system.

Failure to Capture Rate (FCR)

The failure to capture rate is the measure to evaluate the capacity of the device to properly to read the gait characteristics. The failure in the capture device is called as capture error.

False Acceptance Rate (FAR)

The false acceptance rate is defined as the probability that the system incorrectly identifies a non-authorized person, due to incorrect matching of the gait input with a template. The FAR is normally expressed as a percentage of invalid inputs which are incorrectly accepted.

False Rejection Rate (FRR)

The false rejection rate (FRR) is the probability that the system incorrectly rejects access to an authorized person, failing to match the biometric input with a template. The FRR is normally expressed as a percentage of valid inputs which are incorrectly rejected. Also, FRR might increase due to environmental conditions or incorrect gait sequences.

Equal Error Rate (EER) or Crossover Error Rate (CER)

The EER is defined as the performance measure of the system against another system. If the EER is low, the performance of the system is good.

Mean Time to Detect (MTTD)

MTTD is defined as the mean time to detect the identity. It is also defined as time taken to capture the biometrics, process the sample, create a template, compare the reference templates, and generate the decision.

Identification Rate (IDR)

The IDR is defined as the rate at which a biometric subject in the database is correctly identified.

Occlusion Success Rate (OSR)

Occlusion success rate is defined as the proportion of number of successful occlusions to the total number of occlusions.

OSR = Number of successful occlusions/total number of occlusions.

Reliability (R)

Reliability is a measure used to characterize the identification systems.

$$R = 1 - FRR$$

7   Gait Datasets

USF Gait Dataset

The gait data were collected at the University of South Florida, Tampa. The USF dataset [64] consists of about 33 subjects, and it can be partitioned into 32 subsets, based on the various combinations of the covariate factors such as surface type, shoe type, viewpoint, carrying condition, and time. The largest subset is chosen as the gallery set (G, A, R, NB, t_1) (i.e., grass, shoe type A, right camera, no briefcase, and time t_1). The Rest of the subsets form the testing set of various covariates.

CASIA Gait Dataset

The CASIA dataset [3] consists of Dataset A, Dataset B, Dataset C, and Dataset D. Dataset A formerly called as NLPR database. Dataset A was created with 20 persons. Each person has 12 image sequences, 4 sequences for each of the three directions such as parallel, 45°, and 90° to the image plane. The length of each sequence will vary as there is variation in the walker's speed, but it must ranges from 37 to 127. The size of Dataset A is about 2.2 GB, and the database includes 19,139 images. Dataset B is the multiview gait database and consists of 124 subjects from 11 views. Three different scenarios are considered in the database (view angle, clothing, and carrying condition changes). Dataset C was collected by an infrared (thermal) camera, and it is captured in the nighttime. It contains 153 subjects and considers four walking conditions: normal walking, slow walking, fast walking, and normal walking with a bag. Dataset D contains 88 subjects, the videos are collected synchronously, and it is the collection of real surveillance scenes in the indoor environment (Fig. 8).

(a)　　　　　　　　**(b)**　　　　　　　　**(c)**

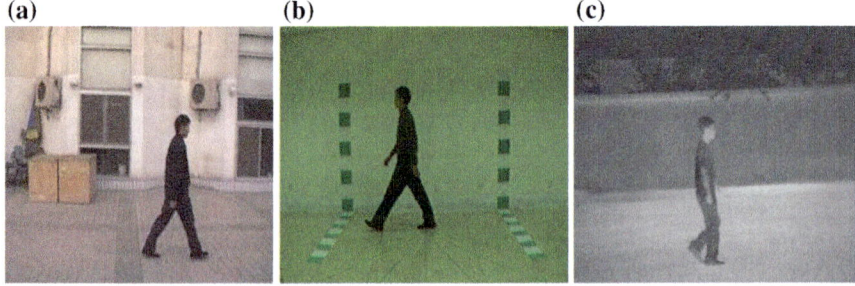

Fig. 8 Sample frames from CASIA Datasets A, B, and C [3]

TUM-IIT KGP Gait Dataset

The database includes static and dynamic occlusions.

(1) (i) Dynamic occlusions are caused by people walking in the line of sight of the camera and (ii) static occlusions are caused by people who are occluding the person of interest by standing in the scene.
(2) TUM-IITKGP dataset [65] contains the videos of new scenarios such as normal walking, person carrying bag, and person wearing long coat.

CMU MoCAP Dataset

The CMU dataset [66] contains video of 25 individuals with 824 gait cycles, and subjects are walking on a treadmill under four different conditions: slow walk, fast walk, incline walk, and walking with a ball in hand.

Southampton Gait dataset

The Southampton Human ID [67] at a distance gait database consists of two major segments—a large population and a small population database. The database consists of videos related to the same subject in various common conditions (carrying items and wearing different clothing or footwear).

8 Research Issues in Gait Recognition

The research on gait recognition is encouraging, and there are some factors to drop down the performance rate of the system. Some of the drop-down factors are grouped together as the issues or the challenges in the biometric security system. The factors can be grouped as external factors and internal factors that affect the recognition rate of the system.

- External factors: the factors that are externally affect the recognition approach and that are referred as challenges to the gait recognition

For example, view-invariant gait recognition—various viewing angles such as frontal view and side view, different lighting conditions such as day and night, different environments, i.e., outdoor/indoor, clothing style such as skirts or long coats, various walking surface conditions such as hard/soft, dry/wet grass/concrete, level/stairs, different shoe types such as boots, sandals, and person carrying objects such as backpack, briefcase, and so on.

- Internal factors: the factors that are internally affect the recognition approach and that are referred as challenges to the gait recognition.

For example, the natural gait of the person may get changed due to foot injury, lower limb disorder, Parkinson's disease, or other physiological changes in the body due to aging, drunkenness, pregnancy, weight gaining or losing, and so on.

Some of the external factors may have serious effects on different gait recognition approaches. For example, while carrying an object may influence the dynamics of gait, in both sensor-based and vision-based categories, it may also create additional difficulties in vision-based category during human silhouette extraction. These are the external and internal factors that affect the performance of the system. These research issues have to be considered to design a robust gait recognition system.

9 Applications and Future Scope

Video surveillance system is the most recent and important research in the smart environment. In order to enhance safety and privacy protection, surveillance cameras [68] are deployed in all commercial and public places such as airports, banks, parks, ATMs, and retail and business intelligence analysis. The widespread use of this technology demanded new functionalities [69–71] to become more smarter surveillance systems. The digital watermarked video recording facilities are used to improve the authenticity [72–74] of the provided video data. To realize more robust and smart environment system, an automated gait analysis and tracking system are needed which can analyze the video streams fed by several surveillance cameras in real time. The primary advantage of the gait recognition is based on its application domain.

The main application includes person identification and authentication in video surveillance applications. In remote monitoring applications, gait plays a major role in identification of the individual person in different situations. In healthcare applications, gait recognition becomes important in identifying the normal and abnormal gaits. Normal and abnormal gaits have been identified using many neurological examinations. In forensic applications, person authentication plays a major role in identifying the criminals. Using gait-based methodologies, students can be identified in classrooms, in shopping malls, and in smart environments. In apartments, unwanted person entering can be avoided using the gait-based security system. This system may be a boon for security applications, remote monitoring, etc., to reduce the risk involved in the incidents of burglaries, unattended, emergency situations, which could enhance social stability. In future, the study of internal factors that cause disturbance to the natural gait is explored and the abnormal gait is investigated. Instead of neurological examinations and without any clinical impairment, the system has to identify the person as normal or the person suffering from neurological disorder from the videos.

10   Conclusion

A study on security system using gait recognition approach has been presented, which is significantly different from other biometric researches. The security was based on analyzing the gait signals of the person which are collected using videos or sensors. Gait recognition methodologies with respect to person identification have been described. Feature extraction and learning models play a main role in gait recognition. Feature extraction in model-based and model-free methodologies has been studied. Static features and dynamic features are analyzed to identify the person and improve the recognition rate. In sensor-based gait analysis, acceleration signals from ankle, hip, and arm were considered as features. The analyzed features are in large dimensions. In order to reduce the dimension of the gait features, feature dimensionality reduction techniques are described. Furthermore, learning the gait patterns using SVM, HMM, manifold, and distance-based learning has been discussed. The internal and external factors that affect the performance of the gait recognition have been discussed. In addition, the impact of gait recognition technology in many application domains has been analyzed. In particular, this study revealed the complete gait recognition system with respect to person identification.

References

1. Wang J, She M, Nahavandi S, Kouzani A (2010) A review of vision-based gait recognition methods for human identification. In: International conference on digital image computing: techniques and applications (DICTA), pp 320–327
2. Yun J (2011) User identification using gait patterns on UbiFloorII. Sensors 11:2611–2639. doi:10.3390/s110302611
3. Zheng S, Huang K, Tan T, Tao D (2012) A cascade fusion scheme for gait and cumulative foot pressure image recognition. Pattern Recognit 45:3603–3610
4. Arora P, Srivastava S (2015) Gait recognition using gait Gaussian image. In: 2nd international conference on signal processing and integrated networks (SPIN), pp 791–794
5. Mohammed S, Saméb A, Oukhellou L, Kong K, Huoa W, Amirat Y (2016) Recognition of gait cycle phases using wearable sensors. Robot Auton Syst 75:50–59
6. Yogarajah P, Chaurasia P, Condell J, Prasad G (2015) Enhancing gait based person identification using joint sparsity model and '1-norm minimization. Inf Sci 308:3–22
7. Xing X, Wang K, Yan T, Lv Z (2016) Complete canonical correlation analysis with application to multi-view gait recognition. Pattern Recognit 50.107–117
8. Muramatsu D, Shiraishi A, Makihara Y, Uddin MZ, Yagi Y (2015) Gait-based person recognition using arbitrary view transformation model. IEEE Trans Image Process 24:1
9. Choudhury SD, Tjahjadi T (2015) Robust view-invariant multi scale gait recognition. Pattern Recognit 48:798–811
10. CC Charalambous, AA Bharath (2015) Viewing angle effect on gait recognition using joint kinematics. In: Sixth international conference on imaging for crime prevention and detection (ICDP-15), pp 1–6
11. Zheng S, Zhang J, Huang K, He R, Tan T (2012) Robust view transformation model for gait recognition 7(2):22–26

12. Burhan IM, Nordin MJ (2015) Multi-view gait recognition using Enhanced gait energy image and radon transform techniques. Asian J Appl Sci 8(2):138–148

13. Zhao X, Jiang Y, Stathaki T, Zhang H (2016) Gait recognition method for arbitrary straight walking paths using appearance conversion machine. Neurocomputing 173:530–540

14. Nandy A, Chakraborty P (2015) A new paradigm of human gait analysis with kinect. In: IEEE eight international conference on contemporary computing, pp 443–448

15. Prakash C, Mittal A, Kumar R, Mittal N (2015) Identification of spatio-temporal and kinematics parameters for 2-D optical gait analysis system using passive markers. In: IEEE international conference on computer engineering and applications, pp 143–149

16. Perez-Sala X, Escalera S, Angulo C, Gonzàlez J (2014) A survey on model based approaches for 2D and 3D visual human pose recovery. Sensors 14:4189–4210. doi:10.3390/s140304189

17. Tafazzoli F, Safabakhsh R (2010) Model-based human gait recognition using leg and arm movements. Eng Appl Artif Intell 23(2010):1237–1246

18. Lu W, Zong W, Xing W, Bao E (2014) Gait recognition based on joint distribution of motion angles. J Vis Lang Comput 25(6):754–763

19. Yam CY, Nixon MS, Carter JN (2004) Automated person recognition by walking and running via model-based approaches. Pattern Recognit 37:1057–1072

20. Ioannidis D, Tzovaras D, Damousis IG, Argyropoulos S, Moustakas K (2007) Gait recognition using compact feature extraction transforms and depth information. IEEE Trans Inf Forensics Secur 2(3):623

21. Zhang R, Vogler C, Metaxas D (2004) Human gait recognition. In: Proceedings of conference on computer vision and pattern recognition workshop. doi:10.1109/CVPR.2004.87

22. Cunado D, Nixon MS, Carter JN (2003) Automatic extraction and description of human gait models for recognition purposes. Comput Vis Image Underst 90(1):1–41

23. Zhang R, Vogler C, Metaxas D (2007) Human gait recognition at sagittal plane. Image Vis Comput 25:321–330

24. Bouchrika I (2015) Parametric elliptic Fourier descriptors for automated extraction of gait features for people identification. In: 12th international symposium on programming and systems (ISPS), pp 1–7

25. Vera-Rodrigueza R, Fierreza J, Masonb JSD, Ortega-Garciaa J (2013) A novel approach of gait recognition through fusion with footstep information. In: IEEE international conference on biometrics, pp 1–6

26. Gafurov D, Snekkenes E (2009) Gait recognition using wearable motion recording sensors. EURASIP J Adv Signal Process Article ID 415817. doi:10.1155/2009/415817

27. Tao W, Liu T, Zheng R, Feng H (2012) Gait analysis using wearable sensors. Sensors 12:2255–2283. doi:10.3390/s120202255

28. Ngo TT, Makihara Y, Nagahara H, Mukaigawa Y, Yagi Y (2014) The largest inertial sensor-based gait database and performance evaluation of gait-based personal authentication. Pattern Recognit 47:228–237

29. Sprager S, Juric MB (2015) Inertial sensor-based gait recognition: a review. Sensors 15:22089–22127. doi:10.3390/s150922089

30. Wu F, Zhao H, Zhao Y, Zhong H (2015) Development of a wearable-sensor-based fall detection system. Int J Telemed Appl, Article ID 576364. doi:10.1155/2015/576364

31. Yan Z, Wang Z, Xie H (2008) The application of mutual information-based feature selection and fuzzy LS-SVM-based classifier in motion classification. Comput Methods Programs Biomed 9:275–284

32. Moustakas K, Tzovaras D, Stavropoulos G (2010) Gait recognition using geometric features and soft biometrics. IEEE Signal Process Lett 17(4):367–370

33. Lam THW, Lee RST, Zhang D (2007) Human gait recognition by the fusion of motion and static spatio-temporal templates. Pattern Recognit 40:2563–2573

34. Tafazzoli F, Bebis G, Louis S, Hussain M (2015) Genetic feature selection for gait recognition. J Electron Imaging 24(1):013036. doi:10.1117/1.JEI.24.1.013036

35. Johansson G (1973) Visual perception of biological motion and a model for its analysis. Percept Psychophys 14(2):201–211

36. Lee L, Grimson WEL (2002) Gait analysis for recognition and classification. In: IEEE conference on face and gesture recognition, pp 155–161
37. Begg R, Kamruzzaman J (2005) A machine learning approach for automated recognition of movement patterns using basic, kinetic and kinematic gait data. J Biomech 38(3):401–408
38. Xiao F, Hua P, Jin L, Bin Z (2010) Human gait recognition based on skeletons. In: International conference on educational and information technology (ICEIT 2010), pp 1–5
39. Luo J, Zhang J, Zi C, Niu Y, Tian H, Xiu C (2015) Gait recognition using GEI and AFDEI. Int J Opt Article ID 763908
40. Choudhury SD, Tjahjadi T (2012) Silhouette-based gait recognition using procrustes shape analysis and elliptic Fourier descriptors. Pattern Recogn 45:3414–3426
41. BenAbdelkader C, Cutler R, Davis L (2004) Gait recognition using image self-similarity. EURASIP J Appl Signal Process 4:572–585
42. Choudhury SD, Tjahjadi T (2013) Gait recognition based on shape and motion analysis of silhouette contours. Comput Vis Image Underst 117(12):1770–1785
43. Sivapalan S, Chen D, Denman S, Sridharan S, Fookes C (2011) Gait energy volumes and frontal gait recognition using depth images. In: IEEE international joint conference on biometrics, pp 1–6
44. Kovač J, Peer P (2014) Human skeleton model based dynamic features for walking speed invariant gait recognition. Math Probl Eng Article ID 484320
45. Lu H, Plataniotis KN, Venetsanopoulos AN (2008) A full-body layered deformable model for automatic model-based gait recognition. EURASIP J Adv Signal Process Article ID 261317. doi:10.1155/2008/261317
46. Kusakunniran W (2014) Attribute-based learning for gait recognition using spatio-temporal interest points. Image Vis Comput 32(12):1117–1126
47. Middleton L, Buss AA, Bazin AA, Nixon MS (2005) A floor sensor system for gait recognition. Fourth IEEE workshop on automatic identification advanced technologies, pp 171–176
48. Tafazzoli F, Bebis G, Louis S, Hussain M (2014) Improving human gait recognition using feature selection. In: Bebis G et al (eds) ISVC 2014, part II, LNCS vol 8888, pp 830–840
49. Jiwen Lu, Zhang E (2007) Gait recognition for human identification based on ICA and fuzzy SVM through multiple views fusion. Pattern Recognit Lett 28:2401–2411
50. Narasimhulu GV, Jilani SAK (2012) Fuzzy principal component analysis based gait recognition. Int J Comput Sci Inf Technol 3(3):4015–4020
51. Das SR, Wilson RC, Lazarewicz MT, Finkel LH (2006) Two-stage PCA extracts spatiotemporal features for gait recognition. J Multimed 1(5):9–17
52. Luo C, Xu W, Zhu C (2015) Robust gait recognition based on partitioning and canonical correlation analysis. IEEE
53. D Skoda, P Kutilek, V Socha, J Schlenker, A Ste, J Kalina (2015) The estimation of the joint angles of upper limb during walking using fuzzy logic system and relation maps. In: IEEE 13th international symposium on applied machine intelligence and informatics
54. Fazli S, Askarifar H, Tavassoli MJ (2011) Gait recognition using SVM and LDA. In: International conference on advances in computing, control, and telecommunication technologies, pp 106–109
55. Libin DU, Wenxin SHAO (2011) An algorithm of gait recognition based on support vector machine. J Comput Inf Syst 7(13):4710–4715
56. Taborri J, Rossi S, Palermo E, Patanè F, Cappa P (2014) A novel HMM distributed classifier for the detection of gait phases by means of a wearable inertial sensor network. Sensors 14:16212–16234. doi:10.3390/s140916212
57. Chen C, Liang J, Zhao H, Hu H, Tian J (2009) Factorial HMM and parallel HMM for gait recognition. IEEE Trans Syst Man Cybern Part C Appl Rev 39(1):114–123
58. Hai HX, Thuc HLU (2015) Cyclic HMM-based method for pathological gait recognition from side view gait video. Int J Adv Res Comput Eng Technol 4(5):2171–2176

59. Kale A, Rajagopalan AN, Cuntoor N, Krüger V (2002) Gait-based recognition of humans using continuous HMMs. In: Fifth IEEE international conference on automatic face and gesture recognition, pp 336–341
60. Zeng W, Wang C (2016) View-invariant gait recognition via deterministic learning. Neurocomputing 175:324–335
61. Huang S, Elgammal A, Lu J, Yang D (2015) Cross-speed gait recognition using speed-invariant gait templates and globality-locality preserving projections. IEEE Trans Inf Forensics Secur 10(10):2071
62. Boulgouris NV, Chi ZX (2007) Gait recognition using radon transform and linear discriminant analysis. IEEE Trans Image Process 16(3):731–740
63. Okumura M, Iwama H, Makihara Y, Yagi Y (2010) Performance evaluation of vision-based gait recognition using a very large-scale gait database. In: Proceedings of the fourth IEEE international conference on biometrics: theory applications and systems (BTAS), pp 1–6. doi:10.1109/BTAS.2010.5634525
64. Sarkar S, Jonathon Phillips P, Liu Z, Robledo I, Grother P, Bowyer KW (2005) The human ID gait challenge problem: data sets, performance, and analysis. IEEE Trans Pattern Anal Mach Intell 27(2):162–177
65. Hofmann M, Sural S, Rigoll G (2011) Gait recognition in the presence of occlusion: a new dataset and baseline algorithms. In: 19th international conference on computer graphics, visualization and computer vision, pp 99–104
66. Gross R, Shi J (2001) The CMU motion of body (MoBo) database. Tech report CMU-RI-TR-01-18, Robotics Institute, Carnegie Mellon University
67. http://www.gait.ecs.soton.ac.uk/
68. Ran Y, Zheng Q, Chellappa R, Thomas M (2010) Applications of a simple characterization of human gait in surveillance. IEEE Trans Syst Man Cybern Part B Cybern 40(4):1009–1020
69. Dey N, Samanta S, Yang XS, Das A, Chaudhuri SS (2013) Optimisation of scaling factors in electrocardiogram signal watermarking using cuckoo search. Int J Bio Inspir Comput 5 (5):315–326
70. Dey N, Mukhopadhyay S, Das A, Chaudhuri SS (2012) Analysis of P-QRS-T components modified by blind watermarking technique within the electrocardiogram signal for authentication in wireless telecardiology using DWT. Int J Image Graph Signal Process 4 (7):33
71. Dey N, Pal M, Das A (2012) A session based watermarking technique within the NROI of retinal fundus images for authentication using DWT, spread spectrum and Harris corner detection. Int J Mod Eng Res 2(3):749–757
72. Pal AK, Das P, Dey N (2013) Odd–even embedding scheme based modified reversible watermarking technique using Blueprint. arXiv preprint arXiv:1303.5972
73. Dey N, Dey M, Mahata SK, Das A, Chaudhuri SS (2015) Tamper detection of electrocardiographic signal using watermarked bio–hash code in wireless cardiology. Int J Signal Imaging Syst Eng 8(1–2):46–58
74. Acharjee S, Chakraborty S, Samanta S, Azar AT, Hassanien AE, Dey N (2014) Highly secured multilayered motion vector watermarking. In: Advanced machine learning technologies and applications. Springer International Publishing, pp 121–134

Face Recognition Under Dry and Wet Face Conditions

K. Dharavath, F.A. Talukdar, R.H. Laskar and N. Dey

Abstract Research fraternities in face recognition were successful in addressing intraclass (intra-personal) and interclass (inter-personal) variations effectively. However, none of the works has made an attempt to address a special case of traditional face recognition, i.e., face recognition under dry and wet face conditions. Here, the gallery face is supposed to be dry face and the probe is a wet face, which comes into picture if the face recognition is employed for an automatic person authentication or in any other intelligent access control application. We name this scenario as wet face recognition (WFR). In such scenario, face remains wet due to several factors including adverse weather conditions such as rain, sweat, high humidity, snow fall, and fog. Essentially, face gets wet due to rain and sweat most commonly. Focus of the current work is to deal with wet image normalization and study its impact on face recognition rate. A framework based on modified bilateral filter is proposed to improve recognition performance in WFR scenario. Sparse representation-based classification over texture features results in impressive recognition performance. Extensive experiments on NITS-DWFDB database demonstrate the efficacy of the proposed wet normalization scheme and WFR framework.

Keywords Face recognition · Wet faces · Bilateral filter · Sparse representation classifier · Fisher criterion · Face database

K. Dharavath (✉) · F.A. Talukdar · R.H. Laskar
Department of ECE, National Institute of Technology Silchar, Silchar,
Assam 788010, India
e-mail: dkrishna@ymail.com

F.A. Talukdar
e-mail: fatalukdar@gmail.com

R.H. Laskar
e-mail: rabul18@gmail.com

N. Dey
Department of IT, Techno India College of Technology, Kolkata, India
e-mail: neelanjandey@gmail.com

© Springer International Publishing Switzerland 2017
N. Dey and V. Santhi (eds.), *Intelligent Techniques in Signal Processing for Multimedia Security*, Studies in Computational Intelligence 660,
DOI 10.1007/978-3-319-44790-2_12

1 Introduction

Expeditious development of face recognition technology in recent years has made it easier for industrial establishments ranging from small scale to large scale to adopt face recognition technology for authentication purposes [1]. However, these systems could perform well only in certain circumstances such as partially controlled environments to fully control. Performance of facial recognition systems in general depends on several factors such as some extreme variations in probe face and adverse weather conditions. Variations in probe image such as pose, expressions, age, and cosmetics and bad weather condition such as illumination changes were well addressed in the literature [2–8]. However, there exists no work to address the issue of wet face recognition. Faces become wet due to many factors such as rain, sweating, snowfall, or by making a face wet necessarily. Image of a wet face often exhibits an illumination changes; therefore, it is being considered as a special case of illumination problem. Numerous techniques have been developed literally to overcome illumination problem [9, 10], but none of them considers the special case, the wet face recognition. This is perhaps because of non-availability of public wet face databases. We believe that this is the first work to address wet face recognition in literature to the best of our knowledge.

To better understand, consider images shown in Fig. 1. It is easy for any person to differentiate wet and dry surfaces by their naked eyes. It is because people retain fundamental cue about wet surfaces. The common behavior of certain rough materials (natural or human made) such as wood, rust, stone, brick, and sand is that they look more specular and darker and exhibit elusive variations in saturation and hue. But look does not change in case a smooth material (such as glass and marble) is wet. For example, the difference between wet and dry bricks or highly polished stones strengthens this statement. Similar is the case with human faces.

Fig. 1 Illustration of dry and wet surface

The look changes when face is wet or sweat (see Fig. 2). Although naked eye cannot observe the difference directly, one can easily notice the illumination difference between dry image and a wet image captured using any digital camera. The difference is that wet faces are specularly brighter which in turn exhibits illumination variation leading to poor matching rates against dry faces. Compared to the particles caused by adverse weather conditions, water droplets are large enough to be visible to the naked eye. In addition, they reflect and refract both environmental lighting and scene radiance which results in a complex spatial and temporal intensity fluctuations in digital images that is random noise. More discussion on vision and rain can be found in [11]. The problem is significant when these images are used for matching purposes. Machines find difficulty in comparing wet image to dry image and vice versa due to intensity fluctuations and give a wrong decision even if they belong to one person.

In general, wet image normalization is a kind of noise removal from a digital image. The same issue can also be considered as the subcategory of image normalization or noise removal. Image-based wet and water droplets removal is a major issue and challenging one. The primary focus here is to design an efficient filter that can abolish structured or unstructured noise caused by various factors from an image. In fact, the structured and unstructured noise appears in an image due to bad image acquisition conditions. Literally, numerous contributions were made by research community to address image de-noising in different aspects. For example, partial difference equations, statistical analysis, transform domain methods, spatial filters, approximation theory methods, and histogram techniques are some of the explored directions [12, 13].

So far the research works on bad weather conditions, essentially noise due to rain and rain streaks found in literature, focused on removing rain streaks or water droplets from videos. Most of the methods utilize the temporal correlation in several video frames consecutively. On the other hand, if wet is present even in the absence of rain, a single image-based wet and water droplets removal technique is required. The current work is on single image-based wet normalization scheme only.

The remainder of the manuscript is organized as follows. In Sect. 2, a brief review of the background behind the employed techniques in proposed framework

Fig. 2 Illustration of dry and wet faces

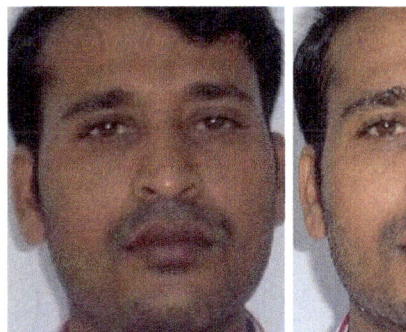

is presented. Initially, we briefly discuss wet normalization schemes followed by robust face recognition methods. Wet normalization schemes and performance analysis of adapted approach are presented in Sect. 3. Technical details of adapted face recognition algorithm and the method of sparse dictionary learning are given in Sect. 4. Section 5 introduces our in-house DWF database and corresponding technical details. Section 6 presents extensive experimental results and analysis, and finally, the work is concluded in Sect. 7.

2 Background Work

Automatic facial recognition is one of the most challenging research topics in the field of pattern recognition and multimedia technology which analyzes the face image and extracts effective information to identify the face image via the computer [1, 2]. It has many applications such as security, authentication, digital surveillance, and fusion, but the change in lighting condition is one of the key factors affecting the performance accuracy of a facial recognition system [14–16]. The ambient illumination most commonly changes with time and locations often. Essentially, face recognition if used for intelligent access control and secure authentications, the adverse weather conditions such as rain, cloudy day, high humidity, fog, and snow make system unreliable. In general, human face gets wet due to rain or sweating or by intentional acts. Matching a probe (wet) face against a gallery (dry) face becomes a challenging task for a recognition system due to the fact that wet on face becomes a noise in image. Pixel intensities vary due to reflection and refraction properties of water while capturing of image. Also, presence of heavy water droplets makes the situation worse. For accurate matching of wet to dry and vice versa, it is essential to normalize the noisy (wet) image or reconstruct the dry image.

The conventional linear filter for nose removal not only performs effectively in smooth areas, but also blurs the image edge structures significantly. Several researchers have made attempt to address edge preserving noise removal. For example, the nonlinear filter introduced by Tomasi and Manduchi [17] combines benefits of both the spatial filter and the range filter and hence is denoted as bilateral filter (BF). A low-pass Gaussian filter is adopted for both the space and range kernels. The Gaussian kernel of spatial filter applies higher weights to pixels adjacent to the center pixel and on the other hand, the Gaussian kernel of range filter applies higher weights to the nearby pixels of the center pixel. At an edge pixel, BF behaves as an edge-oriented elongated Gaussian filter, which ensures proper averaging along the edge and gradient directions (mostly along the edge and lightly in the gradient). This ensures BF to suppress the noise while preserving the edge structures. Figure 3 illustrates the weight computation process on simple input image [17].

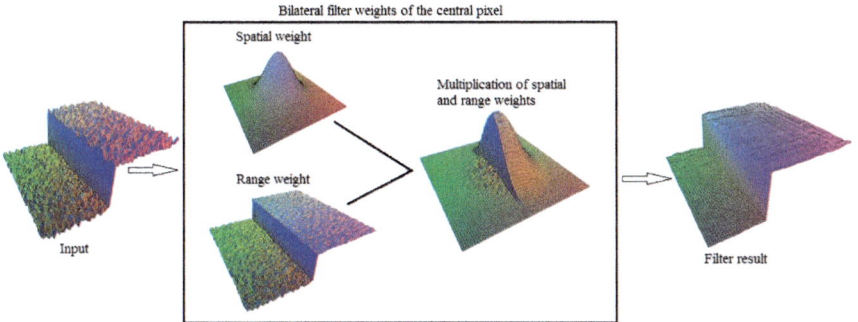

Fig. 3 Illustration of weight computation process in bilateral filter for a central pixel

The bilateral filter proves to be computationally intensive for practical applications if employed directly (as it is). Literature has witnessed several efficient schemes that improved efficiency of the same. Recently, Chaudhury et al. [18] implemented the bilateral filter, using raised-cosine range kernel. Their implementation has demonstrated the efficacy of raised-cosine range kernel. It has exhibited superior approximation quality compared to a Taylor polynomial with equal number of terms.

In case of image sharpening, the unsharp masking (USM) technique [19] is widely applied because of its simplicity. The input-degraded image is high-pass filtered under the guidance of USM which is obtained from the difference image (i.e., by subtracting the test image and its blurred version). This ensures the sharpened output as well as increases contrast along the edges. Unsharp masking technique, however, has two important drawbacks. First, because of addition of undershoot and overshoot to the edges during image sharpening, USM produces halo around the same. Second, it results in a diminished image quality for a given noisy input image as it also amplifies the noise in smooth areas. To overcome the same, Kim et al. [20] presented an optimal unsharp mask (OUM)-based framework to remove the noise in congruent regions, while preserving the equivalent sharpening as USM does. However, OUM has not also been able to overcome halo artifacts completely. In this framework, we adapt the adaptive bilateral filter (ABF) [21] for image sharpening and noise removal due to the fact that ABF outperforms in sharpening degraded image with increase in the slope of edges. It is also able to overcome halo artifacts by not producing any overshoot and undershoot. Unlike USM, it is a most successful and fundamentally different approach to enhance the image sharpness. Compared to bilateral filter, ABF produces significantly sharper images.

The sparse representation-based face recognition is basically based on this hypothesis where all the training images are used to span a face subspace. To solve the issues such as uncontrolled illumination, the pose, and the expression,

Wright et al. [22] presented a sparse representation-based classifier framework which regards the identification problem as classification problem of multiple linear regression models. Firstly, the image is represented according to the compacted sparse; then, the reconstruction error of test samples is computed; finally the minimum residual error is classified and recognized. Figure 4 illustrates the sparse representation-based robust face recognition approach in [22]. The robustness of the algorithm was mainly decided by the sparse dictionary and the method of solving sparse solution. But if the training images are affected by many uncertain factors, the dictionary may become ineffective in representing a probe image.

Yang et al. [23] presented a Gabor features based on sparse representation algorithm. It combines the Gabor features and the SRC algorithm to reduce the computation complexity and improves the human face recognition rate. Ahonen et al. [24] proposed local binary pattern (LBP)-based face recognition with the algorithm's robustness improved rapidly as the LBP histogram is not sensitive to light. However, the coding complexity of the algorithm is increased due to large number of atoms of the dictionary learned. Thus, several dictionary learning methods were proposed in literature.

Compared with the traditional dictionary, the discriminative dictionary contains not only the representational power, but also the discriminative power which is good for object classification by adding a discriminative term into the objective function. Zheng et al. [25] proposed a Fisher discrimination-based dictionary learning method via kernel singular value decomposition (K-SVD) algorithm, whereas Yang et al. [26] proposed an improved dictionary learning method via Fisher discrimination dictionary learning (FDDL) for sparse representation. It can not only improve face recognition rate, but also reduce the computational complexity via applying the Fisher discrimination criterion into sparse representation. The weights here are computed by the help number of samples in each class and the total number of gallery images. In the current work, we employ a face recognition algorithm based on discriminative dictionary learning and sparse representation [27].

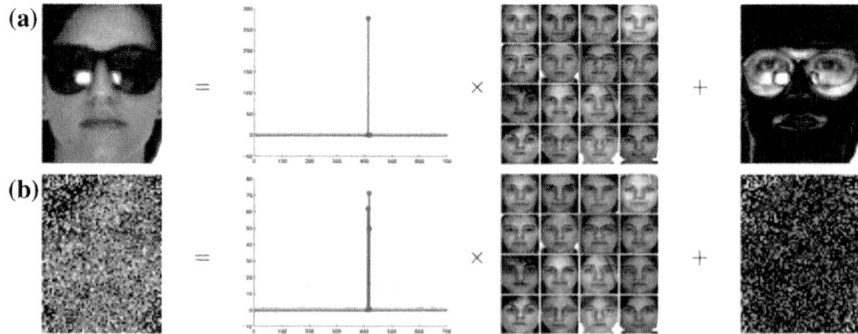

Fig. 4 Face recognition approach presented in [19]. The test image is represented as a sparse linear combination of all the gallery images plus sparse errors. Here test image is occluded (*top*) or corrupted (*bottom*). The algorithm was tested on the standard AR face database

3 Wet Normalization

In this section, initially, bilateral filter with Gaussian kernel is described, followed by wet normalization method employed in the framework; ABF is presented.

3.1 Gaussian Bilateral Filter

Bilateral filter (BF) was first proposed by Tomasi and Manduchi [17] in 1998 as a noniterative method for noise reduction and edge preserving smoothing. A low-pass Gaussian filter is adopted for both the space and range kernels. The Gaussian kernel of spatial filter applies higher weights to pixels adjacent to the center pixel and on the other hand, the Gaussian kernel of range filter applies higher weights to the nearby pixels of the center pixel. A simple and important case of BF is shift invariant Gaussian filtering. Mathematically, it is given by the Eq. 1.

$$\tilde{f}[u,v] = \sum_u \sum_v h[u,v;p,q]i[p,q] \tag{1}$$

where $i[p,q]$ is the degraded image, $\tilde{f}[u,v]$ is the filtered image, and $h[u,v;p,q]$ indicates the response at $[u,v]$ to an impulse at $[p,q]$. The impulse response at $[u_0,v_0]$ is given by Eq. 2.

$$
\begin{aligned}
h[u_0,v_0;u,v] = r_{u_0,v_0}^{-1} \exp &\left(-\frac{(u-u_0)^2 + (v-v_0)^2}{2\sigma_d^2} \right) \\
\times \exp &\left(-\frac{(i[u,v] + i[u_0,v_0])^2}{2\sigma_r^2} \right), [u,v] \epsilon \, \Omega_{u_0 v_0}
\end{aligned}
\tag{2}
$$

where $[u_0,v_0]$ is the center pixel of the window, $\Omega_{u_0 v_0} = \{[u,v] : [u,v] \epsilon [u_0 - N, u_0 + N] \times [v_0 - N, v_0 + N]\}$, σ_r, and σ_d are the standard deviations of the range and domain filters, respectively. r_{u_0,v_0} is the normalization constant and is given by Eq. (3) ensures that filter preserves image energy.

$$
\begin{aligned}
r_{u_0,v_0} = \sum_{u=u_0-N}^{u_0+N} \sum_{v=v_0-N}^{v_0+N} \exp &\left(-\frac{(u-u_0)^2 + (v-v_0)^2}{2\sigma_d^2} \right) \\
\times \exp &\left(-\frac{(i[u,v] + i[u_0,v_0])^2}{2\sigma_r^2} \right)
\end{aligned}
\tag{3}
$$

The bilateral filter combines benefits of both the spatial filter and the range filter. A low-pass Gaussian filter is adopted for both the space and range kernels. The Gaussian kernel of spatial filter applies higher weights to pixels adjacent to the center pixel and on the other hand, the Gaussian kernel of range filter applies higher weights to the nearby pixels of the center pixel. At an edge pixel, BF behaves as an edge-oriented elongated Gaussian filter, which ensures proper averaging along the edge and gradient directions (mostly along the edge and lightly in the gradient). This ensures BF to suppress the noise while preserving the edge structures.

3.2 Adaptive Bilateral Filter

We obtain an adaptive bilateral filter (ABF) from classical bilateral filter by introducing a locally adaptive offset to the range filter [21]. For the ABF, the impulse response at $[u_0, v_0]$ is given by Eq. 4.

$$
h[u_0, v_0; u, v] = r_{u_0, v_0}^{-1} \exp\left(-\frac{(u - u_0)^2 + (v - v_0)^2}{2\sigma_d^2}\right)
$$
$$
\times \exp\left(-\frac{(i[u, v] - i[u_0, v_0] - \eta[u_0, v_0])^2}{2\sigma_r^2}\right), [u, v] \in \Omega_{u_0 v_0}
$$
(4)

where $[u_0, v_0]$ is the center pixel of the window, $\Omega_{u_0 v_0} = \{[u, v] : [u, v] \in [u_0 - N, u_0 + N] \times [v_0 - N, v_0 + N]\}$, σ_r, and σ_d are the standard deviations of the range and domain filters, respectively. r_{u_0, v_0} is the normalization constant and is given by Eq. 5 ensures that filter preserves image energy.

$$
r_{u_0, v_0} = \sum_{u=u_0-N}^{u_0+N} \sum_{v=v_0-N}^{v_0+N} \exp\left(-\frac{(u - u_0)^2 + (v - v_0)^2}{2\sigma_d^2}\right)
$$
$$
\times \exp\left(-\frac{(i[u, v] - i[u_0, v_0] - \eta[u_0, v_0])^2}{2\sigma_r^2}\right)
$$
(5)

Since range filter can be considered as a 1D histogram processing filter, it is introduced with an offset η in the ABF in order to be able to shift the range filter on histogram unlike conventional bilateral filter wherein the range filter is located at current pixel's gray value on the histogram. On the other hand, both σ_r and η (i.e., width of the range filter and offset) are made locally adaptive in order to ensure a more powerful and versatile bilateral filter. Even after these two modifications, ABF preserves the general form of conventional bilateral filter. ABF becomes conventional bilateral filter with $\eta = 0$ and σ_r is fixed. ABF adopted a fixed Gaussian filter with $\sigma_d = 1.0$ for the domain filter. Refer [21] for the detailed description about the role of η and σ_r in ABF.

4 Robust Face Recognition Method

In this section, Gabor filter is discussed in the Sect. 4.1. Description of uniform local binary pattern technique and ULBP dictionary learning are presented in 4.2. SRC-based classification is discussed in last Sect. 4.3. The block diagram of classification process is illustrated in Fig. 5.

4.1 Gabor Wavelet

One of the main features of Gabor filters is that its spatial frequency and orientation representations resemble the cells in human visual system. The response of a simple cell in visual cortex could be approximated using 2D filter. Also spatial and structural characteristics at multiple directions can be extracted from the face images by using Gabor filters because of their efficiency in spatial locality and orientation selectivity. They also have certain tolerance on the variation in rotation, displacement, scaling, deformation, and illumination. Features constructed from the response of Gabor filters (also known as Gabor features) have been successful in image processing and computer vision applications. They are the top-performing features among many others essentially in face recognition technology. In general, the local pieces of information is extracted by using Gabor features and then combined to recognize an object.

Gabor filter in general is a band-pass filter. The impulse response of the same is given by the product of a Gaussian function and a harmonic function. A two-dimensional Gaussian filter thus comprises a sinusoidal plane of spatial frequencies and orientations and is modulated by a Gaussian envelop.

Mathematically, a 2D Gabor filter is represented as in Eq. 6.

$$\mathcal{G}_{\theta_k, f_i, \sigma_x, \sigma_y}(x, y) = \exp\left(-\left[\frac{x_{\theta_k}^2}{\sigma_x^2} + \frac{y_{\theta_k}^2}{\sigma_y^2}\right]\right)\cos(2\pi f_i \theta_k + \varphi) \tag{6}$$

where $x_{\theta k} = x\cos\theta_k + y\sin\theta_k$ and $y_{\theta k} = y\cos\theta_k - x\sin\theta_k$
f_i is the frequency of sinusoidal at an angle θ_k with reference to x-axis. σ_x is the standard deviation (of Gaussian envelope) along the x-axis. Similarly σ_y represents

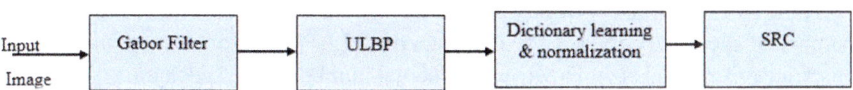

Fig. 5 Flow of classification process

standard deviation along y-axis. We set the phase $\varphi = \frac{\pi}{2}$ and compute each orientation as $\theta_k = \frac{k\pi}{n}$ where $k = 1, 2, 3, \ldots, n$. Equation 6 captures spatial frequency and orientation information from image and is called as the mother wavelet. Each face image is filtered with above Gabor filter at various frequencies and orientations. Mathematical representation of Gabor feature extraction process is given by Eq. 7.

$$\mathcal{O}_{\theta_k, f_i, \sigma_x, \sigma_y}(x, y) = I(x, y) * \mathcal{G}_{\theta_k, f_i, \sigma_x, \sigma_y}(x, y) \tag{7}$$

where $\mathcal{O}_{\theta_k, f_i, \sigma_x, \sigma_y}(x, y)$ is the Gabor representation of image $I(x, y)$ and $\mathcal{G}_{\theta_k, f_i, \sigma_x, \sigma_y}(x, y)$ is the 2D Gabor filter. Alternatively, Eq. 2 can be written as given in Eq. 8.

$$\mathcal{O}_{\theta_k, f_i, \sigma_x, \sigma_y}(I) = \mathbf{FFT}^{-1}[\mathbf{FFT}(I) \cdot \mathbf{FFT}(\mathcal{G})] \tag{8}$$

Here $\mathcal{O}_{\theta_k, f_i, \sigma_x, \sigma_y}(I)$ represents the feature vector of input image.

4.2   Uniform Local Binary Pattern

An efficient texture feature operator, local binary pattern (LBP), was introduced by Ojala et al. in the year 1996 [29]. Because of its computational simplicity and high discriminative power, it has become one of the most popular feature operators. Also it is robust to monotonic grayscale variations occurred. Using LBP operator, the center pixel in an 8-neighborhood is used as threshold value to label every pixel of an image and considers the result as a binary number. With reference to central pixel, a value 1 is assigned to neighboring pixels that has higher gray (intensity) value and a value 0 is assigned if they have lower value. Mathematically, calculation of LBP value is given as in Eq. 9

$$\mathrm{LBP}_{P,R} = \sum_{p=0}^{P-1} s(g_p - g_c)2^p \tag{9}$$

where $s(x) = \begin{cases} 1, & x \geq 0 \\ 0, & \text{Otherwise} \end{cases}$ and g_c and g_p represent the gray level values of center pixel and neighboring pixels, respectively.

The LBP histogram is mostly calculated for each partition of the image, so the number of the binary modes by using standard LBP operator is very large for the block area of normal size. However, the actual number in the block area is relatively little; this means that the histogram is too sparse. For LBP, if the number of transition, such as 0–1 or 1–0, is no more than 2, then the LBP is the uniform LBP (ULBP). We use ULBP pattern in the current work to learn sparse dictionary.

The procedure of dictionary learning

Assuming that the size of an image $f(x)$ is $p \times q$, we can get the Gabor features by convolving it with each of the 40 Gabor filters. Every Gabor image's histogram features are computed according to the order from left to right and from top to bottom. Finally, all subblocks' histogram features are connected to a composite feature vector, that is the ULBP features of the image.

The dictionary of ULBP is denoted as $G = [G_1, G_2, \ldots, G_c]$ where G_i is the subset of the ULBP dictionary from the class i, and c is the total number of classes. We learn a structured dictionary $D = [D_1, D_2, \ldots, D_c]$, where D_i is the class-specified sub-dictionary associated with the class i. Z denotes the coding coefficient matrix of G on D, i.e., $G = ZD$ and $Z = [Z_1, Z_2 \ldots, Z_c]$, where Z_i is the subset of the coding coefficient matrix Z. We use the Fisher criterion to obtain the better dictionary D.

4.3 SRC-Based Classification Algorithm

For a given probe image y, the sparse representation $\bar{x}_1$ is computed as given in Eq. 10

$$(\ell^1) : \quad \bar{x}_1 = \arg \min \|x\|_1 \quad \text{subject to } Dx = y \tag{10}$$

Let $\delta_i : \mathbb{R}^n \rightarrow \mathbb{R}^n$ be the characteristic function for each class i and δ_i selects the coefficients associated with the ith class. For any $x \in \mathbb{R}^n$, $\delta_i(x) \in \mathbb{R}^n$ is a new vector whose only nonzero elements are the elements in x that are associated with class i. Given test sample y can be approximated as $\bar{y}_i = D\delta_i(\bar{x}_1)$ once the coefficients

Table 1 Robust face recognition based on Gabor LBP dictionary learning and sparse representation

Step	Action
1	Convolve the facial image function with the Gabor filter function to get the Gabor images
2	Calculate the ULBP histograms of the 40 Gabor images
3	Gain the better dictionary for sparse representation by dictionary learning
4	Normalized the training dictionary
5	Solve the $l1$-minimization problem $\bar{x}_1 = \arg \min \|x\|_1$ subject to $Dx = y$
6	Compute the residuals $r_i(y) = y - D\delta_i(\bar{x}_1)_2$
7	Identify based on the approximate information $\mathbf{Identify}(y) = \arg \min_i r_i(y)$

associated with the ith class. Test sample y then be classified based on these approximations and by assigning it to the training class which has minimal residual between y and $\bar{y}_i$:

$$\min_i r_i(y) = \|y - D\delta_i(\bar{x}_1)\|_2 \qquad (11)$$

Table 1 presents the summary of the complete face recognition procedure [19].

5 The NITS-DWFDB Database

Automatic facial recognition is one of the most challenging research topics in the field of pattern recognition and multimedia technology which analyzes the face image and extracts effective information to identify the face image via the computer. Along with the development of several face recognition algorithms, literature has witnessed a large number of dry face databases [30, 31]. But, many of them are collected to the specific needs of the corresponding algorithm under development. Existing databases are based on dry faces only. As such, to the best of our knowledge, no standard wet face database is currently available for benchmarking. Therefore, we created a first of its kind database called a dry and wet face (DWF) database at our institute, National Institute of Technology Silchar (NITS), India, in 2014 [32] and is denoted as NITS-DWFDB database.

NITS-DWFDB contains images of 75 individuals. The imaging conditions such as camera parameters, distance, and illumination settings were carefully controlled and the settings were ensured to be identical across subjects by constant recalibration. The resulting face images (RGB Color) are 1024×768 pixels in size. The images are cropped and resized manually to 250×250 pixels to have clear facial features. The subjects were recorded twice at a 2-week interval. During first session, 25–30 conditions with variable facial expressions, pose variations (yaw, pitch, and roll), occlusion, and illumination were captured. Database contains two sets of these variations, wet and dry. There are 25–65 images of each subject in a set and total of 6750 images in the database. Each subject has pose variations (head rotations) in yaw ($\pm90°$, $\pm75°$, $\pm60°$, $\pm45°$, $\pm30°$, $\pm15°$ and $0°$) and pitch ($\pm45°$, $\pm30°$, $\pm15°$ and $0°$), 6–7 random expressions such as smile, angry, disgusting, open mouth, closed eye, and occlusions such as spectacles. Sample images of dry and wet faces are provided in Fig. 6. NITS-DWFDB database includes Indian native faces with different complexions which makes it a unique.

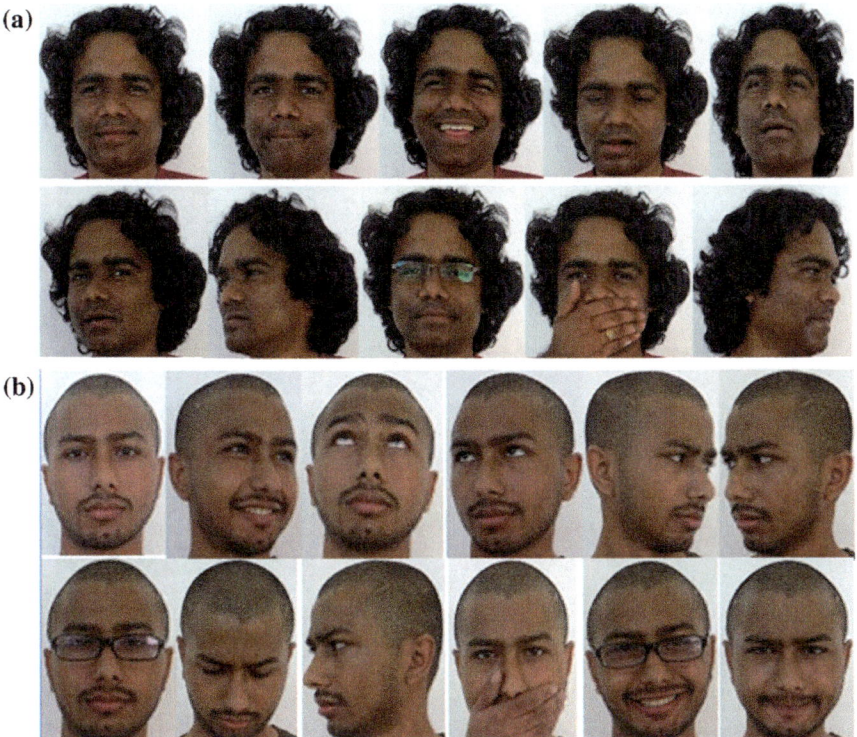

Fig. 6 Sample images of **a** wet and **b** dry faces in NITS-DWFDB database

6 Experimental and Analysis

The experimental procedure and result analysis are presented in this section. Figure 7 illustrates the bilateral filter performance with $\sigma_d = 2$ and $\sigma_r = 20$. Figure 7b shows that bilateral filter removes much of the noise present in the wet image shown in Fig. 7a. Edge structures are also preserved efficiently. From this, one can understand that bilateral filter is able to preserve edge structures while removing noise. The results of the bilateral filtering are a significant improvement over a conventional linear low-pass filter.

Figure 8 illustrates the performance comparison of several photometric normalization techniques presented in [28]. Compared to conventional linear low-pass filters, bilateral filter exhibited significant improvement in results. Excellent visual quality is obtained only with the help of bilateral filter (see Fig. 7) as compared to other methods. However, the result of the bilateral filter confirms that it is essentially an image smoothing filter, as it does not sharpen edges. Therefore, some modifications are necessary in the filter implementation in order to achieve enhanced sharpness.

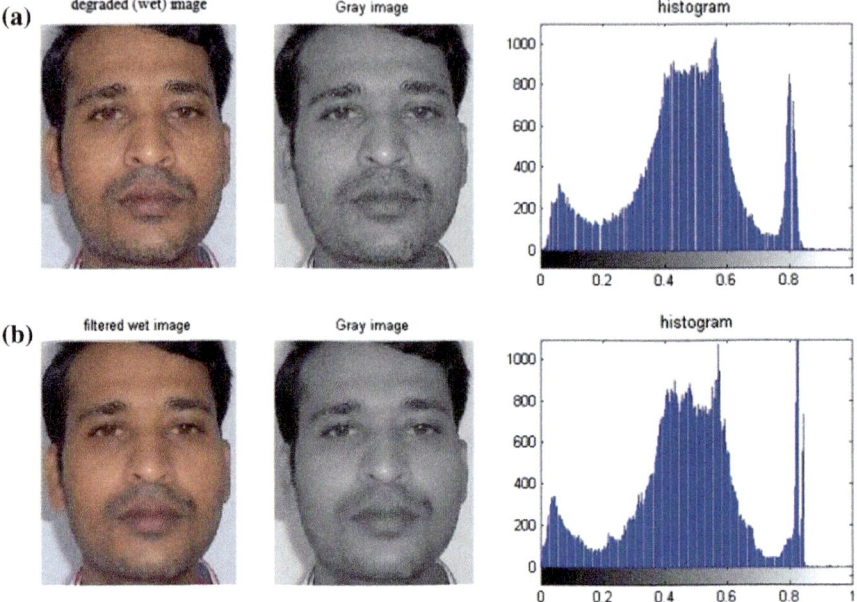

Fig. 7 Illustration of bilateral filter performance **a** wet, **b** filtered image

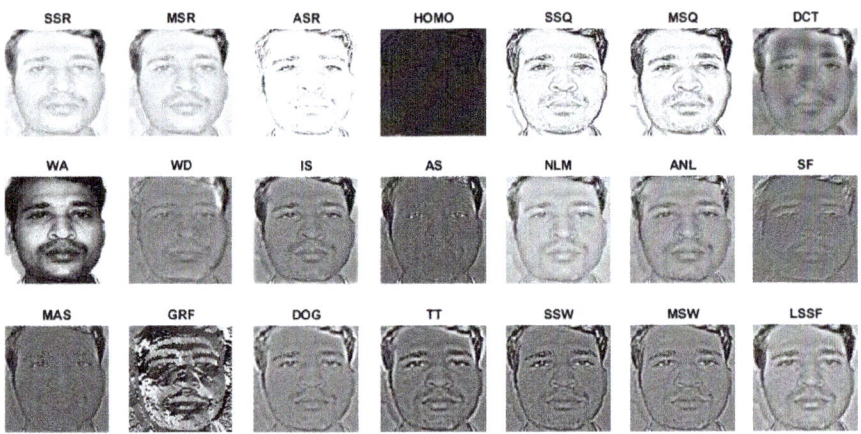

Fig. 8 Performance comparison of photometric normalization techniques (from *left* to *right* and *top* to *bottom*): single scale retinex (SSR), multiscale retinex (MSR), adaptive single scale retinex (ASR), homomorphic filtering (HOMO), single scale self-quotient image (SSQ), multiscale self-quotient image (MSQ), discrete cosine transform (DCT), wavelet-based normalization (WA), wavelet de-noising (WD), isotropic diffusion (IS), anisotropic diffusion (AS), non-local-means (NLM), adaptive non-local-means (ANL), steerable filter (SF), modified anisotropic smoothing (MAS), gradient faces (GRF), difference of Gaussians (DOG), Tan and Triggs (TT), single scale Weber faces (SSW), multiscale Weber faces (MSW), least-squared spectrum fitting (LSSF). Technical details can be found in [28]

Figure 9 demonstrates the performance comparison of ABF against classical BF and its fast version FBF (see [18]). The output of bilateral filter infers that it is able to remove noise in the wet-degraded image, but the edges in the output image are still blurred. The similar is the case even with fast bilateral filter. Though FBF resulted in faster output, it has worsened the visual quality of the output image. Compared to competing filters, the ABF outperforms. Visual quality is improved significantly. It is also able to produce sharper edges/textures and less noisy in smooth regions. As a result, overall quality of the ABF restored image is improved significantly.

Algorithms described in Sects. 3 and 4 are applied on NITS-DWFDB database. In order to demonstrate the effectiveness of the adopted face recognition method, the algorithm is compared with Gabor PCA, Gabor LBP, and SRC face recognition methods. The experimental platform is Intel(R) Core(TM) i5-650CPU, 3.33 GHz frequency, 4.00 GB memory, and Windows 7 Ultimate operating system. MATLAB R2015a version is used for entire experimentation. The face recognition process is shown in Fig. 10. The normalized image was obtained by filtering the test image via

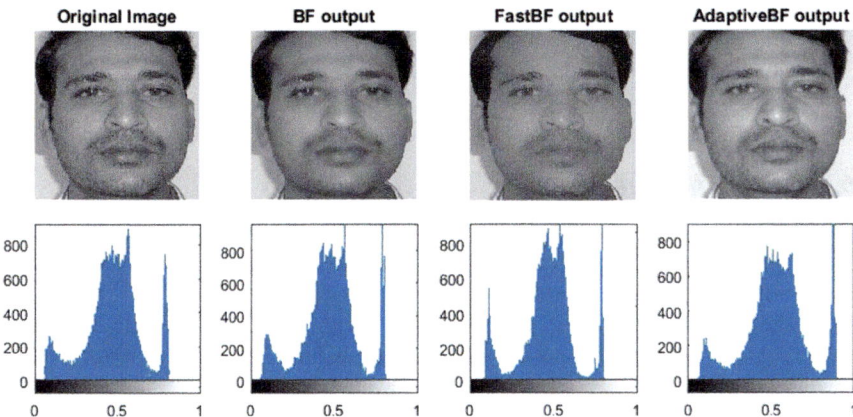

Fig. 9 Illustration of filter outputs. *Upper row* illustrates original wet image, BF output, FBF output, and ABF output, respectively, from *left* to *right* and *lower row* illustrates corresponding histograms

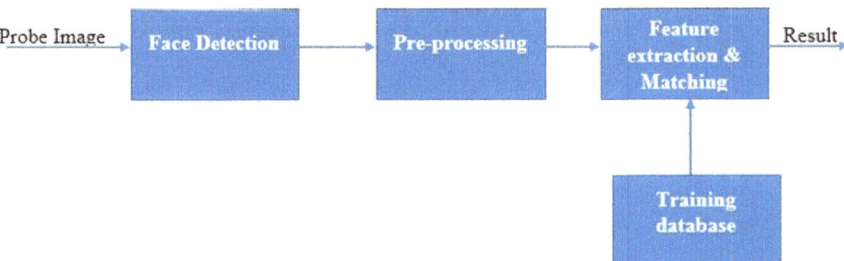

Fig. 10 Block diagram of general face recognition

ABF as described in 3.2. Then, the Gabor amplitude images of a normalized test image are obtained via using Gabor filter at first; then, we extract the uniform local binary histogram and use Fisher criterion to gain a new dictionary, and finally the test image is classified as the existing class via sparse representation coding.

6.1 Assessing the Sparse Representation-Based Classification

In our first series of face recognition experiments, we assess the performance of face recognition techniques, namely Gabor principle component analysis (GPCA), Gabor local binary patterns (GLBP), down-sampling sparse representation classifiers (SRC), and Gabor feature-based SRC. No normalization technique was employed during first series of experiments. Results of the assessment on WDF database in terms of rank one recognition rate (in %) are demonstrated in Table 2. The results listed in the table suggest that performance of the adopted face recognition algorithm, i.e., Gabor ULBP-based discriminative dictionary learning and SRC-based classification, outperforms the other techniques.

6.2 Assessing the Normalization Techniques

Our second series of face recognition experiments aimed at evaluating the performance of the photometric normalization techniques presented in Sect. 3. In Table 3, the results of the assessment are presented. Some of the photometric techniques mentioned in [28] are also considered for assessment. The face recognition process outlined in Sect. 4 is used for the assessment as it has reported better recognition results over its competing techniques as in Table 2.

The results of the experiments suggest that while the majority of photometric normalization techniques ensure significant improvements in the recognition rate, adaptive bilateral filter outperforms significantly the other normalization techniques such as WD, BF, and FBF. Compared to classical bilateral filter (BF), the face recognition rate of ABF is improved by about 11 %. Overall, the proposed face recognition method and ABF normalization technique has improved the performance in terms of recognition rate.

Table 2 Rank one recognition rates obtained before normalization

Algorithm	Rate (%)
Gabor + PCA	43.1
Gabor + LBP	45.8
Sparse representation (SRC)	52.7
Gabor + ULBP + SRC (our approach)	61.2

Table 3 Rank one recognition rates comparison

Normalization techniques	Rate (%)
Single scale retinex (SSR)	62.4
Multiscale retinex (MSR)	63.8
Single scale self-quotient (SSQ)	64.2
Discrete cosine transform (DCT)	62.7
Wavelet-based normalization (WA)	65.3
Wavelet de-noising (WD)	70.1
Bilateral filter (BF)	69.9
Fast bilateral filter (FBF)	65.9
Adaptive bilateral filter (ABF)	81.4

7 Conclusions

In this work, investigation into a special case of classical face recognition, wet face recognition, is presented. Wet faces have greater impact on recognition rates if face recognition is employed in automated face identification. The impact of wet face on recognition rate has so far not been studied. In the current work, an investigation into how wet faces impact a recognition rate is presented. The dry and wet face (DWF) database has also been introduced. The wet normalization technique adaptive bilateral filter (ABF) is presented in this work. Compared to competing filters, the ABF outperforms. Visual quality is improved significantly. It is also able to produce sharper edges/textures and less noisy in smooth regions. As a result, overall quality of the ABF restored image is improved significantly. The quality is compared with several photometric normalization techniques. On the other hand, a robust face recognition algorithm based on discriminative dictionary learning and regularized robust coding is used in the assessment. Compared to classical bilateral filter (BF), the face recognition rate of ABF is improved by about 11 %. Overall, the proposed face recognition method and ABF normalization technique have improved the performance in terms of recognition rate. Removing water droplets from wet image will be the future scope of this work.

References

1. Zhao W, Chellappa R, Phillips PJ (2003) Face recognition: a literature survey. ACM Comput Surv 35:399–458
2. Jafri R, Arabnia HR (2009) A survey of face recognition techniques. J Inf Process Syst 5:41–68
3. Cai J, Chen J, Liang X (2015) Single-sample face recognition based on intra-class differences in a variation model. Sensors 15:1071–1087
4. Hermosilla G, Gallardo F, Farias G, Martin CS (2015) Fusion of visible and thermal descriptors using genetic algorithms for face recognition systems. Sensors 15:17944–17962
5. Yin S, Dai X, Ouyang P, Liu L, Wei S (2014) A multi-modal face recognition method using complete local derivative patterns and depth maps. Sensors 14:19561–19581

6. Xia W, Yin S, Ouyang P (2013) A high precision feature based on LBP and Gabor theory for face recognition. Sensors 13:4499–4513
7. Qin H, Qin L, Xue L, Li Y (2012) A kernel Gabor-based weighted region covariance matrix for face recognition. Sensors 12:7410–7422
8. Lee W, Cheon M, Hyun C-H, Park M (2013) Best basis selection method using learning weights for face recognition. Sensors 13:12830–12851
9. Makwana RM (2010) Illumination invariant face recognition: a survey of passive methods. Procedia Comput Sci 2:101–110
10. Zhichao L, Joo MEr Face recognition under varying illumination. In: Joo MEr (ed) New trends in technologies: control, management, computational intelligence and network systems. InTech, pp 209–226. ISBN: 978-953-307-213-5
11. Garg K, Nayar SK (2007) Vision and rain. Int J Comput Vis 75:3–27
12. Elad M, Aharon M (2006) Image denoising via sparse and redundant representations over learned dictionaries. IEEE Trans Image Process 15:3736–3745
13. Dey M, Dey N, Mahata SK, Chakraborty S, Acharjee S, Das A (2014) Electrocardiogram feature based inter-human biometric authentication system. In: International Conference on Electronic Systems, Signal Processing and Computing Technologies (ICESC). IEEE explore, pp 300–304
14. Dey N, Nandi B, Dey M, Das A, Chaudhuri SS (2013) BioHash code generation from electrocardiogram features. In: 3rd IEEE International Advance Computing Conference (IACC-2013) February 22–23, 2013, Ghaziabad (UP), IEEE Xplore
15. Dey N, Nandi B, Dey M, Das A, Chaudhuri SS (2013) BioHash code generation from electrocardiogram features. In: 3rd IEEE International Advance Computing Conference (IACC-2013) February 22–23, 2013, Ghaziabad (UP), IEEE Xplore
16. Buades A, Coll B, Morel JM (2005) A review of image denoising algorithms, with a new one. Multiscale Model Simul 4:490–530
17. Tomasi C, Manduchi R (1998) Bilateral filtering for gray and color images. In: Proceedings of the IEEE International Conference on ComputerVision, pp 839–846
18. Chaudhury KN, Sage D, Unser M (2011) Fast O(1) bilateral filtering using trigonometric range kernels. IEEE Trans Image Process 20:3376–3382
19. Arcel GR, Bacca J, Paredes JL (2000) Nonlinear filtering for image analysis and enhancement. In: Bovik AC (ed) Handbook of image and video processing. San Diego, pp 95–95
20. Kim S, Allebach JP (2005) Optimal un-sharp mask for image sharpening and noise removal. J Electron Imaging 14:023007–1
21. Zhang B, Allebach JP (2008) Adaptive bilateral filter for sharpness enhancement and noise removal. IEEE Trans Image Process 17
22. Wright J, Yang AY, Ganesh A, Sastry SS, Ma Y (2009) Robust face recognition via sparse representation. IEEE Trans Pattern Anal Mach Intell 31:210–227
23. Yang M, Zhang L (2010) Gabor feature based sparse representation for face recognition with Gabor occlusion dictionary. In: Computer Vision-ECCV2010, Springer, Berlin, pp 448–461
24. Ahonen T, Hadid A, Pietikäinen M (2004) Face recognition with local binary patterns. In: Computer Vision-ECCV. Springer, Berlin, pp 469–481
25. Zheng H, Tao D (2015) Discriminative dictionary learning via Fisher discrimination K-SVD algorithm. Neurocomputing 162:9–15
26. Yang M, Zhang L, Feng X, Zhang D (2011) Fisher discrimination dictionary learning for sparse representation. In: Proceedings of the 2011 IEEE International Conference on Computer Vision (ICCV), IEEE, pp 543–550
27. Lu Z, Zhang L (2015) Face recognition algorithm based on discriminative dictionary learning and sparse representation. Neurocomputing. doi:10.1016/j.neucom.2015.09.091

28. Štruc V, Pavešiæ N (2011) Photometric normalization techniques for illumination invariance. In: Zhang YJ (ed) Advances in face image analysis: techniques and technologies. IGI Global, pp 279–300
29. Ojala T, Pietik MA, Harwood D (1996) A comparative study of texture measures with classification based on feature distributions. Pattern Recogn 29:51–59
30. Gross R (2005) Face databases. In: Li S, Jain A (eds) Handbook of face recognition. Springer, Berlin
31. Dharavath K, Laskar RH, Talukdar FA (2013) Qualitative study on 3D face databases: a review. In: Annual IEEE India Conference (INDICON), Bangalore, pp 1–6
32. Dharavath K (2014) WDF: a wet and dry face database. Technical Report, National Institute of technology, July

Improved Approach for 3D Face Characterization

S. Sghaier, C. Souani, H. Faeidh and K. Besbes

Abstract Representing and extracting good quality of facial feature extraction is an essential step in many applications, such as face recognition, pose normalization, expression recognition, human–computer interaction and face tracking. We are interested in the extraction of the pertinent features in 3D face. In this paper, we propose an improved algorithm for 3D face characterization. We propose novel characteristics based on seven salient points of the 3D face. We have used the Euclidean distances and the angles between these points. This step is highly important in 3D face recognition. Our original technique allows fully automated processing, treating incomplete and noisy input data. Besides, it is robust against holes in a meshed image and insensitive to facial expressions. Moreover, it is suitable for different resolutions of images. All the experiments have been performed on the FRAV3D and GAVAB databases.

Keywords 3D face · Anthropometric · Feature extraction · Salient point · Euclidean distance · Angle

S. Sghaier (✉) · K. Besbes
Microelectronics and Instrumentation Laboratory, Faculty of Sciences
of Monastir, Monastir, Tunisia
e-mail: souhir.sghaier.zouari@gmail.com

K. Besbes
e-mail: kamel.besbes@fsm.rnu.tn

C. Souani · H. Faeidh
Microelectronics and Instrumentation Laboratory, Higher Institute
of Applied Sciences and Technology, Sousse, Tunisia
e-mail: chokri.souani@gmail.com

H. Faeidh
e-mail: faiedh_h@yahoo.fr

© Springer International Publishing Switzerland 2017
N. Dey and V. Santhi (eds.), *Intelligent Techniques in Signal Processing
for Multimedia Security*, Studies in Computational Intelligence 660,
DOI 10.1007/978-3-319-44790-2_13

1   Introduction

Automatic recognition and authentication of individuals is becoming an important research field of computer vision. The applications are varied and are used for the safety of goods and lives as they are utilized in airports, military places and banks as well as by anti-criminal police.

Other usages of 3D authentication are used for video surveillance, biometrics, autonomous robotics, intelligent human–machine interface, photography and image and video indexing. Therefore, different biometric characteristics such as fingerprint, vein, hand geometry, iris, can be used for automatic authentication. The drawbacks of these modalities are that the cooperation of the candidate is usually essential. Among these, the face recognition is a very important biometric modality. The main advantages are that the candidate does not need direct physical contact with the system, and it is a transparent process since hidden cameras can be used without the candidate's knowledge.

It has been shown that 2D face recognition is sensitive to illumination changes and variation of pose and expression. 3D face recognition is likely to solve these problems. 3D face recognition is more robust against pose changes and can better overcome illumination [1]. In the last 5 years, a rapid increase in the usage of 3D techniques for applications in video, face recognition and virtual reality has taken place both in academy and industry applications. The acquisition technology of 3D images has become simpler and cheaper. The main advantage of the 3D data usage is that it conserves all the geometric information of the object, which is close to the reality representation.

In this paper, we work in unaffected region by the pose changes and the variation of facial expressions. Indeed, we focus on the anthropometric landmarks that play an important role to perform quantitative measurements. Indeed, we extract the most important features that help us to recognize the 3D face. Our features vector is composed of the Euclidean distances and angles between these keypoints, the entropy of the region of interest and the width of the forehead.

The paper is organized as follows. In Sect. 2, an overview of some research works is presented. The 3D face databases are shown in Sect. 3. In Sect. 4, we detail our contribution in 3D face characterization. The achieved experiments and the results are discussed in Sect. 5. The conclusion and the perspectives are presented in Sect. 6.

2   Related Works

Recently, many authors worked on 3D face. Among them, we may cite: the algorithm developed in [2] described differential geometry descriptors such as derivatives, coefficients of the fundamental forms, different types of curvatures and Shape Index. After that they computed the geodesic and Euclidean distances between landmarks, nose volume and ratios between geodesic and Euclidean distances.

The method in [3] based on a new triangular surface patch (TSP) descriptor to localize the landmark in the 3D face. This descriptor represents a surface patch of an object, along with a fast approximate nearest neighbour lookup to find similar surface patches from training heads.

In the research work of salzar et al. [4], each landmark is located automatically on the face surface by means of a Markov network. The network captures the statistics of a property of the surface around each landmark and the structure of the connections. The estimation of the location of landmark on a test model is carried out by using probabilistic inference over the Markov network. They performed inference using the loopy belief propagation algorithm.

Another work proposed in [5] represented a linear method, namely linear discriminant analysis, and a nonlinear method, namely AdaBoost to characterize the 3D face.

The work of Böer et al. [6] consisted in the Discriminative Generalized Hough Transform (DGHT) to describe the 3D face scans.

Ballihi et al. [7] are interested in finding facial a curve of level sets (circular curves) and streamlines (radial curves). Indeed, they used a geometric shape analysis framework based on Riemannian geometry to compare pairwise facial curves and a boosting method to highlight geometric features according to the target application.

Ramadan and Abdel-Kader [8] suggested to use spherical wavelet parameterization of the 3D face image.

In [9], feature extraction is done by geodesic distances and linear discriminant analysis (LDA).

Han et al. [10], based on the mapping of a deformable model to a given test image involves two transformations, rigid and nonrigid transformation. They utilized two intrinsic geometric descriptors, namely geodesic distances and Euclidean distances, to represent the facial models by describing the set of distance-based features and contours.

Berretti et al. [11] in their work proposed three different signatures to locally describe the 3Dface at the keypoints, namely the histogram of gradients (HOG), the histogram of orientations (SHOT) and the geometric histogram (GH).

In [12], Fang et al. performed the principal component analysis (PCA) method the feature space to reduce its dimensionality. Then, they applied LDA to find the optimal subspace that preserves the most discriminant information. In fact, these two methods are applied sequentially to reduce the feature dimension and find the optimal subspace. The feature vectors combine geometric information of the landmarks and the statistics on the density of edges and curvature around the landmarks.

The method introduced in [13], based on wavelet descriptors as a multiscale tool to analyze 3D face. Also, they used an ellipsoidal cropping through the detection of facial landmarks to detect and crop the facial region. In the step of the preprocessing of the facial region, they applied a procedure called Sorted Exact Distance Representation in order to fill holes. Gaussian filter is used to the range image to smooth the surface. Then, they are based on the Gaussian function as mother wavelets to extract the landmarks.

Therefore, in our work, we prefer to improve geometric descriptors which are based on an anthropometric measurements and Euclidean distances between feature points of a 3D meshed face.

Experimental results are tested on GAVAB and FRAV 3D databases. Our technique outperforms other latest methods in the state-of-the-art.

3   3D Face Databases

3.1   GAVAB Database

GAVAB is a database that consists of 549 3D facial range scans (Fig. 1) of 61 different subjects (45 are male and 16 are female) captured by a Minolta Vivid 700 scanner. The faces were placed at 1.5 ± 0.5 m of distance from the scanner. Although the chair (with wheels) was at 1.5 m of distance, the individuals had the flexibility to move their head normally. This fact produces any resolution changes from some images to others. The height of the chair could be changed in order to make the head always visible to the scanner [14].

Each image is a mesh of connected 3D points of the facial surface. Textured information of each vertex has been eliminated to reduce the size of all these face models. The subjects are all Caucasian, and most of them are aged between 18 and 40 [15]. Some examples in the GAVAB database are shown in Fig. 2.

Each subject has been scanned 9 times with many poses and different facial expressions. The scans with pose variations contain 1 facial scan while looking up (+35°), 1 while looking down (−35°), 1 for the right profile (+90°) and 1 for the left profile (−90°). Figure 3 shows some examples of pose variation [15].

The facial scan without pose changes include 4 different close frontal facial scans: 2 of them are with a neutral expression, 1 with a smile and 1 with an accentuated laugh. In Fig. 4, we find a different facial expression of some subjects in the database [15].

Fig. 1 3D face scan

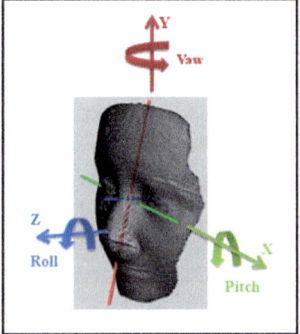

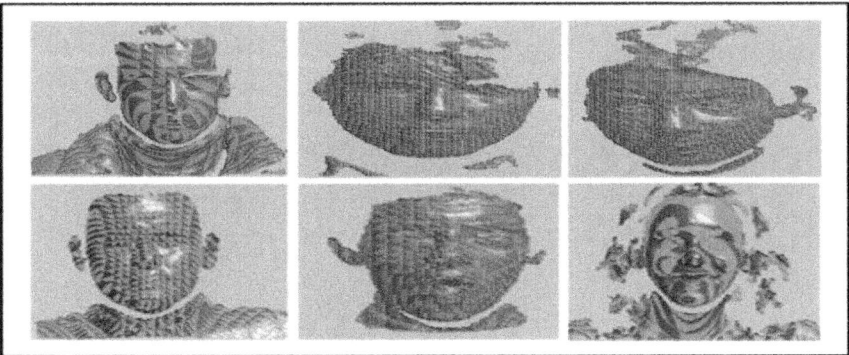

Fig. 2 Example of scans from GAVAB database

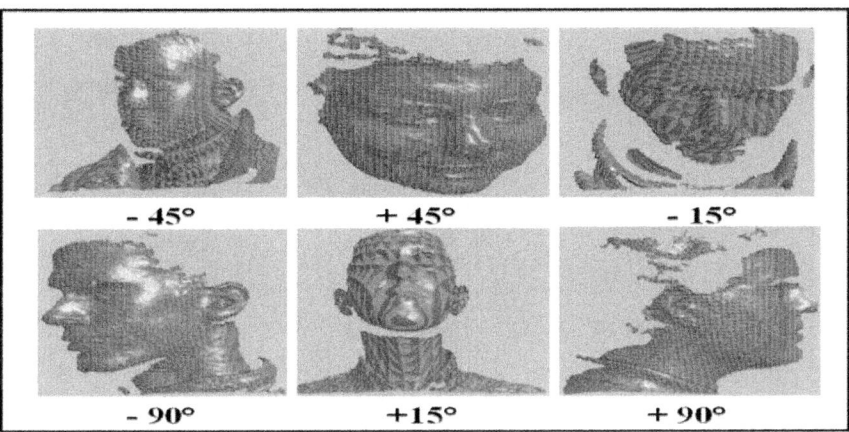

Fig. 3 Example of pose variation

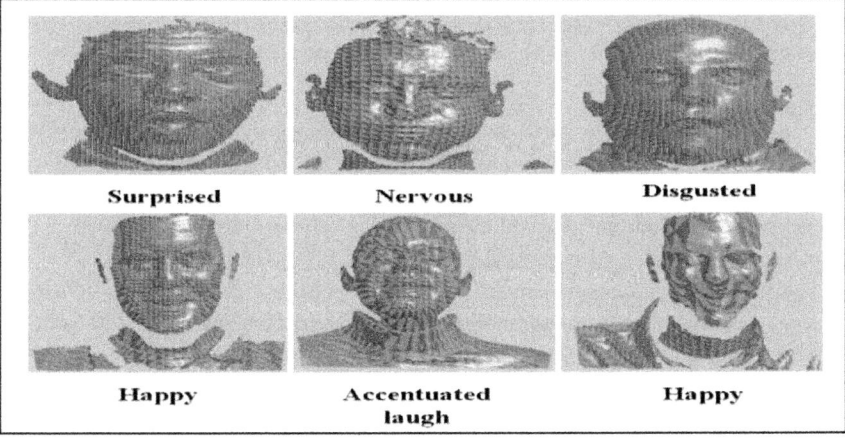

Fig. 4 Variation in facial expression

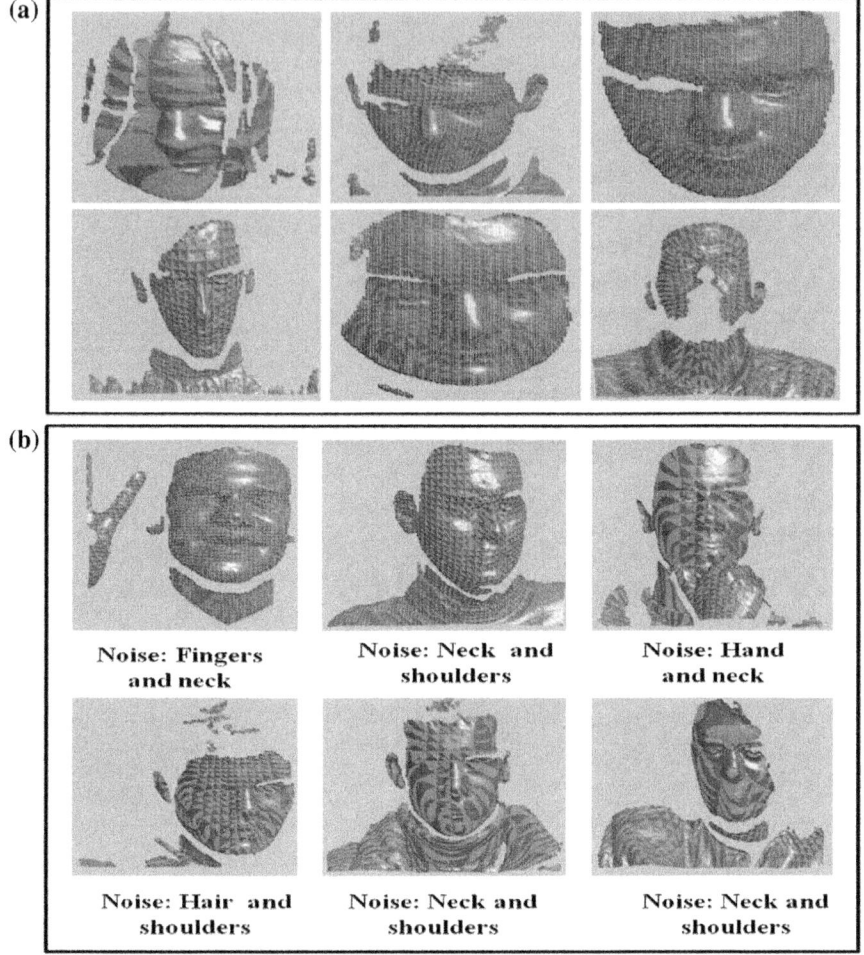

Fig. 5 **a** Scans from Gavab database showing varying amounts of holes (missing data). **b** Scans from Gavab database showing varying amounts of noise and presence of nonfacial regions

In addition, GavabDB is the noisiest dataset currently available to the public, showing many holes and missing data especially in the eyebrows (Fig. 5a). Also, the images of this database have much noise and nonfacial regions, such as neck, hands, shoulders and hair.(Fig. 5b). Moreover, each person has several scans with different poses and facial expressions.

The most important problem in this database is that all the scans of a given person do not have the same number of faces and vertexes (Fig. 6), which can cause a problem to the extraction process in the region of interest. Besides, this database shows a high interpersonal variation.

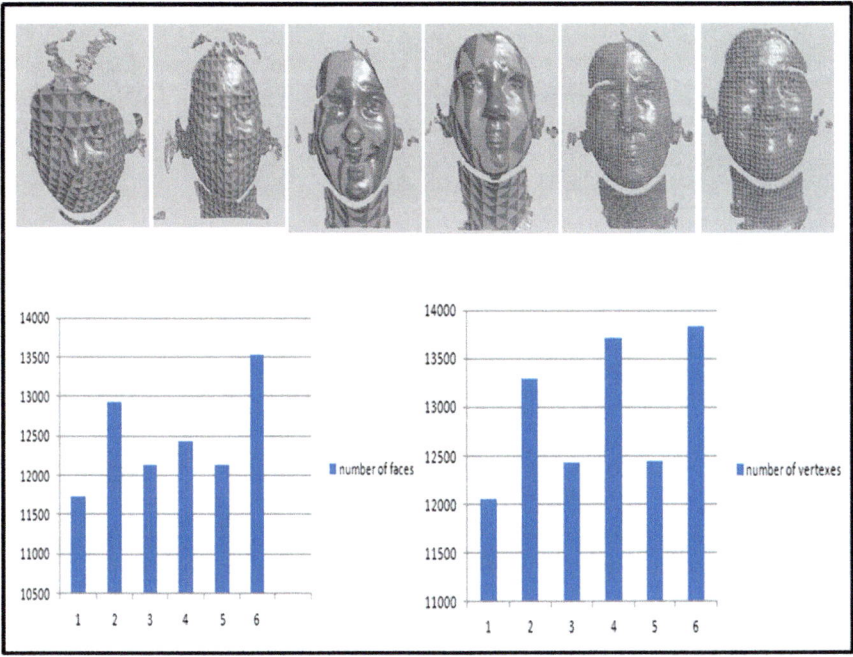

Fig. 6 Different scans of the same person showing an intrapersonal variation of number of vertexes and faces

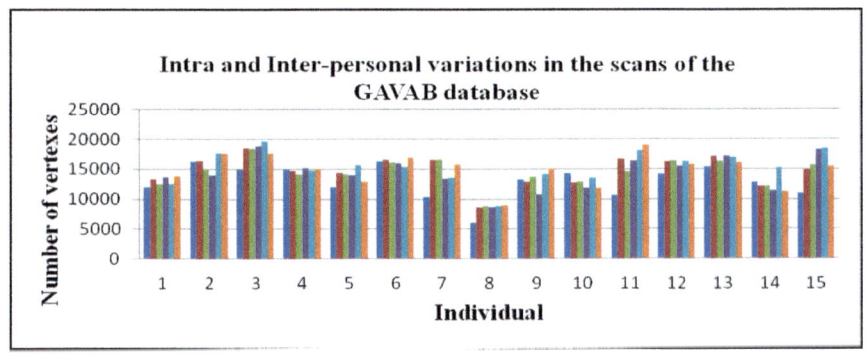

Fig. 7 Intrapersonal and interpersonal variations in the scans of GAVAB database

As a matter of fact, in this database we find all types of scans with pose variations, many facial expressions, occlusions, fat and slim persons, etc. Fig. 7 presents the variation between some individuals.

Therefore, we will use this database to evaluate our technique, since it has been used for automatic face recognition experiments and other possible facial applications such as pose correction or register of 3D facial models.

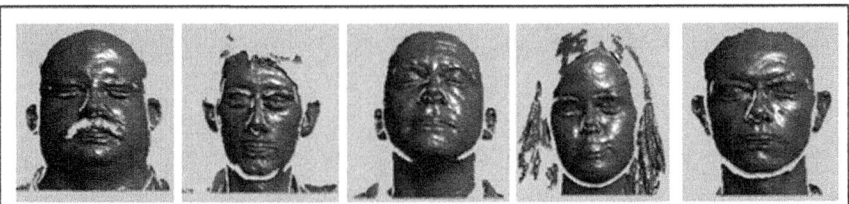

Fig. 8 Example of scans from FRAV 3D database

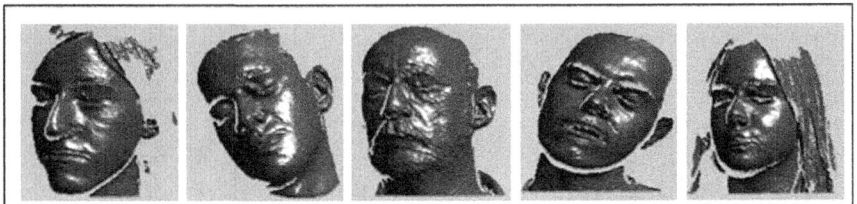

Fig. 9 Example of scans from FRAV 3D database showing poses variation

3.2 FRAV 3D Database

This database contains 106 subjects, with approximately one woman every three men. The data were acquired with a Minolta VIVID 700 scanner, which provides texture information and a VRML file. Some examples in the FRAV V3 database are shown in Fig. 8 [16].

Each person has 16 captures with different poses and lighting conditions, trying to cover all possible variations, including turns in different directions as it is shown in Fig. 9 [16].

Moreover this database provides scans with different facial expression as it is presented in Fig. 10.

In every case only one parameter was modified between two captures. This is one of the main advantages of this database, respect to others [16].

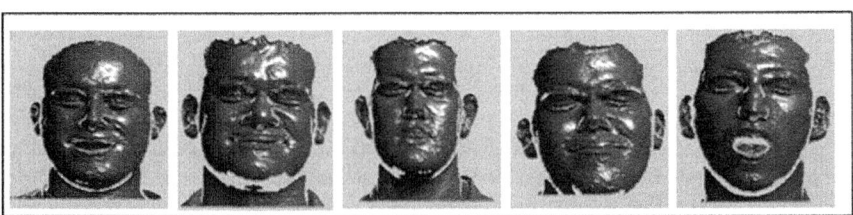

Fig. 10 Example of scans from FRAV 3D database with different facial expression

Fig. 11 Organization chart
of our approach

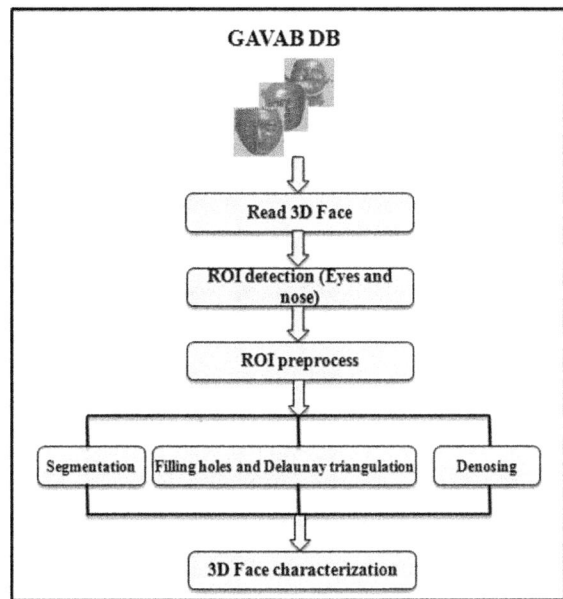

4   Proposed Approach

In this paper, a novel approach based on landmark localization and characterization
of a 3D face is presented. It contains five essential steps (Fig. 11). Indeed, we begin
with the lecture of the 3D mesh. Then we extract the region of interest (ROI) using
the anthropometric proportions and a cropping filter. After that, a pre-treatment of
the extracted part is needed. Finally, after the detection of the salient points, we use
the Euclidean distance and the angle between them to determine the features vector.

4.1   Region of Interest Detection

We all know that any face is divided in three parts in the anthropometric proportions; the first part is between the higher forehead and eyebrows, the second is
between the eyebrows and nose tip and the third region is between the nose tip and
the chin.

We focus on the second part of the face where the number of vertexes is in the
interval [1/3 2/3] of the image (Fig. 12). After that, we have worked exactly in the
region of the nose tip that is located in the end of the second part.

The regions of interest (eyes + nose) are automatically detected for frontal scans
and scans with pose variation. We choose to work on these regions above the mouth
to avoid the expression variations. It is named a static part because it is weakly
affected by the variations of facial expression while the lower part includes the

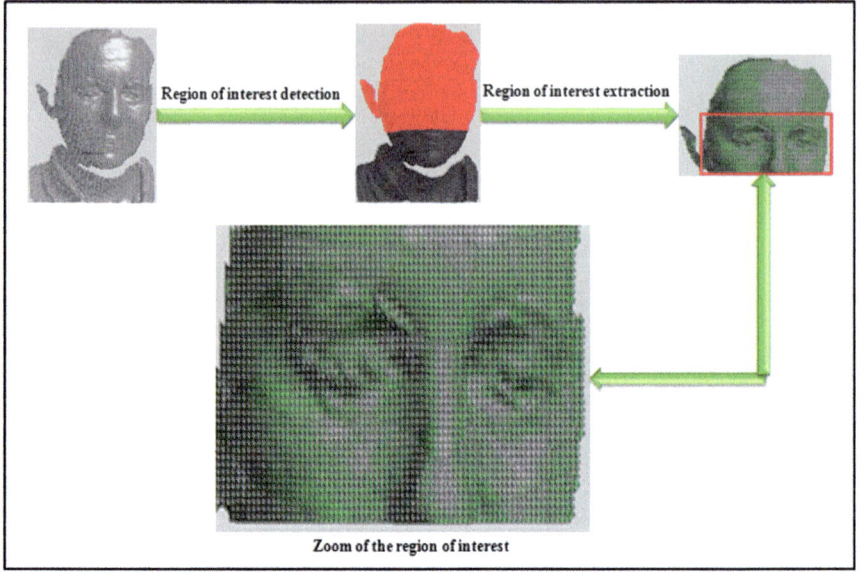

Fig. 12 Detection and extraction of the region of interest

mouth is strongly affected by the change of facial expression. In this part we can detect the keypoints that are used as the foundation to detect the most prominent features to compose the characteristic vector.

4.2 3D Face Preprocessing

Preprocessing data is a very important and nontrivial step of biometric recognition systems. We preprocess a 3D face to minimize the effects of the intrinsic noise of a face scanner, such as holes and spikes, including some undesired parts such as clothes, neck, ears and hair. This part is common to all situations including 3D scan preprocessing. Our preprocessing has three stages: segmentation, filling holes and Delaunay triangulation, median filter and denosing. The details of these stages are given in the following.

- *Segmentation*

The first step of the pretreatment is the segmentation in which we split the 3D face in two areas. We are based on the anthropometric proportions of the face to detect the region of interest that we need (the eyes and the nose). Indeed, we are interested in the second part of the face. Then, we use a cropped filter to extract the ROI. Figure 13 shows the regions of interest that we work on. These regions are unaffected by the variations of facial expression.

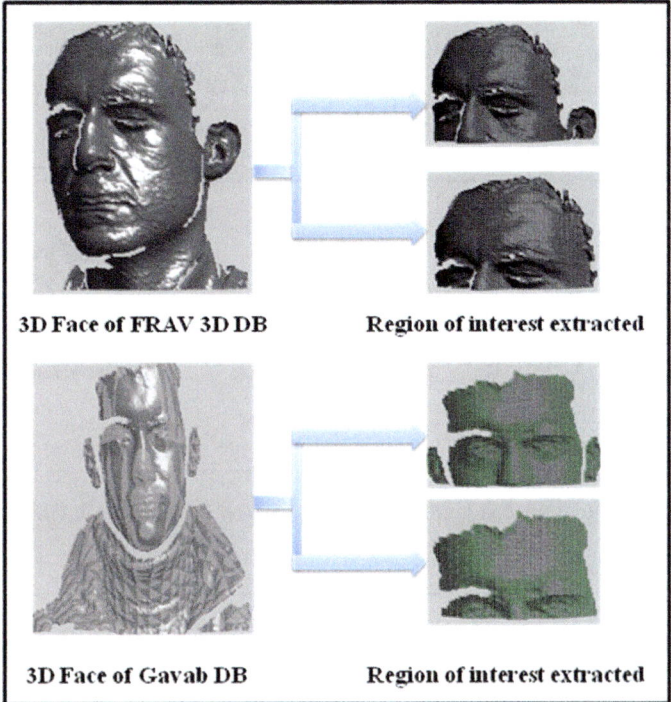

3D Face of FRAV 3D DB   Region of interest extracted

3D Face of Gavab DB   Region of interest extracted

Fig. 13 3D face segmentation

- *Filling holes and Delaunay triangulation*

The holes are created either because of the absorption of laser in dark areas, such as eyebrows and mustaches, or self-occlusion or open mouths [17]. These holes and noises in a 3D scan reduce the recognition system accuracy, that is why it is important to fill holes in the input mesh.

Though, building triangular mesh costs too much time and raises many issues; so we have applied Delaunay triangulation (Fig. 14) to fill holes in the mesh of the extracted face region so as to place them on a regular grid. This technic is not only very fast and easy, but also it solves this problem more efficiently.

- *Denosing*

In this stage, we have used median filter to remove the undesirable spikes and clutter due to the 3D scanner (Fig. 15). This filter may also unnecessarily suppress actual and small variations on the data which appear in the 3D surface.

Finally, we perform a mesh heat diffusion flow in order to eliminate the maximum of the undesirable noise (Fig. 16).

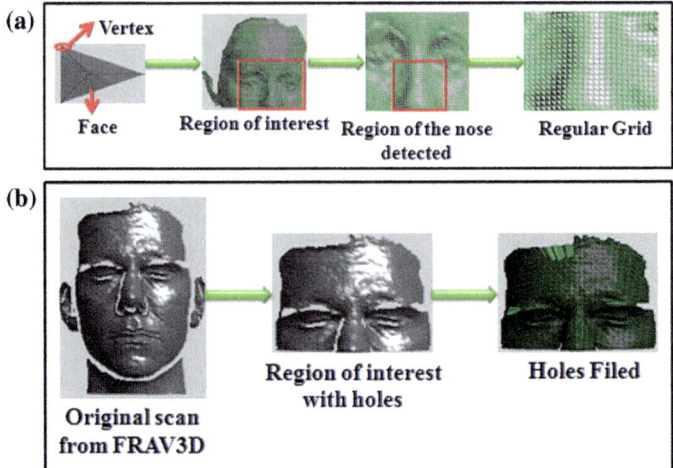

Fig. 14 Delaunay triangulation to fill holes. **a** Triangulation de Delaunay. **b** Filling holes

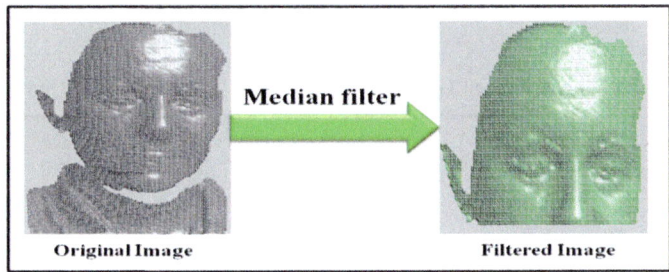

Fig. 15 Median filter

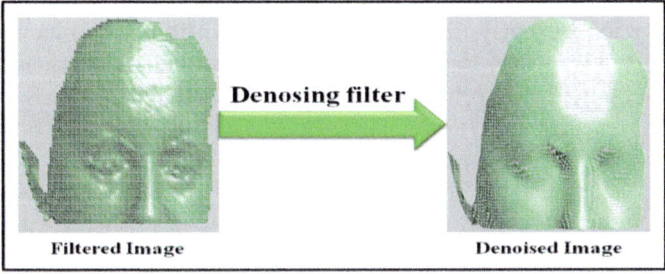

Fig. 16 Denosing the 3D face

4.3 3D Face Characterization

The Cranio-facial measurements are used in different fields: in sculpture to create well-proportioned facial ideals, in anthropology to analyze prehistoric human remains and more recently in computer vision (estimation of the orientation of the head, detection of characteristic points of a face) and computer graphics to create parametric models of human faces.

In this paper, we propose an approach of 3D face characterization using anthropometric measures based on seven salient landmarks: (left inner eye corner, left outer eye corner, right inner eye corner, right outer eye corner, centre of the root of the nose, nose tip and forehead point).

We extract the desired region of interest (eyes and nose) that is unaffected by the variations of pose, occlusions and facial expressions. We focus on the third phase of facial recognition system: the characterization. This step is the most important and interesting step in any recognition system. It takes place after the detection and preprocessing phase. This phase is based on an analysis of a 3D shape image.

After detecting the points of interest, we built our feature vector which is composed by:

- *Entropy oft he region of interest;*
- *Euclidean distances;*

Figure 17 shows the different Euclidean distances suggested in our approach as a feature to characterize the 3D face.
where as follows:

$$A = \begin{pmatrix} x_1 \\ y_1 \\ z_1 \end{pmatrix}; B = \begin{pmatrix} x_2 \\ y_2 \\ z_2 \end{pmatrix} L1 = \sqrt{(x_2 - x_1)^2 + (y_2 - y_1)^2 + (z_2 - z)^2} \qquad (1)$$

$$C = \begin{pmatrix} x_3 \\ y_3 \\ z_3 \end{pmatrix}; D = \begin{pmatrix} x_4 \\ y_4 \\ z_4 \end{pmatrix} L2 = \sqrt{(x_4 - x_3)^2 + (y_4 - y_3)^2 + (z_4 - z_3)^2} \qquad (2)$$

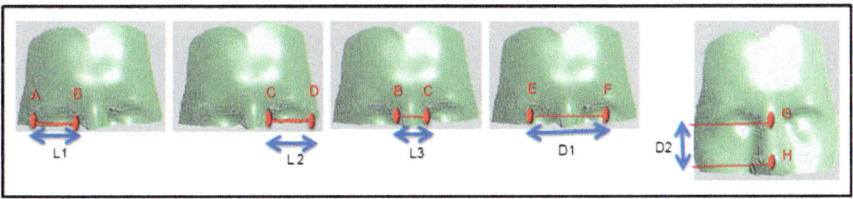

Fig. 17 Euclidean distances

$$B = \begin{pmatrix} x_2 \\ y_2 \\ z_2 \end{pmatrix} ; C = \begin{pmatrix} x_3 \\ y_3 \\ z_3 \end{pmatrix} L3 = \sqrt{(x_3 - x_2)^2 + (y_3 - y_2)^2 + (z_3 - z_2)^2} \quad (3)$$

$$E = \begin{pmatrix} a_1 \\ a_2 \\ a_3 \end{pmatrix} ; F = \begin{pmatrix} b_1 \\ b_2 \\ b_3 \end{pmatrix} D1 = \sqrt{(b_1 - a_1)^2 + (b_2 - a_2)^2 + (b_3 - a_3)^2}$$
(4)

$$a_1 = \frac{x_1 + x_2}{2} ; a_2 = \frac{y_1 + y_2}{2} ; a_3 = \frac{z_1 + z_2}{2} ; b_1 = \frac{x_3 + x_4}{2} ; b_2 = \frac{y_3 + y_4}{2}$$

$$G = \begin{pmatrix} x_2 \\ y_2 \\ z_2 \end{pmatrix} ; H = \begin{pmatrix} x_5 \\ y_5 \\ z_5 \end{pmatrix} D2 = \sqrt{(x_2 - x_5)^2 + (y_2 - y_5)^2 + (z_2 - z_5)^2} \quad (5)$$

The Eqs. (1)–(5) determined, respectively, the length of the right eye, the length of left eye, the length of the root of the nose, the distance between the eyes and the distance between the eyes and nose. These features are calculated using Euclidean distance between the keypoints detected in the region of interest. We extract the coordinates of each point in the three axes (x, y, z). So, the Euclidean distance between two points implicates computing the square root of the sum of the squares of the differences between corresponding values.

- *Angles between keypoints*

Figure 18 shows the angles between the keypoints. When (a) is the angle between nose tip, left outer eye corner and right outer eye corner, (b) is the angle between nose tip, left inner eye corner and right inner eye corner, (c) is the angle between nose tip, right inner eye corner and right outer eye corner, (d) is the angle between nose tip, left inner eye corner and left outer eye corner and (e) is the angle between the nose tip, centre of the root of the nose and centre of the forehead.

We use the trigonometric function to determine the degrees of angles between the landmarks. We use exactly the cosine theta that returns the angle whose cosine is the specified number.

This function calculates the angle between two three-dimensional vectors as it is shown above.

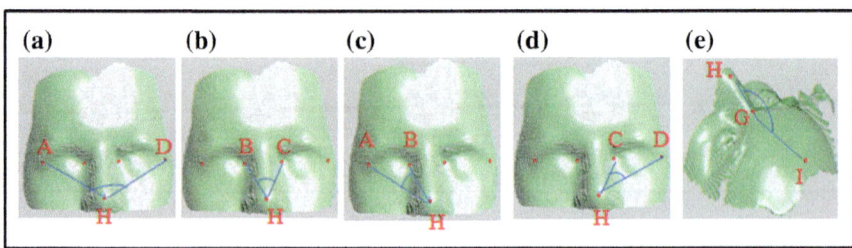

Fig. 18 Angles between keypoints

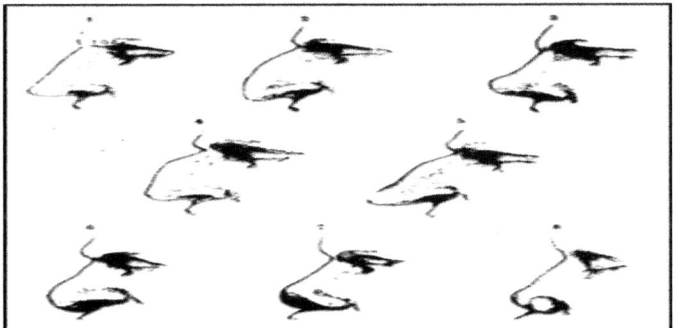

Fig. 19 Different shapes of noses

As it is shown in Fig. 19, there are lots of different shapes that noses can take (dorsal hump, long nose, tension tip, over protected tip, bulbous tip and under projected tip). Therefore, we suggest adding another primitive that is the angle between nose tip, centre of the root of the nose and centre of the forehead (Fig. 18e). This angle is highly differing from an individual to another, and it depends on the nose of people.

The principal keypoints detected are as follows:

A: Right outer eye corner.
B: Right inner eye corner.
C: Left inner eye corner.
D: Left outer eye corner.
H: Nose tip.
I: Centre of the forehead

The keypoints E, F and G are calculated from the principal keypoints. It indicated that:

E: Centre of the right eye.
F: Centre of the left eye.
G: Centre of the root of the nose.

- *Width of forehead;*

Both the GAVAB and the FRAV 3D databases contain variety of scans between people. Each person has its properties and a specific form of the face (see Fig. 20). We propose to add a specific characteristic which is the width of the face to the feature vector. This feature helps us to determine the form of the face.

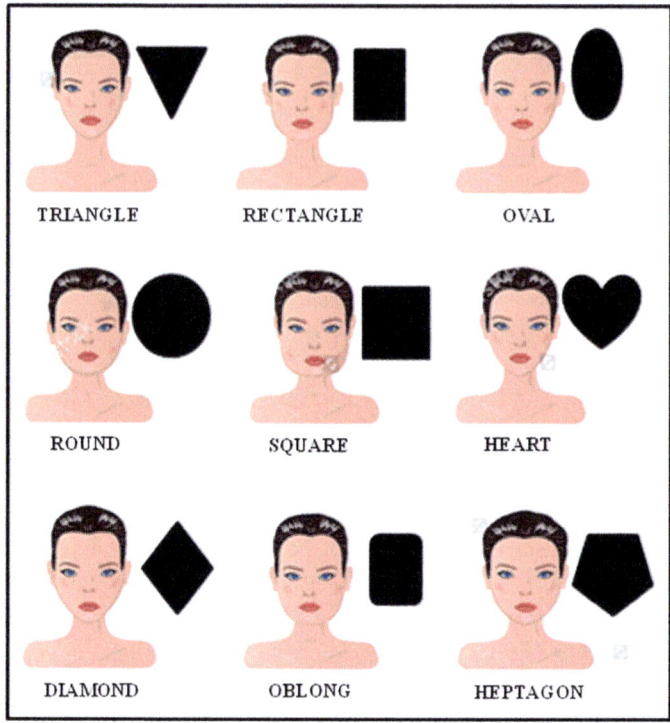

Fig. 20 Different forms of human face

Fig. 21 Arc of the circle

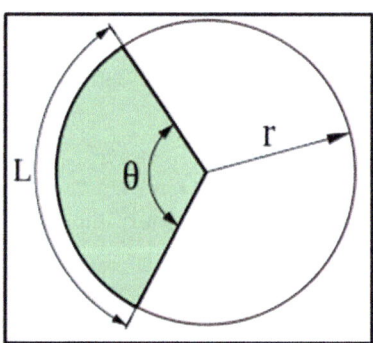

We begin by the extracting of the face in a circle inwhich the nose tip is its centre (Fig. 22). Hence, applying the Eqs. 6 and 7, we can calculate the width of the forehead as an arc of the circle. In Fig. 21, L is the arc of the circle.

Fig. 22 Measure of forehead
width

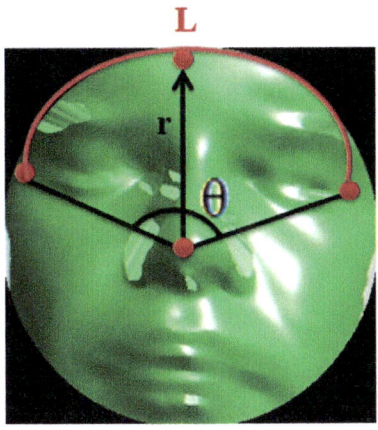

$$\alpha = \frac{180 * \theta}{\pi} \tag{6}$$

$$L = \frac{\alpha \pi r}{180} \tag{7}$$

where:

- **r** = Radius: distance between the nose tip and the first vertex of the 3D face.
- θ = Angle (in radians) between nose tip, left outer eye corner and right outer eye corner (Fig. 17a).
- **L** = Length of the arc = Width of the forehead.

The advantage of our approach is that we work only in the upper part of the face so we work just in the half circle to extract the pertinent characteristics of the 3D face. These characteristics were then fed into a standard classifier such as support vector machines (SVM) to handle the classification.

5 Experiments and Results

We test our algorithm on the GAVAB and FRAV 3D face databases. Experimental results show that the proposed method is comparable to state-of-the-art methods in terms of its robustness, flexibility and accuracy (Table 1).

We suggest working with the geometric measurements because they have a permanent and high subjective perception. These signatures are the most discriminate and consistent that deal with pose variation, occlusions, missing data and facial expressions. Furthermore, the literature shows the absence of 3D biometric systems based on the fusion of these features.

Table 1 Example of
measures of characteristics of
scans from both GAVAB and
FRAV 3D databases

	Scan from GAVAB	Scan from FRAV 3D
L (cm)	10.361	13.756
L1(cm)	2.368	3.329
L2(cm)	2.751	3.391
L3(cm)	1.492	1.645
D1(cm)	3.986	4.809
D2(cm)	4.036	5.117
(a) (°)	97.2967	127.3641
(b) (°)	35.7789	38.7200
(c) (°)	30.5543	43.8591
(d) (°)	30.9635	44.7850
(e) (°)	135.2531	171.8377
Entropy	4.6174	3.5732

Purposely, the major contributions of this paper consist of a new geometric features approach for 3D face recognition system handling the challenge of facial expression.

Moreover, in our fully automatic approach, we work just in the upper part of the 3D face. We success 100 % to extract this region. Therefore, it is not only permanent, but also fast in time of execution.

6 Conclusion and Perspectives

In our approach, we use the anthropometric measurements to extract the pertinent primitives of 3D face. We focus on the upper face because this region is unaffected by the variations of facial expressions. We use seven salient points that are unaffected by the variations of pose, occlusions and facial expressions. We use the Euclidean distances and the measure of angles between these points to extract the features vector of the 3D face. This vector is necessary in 3D face recognition system.

Future evaluations should be carried out on more challenging datasets with an increased number of images, also incorporating rotations and translations to improve the success rate.

References

1. Bockeler M, Zhou X (2013) An efficient 3D facial landmark detection algorithm with haar-like features and anthropometric constraints. In: International conference of the Biometrics Special Interest Group (BIOSIG), pp 1–8. IEEE
2. Vezzetti E, Marcolin F, Fracastoro G (2014) 3D face recognition: An automatic strategy based on geometrical descriptors and landmarks. Robot Auton Syst 62(12):1768–1776
3. Papazov C, Marks TK, Jones M (2015) Real-time 3D head pose and facial landmark estimation from depth images using triangular surface patch features. In: Proceedings of the IEEE conference on computer vision and pattern recognition, pp 4722–4730
4. Papazov C, Marks TK, Jones M (2015) Real-time 3D head pose and facial landmark estimation from depth images using triangular surface patch features. In: Proceedings of the IEEE conference on computer vision and pattern recognition, pp 4722–4730
5. Creusot C, Pears N, Austin J (2013) A machine-learning approach to keypoint detection and landmarking on 3D meshes. Int J Comput Vision 102(1–3):146–179
6. Böer G, Hahmann F, Buhr I et al (2015) Detection of facial landmarks in 3D face scans using the discriminative generalized hough transform (DGHT). Bildverarbeitung für die Medizin. Springer, Berlin, pp 299–304
7. Ballihi L, Ben Amor B, Daoudi M et al (2012) Boosting 3-D-geometric features for efficient face recognition and gender classification. IEEE Trans Inf Forensic Secur 7(6):1766–1779
8. Ramadan RM, Abdel-Kader RF (2012) 3D Face compression and recognition using spherical wavelet parametrization. Int J Adv Comput Sci Appl 3(9)
9. Hiremath PS, Hiremath M (2013) 3D face recognition based on deformation invariant image using symbolic LDA. Int J 2(2)
10. Han X, Yap MH, Palmer I (2012) Face recognition in the presence of expressions. J Softw Eng Appl 5:321–329
11. Berretti S, Werghi N, Del Bimbo A, Pala P (2013) Matching 3D face scans using interest points and local histogram descriptors. Comput Graph 37(5):509–525
12. Fang T, Zhao X, Ocegueda O et al (2011) 3D facial expression recognition: a perspective on promises and challenges. In: IEEE international conference on automatic face and gesture recognition and workshops, pp 603–610
13. Pinto SCD, Mena-Chalco JP, Lopes FM et al (2011) 3D facial expression analysis by using 2D and 3D wavelet transforms. In: 18th IEEE international conference on image processing (ICIP), pp 1281–1284
14. Hatem H, Beiji Z, Majeed R et al (2013) Nose tip localization in three-dimensional facial mesh data. Int J Adv Comput Technol 5(13):99
15. Zhang Y (2014) Contribution to concept detection on images using visual and textual descriptors (Doctoral dissertation, Ecully, Ecole centrale de Lyon)
16. Grgic M, Delac K (2013) Face recognition homepage. Zagreb, Croatia (www.face-rec.org/databases), 324. Accessed 1 May 2016
17. Drira H, Ben Amor B, Srivastava A et al (2013) 3D face recognition under expressions, occlusions, and pose variations. IEEE Trans Pattern Anal Mach Intell 35(9):2270–2283

Attendance Recording System Using Partial Face Recognition Algorithm

Borra Surekha, Kanchan Jayant Nazare, S. Viswanadha Raju
and Nilanjan Dey

Abstract In today's world, recording the attendance of a student plays an important role in improving the quality of educational system. The manual labor included in the maintenance and management of the traditional attendance sheets is tedious as it costs quite a time for the lecturer. Thus, there is a requirement for robust computerized biometric-based attendance recording system (ARS). Face recognition-based methods are a potential replacement for conventional systems, in case if the students to be addressed are more. This chapter gives an overview of the existing attendance recording systems, their vulnerabilities, and recommendations for future development. A smart attendance capturing and management system based on Viola–Jones algorithm and partial face recognition algorithms is introduced for two environments: controlled and uncontrolled. While the proposed system proved 100 % accurate under controlled environment, the efficiency under uncontrolled environment is quite low (60 %). It is observed that the face recognition rate varies from frame to frame. Further, the performance of the proposed attendance system completely depends upon the database collected, the resolution of the camera used and the capacity of students. Further work can be carried out to make the system more efficient in the real-time scenario.

Keywords Biometrics · Face recognition · Face detection · Viola–Jones algorithm · Gabor ternary pattern · Sparse representation

B. Surekha (✉) · K.J. Nazare
K S Institute of Technology, Bangalore, India
e-mail: borrasurekha@gmail.com

K.J. Nazare
e-mail: hrl.lakshmi@gmail.com

S. Viswanadha Raju
JNTUHCEJ, Karimnagar, Telangana, India
e-mail: svraju.jntu@gmail.com

N. Dey
Techno India College of Technology, Kolkata, India
e-mail: neelanjandey@gmail.com

© Springer International Publishing Switzerland 2017   293
N. Dey and V. Santhi (eds.), *Intelligent Techniques in Signal Processing
for Multimedia Security*, Studies in Computational Intelligence 660,
DOI 10.1007/978-3-319-44790-2_14

1 Introduction

Every educational institution or organization employs attendance recording system. Some continue with the traditional method for taking attendance manually while some have adopted biometric techniques [1]. The traditional method makes it hard to authenticate every student in a large classroom environment. Moreover, the manual labor involved in computing the attendance percentage becomes a major task.

The radio frequency identification (RFID) helps to identify a large number of crowds using radio waves. It has high efficiency and hands-free access control. But it is observed that it can be misused. In radio frequency identification (RFID)-based automatic attendance recording system [2] uses RFID tags, transponders, and RFID terminals for attendance management of employees and students. The captured data is processed by the server to update the database. Attendance recording using Bluetooth technology proposed by Bhalla et al. [3] demands the students to carry their mobile phones to classroom, so that the software installed in the instructors' mobile phone can detect it via Bluetooth connection and MAC protocols. The common drawback of these two approaches is proxy attendance, as there is no provision for verification.

A biometric-based system indeed provides the solution as they measure characteristics that are unique to every human being and hence making it impossible to duplicate the biometric characteristics of a person. Therefore, there is an extremely low probability of two humans to share the same biometric data and it can only be lost due to a serious accident. It has proven it usefulness and reliability in many organizations, government bodies, and commercial banks. Biometric measurements can be subdivided into physiological and behavioral. However, each biometric method has its own advantages and disadvantages. A physiological method derives its data directly using the body parts of human beings. They include fingerprint scan, iris scan, retina scan, hand scan, and facial recognition. A behavioral method derives its data from an action done by human beings. It includes the following: voice scan, signature scan [4, 5], and keystroke scan.

Various biometric-based authentication systems have been developed and implemented in the past to yield maximum efficiency. These methods include fingerprints, eye retina, voice etc.

1.1 Related Work

Rashid et al. [6] proposed biometric voice recognition technology using voiceprints of an individual to authenticate. This system is useful for people having difficulty in using hands and other biometric traits. However, this system is sensitive to background noise. Also, the voice of the person tends to change with age. The voice recognition system may not accurately identify the person when he/she is suffering from throat infection or flu. Hence, this system is not reliable.

Retina scanning [7]-based methodology uses a blood vessel pattern to authenticate. The pattern remains the same and is not affected by aging as well. However, this device can be used by only one person at a time. It proves to be time-consuming for a large crowd. This equipment also requires the person to be in close contact with it for authenticating. Since it is open for the public, it is susceptible to be vandalized. Alternately, optical sensors are used for scanning the fingerprints of an individual [8]. This system is most commonly used in every organization because of its high reliability. However, the optical sensor can be used only one at a time which tends to waste a considerable amount of time for large crowds. The optical sensor comes in direct contact with the student. It is exposed to a high risk of getting dirty or damaged.

To overcome the disadvantages of existing ARS, face recognition-based attendance authentication techniques are being developed. Joseph et al. [9] proposed a face recognition-based ARS using principal component analysis (PCA) [10]. Roshan Tharanga et al. [11] proposed a smart ARS based on PCA and Haar transform. Shireesha et al. [12] have used PCA, LBPH and LDA for face recognition in their research [13]. Yohei et al. [14] proposed attendance management system which can estimate the position of each student and attendance by continuous observation and recording.

The biometric systems defined above are efficient and reliable and provides immense security when compared to the traditional method. However, these systems offer some disadvantages as well. Most of the devices are unable to enroll some small percentage of users, and the performance of the system can deteriorate over time. Table 1 gives the advantages and disadvantages of various biometric traits.

1.2 Advantages of Face Recognition-Based ARS

Face recognition-based ARS has proven [15] to be a promising due to the following advantages:

- No physical interaction is required from the user.
- It is very accurate and provides a high level of security.
- It has an advantage of ubiquity and of being universal over other biometrics, i.e., everyone has a face and everyone readily displays the face.
- Non-intrusive nature.
- Use of one biometric data in different environments.
- Ease of use of any camera to capture the biometric data of the faces.

This chapter has its focus on development of face recognition-based attendance recording and management system under both controlled and uncontrolled environments.

Section 2 gives an overview of face recognition-based ARS, elaborate various requirements, and the state of the art. Section 3 describes the proposed ARS system

Table 1 A comparison of various biometric technologies

Biometric trait	Advantages	Disadvantages
Iris recognition	1. Very high accuracy 2. Verification time is less	1. Intrusive 2. Very expensive
Retinal scan	1. Reliable 2. Very accurate and efficient for identifying individuals	1. Device can be used one at a time 2. Time-consuming for large crowd 3. Susceptible to be vandalized
Voice recognition	1. Helps people having trouble with working hands	1. Person needs to be in close contact with the device 2. Not reliable 3. Less accurate with background voice
Hand geometry	1. Easy integration into devices and systems 2. Amount of data required to uniquely identify a user in a system is small	1. Very expensive 2. Considerable size 3. It is not valid for arthritic person, since they cannot put the hand on the scanner properly
Fingerprint	1. Reliable Very accurate and efficient for identifying individuals	1. Device can be used one at a time 2. Time-consuming for large crowd 3. Direct contact with instrument
Signature	1. Non-intrusive 2. Less verification time 3. Cheap technology	1. Not reliable as signatures can be copied if easy 2. Non-consistency in every signature made by an individual
Facial recognition	1. Non-intrusive 2. Cheap technology 3. Less time-consuming	1. Cannot detect partial faces 2. Cannot detect faces with improper illumination, pose variations, occlusion

which is based on partial face recognition algorithms. Section 4 discusses about the experiments conducted and the associated results for both the controlled and uncontrolled environments. Finally, Sect. 5 concludes the chapter.

2 Overview of Face Recognition-Based ARS

Face recognition is defined as a process of identifying a person with the help of his facial features [16]. Face recognition technology involves scanning the distinctive features of the human face to authorize the student.

2.1 Overview

The general block diagram of face recognition-based ARS is shown in Fig. 1. A database of students' personal information along with their face image is to be created first. The existing images in the database are known as the "standard

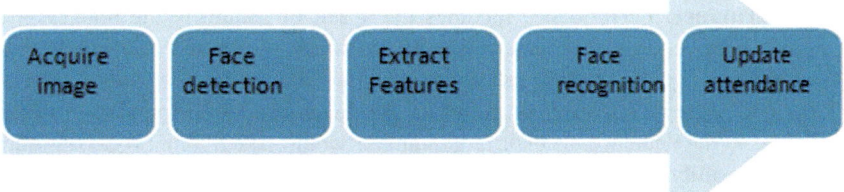

Fig. 1 Steps involved in face recognition-based ARS system

images." The image or the video of students is acquired by using a digital camera or video recorder placed in classroom. Detection of faces from the images or video frames is then performed. It is then required to locate the faces of the students. Locating the faces is a challenging job in real-time applications. The spatial features of the detected faces are then extracted as part of dimensionality reduction. The features also define the behavior of the image. Recognition algorithm is then applied to identify the real-time face image with the database created. A matching score is used to obtain how well the two images are matched. This score matched against the database image reveals the identity of the student.

2.1.1 Requirements

The factors to be considered while selecting face recognition-based ARS are as follows [17]:

- **Uniqueness**: Every person has unique features when compared with every other person. However, this is not true for twins. They have identical features. Thus, a face recognition system must have the ability to identify every person.
- **Universality**: Every person's appearance differs from another person. Due to this reason, the face recognition algorithm might not work for some people while it might work for another set of people. Factors such as long hair, beard, and spectacles might create an extra difficulty to recognize the faces. The resulting solutions for these problems might not equally work well with the others.
- **Permanence**: The human face appearance changes with age. The face might not look the same for a long period of time. It is also subject to permanent changes such as plastic surgery or to temporary changes such as veil or sunglasses.
- **Collectability**: Collectability account for the biometric features that can be determined quantitatively. This biometric system does not require the direct physical contact with the individual whose biometrics are to be captured. Capturing facial images is easy. In fact, a person photograph can be taken without his notice. However, the facial recognition system requires proper

lighting; a correct positioning of the person, long scanning time. Thus, a facial recognition system is a highly professional system.

- **Performance**: Performance includes the speed of acquiring the images, their processing times; which determines the accuracy of correctly recognizing the right faces against the images in the database. The speed of operation depends upon the face recognition algorithm. It also depends upon the number of images stored in the database, as their large number would take a long processing time.
- **Acceptability**: Acceptability defines the user friendliness of the system used in daily lives. Face recognition technique is highly user friendly as it involves a non-intrusive way of capturing the biometric information of the person. It provides an easier access control as compared with the other biometric solutions.
- **Circumvention**: Circumvention states whether the biometric system can be fooled or hacked by other fraud people. It depends on technical implementation, quality of the camera, surrounding background, and algorithm.

2.2   Performance Metrics

The accuracy of any biometric-based system is determined by measuring two kinds of error rates.

- **False Acceptance Rate (FAR)**: FAR is the measure of number of unknown students being falsely accepted into the ARS system as known students. This is called "Type-I error."
- **False Rejection Rate (FRR)**: FRR is the measure of how many known students are falsely rejected by the ARS system as unknown students. This is called "Type-II error."

The authentication [18] procedure requires low false acceptance rate (FAR) which says the score must be high enough before matching. It also requires a low false rejection ratio (FRR) to avoid the unknown students being missed out from marking their attendance.

2.3   State of the Art

Attendance management systems (ARS) using facial recognition techniques have evolved tremendously since the past decade. Various methods such as principal component analysis (PCA) [19], local binary pattern [20], eigenface [21], AdaBoost [22], Haar classifier [23], two-dimensional Fisher's linear discriminate (2DFLD) [24], and 3D modeling [25] have been used for the same.

In the method proposed by Jha et al. [26], color-based technique is used for face detection. This method detects the skin color of humans and its variations. The skin area is then segmented and fed as an input to the recognition process. For face recognition and feature extraction, principal component analysis [27] is used. PCA [28] technique is based on a statistical approach which deals with pure mathematical matrixes. The entire system is implemented in MATLAB. However, the skin tones vary dramatically within and across individuals. Also, due to the changes in the ambient light and shadows, it changes the apparent color of the image. The movement of objects causes blurring of colors.

Shehu and Dika [29] in their work used real-time face detection algorithm which is integrated on the learning management system (LMS). This system automatically detects and registers students present in the classroom. Their approach uses a digital camera installed on a classroom scanning the room every 5 min to capture the images of the students. HAAR classifier is used for face detection. However, the students are required to pay attention to the camera while capturing images. This method, however even detects objects as faces creating a large number of false positives. For face recognition, eigenface methodology is implemented. A drastic change in the student's appearance causes false recognition of the student. A manual method of cropping the region of interest is done to increase its efficiency.

Balcoh [30] proposed a method that uses Viola–Jones algorithm for face detection algorithm and eigenface methodology for face recognition. However, cropping of images is required after the face detection process, in order to recognize the faces of the students. Mao et al. [31] performed the multiobject tracking to convert detected faces into tracklets. This method uses spare representation to clutter the face instances each tracklet into a small number of clusters. Experiments have been performed on Honda/UCSD database. Real-time face detection and recognition is not used.

Tsun et al. [32] performed the experiment by placing the Webcam on the laptop to continuously capture the video of the students. At regular time intervals, frames of the video are captured and used for further processing. Viola–Jones algorithm is used for face detection algorithm due to high efficiency and eigenface methodology is used for face recognition. However, the students are required to remain alert as the eigenface methodology is not capable to recognize tilted faces captured in the frames. Also, a small classroom was used due to the limited field of view of the Webcam used on the laptop.

Yet another method proposed for ARS by Fuzil et al. [33] used HAAR classifier for face detection and eigenface methodology for face recognition. This system is intended only for frontal images. The different facial poses cannot be recognized. Moreover, the faces which are not detected by the HAAR classifier require manual cropping of the facial features which results in lower efficiency of the overall system. Rekha and Chethan, in their proposed method [34], use Viola–Jones algorithm face detection and a correlation method for recognition. This method also uses the manual cropping of the region of interest where is it further compared with the existing database. However, the images where multiple people are captured in the same or different sequence, the face recognition efficiency is very low.

Shirodkar et al. [35] have proposed an ARS using a Webcam to capture the facial features of the students. Facial detection using Viola–Jones algorithm basically including HAAR features, Integral image, AdaBoost, cascading, and local binary pattern (LBP) was used for facial recognition. In this algorithm, the image is divided into several parts, where on each part LBP is applied. An accuracy of 83.2 % was achieved in this system. However, this system overlooks the pose variations which can occur during the lecture hours. Muthu Kalyani et al. [36] proposed a methodology which uses 3D face recognition to provide more accuracy for recognizing the faces from the images stored in the database. It uses a CCTV camera fixed at the entrance of the classroom. However, the facial recognition from still images is a problem under various illuminations, pose, and expression changes. Moreover, the installation cost of CCTV camera is expensive.

Kanti et al. [37] in their paper for ARS use Viola–Jones algorithm for face detection and principal component analysis for face recognition. PCA can be used for holistic faces and not for partial faces. Thus, the facial recognition efficiency reduces for the overall system. In the method proposed by Rode et al. [38], faces are detected using the skin classification method [39]. Eigenfaces are generated for facial recognition. However, the above algorithm is not implemented for real-time images. There is a need for the selection of region of interest in the images for further processing. The comparison of different face recognition-based ARS is given in Table 2.

Table 2 indicates that the existing automated attendance systems are proved effective only for frontal faces. This chapter focuses on these issues and presents an ARS that can improve the performance of existing automated systems, using modified Viola–Jones algorithm to detect faces and alignment-free partial face recognition algorithm for face recognition.

3   Proposed ARS

A database created with the student's personal data along with their face images. Figure 2 shows the block diagram of the system. A camera is used to capture the images of the faces or to capture the real-time video. Optical devices such as the camera or video recorder are used to accomplish this task. The students face images to be recognized are fed to the image processing block where it performs preprocessing, face detection, and face recognition tasks. Preprocessing includes tasks such as cropping of image and enhancement procedures. These processed images are fed to the face recognition algorithm. These database images are then compared with the real-time recognized faces to identity the student.

Table 2 Summary of face recognition-based ARS

Author	Face detection algorithm	Face recognition algorithm	No. of cameras	Face type	Efficiency of face detection	Efficiency of face recognition
Naveed Khan et al. [29]	Viola–Jones	Eigenface	One (top of blackboard)	Holistic	40 % veil 95 % unveil 75 % beard	2 % veil 85 % unveil 63 % beard
Muhammad et al. [32]	HAAR classifier	Eigenface	One (at the entrance of classroom)	Holistic	<50 %	<30 %
Visar Shehu et al. [26]	HAAR classifier	Eigenface	One (top of blackboard)	Holistic	70 %	56 % (Due to position of camera)
Jonathan et al. [31]	Viola–Jones	Eigenface	Laptop camera	Holistic	100 %	66 % (Assumption: object has no variations in face pose)
Rekha et al. [33]	Viola–Jones	Correlation technique	One (on top of the blackboard)	Holistic	70 %	90 % (Due to cropping ROI)
Yunxiang et al. [30]	HAAR classifier	Global Tracklet Recognition	One (top of blackboard)	Holistic	95 % (Assumption: all students are sitting upright)	75 % (Due to tracking of the faces)
Mrunmayee et al. [34]	Viola–Jones	Local binary pattern	One (top of blackboard)	Holistic	78 %	83.33 %
Muthu et al. [35]	Viola–Jones	3D methodology	One (entrance of classroom)	Holistic	100 %	70 %
Jyotshana et al. [36]	Viola–Jones	Principal component analysis	Webcam attached to the PC	Holistic	100 %	100 %
Rahus et al. [37]	Skin classification	Eigenfaces	NA	Holistic	100 %	100 %

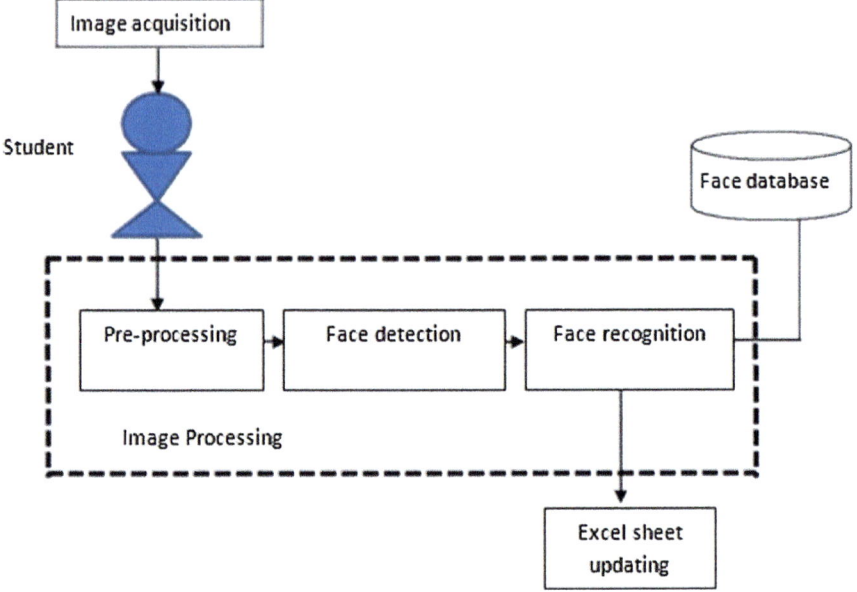

Fig. 2 Block diagram of the proposed ARS

3.1   Face Detection Using Viola–Jones Algorithm [40]

The face detection in the proposed ARS is performed by employing modified Viola–Jones algorithm. The images of the students are captured from the camera placed at the top center of the blackboard at fixed intervals. These images are then preprocessed and are converted to gray scale before performing the face detection. The Viola–Jones method uses integral images to compute the features of the faces. The advantages of Viola–Jones algorithm are as follows:

- High accuracy,
- Low false detection rate,
- Fast feature extraction is possible,
- Location and scale invariant feature detector,
- Features are scalable.

A subwindow is swept across the selected real-time image for catching the faces. Before this step, the rescaling of the image is done to different sizes and then run the size locator across these images. However, in the Viola–Jones algorithm, the detector is rescaled rather than the images. The locator is operated every time with a different size through one image at a time. Viola–Jones algorithm uses a scale invariant detector. The locator is established using a fundamental picture and HAAR wavelets. It uses AdaBoost algorithm to select important features. The background region present in the images is eliminated. The majority of the computational time is spent on face regions. Boundary boxes are then inserted for the detected faces. The size of the boundary box depends upon the size of the face detected in an image or video frame.

3.2   Face Recognition Using MKD-SRC Representation

The proposed system employs the MKD-SRC method of partial face recognition proposed by Liao et al. [41] irrespective of whether the face detected is holistic or partial. This approach represents database images and real-time images as multi-keypoint descriptor (MKD) and then applies sparse representation-based classification (SRC) for face recognition. This kind of representation performs well in cases where the training data available is small. Figure 3 shows the block diagram of MKD-SRC-based face recognition employed in the proposed system.

Each face in the database is represented by a set of descriptors, where the size of descriptors depends upon the information available about the face. Therefore, the descriptor size for frontal images is large when compared to partial faces in the database. A scale and affine invariant detector namely CanAff is first used to detect the key points. This detector is robust to viewpoint changes and, hence, works best in uncontrolled environments, where the captured images are composed of more partial faces.

Each detected key point region is first enclosed by ellipse and then is normalized to a uniform size circles using affine transformations. After normalizing the detected key point regions to a fixed size, a Gabor ternary pattern (GTP) descriptor is constructed for each key point region. GTP features are robust to illumination changes and noise and are proved as best local features. The procedure is as follows:

Apply Gabor filter to each detected and normalized region. Only odd Gabor kernels are processed with single scale and with just four different orientations

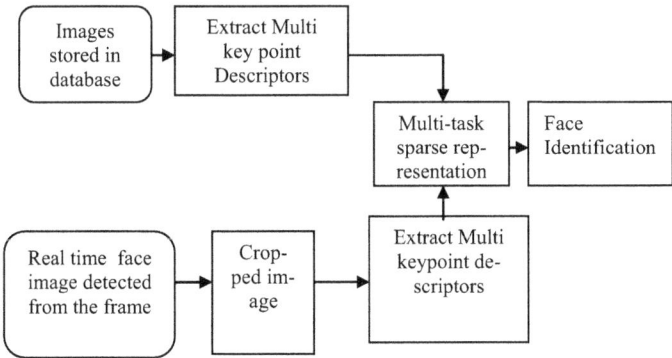

Fig. 3 Block diagram of face recognition algorithm

($0°$, $45°$, $90°$, and $135°$) as they are efficient at detecting edges in these directions. The results of these four Gabor filters are combined to form a ternary pattern called Gabor ternary pattern (GTP), which has the same size as the original detected region. Each GTP region is then divided into subregions and a histogram is extracted for every subregion. The resultant histograms are all concatenated to form a feature vector. To eliminate the extremes and outliers and to form fixed size a normalization step is applied. Finally, principal component analysis (PCA) is applied to reduce the dimensionality of feature vector to M. The resultant feature vector is called multikeypoint descriptor (MKD).

The MKD's obtained from each image in a class forms a subdictionary. A class here is defined as a set of images with different poses collected for the same student. A dictionary is composed of all subdictionaries and represents features from all images in a class.

Let C denote the total number of classes. Then, the class dictionary

$$D = (D_1, D_2, \ldots, D_c) \tag{1}$$

where each dictionary D_i corresponds to one class and is given by

$$D_i = (di_1, di_2 \ldots di_n) \tag{2}$$

where n is the total number of images in a class, and di_n is a subdictionary for nth class image, which represents one MKD. The dictionary D gives complete description about the database of images. As the dictionary size depends on the size of the input real-time detected face image, filtering is adapted to keep only the largest values of the descriptors.

Application of sparse representation-based classification (SRC) is preferred in this paper for its effectiveness in the classification. From the theory of compressed sensing (CS) [41], "a sparse solution is possible for an over complete dictionary and hence any descriptor from real-time image can be expressed by a sparse linear combination of the dictionary **D**, with a high probability using $\ell 1$ minimization." Inspired by this statement, a multitask SRC based on least residual, proposed by Laio et al. [41] is employed directly, to determine the identity of the real-time partial face image.

3.3   Controlled Environment

In the controlled environment, the camera is fixed to the wall. Each student is required to come in front of this camera to get his/her image to be captured. In the controlled environment, motion detection algorithm (in built in MATLAB) is used to detect if a student came in front of the camera.

3.4 Uncontrolled Environment

In the uncontrolled environment, the camera is placed on top of the blackboard to capture the video of the students during the lecture. The wide capturing angle of the camera enables to capture all the students present in the classroom. Images can be captured at regular intervals or video can be taken by using a high definition camera. Figure 4 shows the camera arrangement in the classroom.

Figure 5 shows flowchart for the proposed ARS for one lecture. At the start of the lecture, camera is initiated to capture the video of the students attending the lecture. One second of the video consists of approximately 30 frames. These frames are further processed one at a time.

The faces of students are detected using the Viola–Jones algorithm. The detected faces are fed as an input to the face recognition algorithm. If all the students are not recognized in the first frame, then another frame is given as an input to the face detection algorithm. Further, these detected faces are given to the face recognition algorithm. This process continues till all students are detected and recognized. At the end of the lecture, recognized students are awarded with points. This data is stored in the attendance sheet, and their respective percentage of attendance is calculated.

Fig. 4 Demonstration of camera arrangement in the classroom

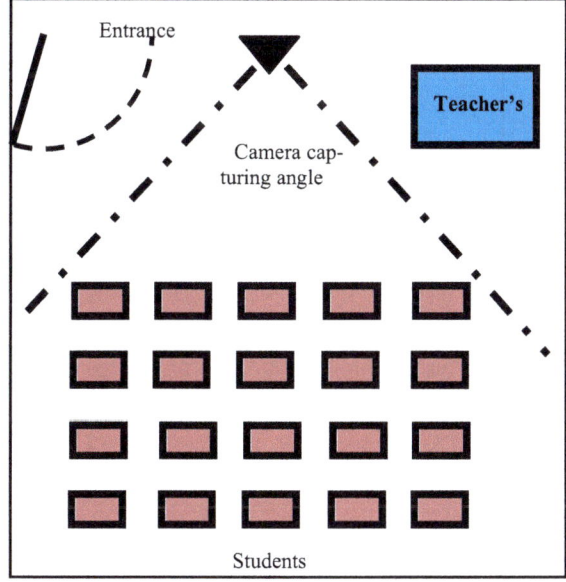

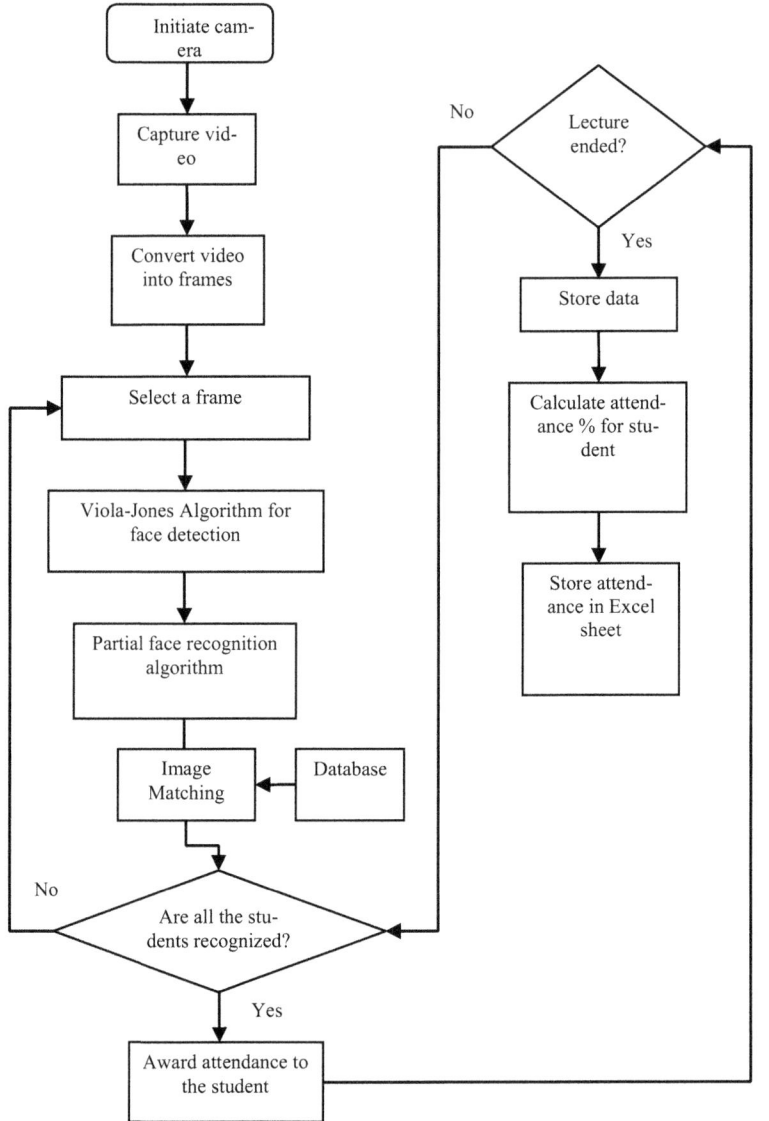

Fig. 5 Flowchart of the uncontrolled environment

4 Results and Discussions

MATLAB 2012b version is used for implementation of software algorithm. The computer vision toolbox is used for implementations of various algorithms related to feature detection, feature extraction, feature matching, object detection and tracking, motion estimation, and video processing. A DSLR 3100 (digital single-lens reflex) camera is used for image acquisition and video capturing. The proposed algorithms and systems are defined, implemented, tested, and performance evaluated for two scenarios: controlled and uncontrolled environments.

4.1 Database Creation

Initially, a database is created for 20 students with 10 different poses. Table 3 shows the database images along with their roll numbers.

4.2 Results of Controlled Environment

Table 4 presents the results of face recognition in controlled environment. A test image is the real-time image captured by the camera when a student comes in front of it. A recognized image is a photograph of a student that scores high when compared with the test image in features. After recognition algorithm updates excel sheet accordingly. Table 4 shows that all the 20 students' faces were successfully detected and recognized. Hence, for the proposed attendance recording system, the face detection efficiency is 100 % and face recognition efficiency is 100 %.

4.3 Results of Uncontrolled Environment

Figure 6 shows a sample frame from the real-time video captured by the camera, placed above the blackboard. Table 5 shows the results of the proposed ARS under uncontrolled environment. It shows the number of faces detected of the students and whether these detected faces are recognized. Figure 6 shows one frame from the video.

The proposed system is trained with a class of frontal and partial face images. For each class, a dictionary is created which contains the biometric information at the training stage. During face recognition, the features extracted from every real input image matched against the trained images dictionary. The sparse representation is used in calculating the minimum distance between them.

However, there are chances of an unknown student being identified correctly against the known student. Thus, a threshold level is maintained to avoid the unknown student being correctly recognized. Ideally, the minimum distance of the unknown must be lower than the known student. Achieving this ideal case in the real world scenario is quite a challenge. A threshold is chosen such high that no unknown student's results will exceed the result of the known student. This will reduce the false acceptance of the system. On the other hand, the known student's

Table 3 Database of 20 students

Sl. No	Roll No.	Database images of students
1	1KS14EC01	
2	1KS14EC02	
3	1KS14EC03	
4	1KS14EC04	
5	1KS14EC05	
6	1KS14EC06	
7	1KS14EC07	

(continued)

Table 3 (continued)

8	1KS14EC08					
9	1KS14EC09					
10	1KS14EC10					
11	1KS14EC11					
12	1KS14EC12					
13	1KS14EC13					
14	1KS14EC14					

(continued)

Table 3 (continued)

15	1KS14EC15	
16	1KS14EC16	
17	1KS14EC17	
18	1KS14EC18	
19	1KS14EC19	
20	1KS14EC20	

results are lower than the highest unknown student's results. To avoid this situation, we can choose a threshold so low that no known student's images are falsely rejected. This also results in the false acceptance of the unknown students. Therefore, choosing a threshold as a compromise between them is necessary.

In the proposed system, in the uncontrolled environment, only 16 students out of 20 were detected. Thus, the face detection efficiency is 80 %. Out of which only 12 students were recognized. The face recognition efficiency is 60 %. Note that, the output differed from frame to frame. This is because, in each frame, the number of

Table 4 Results of proposed ARS under controlled environment

Sl. No.	USN	Test image	Recognized image	Impression
1	1KS14EC01			√
2	1KS14EC02			√
3	1KS14EC03			√
4	1KS14EC04			√
5	1KS14EC05			√
6	1KS14EC06			√
7	1KS14EC07			√
8	1KS14EC08			√
9	1KS14EC09			√
10	1KS14EC10			√
11	1KS14EC11			√
12	1KS14EC12			√

(continued)

Table 4 (continued)

13	1KS14EC13		√
14	1KS14EC14		√
15	1KS14EC15		√
16	1KS14EC16		√
17	1KS14EC17		√
18	1KS14EC18		√
19	1KS14EC19		√
20	1KS14EC20		√

Fig. 6 A frame captured from the video

Table 5 Results of proposed ARS under uncontrolled environment

Sl. No.	Students detected	Students recognized	Impression
1			√
2			√
3			x
4			√
5			√
6			x

(continued)

Table 5 (continued)

7		√
8		√
9		√
10		√
11		x
12		√

(continued)

Table 5 (continued)

13	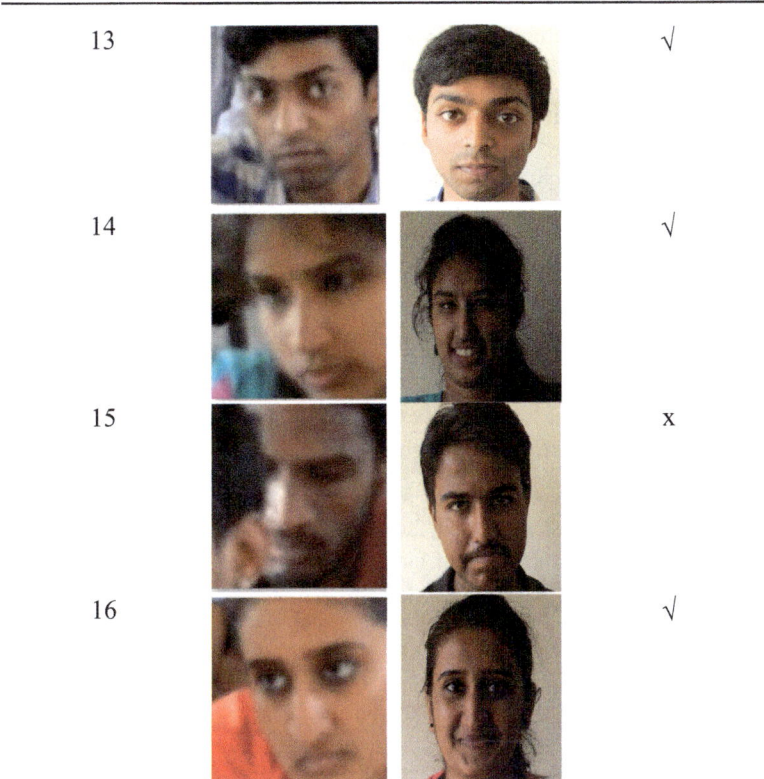	√
14		√
15		x
16		√

Table 6 Consolidated results of proposed ARS

	Face detection efficiency (%)	Face recognition efficiency (%)
Controlled environment	100	100
Uncontrolled environment	80	60

students being detected varies. This affects the face recognition performance. Comparing the Tables 4 and 5, it can be shown that the results of the controlled environment are more accurate than the ones for uncontrolled environment. The consolidated results for the controlled and uncontrolled environment are shown in Table 6. Note that the image quality affects the output results. The image quality in the controlled environment is better than in the uncontrolled environment.

4.3.1 Advantages

The face recognition method employed in the system is alignment free [41]. This method is capable of recognizing the partial faces under the following conditions:

- Self-occlusion: this includes blocking the face due to non-frontal poses.
- Facial accessories: this includes blocking of the face due to facial accessories
- Limited view: this includes the faces which lie out of the camera's field of view.
- Extreme illumination: this includes images in which the facial area is gloomy or high-lightened.
- Sensor saturation: this includes the underexposure or overexposure of the facial areas in the images.
- External occlusion: this includes blocking of the faces due to other objects or faces in the image.

Further, the alignment-free partial face recognition algorithm has the following advantages over the other existing algorithms:

- No prealignment of the images is required,
- No presence of eye or any other facial component required,
- No prior knowledge of the input face required, i.e., whether the face is holistic or partial.

4.4 Limitations

It is also observed that the proposed ARS system works under some limitations. Due to the low image quality in the uncontrolled scenario, the efficiency of the system is reduced. Also the computation time required is large when the number of images stored in the database is large. Further work can be carried out to make the system more efficient in the real-time scenario. Also, the training images for each student can be increased to make the system more robust to the recognition problem. Moreover, different ways can be explored to reduce the computational time. This can be achieved by using much more efficient algorithms.

The camera height can be raised to prevent the blockage created by tall students. The use of more number of cameras mounted across the classroom, and by employing image stitching algorithms can enhance the capacity.

An independent system can be implemented by extending the hardware to Internet Protocol (IP) camera. This effort can make the system more accurate as it can continuously monitor the students. An FPGA based system can be implemented as a future scope.

5 Conclusion

A smart attendance capturing and management system is proposed to overcome the existing drawbacks present in the biometric-based attendance management systems, based on existing partial face recognition algorithms. The proposed system was implemented in two phases: controlled environment and uncontrolled environment. The facial features of the students are captured and recognized. The names of the recognized students are then updated in the excel sheet. This system could be used for attendance marking of the students and staff in any organization. This system saves time and manual effort otherwise required to put by the lecturer. This system helps lecturers' to efficiently manage a large number of students present in the classroom. The system will also help prevent a large number of students from skipping the daily classes.

The proposed system proved accurate under controlled environment. While the efficiency of uncontrolled environment is quite low, it is user friendly. The performance of the proposed attendance system completely depends upon the resolution of the camera used, the number of students detected and hence recognized. Further, the large number of images stored in the database increases the computation time. The techniques employed to make the existing ARS system more efficient and user friendly is never ending.

References

1. Dey M, Dey N, Mahata SK, Chakraborty S, Acharjee S, Das A (2014) Electrocardiogram feature based inter-human biometric authentication system. In: 2014 International conference on electronic systems, signal processing and computing technologies (ICESC). IEEE, pp 300–304
2. Sharma S, Shimi SL, Chatterji S (2014) Radio frequency identification (rfid) based employee monitoring system (EMS). Int J Curr Eng Technol 4(5):80–83
3. Bhalla V, Singla T, Gahlot A, Gupta V (2013) Bluetooth based attendance management system. Int J Innovat Eng Technol 3:2319–1058
4. Maji P, Chatterjee S, Chakraborty S, Kausar N, Samanta S, Dey N (2015) Effect of euler number as a feature in gender recognition system from offline handwritten signature using neural networks. In: 2015 2nd International conference on computing for sustainable global development (INDIACom). IEEE, pp 1869–1873
5. Sirshendu II, Sankhadeep C, Santhi V, Fuqian S (2015) Indian sign language recognition using optimized neural networks. In: International conference on information technology and intelligent transportation systems (ITITS 2015)
6. Rashid RA, Mahalin NH, Sarijari MA, Abdul Aziz A (2008) Security system using biometric technology: design and implementation of Voice Recognition System (VRS). In: International conference on computer and communication engineering, 2008. ICCCE 2008. IEEE, pp 898–902
7. Dehghani A, Ghassabi Z, Moghddam HA, Moin MS (2013) Human recognition based on retinal images and using new similarity function. EURASIP J Image Video Process 1:1–10
8. Cappelli R, Maio D, Maltoni D, Wayman JL, Jain AK (2006) Performance evaluation of fingerprint verification systems. IEEE Trans Pattern Anal Mach Intell 28(1):3–18
9. Joseph J, Zacharia KP (2013) Automatic attendance management system using face recognition

10. Zhu J, Yu YL (1994) Face recognition with eigenface. IEEE international conference on industrial technology, pp 434–438
11. RoshanTharanga JG, Samarakoon SMSC, Karunarathne TAP, Liyanage KLPM, Gamage MPAW, Perera D (2013) Smart attendance using real time face recognition (smart-fr). SAITM Research Symposium on Engineering Advancements 2013
12. Chintalapati S, Raghunadh MV (2013) Automated attendance management system based on face recognition algorithms. In: 2013 IEEE international conference on computational intelligence and computing research (ICCIC). IEEE, pp 1–5
13. Javed A (2013) Face recognition based on principal component analysis. Int J Image Graph Sig Process 5(2):38
14. Kawaguchi Y, Shoji T, Weijane LIN, Kakusho K, Minoh M (2005) Face recognition-based lecture attendance system. In: The 3rd AEARU workshop on network education, pp 70–75
15. Selvi KS, Chitrakala P, Jenitha AA (2014) Face recognition based attendance marking system
16. Gunes H (2010) Automatic, dimensional and continuous emotion recognition
17. Virmani J, Dey N, Kumar V (2016) PCA-PNN and PCA-SVM based CAD systems for breast density classification. In: Applications of intelligent optimization in biology and medicine. Springer, New York, pp 159–180
18. Dey N, Nandi B, Das P, Das A, Chaudhuri SS (2013) Retention of electrocardioGram features insiGnificantly devalorized as an effect of watermarkinG for. Advances in biometrics for secure human authentication and recognition, p 175
19. Yang J, Zhang D, Frangi AF, Yang JY (2004) Two-dimensional PCA: a new approach to appearance-based face representation and recognition. IEEE Trans Pattern Anal Mach Intell 26 (1):131–137
20. Rongbao C, Shibing L (2009) A fast face recognition based on LDA and application on auto-guard against theft. J Univ Sci Technol China 39:181–185
21. Turk MA, Pentland AP (1991) Face recognition using eigenfaces. In: Proceedings CVPR'91, IEEE computer society conference on computer vision and pattern recognition. IEEE, pp 586–591
22. Huang D, Wang YH, Wang YD (2007) A robust infrared face recognition method based on adaboost gabor features. In: International conference on wavelet analysis and pattern recognition, 2007. ICWAPR'07, vol 3. IEEE, pp 1114–1118
23. Nishimura J, Kuroda T (2010) Versatile recognition using Haar-like feature and cascaded classifier. IEEE Sens J 10(5):942–951
24. Cao L, Chen D, Fan J (2012) Face recognition using the wavelet approximation coefficients and fisher's linear discriminant. In: 2012 5th International congress on image and signal processing (CISP). IEEE, pp 1253–1256
25. Sun Y, Yin L (2008) 3D Spatio-Temporal face recognition using dynamic range model sequences. In: IEEE computer society conference on computer vision and pattern recognition workshops, 2008. CVPRW'08. IEEE, pp 1–7
26. Jha A (2007) Class room attendance system using facial recognition system. Int J Math Sci Technol Manage 2(3):4–7
27. Kumar P, Agarwal MM, Nagar MS (2013) A survey on face recognition system-a challenge. Int J Adv Res Comput Commun Eng, 2(5):2167–2171
28. Nandi D, Ashour AS, Samanta S, Chakraborty S, Salem MA, Dey N (2015) Principal component analysis in medical image processing: a study. Int J Image Mining 1(1):65–86
29. Shehu V, Dika A (2010, June) Using real time computer vision algorithms in automatic attendance management systems. In: 2010 32nd International conference on information technology interfaces (ITI). IEEE, pp 397–402
30. Balcoh NK, Yousaf MH, Ahmad W, Baig MI (2012) Algorithm for efficient attendance management: Face recognition based approach. IJCSI Int J Comput Sci Issues 9(4):146–150
31. Mao Y, Li H, Yin Z (2014) Who missed the class? Unifying multi-face detection, tracking and recognition in videos. In: 2014 IEEE international conference on multimedia and expo (ICME). IEEE, pp 1–6
32. Tsun JCE, Jen CW, Mei FCC (2014) Automated attendance capture system 2nd Eureca 2014, pp 1–2

33. Fuzail M, Nouman HMF, Mushtaq MO, Raza B, Tayyab A, Talib MW (2014) Face detection system for attendance of class' students. Int J Multidiscip Sci Eng, 5(4), pp 6–10
34. Rekha AL, Chethan HK (2014) Automated attendance system using face recognition through video surveillance. Int J Technol Res Eng 1(11):1327–1330
35. Shirodkar M, Sinha V, Jain U, Nemade B (2014) Automatic attendance management system using face recognition. Int J Comput Appl 3:23–28
36. MuthuKalyani K, VeeraMuthu A (2013) Smart application for AMS using face recognition. Comput Sci Eng 3(5):13
37. Kanti J, Papola A (2014) Smart attendance using face recognition with percentage analyzer. Database, Int J Adv Res Comput Commun Eng, 3(6):7321–7324
38. Rode RS, Dahelkar NA, Nagdive NR, Gajbhiye AS, Shriwas MK (2015) Review of automatic attendance using facial recognition. Int J Innovat Res Comput Commun Eng 3(3):1716–1723
39. Dey N (ed) (2016) Classification and clustering in biomedical signal processing. IGI global
40. Viola P, Jones MJ (2004) Robust real-time face detection. Int J Comput Vision 57(2):137–154
41. Liao S, Jain AK, Li SZ (2013) Partial face recognition: Alignment-free approach. IEEE Trans Pattern Anal Mach Intell 35(5):1193–1205

Automatic Human Emotion Recognition in Surveillance Video

J. Arunnehru and M. Kalaiselvi Geetha

Abstract Recognition and study of human emotions have fascinated a lot of attention in the past two decades and have been researched broadly in the field of computer vision. The recognition of complete-body expressions is significantly harder, because the pattern of the human pose has additional degrees of self-determination than the face alone, and its overall shape varies robustly during articulated motion. This chapter presents a method for emotion recognition based on the gesture dynamics features extracted from the foreground object to represent various levels of a person's posture. The experiments are carried out using publicly available emotion recognition dataset, and the extracted motion feature set is modeled by support vector machines (SVM), Naïve Bayes, and dynamic time wrapping (DTW) which are used to classify the human emotions. Experimental results show that DTW is efficient in recognizing the human emotion with an overall recognition accuracy of 93.39 %, when compared to SVM and Naïve Bayes.

Keywords Emotion recognition · Dynamic time wrapping · Support vector machines · Naïve Bayes · Object detection

1 Introduction

Human–Computer interaction is mounting its attention nowadays. In order to put some prominence on socializing computer with human, understanding the human body gestures and visual cues of an individual is a need [1]. It allows a system to understand the expressions of humans in turn, enhancing its effectiveness in performing various tasks. It serves as a measurement system for behavioral science,

J. Arunnehru (✉) · M. Kalaiselvi Geetha
Department of CSE, Annamalai University, Chidambaram, India
e-mail: arunnehru.aucse@gmail.com

M. Kalaiselvi Geetha
e-mail: geesiv@gmail.com

© Springer International Publishing Switzerland 2017
N. Dey and V. Santhi (eds.), *Intelligent Techniques in Signal Processing for Multimedia Security*, Studies in Computational Intelligence 660, DOI 10.1007/978-3-319-44790-2_15

and socially intelligent software tools can be accomplished. Recognition of gesture-based human emotions has attracted a lot of attention in the precedent two decades and has been researched broadly in the field of computer vision. The recognition of complete-body expressions is significantly difficult to express, because the semantic pattern of the human pose has additional degrees of range than the face alone, and its overall shape varies robustly during articulated motion [2, 3].

Human emotion recognition is having a wide range of human–computer interaction (HCI)-related applications. Future shopping is going to be more analytical in the near future using video to track emotions of the buyers. Future retail will avoid silly queries from the customers and combine business, technology, human behavior, and psychology using HCI. For example, if a customer looks frustrated in interactive monitors, then the retailer could understand that something needs to be done for him. Thus, HCI enables the marketers to analyze the actual customer behavior in emotions in real time than to process-biased answers or survey questions.

This chapter investigates recognizing emotions from gestures which is a challenging task. Automatic human emotion recognition system could offer the marketers the ability to interact with the customers in real time, facilitate better decisions, and to be effective in providing service to customers.

Human emotions emerge differently on the same external stimulus through individual standards in the individual style [4]. According to gender, age, society, or residential areas, the emotion expression can be different on the same multimedia information. The most important challenge in HCI-based system is to provide the capability to computers to evaluate human emotions strongly. The immense potential applications of human emotion recognition, such as interactive learning systems, consumer care, web cinema, safety and video surveillance, just to name a few, provide increased need of a strong human emotion recognition. A massive amount of research has been done on human emotions, and it is almost impossible to identify all of it. One can categorize the proposed methods based on features extraction. Action recognition systems can be divided into three categories: (1) walking, (2) sitting, and (3) jumping.

Generally, human emotions consist of 3 principal emotions: happy, angry, and fearful. Remaining emotions are considered to be combinations of these primary emotions. The outcome reported the efficacy of combining the visual information into single framework. The majority of the multimodal systems better the modal approaches for emotion recognition applications. To understand the emotions from human pose motion in normal environments can be extremely hard as body movements are almost unconstrained. This makes it complicated to train emotion recognizer's which are strong as much as necessary to endure this kind of real-world inconsistency problems.

Further, an emotion recognition approach must be able to discriminate between emotions like happy, angry and fearful, combined with activities like walking, sitting, and jumping performed by an individual person. This is not an easy task, since certain movements such as going from happy-walking posture to angry-walking posture have strong similarities. Hence, the problem of human emotion recognition is viewed as motion gesture in this proposed work. This chapter analyzes the emotion recognition problem as vision-based gesture recognition. In vision-based gesture recognition, the procedure is carried at four steps, viz. human detection, human tracking, emotion recognition, and then a complex activity assessment to evaluate happy, angry, and fearful emotions.

Automatic emotion recognition has turned into a significant research area in computer vision for last few decades with applications in video surveillance, sport event analysis, human–computer interaction, computer-aided games, etc. Recent studies on visual analysis of human body pose movement reveal that the human activities vary from other motion movements. Speech emotion recognition is achieved through untainted sound processing without linguistic information. Anagnostopoulos et al. [5] proposed speech emotion recognition through processing approaches that include the separation of the speech signal, and speech features are extracted for the emotion classification. For capturing the information about emotion from audio and facial features, auto associative neural network (AANN) models [6] are considered in video for emotion recognition. Although several automatic action recognition systems have considered the use of both gesture and facial expressions, relatively few efforts have focused on emotion recognition using both modalities. The emotion recognition is based on the combination of facial expressions and speech data [7] and facial expressions and gesture [8]. Karpouzis et al. [9] presented a multi-cue framework based on facial, oral, and bodily expressions to model emotional states, but the synthesis of modalities is modeled at the point of facial expressions and speech information only. The effort–shape analysis [10] was used to illustrate the movement style distinctiveness connected with each of the objective emotions on knocking movement, and the target emotions are angry, anxious, content, joyful, proud, and sad. Castellano et al. [1] proposed human emotion recognition of four performed emotional states (angry, joy, fearful, and sadness) based on the psychiatry of body pose movement and gesture expressivity. The various methods from psychology focus on the relationships between action and emotion behavior, investigating expressive body movements [11, 12]. The automatic analyses of emotions in multimedia records are helpful for indexing and retrieving the multimedia information based on emotion-specific information [13].

Kapoor et al. [14] presented a method based on correlation between body posture and aggravation in a computer-based training environment. Kapur et al.'s [15] four basic emotions may possibly be automatically distinguished from statistical measures of motion's dynamics. Balomenos et al. [16] proposed a technique by fusing the facial expressions and body gestures for the recognition of six classical emotions. Camurri et al. [17] presented a method based on human full-body movement to discriminate the expressive gestures. In particular, they recognized motion cues similar to time duration, contraction index, magnitude of motion, and motion confidence. From these motion cues, they defined an intelligent classifier that has the ability to discriminate four emotions, namely angry, fear, sorrow, and joy. Zhaojun Yang et al. [18] presented a graph-based approach to identify the gesture's dynamic pattern and emotion from body motion cues in common interpersonal interactions, where dynamic patterns for particular emotions from a weighted graph-based method improved separation among distinct emotion classes in order to maintain fewer inconsistencies within the similar emotion class. Zaboleeva-Zotova et al. [19] presented a new methodology for recognition of human emotions based on analysis of body distinguishing gestures and poses; the characteristics of body pose movements are modeled with linguistic variables for sequential activities.

From the literature, it is clear that emotions are mostly expressed as mental or psychological states [20], which are the most important human cognitive features that attract life by relations processed in the human race [21]. Mental states like happy, angry, jealousy, fear, envy, indignation, embarrassment are classified and referred to "Emotion." Recently, artificial intelligence researchers have measured the significance of adding emotion to computers, which prioritize the primary human motive and direct their processing for full expression of emotions [22]. However, one can easily see emotion from an image sequence than from a still image. One reason for that is the information from the image sequence contains appearance and motion feature. Due to this reason, the analysis of dynamic image sequences has become very attractive in computer vision, virtual reality [23], and cyborgs [24] field; this has motivated to analyze the gesture expressions from a dynamic image sequence in the proposed work.

2   Emotion Recognition Framework—Overview

For better understanding of the underlying study of this chapter, a real-life scenario for emotion recognition is experimented. Dataset comprising human emotion performed by several actors in static environment is used for experimental purpose. The proposed work employs ideal features, extraction approaches, and classifier which reveal promising outcomes. The overview of the proposed approach is shown in Fig. 1.

This chapter deals with human emotion recognition, which aims to discriminate different emotions based on actions from the video sequences. The proposed method is evaluated using emotion recognition dataset [25], considering emotions such as happy, angry, and fearful with activities like walking, sitting, and jumping. Background subtraction technique is applied to current frame in order to obtain the foreground object. Thus, the motion and shape features are computed and chosen as

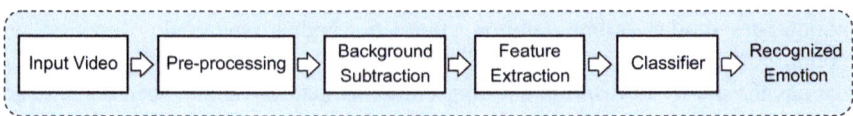

Fig. 1 Overview of the proposed emotion recognition approach

a feature set. The extracted feature set is fed to the SVM, Naïve Bayes, and DTW classifiers for classification.

2.1 Feature Description

The extraction of selective characteristic is the most essential crisis in human emotion and activity recognition which represents the momentous information that is necessary for further study. To identify a person's movement across image sequences, background subtraction approach is widely used. The foreground object is detected by subtracting the current frame from the reference frame. The foreground images are obtained by applying thresholds to reduce pixel modification due to camera noise and changes in illumination conditions. This process is really adaptive to perceive the motion region equivalent to moving objects in static scenes and better quality for extracting significant feature pixels.

The foreground object is obtained by simply subtracting the current frame at time $I(x, y, t)$ from background reference frame at time $B(x, y, t)$ on a pixel-by-pixel basis. The extracted foreground object $F(x, y, t)$ is considered as the region of interest (ROI) or minimum bounding box. Figure 2a and Fig. 2b illustrate the background frame and current frame of the happy-walk emotion dataset. The resulting foreground object is shown in Fig. 2c. The sample foreground objects for different human emotions are extracted from the emotion dataset [25] and are shown in Fig. 3. From the figure, it is clearly indicated that the foreground object (shape) represents the different gesture dynamics of action and emotion. The foreground object $F(x, y, t)$ is calculated using

$$F(x, y, t) = |I(x, y, t) - B(x, y, t)| \tag{1}$$
$$1 \le x \le w, 1 \le y \le h$$

$$F(x, y, t) = \begin{cases} 1, & \text{if } I(x, y, t) < t \\ 0, & \text{otherwise} \end{cases} \tag{2}$$

where w—width of the foreground object, h—height of the foreground object, and threshold $t = 30$ is to be considered in this work.

(a)   **(b)**   **(c)**

Fig. 2 **a** Background model, **b** current frame, **c** foreground object obtained from (**a, b**)

Fig. 3 Sample foreground objects from emotion recognition dataset

To recognize the human emotion, motion and shape information is an essential signal usually extracted from video sequences. Distance, speed, orientation, elongation, solidity, and rectangularity measures are compact representation of motion and shape information, since much valuable information is retained in this measure. The distance and speed are extracted from the successive frame that consists of foreground object motion information only. The orientation, elongation, solidity, and rectangularity are extracted from the foreground object that consists of shape data only.

Distance is a measure between successive frame objects by finding the centroid of the object.

$$\text{Centroid} = \left(\text{ROI}_{\text{width}/2}, \text{ROI}_{\text{height}/2}\right) \tag{3}$$

Euclidean distance measure is used to calculate the distance between (x_1, y_1) and (x_2, y_2) centroid points.

$$D = \sqrt{(x_2 - x_1)^2 + (y_2 - y_1)^2} \tag{4}$$

where D—is the distance, (x_1, y_1)—centroid point on frame t, (x_2, y_2)—centroid point on frame $t + 1$.

Speed is an essential cue to determine the object's motion as fast or slow. Speed is defined as distance divided by time. Distance is directly proportional to velocity when time is constant. After finding the displacement of the object, speed is calculated using

$$S = \frac{D}{t} \qquad (5)$$

where S—is the speed in m\s, D—is the distance travelled in pixels, t—is the time taken, $t = 0.08$ (25 fps\2—two frames used for distance measure).

Orientation is a measure of observant angle between the x-axis and the y-axis of the ellipse that has the similar second-moments as the region. The obtained cost is in degrees, ranging from $-90°$ to $90°$. Figure 4 illustrates an image area, and its corresponding ellipse and same ellipse with the solid red lines represents the axes; the orientation is the angle between the horizontal line axis and the vertical line axis.

Elongation (Elo) is also called as minimum bounding rectangle or minimum bounding box of an arbitrary shape. In arbitrary shape, eccentricity is the relation of the length L and width W of minimum bounding rectangle of the shape at some set of orientations as shown in Fig. 5 Elongation is a measure of values range from [0,1].

$$Elo = 1 - W/L \qquad (6)$$

A regular shape in all axes such as a circle or square will have an elongation value of 0, whereas shapes with large aspect ratio will have an elongation closer to 1.

Solidity characterizes the extent corners to which the shape can be represented by convex or concave, and it is defined by

Fig. 4 Orientation of the angle between x-axis and y-axis moments

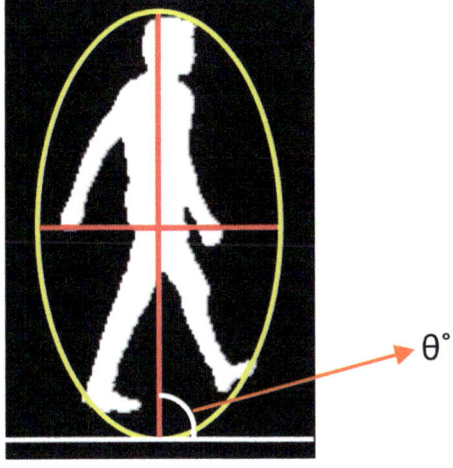

Fig. 5 Minimum bounding
box and corresponding
parameters for elongation

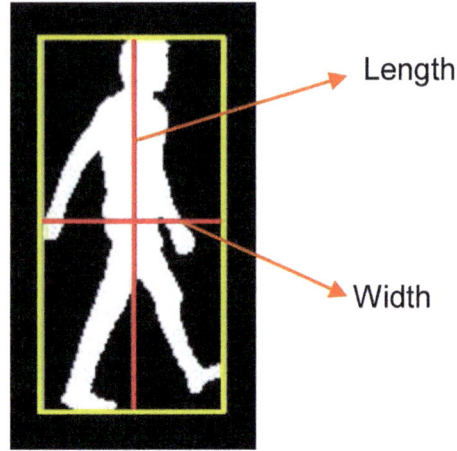

$$\text{Solidity} = A_S / H \tag{7}$$

where A_s is the shape region area and H is the shape area of convex hull. The solidity of a convex shape is always 1 as shown in Fig. 6.

Rectangularity describes the rectangular shape is how much it fills its minimum bounding rectangle as shown in Fig. 7

$$\text{Rectangularity} = A_s / A_R \tag{8}$$

where A_S is the shape area, A_R is the minimum bounding rectangle area.

The motion and shape cues such as distance, speed, orientation, elongation, solidity, and rectangularity are represented as six-dimensional feature vectors to depict the emotion and action. The extracted features are fed to the SVM, Naïve Bayes, and DTW for human emotion recognition.

Fig. 6 Solidity of the shape
is represented by shape area
and convex hull area

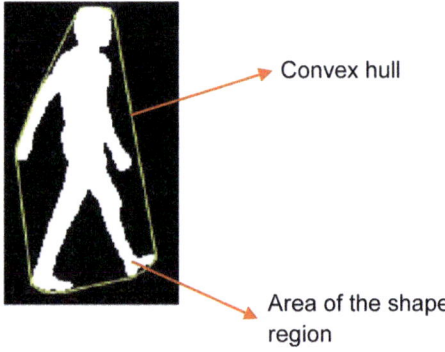

Fig. 7 Rectangularity of the shape is represented by shape area and minimum bounding rectangle area

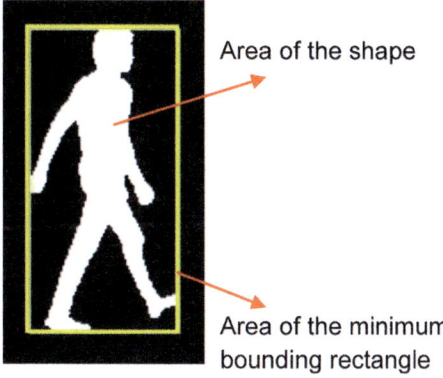

Area of the shape

Area of the minimum bounding rectangle

3 Support Vector Machines (SVM)

Support vector machine (SVM) is a well popular technique for classification in visual pattern recognition [27]. The SVM is most widely used in kernel learning algorithm. It achieves reasonable vital pattern recognition performance in optimization theory. Classification tasks are typically involved with training and testing data. The training data are separated by $(x_1, y_1), (x_2, y_2), \ldots (x_m, y_m)$ into two classes, where $x_i \in \mathbb{R}^n$ contains n-dimensional feature vector, and $y_i \in \{+1, -1\}$ are the class labels. The aim of SVM is to generate a model which predicts the target value from testing set. In binary classification, the hyper plane $w.x + b = 0$, where $w \in \mathbb{R}^n$, $b \in \mathbb{R}^n$ is used to separate the two classes in some space $\mathbb{Z}$ [28]. The maximum margin is given by $M = 2/\|w\|$ as shown in Fig. 8. The minimization problem is solved by using Lagrange multipliers $\alpha_i (i = 1, \ldots m)$ where w and b are optimal values obtained from Eq. 9.

$$f(x) = \text{sgn}\left(\sum_{i=1}^{m} \alpha_i y_i K(x_i, x) + b \right) \tag{9}$$

Fig. 8 Illustration of hyperplane in linear SVM

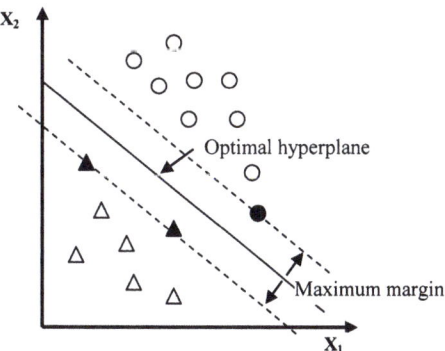

Table 1 Types of SVM inner product kernels

Types of kernels	Inner product kernel $K(x^T, x_i)$	Details
Polynomial	$(x^T x_i + 1)^p$	Where x is input patterns
Gaussian	$\exp\left[-\frac{\|x^T - x_i\|^2}{2\sigma^2}\right]$	x_i is support vectors σ^2 is variance $1 \leq i \leq N_s$
Sigmoid	$\tanh(\beta_0 (x^T x_i) + \beta_1)$	N_s is number of support vectors β_0, β_1 are constant values p is degree of the polynomial

The non-negative slack variables ξ_i are used to maximize margin and minimize the training error. The soft margin classifier is obtained by optimizing by Eqs. 10 and 11.

$$\min_{\omega, b, \xi} \frac{1}{2} w^T w + C \sum_{i=1}^{l} \xi_i \tag{10}$$

$$y_i \left(w^T \phi(x_i) + b \right) \geq 1 - \xi_i, \ \xi_i \geq 0 \tag{11}$$

If the training data are not linearly separable, the input space mapped into high-dimensional space with kernel function $K(x_i, x_j) = \phi(x_i) . \phi(x_j)$ is explained in [29]. There are several SVM kernel functions as given in Table 1.

4 Naive Bayes

In recent years, researchers in pattern recognition, machine learning, and classification have been concerned with naive Bayesian classifiers. The naive Bayes algorithm makes use of Bayes theorem, which is a formula that determines probability by estimating the frequency of values and mixture of values. A naive Bayes is a simple probabilistic-based classifier, which can able to predict the probabilities of the membership class [28]. The naive Bayesian classifier is simple and computationally efficient learning algorithm with theoretical roots in the Bayes theorem. The Bayes theorem states:

- Let $A_1, A_2, \ldots, A_L$ be mutually exclusive events whose union has probability one. That is, $\sum_{i=1}^{L} P(A_i) = 1$
- Let the probabilities $P(A_i)$ be known.
- Let B be an event for which the conditional probability of B given A_i, $P(B|A_i)$ is known for each A_i.

Then:

$$P(A_i|B) = \frac{P(B|A_i)P(A_i)}{\sum_{j=1}^{L} P(B|A_i)P(A_i)} \tag{12}$$

The probabilities $P(A_i|B)$ reflect updated or revised beliefs about A_i, in light of the knowledge that B has occurred. Once the probabilities have been estimated, the class is predicted by identifying the most probable one.

5   Dynamic Time Wrapping (DTW)

Dynamic time warping (DTW) is an essential method for measuring resemblance between two temporal sequences which may vary in time [26]. For example, similarities in consecutive patterns might be identified using DTW, even if single person was running faster than the other one, or if there were accelerations and decelerations through the path of a scrutiny. DTW has been efficient to temporal sequences of video, which can be turned into a linear sequence and analyzed with DTW. Figure 9 shows the time association of two time-dependent sequences.

The theory of DTW is to measure two (time-dependent) sequences $x :=(x_1, x_2, \ldots, x_n)$ of length $N \in \mathrm{N}$ and $Y := (y_1, y_2 \ldots, y_m)$ of length $M \in N$. These sequences possibly will be distinct signals (time series), feature sequences sampled at equidistant points in time, and the feature space is denoted by F. Then $x_n, y_m \in F$ for $n \in [1 : N]$ and $m \in [1 : m]$. To evaluate two diverse feature vectors $x, y \in F$, one desires a partial cost measure, sometimes also referred to as local distance measure, which is defined as a function

$$C : F \times F \to R \geq 0 \tag{13}$$

Fig. 9 Time alignment for two time-dependent sequences

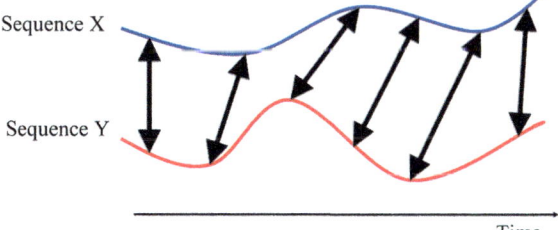

6 Emotion Dataset

Emotion dataset (University of York) is a publicly available dataset [25], containing four different emotions (happy, angry, fear, and sad) performed by 25 actors. The sequences were taken over static (black) background with the frame size of 1920 × 1080 pixels at a rate of 25 fps. For each emotion, actors are performed five different actions: walking, jumping, box picking, box dropping, and sitting, having an approximate length of 15 s of video. In this work, three emotions (happy, angry, and fear) and three actions (walking, jumping, and sitting) of 10 persons (male and female) are used for experimental purpose. Table 2 shows the properties of emotion recognition dataset.

In Figure 10, first row shows the walking action recognition with three different emotions, happy, angry, and fearful. Second row shows the sitting action recognition with three different emotions, happy, angry, and fearful, and the third row shows jumping with three different emotions, happy, angry, and fearful. For each approximate length of 15 s of video obtained, 90 data records are considered for experimental purpose. In this work, 10 persons are taken randomly from emotion dataset for evaluation. The samples are divided into training set of 5 persons and testing set of 5 persons.

7 Performance Measure

Table 3 illustrates a confusion matrix for a human emotion recognition problem having true positive, false positive, true negative, and false negative class values. If the classifier predicts correct response of class at each instance, it is counted as "success," if not, it is an "error." The overall performance of the classifier is obtained by error rate, which is a proportion of the errors made over the whole set

Table 2 Properties of 10 subjects in emotion recognition dataset

Subject	Gender	Happy (s)			Angry (s)			Fearful (s)		
		Walk	Sit	Jump	Walk	Sit	Jump	Walk	Sit	Jump
1	Male	3	5	7	2	5	6	5	2	4
2	Female	3	4	4	2	3	2	5	4	3
3	Male	3	7	6	2	8	4	2	6	7
4	Female	4	3	3	4	3	4	3	3	6
5	Male	4	6	4	5	8	4	5	7	3
6	Female	3	5	5	3	4	4	6	5	3
7	Female	4	7	4	3	2	3	2	6	8
8	Male	3	8	4	2	3	4	2	2	4
9	Female	3	5	3	3	4	3	2	6	5
10	Male	3	6	4	3	3	4	2	6	5

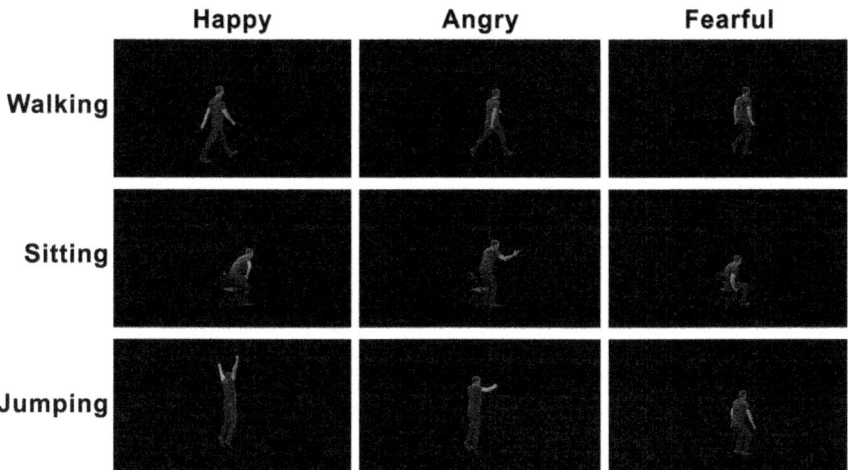

Fig. 10 Sample frames from the emotion dataset

Table 3 Confusion matrix

Actual	Predicted	
	Positive	Negative
Positive	TP	FP
Negative	FN	TN

of instances. From the confusion matrix, it is possible to extract a statistical metrics (recall, precision, F-measure, and accuracy) for measuring the performance of classification systems and is defined as follows:

Recall (R) or sensitivity is a ratio between correctly labeled instances and total instances in the class. It has an ability to measure the prediction model and is also called as true positive rate. It is defined by:

$$\text{Recall (R)} = \frac{TP}{TP + FN} \tag{14}$$

where TP and FN are the numbers of true positive and false negative predictions for the particular class. TP + FN is the total number of test examples of the particular class.

Precision (P) or detection rate is a ratio between correctly labeled instances and total labeled instances. It is a percentage of positive predictions in specific class that are correct. It is defined by:

$$\text{Precision (P)} = \frac{TP}{TP + FP} \tag{15}$$

where TP and FP are the number of true positive and false positive predictions for the particular class.

The F-measure is the harmonic mean of precision and recall and it attempts to give a single measure of performance. A good classifier can provide both recall and precision values high. The F-measure is defined as:

$$F_\beta = \frac{(1+\beta)^2 \cdot \text{TP}}{(1+\beta)^2 \cdot \text{TP} + \beta^2 \cdot \text{FN} + \text{FP}} \qquad (16)$$

where β is the weighting factor. Here, $\beta = 1$, that is, precision and recall are equally weighted and used to measure the F_β-score which is also called as F1-measure.

The most common metric accuracy is defined as the ratio between sum of correct classifications and total number of classifications.

$$\text{Accuracy (A)} = \frac{\text{TP} + \text{TN}}{\text{TP} + \text{TN} + \text{FP} + \text{FN}} \qquad (17)$$

8   Experimental Results

The experiments are carried out in MATLAB R2013a in Windows 7 operating system on computer with Intel Pentium i3 processor 2.10 GHz with 3 GB of RAM. As explained in feature extraction section, the six-dimensional features are extracted. The performance of the proposed feature method is tested on SVM, Naïve Bayes, and DTW classifiers to test the performance.

8.1   Results Obtained with SVM

The extracted features are fed to SVM with polynomial kernel. In polynomial kernel, different degrees (2, 3, 4, 5, and 6) were tested. Based on the classification results, degree 4 performs superior than the other kernel degrees. Further, it has been observed that increase in kernel degree does not give any improvement in performance.

The confusion matrix of the SVM with polynomial classifier on emotion dataset is shown in Table 4, where correct responses define the main diagonal, and most of the emotion classes like happy-walk, happy-sit, happy-jump, angry-sit, angry-jump, and fearful-walk are almost predicted well. An average recognition rate of SVM with polynomial kernel classifier on emotion dataset is 91.98 %. From this, some of the fearful-sit emotions are misclassified as happy-sit. Fearful-jump emotion is misclassified as happy-jump and angry-jump, respectively. Angry-walk is partly confused with happy-walk and fearful-walk, respectively.

Table 4 Confusion matrix for emotion and action recognition using SVM with polynomial classifier

Emotion/action	Happy (walk)	Happy (sit)	Happy (jump)	Angry (walk)	Angry (sit)	Angry (jump)	Fearful (walk)	Fearful (sit)	Fearful (jump)
Happy (walk)	**93.10**	0	0	6.90	0	0	0	0	0
Happy (sit)	0	**94.21**	0	0	2.51	0	3.28	0	0
Happy (jump)	0	0	**93.36**	0	0	6.64	0	0	0
Angry (walk)	4.21	0	0	**90.16**	0	0	5.63	0	0
Angry (sit)	0	6.79	0	0	**93.21**	0	0	0	0
Angry (jump)	0	0	8.58	0	0	**91.42**	0	0	0
Fearful (walk)	0	0	0	6.77	0	0	**93.23**	0	0
Fearful (sit)	0	10.68	0	0	0	0	0	**89.32**	0
Fearful (jump)	0	0	7.17	0	0	3.05	0	0	**89.78**

Table 5 Performance measure of the emotion-based actions using SVM with polynomial classifier

Emotion/action	Recall (%)	Precision (%)	F-measure (%)
Happy (walk)	93.10	95.67	94.37
Happy (sit)	94.21	84.36	89.01
Happy (jump)	93.36	85.57	89.29
Angry (walk)	90.16	86.83	88.47
Angry (sit)	93.21	97.38	95.25
Angry (jump)	91.42	90.42	90.92
Fearful (walk)	93.23	91.28	92.24
Fearful (sit)	89.32	100	94.36
Fearful (jump)	89.78	100	94.61
Average	91.98	92.39	92.06

Table 5 shows the average performance metrics of SVM with polynomial classifier; from the results, it is clearly indicated that proposed method gives higher recall = 91.98 %, precision = 92.39 %, and F-measure = 92.06 % (trade-off between precision and recall), where high recall value indicate that an SVM with polynomial classifier returned most of the relevant samples correctly.

8.2 Results Obtained with Naïve Bayes

The confusion matrix of the Naïve Bayes classifier on emotion dataset is shown in Table 6, where correct responses define the main diagonal, and most of the emotion classes like happy-sit, happy-jump, angry-sit, and fearful-walk are almost predicted well. An average recognition rate of Naïve Bayes classifier on emotion dataset is 89.91 %. From this, some of happy-walk and angry-walk emotions are misclassified as angry-walk and fearful-walk, respectively. Fearful-sit is mostly confused with happy-sit. In contrast, angry-jump and fearful-jump emotion are misclassified as happy-jump.

Table 7 shows the average performance metrics of Naïve Bayes classifier; from the results, it is evidently indicated that proposed method gives good recall = 89.91 %, precision = 90.16 %, and F-measure = 89.96 %. From the quantitative evaluation results, the proposed method has good recall, precision, F-measure, and accuracy for the Naïve Bayes classifier on emotion recognition dataset. It is found that the overall performance is dropped to 2 % on Naïve Bayes classifier, when compared to SVM with polynomial classifier.

Table 6 Confusion matrix for emotion and action recognition using Naïve Bayes classifier

Emotion/action	Happy (walk)	Happy (sit)	Happy (jump)	Angry (walk)	Angry (sit)	Angry (jump)	Fearful (walk)	Fearful (sit)	Fearful (jump)
Happy (walk)	**89.30**	0	0	8.10	0	0	2.60	0	0
Happy (sit)	0	**91.11**	0	0	4.21	0	2.61	2.07	0
Happy (jump)	0	0	**90.25**	0	0	8.26	0	0	1.49
Angry (walk)	4.81	5.49	0	**89.56**	0	0	5.63	0	0
Angry (Sit)	0	0	0	0	**91.41**	0	0	3.10	0
Angry (jump)	0	0	9.18	0	0	**89.36**	0	0	1.46
Fearful (walk)	2.39	0	0	7.42	0	0	**90.19**	0	0
Fearful (sit)	0	10.28	0	0	0	0	0	**89.72**	0
Fearful (jump)	0	0	8.53	0	0	3.21	0	0	**88.26**

Table 7 Performance measure of the emotion-based actions using Naïve Bayes classifier

Emotion/action	Recall (%)	Precision (%)	F-measure (%)
Happy (walk)	89.30	92.54	90.89
Happy (sit)	91.11	85.25	88.08
Happy (jump)	90.25	83.60	86.80
Angry (walk)	89.56	85.23	87.34
Angry (sit)	91.41	95.60	93.46
Angry (jump)	89.36	88.62	88.99
Fearful (walk)	90.19	89.27	89.73
Fearful (sit)	89.72	94.55	92.07
Fearful (jump)	88.26	96.77	92.32
Average	**89.91**	**90.16**	**89.96**

8.3   Results Obtained from DTW

The confusion matrix of the DTW classifier on emotion dataset is shown in Table 8, where correct responses define the main diagonal, and most of the emotion classes like happy-walk, happy-sit, happy-jump, angry-walk, angry-sit and fearful-walk are almost predicted well. An average recognition rate of DTW classifier on emotion dataset is 93.39 %. From this, some of fearful-sit emotions are misclassified as happy-sit. Angry-jump emotion is misclassified as happy-jump. Fearful-jump is mostly confused with happy-jump and fearful-jump, respectively.

Table 9 shows the average performance metrics of DTW classifier; from the results, it is clearly indicated that proposed method gives higher recall = 93.7 %, precision = 93.2 %, and F-measure = 93.4 %, where high recall value indicates that DTW classifier returned most of the relevant samples correctly. From the quantitative evaluation results, the proposed approach has a superior recall, precision, F-measure, and accuracy for the DTW classifier on emotion recognition dataset.

It is found that the overall performance is increased to 3.5 % on DTW classifier, when compared to SVM with polynomial and Naïve Bayes classifiers.

The potential use of automatic visual surveillance application is to detect and analyze abnormal situations. To achieve this, technological support and connected smart devices, which are becoming a reality, are used. The devices can monitor and can act as sensors to sense the environment and in particular to monitor human assistance and to ensure the public safety. They also process information, control traffic lights, lock doors, and remind people to take medications.

This chapter proposes an approach for emotion recognition system in the context of civil safety which raises an alarm when an abnormal emotion is detected. Figures 11, 12, and 13 shows the detection of happy-walking and angry-sit and fearful-jump while monitoring emotion of the person.

Table 8 Confusion matrix for emotion and action recognition using DTW classifier

Emotion/action	Happy (walk)	Happy (sit)	Happy (jump)	Angry (walk)	Angry (sit)	Angry (jump)	Fearful (walk)	Fearful (sit)	Fearful (jump)
Happy (walk)	**94.6**	0	0	5.41	0	0	0	0	0
Happy (sit)	0	**96.08**	0	0	1.96	0	1.96	0	0
Happy (jump)	0	0	**95**	0	0	5	0	0	0
Angry (walk)	2.56	0	0	**92.31**	0	0	5.13	0	0
Angry (sit)	0	5.71	0	0	**94.29**	0	0	0	0
Angry (jump)	0	0	9.38	0	0	**90.63**	0	0	0
Fearful (walk)	0	0	0	5	0	0	**95**	0	0
Fearful (sit)	0	10	0	0	0	0	0	**90**	0
Fearful (jump)	0	0	4.55	0	0	4.55	0	0	**90.91**

Table 9 Performance measure of the emotion-based actions using DTW classifier	Emotion/action	Recall (%)	Precision (%)	F-measure (%)
	Happy (walk)	97.2	94.5	95.8
	Happy (sit)	90.7	96	93.3
	Happy (jump)	88.3	95	91.5
	Angry (walk)	90	92.3	91.1
	Angry (sit)	97	94.2	95.6
	Angry (jump)	87.8	90.6	89.2
	Fearful (walk)	92.6	95	93.8
	Fearful (sit)	100	90	94.7
	Fearful (jump)	100	90.9	95.2
	Average	**93.7**	**93.2**	**93.4**

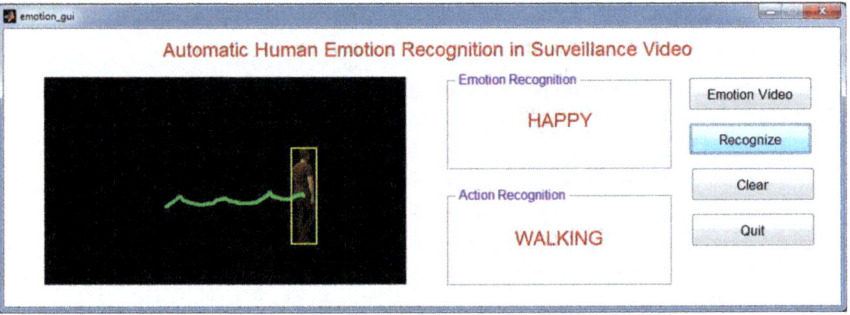

Fig. 11 Snapshot of the happy emotion with walking action using DTW

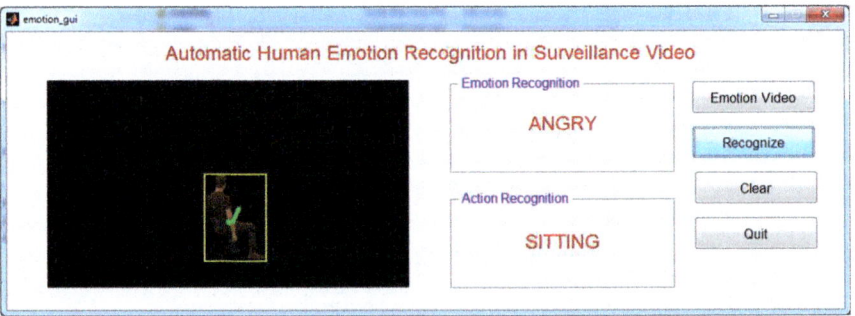

Fig. 12 Snapshot of the angry emotion with sitting action using DTW

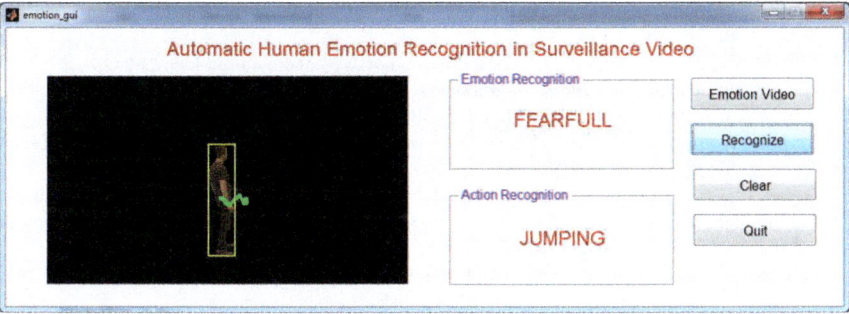

Fig. 13 Snapshot of the fearful emotion with jumping action using DTW

9 Conclusion

This chapter introduces a human emotion recognition using gesture dynamics in surveillance video. Experiments are conducted on emotion dataset considering different persons. The proposed gesture dynamics features are extracted from the emotion video sequences and modeled using SVM with polynomial, Naïve Bayes, and DTW; the performance measures such as recall, precision, F-measure, and accuracy were calculated. Experimental results show the overall accuracy of SVM with polynomial, Naïve Bayes, and DTW results as 91.98, 89.91, and 93.39 %, respectively. The performance results indicate that DTW outperforms SVM and Naïve Bayes classifiers. It is observed from the experiments that the system could not distinguish angry-jump, fearful-sit, and fearful-jump with high accuracy and is of future interest.

Smart screens kept at homes, offices, and shopping centers would provide information about the customer's interest about the products. Further, by dynamically recognizing emotions of buyers and their whereabouts, sellers could send them alerts related to their products, delivered to their mobile. Customer service could be improved without having customers to go through the terrific process of reporting problems to the sellers by recognizing their emotions.

Automatic emotion recognition is addressed in this chapter at the budding level. Lot more issues are to be addressed by the base problem itself, which attracts the researchers nowadays.

References

1. Castellano G, Villalba SD, Camurri A (2007) Recognising human emotions from body movement and gesture dynamics. Affective computing and intelligent interaction. Springer, Berlin, pp 71–82
2. Bernhardt D, Robinson P (2009) Detecting emotions from connected action sequences, visual informatics: bridging research and practice. Springer, Berlin, pp 1–11

3. Yoo H-W, Cho S-B (2007) Video scene retrieval with interactive genetic algorithm. Multimed Tools Appl 34(3):317–336
4. Ke S-R (2013) A review on video-based human activity recognition. Computers 2(2):88–131
5. Anagnostopoulos C-N, Iliou T, Giannoukos I (2015) Features and classifiers for emotion recognition from speech: a survey from 2000 to 2011. Artif Intell Rev 43(2):155–177
6. Rao KS, Koolagudi SG (2015) Recognition of emotions from video using acoustic and facial features. SIViP 9(5):1029–1045
7. Busso C (2004) Analysis of emotion recognition using facial expressions, speech and multimodal information. In: Proceedings of the 6th international conference on Multimodal interfaces. ACM
8. Gunes H, Piccardi M (2007) Bi-modal emotion recognition from expressive face and body gestures. J Netw Comput Appl 30(4):1334–1345
9. Karpouzis K et al (2007) Modeling naturalistic affective states via facial, vocal, and bodily expressions recognition. Artifical intelligence for human computing. Springer, Berlin, pp 91–112
10. Gross MM, Crane EA, Fredrickson BL (2010) Methodology for assessing bodily expression of emotion. J Nonverbal Behav 34(4):223–248
11. Hassan M et al (2014) A review on human actions recognition using vision based techniques. J Image Graph 2(1):28–32
12. Kessous L, Castellano G, Caridakis G (2010) Multimodal emotion recognition in speech-based interaction using facial expression, body gesture and acoustic analysis. J Multimodal User Interfaces 3(1-2):33–48
13. Gunes H(2010) Automatic, dimensional and continuous emotion recognition
14. Kapoor A, Burleson W, Picard RW (2007) Automatic prediction of frustration. Int J Hum Comput Stud 65(8):724–736
15. Kapur A (2005) Gesture-based affective computing on motion capture data. Affective computing and intelligent interaction. Springer, Berlin, pp 1–7
16. Balomenos Themis et al (2004) Emotion analysis in man-machine interaction systems. Machine learning for multimodal interaction. Springer, Berlin, pp 318–328
17. Camurri A, Lagerlöf I, Volpe G (2003) Recognizing emotion from dance movement: comparison of spectator recognition and automated techniques. Int J Hum Comput Stud 59 (1):213–225
18. Yang Z, Ortega A, Narayanan S (2014) Gesture dynamics modeling for attitude analysis using graph based transform. In: 2014 IEEE international conference on image processing (ICIP). IEEE
19. Zaboleeva-Zotova AV (2013) Automated identification of human emotions by gestures and poses. In: 2013 BRICS congress on computational intelligence and 11th Brazilian congress on computational intelligence (BRICS-CCI & CBIC). IEEE
20. Cowie R, McKeown G, Douglas-Cowie E (2012) Tracing emotion: an overview. Int J Synth Emot 3(1):1–17
21. Alvandi EO (2011) Emotions and information processing: a theoretical approach. Int J Synth Emot 2(1):1–14
22. Salovey P, Mayer JD (1990) Emotional intelligence. Imagin Cogn Personal 9(3):185–211
23. Oker A et al (2015) A virtual reality study of help recognition and metacognition with an affective agent. Int J Synth Emot 6(1):60–73
24. Warwick K, Harrison I (2014) Feelings of a cyborg. Int J Synth Emot 5(2):1–6
25. Keefe Bruce D et al (2014) A database of whole-body action videos for the study of action, emotion, and untrustworthiness. Behav Res Methods 46(4):1042–1051
26. Müller M (2007) Dynamic time warping. Information retrieval for music and motion, pp 69–84
27. Cristianini N, Shawe-Taylor J (2000) An introduction to support vector machines and other kernel-based learning methods. Cambridge University Press, Cambridge
28. Mitchell TM, Michell T (1997) Machine learning. Mc-graw-Hill Series in Computer Science
29. Vapnik V (1998) Statistical learning theory

Part III
Medical Security Applications

Watermarking in Biomedical Signal Processing

Nilanjan Dey, Amira S. Ashour, Sayan Chakraborty,
Sukanya Banerjee, Evgeniya Gospodinova, Mitko Gospodinov
and Aboul Ella Hassanien

Abstract Recently, by means of technological innovation in communication networks and information, it has assisted healthcare experts across the world to seek high-quality diagnosis as well as to communicate each other as second opinions via enabling extensive and faster access to the patients' electronic medical records, such as medical images. Medical images are extremely precious owing to its importance in diagnosis, education, and research. Recently, telemedicine applications in telediagnoisis, teleconsulting, telesurgery, and remote medical education play an imperative role in the advancement of the healthcare industry. Nevertheless, medical images are endured security risk, such as images tampering to comprise false data which may direct to wrong diagnosis and treatment. Consequently,

N. Dey (✉) · S. Chakraborty
Department of Information Technology, Techno India College
of Technology, Kolkata, India
e-mail: neelanjan.dey@gmail.com

S. Chakraborty
e-mail: sayan.cb@gmail.com

A.S. Ashour
Department of Electronics and Electrical Communications Engineering,
Faculty of Engineering, Tanta University, Tanta, Egypt
e-mail: amirasashour@yahoo.com

S. Banerjee
Department of CSE, JISCE, Kalyani, West Bengal, India
e-mail: banerjee.sukanya7@gmail.com

E. Gospodinova · M. Gospodinov
Institute of Systems Engineering and Robotics, Bulgarian Academy
of Sciences, Sofia, Bulgaria
e-mail: jenigospodinova@abv.bg

M. Gospodinov
e-mail: mitgo@abv.bg

A.E. Hassanien
Scientific Research Group, CairoUniversity, Giza, Egypt
e-mail: aboitcairo@gmail.com

© Springer International Publishing Switzerland 2017
N. Dey and V. Santhi (eds.), *Intelligent Techniques in Signal Processing
for Multimedia Security*, Studies in Computational Intelligence 660,
DOI 10.1007/978-3-319-44790-2_16

watermarking of medical images offers the compulsory control over the flow of medical information. It is the typically used data hiding technique in the biomedical information security domain and legal authentication. In the field of telemedicine, exchange of medical signals is a very common practice. The signals are transmitted through the web and the wireless unguided media. The security and the authenticity are the matter of concern due to the various attacks on the web. Any type of the signals vulnerability in the biomedical data is not acceptable for the sake of proper diagnosis. A watermark is used to prove the ownership of the exchanged data. The logos of the hospitals or medical centers and electronic patient's report card can be added to the biomedical signals as a watermark to establish the property right. This work provides an extensive view about the existing research works in the field of watermarking techniques on different biomedical signals. It includes the design and evaluation parameters serving as a guideline in the watermarking schemes' development and benchmarking. This work also provides the comparative study between different watermarking methods. It reviews several aspects about digital watermarking in the medical domain. Also, it presented the properties of watermarking and several applications of watermarking. Meanwhile, it discusses the requirements and challenges that the biomedical watermarking process face.

Keywords Watermarking · Telemedicine · Biomedical signal · Watermarking techniques · Watermarking challenges

1 Introduction

Biomedical systems are often sustained by information streams that are represented using mathematical models [1], analysis of spatial structures, and self-organization. Consequently, biomedical signals extract information from the complex measured data, which are characteristically a time series contains both a regular/random components. Solutions endeavor to map broad standards to model the biomedical systems work.

Nowadays, due to the vast progress in the multimedia usage besides our information and communication era, the potential of managing medical information and sharing through telediagnosis to telesurgery techniques is boosted as well as a cooperative remote session is held. Web and wireless media facilitate the diagnosis results to be exchanged for mutual availability among the various diagnostic centers. This assists the inhabitants of rural and remote areas in various countries all over the world who face numerous difficulties accessing proper and timely medical treatments due to the lack of expert personnel. Accordingly, telemedicine becomes a must to serve patient who could not get better treatment due to time and cost difficulties [2]. Thus, one of the common research problems that require investigation is the exchange of medical signals between the hospitals or advanced clinics. Telemedicine provides the ability to interactive health care employing contemporary technology and telecommunications [3]. Essentially, it allows patients to communicate with their physicians over video for instantaneous care or capture video/still images. These patient's data are stored and sent to physicians in order to analyze/diagnose and follow-up treatment over a distance. At the same time, telemedicine benefits are posed to simultaneous risks during shared electronic patient records (EPR); thus, it requires more secure information management. Therefore, transmission of medical signals needs high level of security and authentication. The data transmission uses wireless media that requires a proper technique to maintain the security and authenticity of the data.

Forensics, steganography, and digital watermarking are widely used for secured digital multimedia. Digital forensics uses computational approaches in order to process multimedia content to help in crime investigations. Steganography converts test after encrypting the steganography message. Moreover, watermarking is used for authenticity and can be applied as a proof for ownership. Typically, forensic watermarking is defined as the process through which an invisible, unique serial number can be added to audio or video content. The watermark should be designed to endure with the content regardless of any transcode, resized, or downscale. Typically, according to the multimedia document Digital Rights Management [4], watermarking has attractive properties that fit with the healthcare field. Researchers used watermarking with various applications in the medical domain as in [5, 6]. Generally, digital watermarking [7] is the process of hiding a message related to the actual content of the digital data such as audio, image, and video inside the signal itself. It is a prospective method for ownership rights protection of the digital data. The information is embedded imperceptibly and directly into these digital data as the original (host) data to form watermarked data in the digital watermarking. This embedded information is continually bound to and decodable from the watermarked data, even it is processed, redistributed, or copied. In addition, watermarking can be defined as the inclusion of a watermark message (content) in a host document. It has very important requirements as follows. The watermark information should remain hidden to any unauthorized user, robust to maintain intact facing to the attacks [8], also to be non-interfering with the use of the watermarked document and fragile. Moreover, watermarking is supposed to have supplementary properties such as fidelity, where the watermarked image must be similar to the original host.

Digital watermarking has potential applications including authentication [9], copyright protection [10], authorized access control [11], distribution tracing [12], and in the medical domain [13]. Although the information can be a serial number, user ID, or an authentication information, in the medical domain, it facilitates a security layer at the information level, which endows with authentications and traceability abilities with the security services within and between medical information systems. Consequently, a watermark is used to prove the ownership of the exchanged data. The logos of the hospitals or medical centers and electronic patient's report card can be added to the biomedical signals as a watermark to establish the property right. As a corollary, in the medical domain, there are two major purposes of watermarking which are as follows: (1) hiding data for the intention of inserting meta-data to obtain a more usable image and (2) providing protected information [5].

Table 1 Different watermarking system classifications

According to human perception	According to robustness	According to types of document	According to the access authentication	According to the domain
• Visible • Invisible	• Fragile • Semi-fragile • Robust	• Audio • Text • Image • Video	• Blind • Semi-blind • Non-blind	• Spatial domain • Transform domain

Generally, the watermark techniques can have an assortment of classifications as presented in Table 1, which are as follows:

- According to human perception: visible or invisible.
- According to robustness: fragile, semi-fragile, or robust.
- According to types of document: audio, text, image, or video.
- According to its characteristics: blind (public), semi-blind (semi-private), or non-blind (private).
- According to the working domain: spatial domain or transform domain.

Commonly, the foremost classification for the watermarking structure techniques was suggested in [14], which categorize watermarking in two types, namely visible and invisible. Visible watermark can be seen [15], such as diverse logos either on paper or on television as the logo of a hospital that proves the legal authenticity of the medical signal. However, the invisible (transparent) watermarking techniques [16] cannot be recognized by the human sensory system, as the original signal and the watermark embedded signal are quite similar. An invisible watermark can be divided into either fragile or robust. For actuality requirements when the protected media was interfered, the fragile watermark is to be used. Conversely, robust watermarking is proposed to offer proof of the media ownership. A number of techniques are available to prevent the unauthorized access or illegal copying of the signals. Robust watermarking is a technique which resists the data from several attacks like geometric distortion and signal processing. The fragile watermarking technique mainly deals with integrity issues. A minute change in the original data is easily detected, and the embedded watermark also changes itself accordingly.

Another classification for the watermarking based on its characteristics (access authentication) that can be categorized as follows: (i) blind [17], (ii) semi-blind [18], and (iii) non-blind [19] watermarking. Blind watermarking is the challenging public watermarking scheme that neither uses the cover nor the embedded watermark. In the semi-blind (semi-private) watermarking, only the embedded watermark is required to extract the watermark, while the non-blind watermarking scheme requires the original signal to detect the watermark. In case of public watermarking, all the users of the data are able to extract the embedded watermark, but in case of private watermarking, only the authorized users are able to detect the watermark.

As a complementary, in the domain watermarking classification, the transform domain watermarking is much more effective compared to the spatial domain watermarking. The former does not use the original data for the encoding purposes, while the spatial domain watermarking is embedding the watermark directly with the original data.

There also several applications of watermarking based on the segments, such as copyright protection watermarks, data authentication watermarks, fingerprint watermarks, copy control watermarks, and device control watermarks. Since the watermarking system hides information inside some other data, there are three main criteria to measure the watermarking system performance, namely the embedding effectiveness, fidelity, and data payload [20]. The embedding effectiveness is defined as the probability of immediate detection after embedding. Meanwhile, the perceptual similarity between the host un-watermarked data and watermarked data at the when presented to a user is referred to as the fidelity watermarking system. As a supplementary, data payload ratio is the ratio between the watermark size and its carrier size. Consequently, proper detection algorithms are required for extraction of the watermark.

The watermarking procedure can be generalized as the characteristics of the carrier signal which consist of several information like license, copyright, and authorization. No relation exists between the embedded signal and the digital cover. Thus, watermarking is quite different from stereograph. Stereography [21] is the procedure of data hiding during the data transmission process.

In general, reversible watermarking [22] is the technique which is mostly used in medical data analysis, where the loss of original content may cause severe effects. The original data could be retrieved from the watermarked data in reversible watermarking. However, irreversible watermarking has been just the opposite of reversible where the original signal after discarding the embedded watermark could not extract.

Since the human body contains several systems, including the nervous system, cardiovascular system, and musculoskeletal system. Most physiological processes consist of the signals that reflect their features and activities. Electrical signals are in the form of current or potential, and physical signals are in the form of pressure or temperature. These signals are produced from different modalities used to measure the performance of a certain body part as they produce medical data. Consequently, there are different types of biomedical signals produced from various modalities [23] as follows.

- The electronurogram (ENG) is an electrical signal that observes and records the reaction and the activity of the neurons of both the central nervous system (spinal cord, brain) and the peripheral nervous system (nerves, ganglions). It measures the conduction velocity of a stimulus or reaction in a nerve.
- The electromyogram (EMG) [24, 25] is a biomedical signal that detects the electrical pulses/activities of skeletal muscles. It is used to record the EMG signals.
- The photoplethysmography (PPG) provides signal popular biomedical signal. It is a significant and sensitive tool that diagnoses the function of the human heart.

- The electrocardiogram is the electrical display of the heart activity which is recorded from the body surface [26]. It is mainly used to monitor the heart rate. The waveform and the time interval denote some significant characteristics of ECG signals.
- The electroencephalogram (EEG) is popularly known as brain waves [27]. It represents the electrical activity of the brain waves.
- The PCG signal is the sound or the vibration signal of the cardiohemic system's activity [28]. It is the most traditional biomedical signal that represents the heart sound signal.
- Electrooculography (EOG) endows with an electrooculogram biomedical signal that measures the difference between the electrical potential of cornea and retina. The electrical signal of light-sensitive cells and the eyes' motor nerves are captured by this technique.

These biomedical signals carry the information for accurate and effective disease diagnosis. Embedding watermark with the original signal may cause increased noise or distortion of the data. Changes in the original data may lead to wrong diagnosis as well as wrong treatment. Therefore, it is necessary to select proper watermark embedding and extracting techniques to reduce the noise and distortion. In this work, the main aim is to report various different techniques of watermarking on different biomedical signals. Also, a comprehensive literature review is to be discussed.

The organization of the remaining sections is as follows. Section 2 introduces the general digital watermarking framework in the medical domain, its types, requirements, evaluation as well as the watermarking model. The biomedical watermarking challenges including the distortion and attack besides the security and privacy requirements are discussed in Sect. 3. Consequently, a review on biomedical watermarking applications is given in Sect. 4. Finally, the discussion and concluding remarks and future work are given in Sects. 5 and 6, respectively.

2 General Digital Watermarking Framework in the Medical Domain

Watermarking is the process that inserts (embeds) data (watermark) into an original audio, image, or video. The typical watermarking framework model is shown in Fig. 1. Later, the watermark can be detected/extracted from the cover (carrier), which contains information such as license, copyright, and authorship. Any watermarking algorithm contains three main parts: (i) watermark that is unique to the owner, (ii) encoder for embedding the watermark into the data, and (c) decoder for extraction and verification.

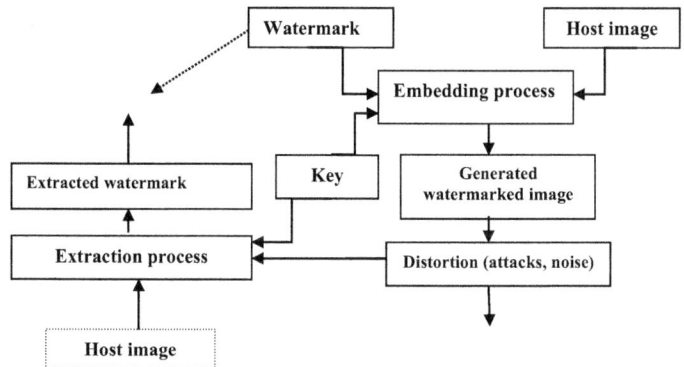

Fig. 1 Digital watermarking system

2.1 Watermarking Properties and Characteristics

The watermarking system has popular properties [29] such as:

- Effectiveness: It is defined as the probability that the message in a watermarked image will be correctly detected.
- Image fidelity: If a minimum change occurs in the image quality during the watermarking process, no clear difference in the image's fidelity can be noticed.
- Payload size: The size of the message is necessary as many systems necessitate a reasonably big payload to be embedded in a covered work.
- False-positive rate: It is the number of digital data that are identified as it has embedded watermark, while they do not have a watermark embedded.

In addition, the major characteristics of the digital watermarking are [30]:

- Robustness: It is the capability of the watermark system to resist after the host image processing operations such as image compression, transformation, and cropping.
- Transparency: Transparent watermark leads to no artifacts or feature loss.
- Imperceptibility: It indicates the similarity between the watermarked image and the host image.
- Capacity: It expresses the amount of the fixed information bits. It shows the possibility of embedding multiple watermarks in one document in parallel. A higher capacity is gained at the expense of either robustness strength or imperceptibility, or both.

Other three compulsory security characteristics can be mentioned on medical patient complex information records and diagnosis explanations from the images. These characteristics for the medical security are preserved through the following security services according to the medical information systems (MIS) [31]:

- Availability: It ensures access to the medical information in usual conditions of access and exercise by the authorized users.
- Confidentiality: It means that only the authorized users have access to the medical data [32].
- Reliability: It signifies both the integrity and authenticity aspects. The integrity means that the information has not been modified by unauthorized users, while authenticity refers to the information that belongs to the correct patient and concerned from the right source.
- Non-repudiation service: It handles delivery verifications as well as of the message sender's identity.

Another watermarking objective rather than the security is the data hiding that is used to prepare the image to be more informative/usable with the insertion of meta-data. However, this aspect is not widely used in the medical application images, due to the message size that can be embedded in a medical image. An example for the data hiding in the medical applications was proposed in [33], which was used for image management. The authors proposed embedding the description of the identified pathology into the image.

2.2   Types of Watermarking in the Medical Domain

Generally, watermarking techniques are designed according to the application framework that compromise between the diverse characteristic requirements mentioned above. For example, more robust and/or of higher capacity, albeit perceptibility requirements, are achieved with higher strength watermark signals. There are three classes of watermarking methods [5] as follows:

(A) Regroups methods that embed information within a region of non-interest (RONI) in order not to compromise the diagnosis capability [34]. Thus, the watermark signal amplitude has to be correctly selected.
(B) Classical watermarking methods with distortion minimization, where the watermark replaces several image details, such as details lost after lossy image compression [35], or the image least significant bit [36].
(C) Reversible watermarking, where once the embedded content is read, the watermark can be removed from the image to permit retrieval of the original image [37].

Another classification for the medical images types of watermarking methods is as follows:

(a) Minimum distortion [38],
(b) Lossless (reversible) watermarking,
(c) Segmentation [39]

The first method uses the classical watermarking methods for minimizing the distortion. Thus, the watermark replaces several image details of the image, for instance the least significant bit. Nevertheless, embedding diverse watermarks may lead to degradation in the watermarked medical images. Through the lossless watermarking, once the embedded data are read, the watermark can be removed from the image, tolerating retrieval of the original image. This approach provides authentication without proof of ownership. The third type is realized via segmenting the medical image into two regions using image segmentation techniques. The first region is known as a region of interest (ROI), and the second region is the region of the background (ROB). The watermarked data are embedded within the ROB in order not to compromise the diagnostic capability.

2.3 Watermarking Requirements for the Medical Imaging Domain

Particularly, medical images have an imperative role in telediagonosis, telesurgery, etc. Privacy and security are significant issues to be considered in teleradiology [40, 41] as well as when images are exchanged through the Internet [42]. Confidentiality, availability, and reliability are the main requirements should be considered in the context of security. Typically, the watermarking techniques for the medical and healthcare applications differ from those for multimedia applications rooted in the medical image properties and usage purpose. Consequently, there exist specific requirements for the medical images authentication schemes as follows [43]:

- Security: It is the ability to guarantee authenticity/integrity of the watermark under attack. This requires the complexity in removing the watermarking.
- Fragility: It means the technique should be sensitive to the attacks, so that the watermark can be easily broken for authentication.
- Tamper detection: It is the mechanism that assists the identification of whether the image is authentic or not [44], wherever may hospitals and even patients modify the image for illegal purpose.
- Reversible: Reversible data hiding supports recovering the cover media lossless subsequent to the extraction of the hidden data [45]. Medical images require good fidelity for diagnosis, which demands recovery of the original image without any loss after removing the watermark [46].
- Imperceptibility: It means the watermark should not be visible under normal vision nor should corrupt the image visual quality. In the medical applications, visible watermarks are not preferable.
- Localization: It refers to the system's ability to locate and recover the tampered region without any loss [47].

- Capacity: It expresses how many information bits can be embedded. It is clear that to achieve higher capacity degradation, other two requirements, namely the imperceptibility and robustness, will occur.
- Robustness: It is used for copyright protection and content tracking, as the watermark must oppose the attempts of removing or destroying.
- Blind detection: It is required to guarantee better medical images security, where the extraction process should not require the original/watermark images [48].
- Time: It refers to the time involved in embedding/extracting that should be kept small for doctors to have a fast access to stored images.

Consequently, the designers can interact with watermarking tools and watermarking requirements according to the application under concern. Depending on the medical application area, the trade-offs among the required parameters such as robustness, invisibility, and capacity varies should be considered. For example, image application concerning with the patient's health care is less tolerant of degradations. For example, when using the regroups watermarking method, the capacity depends on the RONI area size; besides this, the embeddable size varies with the image modality. The RONI watermarking tolerates watermark superimposition for image tracing.

2.4 Watermarking Performance Evaluation for Medical Imaging Domain

In order to calculate the quality performance of the watermarking techniques, the following quality measures are employed.

2.4.1 Image Quality Analysis

For medical imaging, watermarking has been established as a capable technique for security, reliability, integrity, and confidentiality. The key problem is the degradation that the medical images suffer from when secret data are embedded. To quantify image degradation, the following image quality analysis (IQA) metrics are to be used [49]:

1. Mean square error (MSE): It is the mean-squared difference between the original image pixels $x_{j,k}$ and the watermarked image pixels $x'_{j,k}$ that contain M and N pixels. Higher mean square error refers to a greater degradation level.

$$\text{MSE} = \frac{1}{MN} \sum_{j=1}^{M} \sum_{k=1}^{N} \left(x_{j,k} - x'_{j,k} \right)^2 \tag{1}$$

2. Peak signal-to-noise ratio (PSNR): It is the ratio between the maximum power of a signal and the power of the watermarked version that affects the reliability of its representation.

$$\text{PSNR} = 10\log_{10}\left(225^2/MSE\right) \tag{2}$$

3. Maximum difference (MD): It is the difference between any two pixels such that the larger pixel appears after the smallest pixel. The largest value of maximum difference refers to an image which is poor in quality.

$$\text{MD} = \text{Max}\left(\left|x_{j,k} - x'_{j,k}\right|\right) \tag{3}$$

4. The structural content (SC): It measures the similarity between the two images (original and the watermarked) in a number of small image patches the images have in common.

$$\text{SC} = \sum_{j=1}^{M}\sum_{k=1}^{N} x_{j,k}^2 \tag{4}$$

5. Entropy (H): It represents the amount of information in the image and is given by:

$$H(x, x') = \sum_{k} p_k \log_2\left(p_k x_{qk}\right) \tag{5}$$

Here, p and q are the probability distribution of x and x' with k pixel intensities. The image entropy is low for similar image as $H = 0$ indicated both images are identical.

2.4.2 Image Error Analysis

Another measuring metrics to evaluate the watermarked images can be done through the measure of the error in the image as follows:

1. Root-mean-square error (RMSE): It indicates the difference between original and watermarked version.
2. Mean absolute error (MAE): It computes the absolute pixel via pixel differences in original and watermarked images [50].
3. Percentage fit error (PFE): It measures the deviation from original and watermarked images. For no deviation, its value will be (PFE = 0) [51].
4. Bit error ratio (BER): It is the ratio that evaluates how many bits received in error over the number of the total bits received.

5. Image error rate (IER): It indicates the ratio of the number of the images recovered with errors to the total number of under-test images.
6. Normalized cross-correlation (NCC) [52]: It is used to authenticate the robustness of the watermarking systems, and it has values ranging between 0 and 1. The large NCC value indicates the high watermark robustness.

2.4.3 Variety of Image Analysis Metrics in the Medical Domain

- Bits per pixel calculation: It is used to measure the hiding capacity. It refers to how many bits are embedded in a pixel. The higher hiding capacity leads to lower quality.
- Analysis complexity: It is to calculate the average time taken for embedding. As the time taken decreases, the system complexity will be less.

3 Challenges of Biomedical Signal Watermarking

Foremost, an emphasis on the issues and challenges while applying the digital watermarking techniques is discussed as follows:

3.1 General Prospective Challenges of Digital Watermarking

(i)   Need to preserve balance between robustness, imperceptibility, and capacity as rising one factor negatively effects on another. Thus, to realize superior imperceptibility, the watermark should be embedded in the high-frequency component, while robustness arises in low-frequency component [53].
(ii) During spatial domain watermarking, the pixel values are changed, so it is barely to resist against various attacks such as low pass filtering, JPEG compression and cropping, high pass filtering that affects the robustness [54].
(iii) The payload size indicates how much amount is the carried information. Thus, for more payload size, it is required to compromise with the imperceptibility [54].
(iv) The computational cost should be minimized which is the cost of inserting and detecting watermarks [55].
(v)   Another challenging issue is regarding the fragile watermarking where the content recovery against cropping is exigent.
(vi) The major challenging point is to design a robust universal system for all digital media against various types of attacks.

3.2 Distortion and Attack Challenges of Medical Watermarking

Distortions and attacks are the challenges done to weaken the image quality and destroy it [56]. Generally, against a watermark image, there are two reasons for an attack which are hostile (malicious) attacks and coincidental attacks. The foremost is considered as an attempt completed to weaken, alter, or remove the watermark, while the second occurs during common image processing and is not aimed at tampering the watermark.

In the medical domain, picture archiving communication system and hospital information system share medical images through the Internet. During image transmission, attack can occur either intentionally or inadvertently which degrades the image quality affecting the system performance. The intentional attacks are executed to destroy the hidden watermark. Therefore, the watermarking system must be robust enough to endure the attacks.

Possible attacks on medical image watermarking are grouped into the following:

- Passive attacks: The attacker tries to decide whether a watermark exists, but the removal of the watermark is not a goal.
- Active attacks: As the unauthorized user attempts intended to remove the watermark or make it undetectable, it can be divided into three types as follows:

  - Robustness attack: The attacker tries to remove/destroy the watermark. Types of this attack comprise filtering, geometrical, and noise, such as:

    Geometrical attacks: These modify the spatial relationship between the pixels based on geometrical transformations [57], such as scaling, translation, rotation, and cropping.
    Low pass filtering attack: It results in a difference map composed of noise due to low pass filtering the watermarked image.
    Interference attack: It occurs by adding additional noise to the watermarked object.

  - Collusion attacks: The attacker employs some copies of watermarked data (images, video, etc.) to discover the watermark and to create a copy with no watermark.
  - Forgery attacks: The attacker tries to embed a valid watermark, which is a serious implication in authentication

Other types of attacks in the digital watermarking applications are [58, 59]:

- Protocol attacks: These occur when the attacker embeds his watermark in the image and claims ownership of the image by extracting his watermark from it.
- Cryptographic attacks: These deal with the security cracking via, for example, finding the secret watermarking key using comprehensive brute-force search and oracle attack methods. Brute-force search attack endeavors at finding the

embedded secret information, whereas oracle attack produces a non-watermarked signal from watermarked signal.

- Security attack: In particular, if the watermarking algorithm is known, in this type the attacker tries to execute modifications to make the watermark invalid or to estimate/modify the watermark.
- Removal attack: It aims to eliminate the watermark data from the watermarked object.
- Image degradation attacks: It damages robust watermarks via removing parts of the image, as the replaced parts may carry watermark information.

3.3 Security and Privacy Requirements

On behalf of medical applications, medical images involve special confidentiality and safety, as significant decision is made of the information presented by medical images. As a result, they must not be varied in an illegitimate way; otherwise, an unwanted outcome may result due to loss of the vital information. Medical images security is very significant when the images are exchanged through Internet [42]. Consequently, reliability and confidentiality are to be principally considered. For medical healthcare applications, watermarking techniques differ from multimedia applications rooted in the properties of medical image and usage purpose.

4 Review of Digital Watermarking Applications on Medical Images

Generally, forensic watermark (digital watermark) is a character/code sequence that can be embedded in a digital document, image, video, or even computer program in order to exclusively identify its creator and thus authorize the user. Forensic watermarks can be recapped randomly within the content to achieve hard detection and removal. Thus, forensic watermarking is utilized to protect the interested contents against illegal use. Moreover, the forensic watermark can alert the owners in the case of receiving illegitimate image, report, documents, or programs. Since medical images have crucial properties and are very central and important part of the medical information (ROI), the ROI is supportive in providing further diagnosis by the physician. Any bit of distortion in this part may direct for undesirable treatment for the patient. For the sake of securing, medical images through watermarking ROI should be protected and the watermarks can be applied to the residual part of the image (RONI). Consequently, application of watermarking in medical images can be considered as a two-step procedure which comprises the following:

1. Extracting ROI from the medical images
2. Applying watermarking on RONI

Diverse algorithms are available for segmentation of ROI on the different types of medical images. Additionally, there are different algorithms available for applying watermarks.

Digital watermarking is employed to guarantee the integrity of medical images by supplying tamper protection even when the images depart the network. This work mainly focused on diverse attributes and a variety of features on numerous proposed methods. Consequently, a brief review about the medical watermarking techniques and their purpose is given as follows.

Kong et al. proposed three digital watermarking methods [60] for the purpose of security and authentication of the transmission of biomedical signals. It verified the integrity of the EEG signal. Mainly three methods were used for this purpose, namely patchwork, quantization watermarking, and the least significant bit method. These methods were compared on the basis of their noise detection rate. Several parameters such as the robustness of the EEG signal, the change in signal due to brain injury, and response to the noise contamination were used for the evaluation/comparison of these techniques. A watermarking technique was described by Engin et al. [61], which is based on discrete wavelet transformation. This method established the signal integrity of the electrocardiogram (ECG) signal while transmitting it to detect various cardiovascular diseases. The method was evaluated under various noise conditions for the various wavelet functions. Later, He et al. in [62] introduced a self-synchronized watermark technology to achieve the security of electrocardiogram (ECG) signal. The ECG signal was transformed using Harr wavelet transform with 7-level decomposition. The watermark and synchronization code were embedded in the low-frequency sub-band of level 7. The signal-to-noise ratio (SNR) between the embedded ECG and original ECG was greater than 30. The difference between these two ECG signals was very small. To test the robustness of ECG signals, a white noise attack with various strengths was simulated.

In the recent few years, Dey's group has been working on this field and proposed variety of techniques on both blind and reversible watermarking techniques on different biomedical signals. Dey et al. proposed variety of methods of blind watermarking techniques on different signals as follows. Through the year 2012 [63], Dey et al. proposed a low-distortion prediction error expansion reversible watermarked technique to meet the security, integrity, and hiding requirements of the biomedical signal. The authors included the watermark with the original signal as a multimedia content. The method used the low-distortion technique to add and remove the watermark from the EOG signal, which was mainly used to detect the position and movement of the eye. The insertion of the watermark did not violate the original signal qualities proved via the experimental results. The value of the SNR between the original and the extracted signal increased the robustness of the technique compared to the other techniques. Analysis between the original signal, watermarked signal, and the extracted signal was done briefly. Then, a reversible watermarking method was offered by Dey et al. [64] on ECG signals. During this paper, a reversible binary watermarking technique in which the watermark was embedded with the PPG signals as well as the prediction error-based mechanism was used for watermark extraction. As photoplethysmogram (PPG) is a diagnostic tool that measures the electrical activity of the human heart in order to detect a variety of cardiovascular diseases. The distortion and noise generated due to the watermark embedding to the original signal were very low and negligible. Various changes in diagnostic values due to watermark were studied and analyzed. Telecardiology was achieved by a successful transmission of the watermarked PPG Signal.

The above-mentioned methods are mainly based on reversible watermarking techniques, while various other researches on blind watermarking techniques for different biomedical signals are discussed as follows. Nambakhsh et al. in [65] introduced a blind watermarking technique which used a secret key for encryption of the original signal and added the electroencephalogram (EEG) signal to the biomedical image. Using the zero-tree wavelet (EZW) algorithm to perform the embedding procedure, the original signal was compressed. Besides it, a decompression technique was used for extraction. The algorithm had been tested over many computed tomography (CT) scan report and magnetic resonance imaging (MRI) images. The resultant peak signal-to-noise ratio was greater than 35 dB in case of watermarked of 512–8192 bytes signal. It was capable to use 15 % of the carrier signal for embedding the marked signal. In [66], Kaur et al. proposed a digital watermarking method which increased the authenticity and security of the ECG signal while it transmitted through the wireless communication media. A 15-digit code was used as the patient's identification. That code was employed for the embedding of the original signal watermark. This method can completely remove the watermark from the marked signal at the receiver end. A low-frequency chirp signal was used for the embedding purpose. It performed the blind recovery of the watermarked images and detected the noise addition and the filtering attacks. During the same year, Ibaida et al. [67] introduced a method that embedded the patient information with the ECG signal to maintain the owner's authenticity without causing any fatal distortion of the original image. Numerous situations were tested to evaluate the signal modification due to the watermark. The marginal amount of distortion of the original image does not decrease the overall quality of the ECG signal. The size of the signal did not increase due to the added information. The scaling and bandwidth also remained the same. The authors proposed another method in [68], which embedded the patient information with the ECG signal to maintain the owner's authenticity without causing any fatal distortion of the original image. Simulation results were conducted to evaluate the signal modification due to the watermark, which illustrated that the marginal amount of distortion of the original image does not decrease the overall quality of the ECG signal. In addition, the size of the signal did not increase due to the added information; besides, the scaling and bandwidth were remaining the same. The technique recommended by Yina et al. [69] was based on single channel electromyography blind recognition. To reduce the noise, to increase the reliability of the hardware, and to reduce the complexity of the surface electromyography (sEMG) signals, a single channel independent component analysis was adopted. The embedded watermarks assisted to solve the blind source separation disorder problem. For the classification of sEMG features, self-adaptive neural network and some eigenvectors were applied. A different hand position could be recognized for the classification results. The carrier sEMG signal was transforming into wavelet domain embedded with the synchronization code values.

In [70], the researchers introduced a watermark method where various types of gray-scale biomedical images were added as the watermark into the original signal. It mainly preserved the ownership right of biomedical signal. The electrooculgraphy signal was used as the biomedical signal to measure and record the movement and the position of the eye. The 1-D EOG signal was converted into a 2-D signal. Then, discrete wavelet transforms (DWT), discretecosine transforms (DCT), and singular value decomposition (SVD) were applied to add watermark in a 2-D signal. For the watermark extraction, the inverse DWT, inverse DCT, and SVD were used. The PSNR between the original and watermarked signal was comparatively lower than other techniques. Moreover, Dey et al. [71] established another method on ECG signal. The proposed method worked on a blind watermarking technique based on novel session that exploited a secret key for adding the watermark to the electrocardiogram (ECG) signal. The earlier part of this paper consisted with the multiresolution wavelet transform-based system for complexity analysis of the P, Q, R, S, T peaks compared to the original ECG report of the human heart. The second portion proposed a DWT and spread spectrum-based watermarking technique. A comparative study was done in the intervals of two consecutive R-R, P-R, QRS duration, cardiac output duration, cardiac output between original P, QRS, and T components and the watermarked P, QRS, and T components. A further work in [72] was conducted by our group that reported a different technique on the same ECG signal. It provided a binary watermarking method that embedded watermark in the ECG signal. Self-recovery-based watermark extraction method used the stationary wavelet transformation (SWT), spread spectrum and quantization. In this approach, the level of distortion and noise was remarkably low. A comparative analysis of detecting P-QRS-T components was done to measure the diagnostic value changes.

Chen et al. [73] proposed a digital watermarking encryption methodology based on quantization to protect the copyright and maintain security of the electrocardiogram (ECG) signal. Three transformation domains, namely DWT, DCT, and discrete Fourier transform (DFT) were used for quantization. The embedding process was not invertible, though the changes and the distortion of the original signal when compared to the watermarked signal were significantly low and tolerable. The watermarked signal could be used for the diagnosis purpose easily without a devaluation of the signal properties. Without the knowledge of the original ECG signal, the hidden information could be extracted.

Watermark optimization using evolutionary algorithms has also been investigated extensively. By the year 2010 [74], a novel method using genetic algorithms (GA) for adaptive watermark strength optimization in DCT domain were employed in which watermark strength is intelligently selected through the GA. While in [75], a new hybrid particle swarm optimization (HPSO) was suggested develop the performance of fragile watermarking supported DCT. The results proved an enhancement in both the watermarked image quality and the extracted watermark. Based on optimization algorithms, Soliman et al. in [76] designed a watermarking scheme through invoking particle swarm optimization (PSO) algorithm in adaptive quantization index modulation and SVD in conjunction with DWT and DCT in the medical imaging. The proposed approach endorsed the watermarked image quality and robustness. The experimental results showed that the recommended algorithm yields a watermark that is invisible to human eyes, reliable enough for tracing colluders and robust against a wide variety of common attacks.

For medical images, an intelligent reversible watermarking approach using differential evolution (DE) was proposed in [77]. In the same year [78], the authors introduced both GA and PSO as intelligent reversible watermarking technique. Recently, through the year 2015, Venugopal and Siddaiah in [79] utilized a dual security scheme as it considered the medical image as a watermark and then is watermarked inside the natural image. Encryption algorithms are used for supplementary enhancement of the watermarked image security. In addition, a multi-objective optimization approach using different heuristic and metaheuristic approaches like the GA, DE, and bacterial foraging optimization (BFOA) was proposed to preserve the structural integrity of the medical images. The watermarking was implemented using both lifted wavelet transforms (LWT) and SVD technique. The experimental results proved that the proposed optimization approach endows with a high degree of robustness coupled with imperceptibility in addition to preserved structural integrity making it highly suitable for medical application.

5 Discussion

Hospitals and medical centers have enormous databases comprising medical text, images, and patient records. The exchange of these databases through the networks necessitates content organization to index medical record information and a high degree of authenticity as well as security to preserve the privacy of the patients' information. To realize these objectives, diverse techniques of digital watermarking have been employed to hide the patient information in the form of a watermark. As in the field of medical information system, there are numerous formations about the patients which must be kept secret for the privacy concern. The traditional security protection technologies ensure the content integrity, but they are incapable of preventing the internal data access and data disclosure. Typically, there are several effective schemes for the protection of patient's records. The original information is encrypted using some traditional encryption algorithms, and then, the encrypted data are stored into the database of the medical information system. The main advantage of the encryption technique is that it is not possible to get the original data from a normal perceptual event that is disclosed. Watermarking is the mostly used data hiding technique in the field of biomedical information security and legal authentication. Digital watermarking refers to the technique of embedding watermark to the carrier signal. It does not affect the original carrier signal and is also hard to detect by visual or auditory perception. The embedded watermark ensures the authentication of the creator and buyers and also detects any manipulation done during the transmission of the data. The digital watermark could be extracted using numerous extraction procedures in the future. There are some effective watermarking techniques that can be classified into mainly four categories based on their characteristics, which are namely blind watermarking, non-blind watermarking, reversible watermarking, and irreversible watermarking.

In blind (public) watermarking, a secret key is used for the watermark extraction purpose. The carrier signal or the embedded watermark is not required, although the non-blind (private) watermarking is the watermarking, where the original signal is needed to extract the watermark the watermarked data. Meanwhile, reversible watermarking technique is a lossless technique used where distortions of original data are strictly restricted. The original data can be obtained from the watermarked data easily. This technique is widely used in military and medical image processing. Conversely, the irreversible technique is a lossy technique, where the original data cannot extract if discarding the watermark from the watermarked data. For the last couple of years, various researchers working on different watermark techniques on different biomedical signals as a number of original contributions have been made and reported in this work. They are mainly based on blind watermarking and reversible watermarking.

Design and evaluation parameters assist as a guideline in watermarking schemes benchmarking and development. The parameters of an ideal watermarking scheme should be achieved, namely as high embedding capacity, robust, invisible, non-blindness, and reversible. However, a compromise between diverse parameters has to be done depending on the application framework. The higher the watermark signal strength, the more it is robust of higher capacity albeit perceptibility is compromised.

Tamper/interfere localization and recovery ability in watermarking are essential to preserve the medical image's integrity, where any distortion of medical image is unacceptable for diagnosis purposes. It has been found out that there has been a number of works about the integrity and authenticity of medical images. For the severe requirements of medical images, this reluctance is prompted from the incomplete justification of watermarking applicability. To increase the approval of digital watermarking, it is crucial to improve a reversible watermarking with tamper localization and recovery competency into simulating the real-world environment requirements.

Generally, digital watermarking algorithms [43] can be categorized according to the domain of watermark embedding into two groups. The first one is related to the algorithms that uses a spatial domain for data hiding [80, 81], while the algorithms of the second assembly use transform domain, such as DCT [82], DFT [83], and DWT [84] for watermarking purpose. Moreover, biometrics integrate several approaches to recognize unique features of human beings through utilizing their biological features for authentication. Nandi et al. [85] proposed a novel system for constructing a secure biometric recognition system by using an electrocardiogram (ECG)-hash code of two distinct individuals. The ECG-hash code was encrypted using rule vector of cellular automata, which provided better security in terms of randomness of generated cipher text.

Previous works expose that transform domain arrangements are typically more robust to noise and common image processing tasks when compared to spatial transform techniques [86]. Another category for digital watermarking can be considered as visible and invisible, robust and fragile, blind and non-blind with emphasis on authentication, copyright ownership, and the host image availability.

6 Conclusion

In the medical application domain, traditional diagnosis has been replaced by e-diagnosis. This paradigm leads to applications in healthcare productiveness such as telesurgery, teleconsulting, and telediagnosis. Entirely, these applications involve the medical image exchange in the digital format from one environmental location to alternative through the globe through a fast network such as the Internet. Conversely, digital medical image forms can be deployed using image processing software. Patients as well as hospitals might need to adjust the medical images or for any illegal purposes. The healthcare community is becoming responsive with the efficiency of utilizing digital imaging technologies. Substantial research efforts are directed toward supplying integrated digital patient records. This increased usage of the digital image-based application results in huge databases which are hard to use and complex to privacy concerns which is crucial in medical image applications. Specifically, security and privacy controls in images are novel and challenging problems. The utmost adopted solution in hospitals is the encryption of the image/patient data in separate files. In order to recover patient data or analysis details, it is necessary to decrypt both files, so this solution is inefficient.

Consequently, watermarking has become a significant concern in medical image, confidentiality, security, and integrity. Medical image watermarks are used to authenticate and/or integrity investigation of medical images. Authentication is to trace the image origin, while the integrity is to detect whether changes have been done. One of the key problems with medical image watermarking is that medical images have special requirements.

Watermarking techniques for biomedical signals deals with the encryption of the original data and embedding the watermark with it to keep the authenticity, security, and the owner's legal right of the signal. Addition of the watermark may cause distortion into the original signal. The distorted signal leads the doctors to the wrong diagnosis. The main aims of the various watermarking techniques are reducing the level of distortion as much as possible. This work could help us to select a better watermarking technique among the various existing methods.

Improving reversible watermarking is promising for medical image application with compromising with solving the risk of losing the diagnostic accuracy, although computational properties may suffer additional complexity in diverse processing domains. In addition, the watermarking medical signals require a more standard to quantitatively evaluate watermark interference with the diagnosis. Essentially, there is a necessity for distortionless and robust methods adapted to several medical imaging modalities and services such as surgery. Furthermore, attempts need to be made to solve trade-offs between the robustness and capacity of applications in medical information protection, safety, and management, as higher capacity can be achieved at the expense of either robustness or imperceptibility.

References

1. Ledzewicz U, Schattler H (2002) Sufficient conditions for optimality of controls in biomedical systems. In: Proceedings of the 41st IEEE conference on decision and control 3: 3524–3529
2. Al Giakoumaki, Perakis K, Tagaris A, Koutsouris D (2006) Digital watermarking in telemedicine applications-towards enhanced data security and accessibility. Conf Proc IEEE Eng Med Biol Soc. 1:6328–6331
3. Moghadas A, Jamshidi M, Shaderam M (2008) Telemedicine in healthcare system. World Automation Congress (WAC 2008), pp 1–6
4. Barni M, Bartolini FD (2004) Data hiding for fighting piracy. IEEE Signal Process Mag 21 (2):28–39
5. Coatrieux G, Maître H, Sankur B, Rolland Y, Collorec R (2000) Relevance of Watermarking in Medical Imaging. Conf. Proc. IEEE Int. ITAB, USA, pp 250–255
6. Delp E (2005) Multimedia security: the 22nd century approach! Proc Workshop Multimed Syst 11(2):95–97
7. Kumari G, Kumar B, Sumalatha L, Krishna V (2009) Secure and robust digital watermarking on grey level images. Int J Adv Sci Technol 11
8. Miyazaki A, Okamoto A (2002) Analysis and improvement of correlation-based watermarking methods for digital images. Kyushu University
9. Jin C, Zhang X, Chao Y, Xu H (2008) Behavior authentication technology using digital watermark. In: International conference on multimedia and information technology (MMIT '08), pp 7–10
10. Jin C, Wang S, Jin S, Zhang Q, Ye J (2009) Robust digital watermark technique for copyright protection. In: International symposium on information engineering and electronic commerce (IEEC '09), pp 237–240
11. Pan W, Coatrieux G, Cuppens-Boulahia N, Cuppens F, Roux C (2010) Watermarking to enforce medical image access and usage control policy. In: 2010 Sixth international conference on signal-image technology and internet-based systems (SITIS), pp 251–260
12. Li M, Narayanan S, Poovendran R (2004) Tracing medical images using multi-band watermarks. In: 26th Conference of the IEEE annual international engineering in medicine and biology society (IEMBS '04) 2:3233–3236
13. Kishore P, Venkatram N, Sarvya C, Reddy L (2014) Medical image watermarking using RSA encryption in wavelet domain. In: 2014 First international conference on networks & soft computing (ICNSC), pp 258–262
14. Serdean C (2002) Spread spectrum-based video watermarking algorithms for copyright protection. PhD thesis, University of Plymouth
15. Hu Y, Kwong S (2001) Wavelet domain visible watermarking. IEE Electron Lett 37 (20):1219–1220

16. Mohanty S, Guturu P, Kougianos E, Pati N (2006) A novel invisible color image watermarking scheme using image adaptive watermark creation and robust insertion-extraction. In: Proceedings of the 8th IEEE international symposium on multimedia (ISM), pp 153–160

17. Razafindradina H, Karim A (2013) Blind and robust images watermarking based on wavelet and edge insertion. Int J Cryptogr Inf Secur (IJCIS) 3(3)

18. Kumar Y, Vishwakarma P, Nath R (2012) Semi-blind color image watermarking on high frequency band using DWT-SVD. Int J Eng Res Dev 4(3):57–61

19. Dharwadkar N, Amberker B (2010) An efficient non-blind watermarking scheme for color images using discrete wavelet transformation. Int J Comput Appl 2(3)

20. Cox I, Miller M, Bloom J, Fridrich J, Kalker T (2007) Digital Watermarking and Steganography. Morgan Kaufmann Publishers Inc., San Francisco, CA, USA. ISBN 978-0123725851

21. Kumar A, Pooja K (2010) Steganography—a data hiding technique. Int J Comput Appl 9(7)

22. Caldelli R, Filippini F, Becarelli R (2010) Review article reversible watermarking techniques: an overview and a classification. Hindawi Publishing Corporation EURASIP Journal on Information Security 2010:1–19. doi:10.1155/2010/134546

23. Maeder A, Planitz B (2005) Medical image watermarking for multiple modalities. In: 34th Workshop proceedings on applied imagery and pattern recognition (AIPR 2005), 19–21 October 2005, Washington, DC, USA

24. Tsai A, Luh J, Lin T (2012) A modified multi-channel emg feature for upper limb motion pattern recognition. In: 2012 Annual international conference of the IEEE engineering in medicine and biology society 2012:3596–3599

25. Gut R (2000) High-precision EMG signal decomposition using communication techniques. IEEE Trans Signal Process 48(9)

26. Wang Z, Ning X, Zhang Y, Gonghuan (2002) Nonlinear dynamic characteristics analysis of synchronous 12-lead ECG signals. IEEE Eng Mag Med Biol 19(5)

27. Lisha S, Minfen S, Congtao X (2002) Nonlinear analysis of EEG signals based on the parametric bispectral estimation. In: 2002 6th international conference on signal processing, vol 1

28. Velasquez-martinez L, Murillo-rendon S, Castellanos-dominguez G (2012) Relevance and redundancy analysis of PCG signals in the detection of heart murmurs by anova. In: 2012 XVII symposium of image, signal processing, and artificial vision (STSIVA), pp 190–195

29. Abdullatif M, Zeki A, Chebil J, Gunawan T (2013) Properties of digital image watermarking. In: 2013 IEEE 9th international colloquium on signal processing and its applications, 8–10 Mac. 2013, Kuala Lumpur, Malaysia

30. Katariya S (2012) Digital watermarking: review. Int J Eng Innov Technol (IJEIT) 1(2)

31. Cagalaban G, Kim S (2011) Towards a secure patient information access control in ubiquitous healthcare systems using identity-based signcryption. In: 2011 13th international conference on advanced communication technology (ICACT), pp 863–867

32. Alhaqbani F, Fidge C (2008) Privacy-preserving electronic health record linkage using pseudonym identifiers. In: Proceedings of 10th international conference on e-Health, Networking, applications and services, pp 108–117

33. Coatrieux G, Lamard M, Daccache W, Puentes J, Roux C (2005) A low distortion and reversible watermark: application to angiographic images of the retina. In Proceedings of international conference of the IEEE-EMBS, pp 2224–2227

34. Coatrieux G, Sankur B, Maître H (2001) Strict Integrity Control of Biomedical Images. Proc. Electronic Imaging, Security and Watermarking of Multimedia Contents. SPIE, USA, pp 229–240

35. Li M, Poovendran R, Narayanan S (2005) Protecting patient privacyagainst unauthorized release of medical images in a group communication environment. Comput Med Imaging Graph 29(5):367–383

36. Zhou X, Huang H, Lou S (2001) Authenticity and integrity of digital mammography images. IEEE Trans Med Imaging 20(8):784–791

37. Wakatani A (2002) Digital watermarking for ROI medical images by using compressed signature image. In: Proceedings of 35th Hawaii international conference on system sciences, pp 2043–2048
38. Gupta N, Sharma M (2012) Invisible multiple watermarking with minimum distortion using DWT-DCT-CDMA. Int J Comput Sci Eng Technol (IJCSET) 3(8)
39. Coatrieux G, Lecornu L, Sankur B, Roux C (2006) A review of image watermarking applications in healthcare. In: International conference IEEE engineering in medicine and biology society, New York, NY, pp 4691–4694
40. Bouslimi D, Coatrieux G, Roux C (2012) A joint encryption/watermarking system for verifying the reliability of medical images. Comput Methods Biomed 106(1):47–54
41. Poonkuntran S, Rajesh R, Eswaran P (2009) Wavetree watermarking: an authentication scheme for fundus images. In: Proceedings of IEEE international conference on emerging trends in computing, pp 507–511
42. Cao F, Huang X, Zhou Q (2003) Medical image security in a HIPAA mandated PACS environment. Comput Med Imaging Graph 27(2–3):185–196
43. Umaamaheshvari A, Thanuskodi K (2012) Survey of watermarking algorithms for medical images. Int J Eng Trends Technol 3(3)
44. Ni R, Ruan Q, Zhao Y (2008) Pinpoint authentication watermarking based on a chaotic system. Int J Forensic Sci 179(1):54–62
45. Al-Qershi B (2010) Reversible watermarking scheme based on two-dimensional difference expansion (2DDE). In: Second international conference on computer research and development, pp 228–232. doi:10.1109/ICCRD.2010.76
46. Zhang X, Wang S, Qian Z, Feng G (2010) Reversible fragile watermarking for locating tampered blocks in JPEG images. Sig Process 90(12):3026–3036
47. Chaing K, Chang-Chien K, Chang R, Yen H (2008) Tamper detection and restoring system for medical images using wavelet-based reversible data embedding. J Digit Imaging 21(1):77–90
48. Li C, Yang F (2003) One dimensional neighbor forming strategy for fragile watermarking. J Electron Imaging 12(2):284–291
49. Avcibas I, Nasir M, Sankur B (2001) Steganalysis based on image quality metrics. In: 2001 IEEE fourth workshop on multimedia signal processing, pp 517–522
50. Planitz B, Maeder A (2005) Medical image watermarking: a study on image degradation. In: Proceedings of Australian pattern recognition society (APRS), pp 3–8
51. Viswanathan P, Krishna P (2014) A joint FED watermarking system using spatial fusion for verifying the security issues of teleradiology. IEEE J Biomed Health Inform 18(3):753–764
52. Perwej Y, Parwej F, Perwej A (2012) An adaptive watermarking technique for the copyright of digital images and digital image protection. Int J Multimed Its Appl (IJMA) 4(2)
53. Durvey M, Satyarhi D (2014) A review paper on digital watermarking. IJETTCS 3(4):99–105
54. Singh Y, Devi B, Singh K (2013) A review of different techniques on digital image watermarking. IJER 2(3):193–199
55. Mller M, Cox J, kalker T (1999) A review of watermark principal and practices. In: Parhi KK, Nishitani T (eds) Digital signal processing in multimedia system. Marcell Dekar Inc., pp 461–485
56. Nyeem H, Boles W, Boyd C (2014) Digital image watermarking: its formal model, fundamental properties and possible attacks. EURASIP Journal on Advances in Signal Processing 2014EURASIP, 2014:135
57. Raul R, Claudia F, Trinidad-Bias G (2007) Data hiding scheme for medical images. In: Proceedings of the 17th international conference on electronics, communications and computers (CONIELECOMP '07), pp 32
58. Singh P, Chadha R (2013) A survey of digital watermarking techniques, applications and attacks. Int J Eng Innov Technol (IJEIT) 2(9)
59. Voloshynovskiy S, Pun T, Eggers J, Su J (2001) Attacks on digital watermarks: classification, estimation based attacks, and benchmarks. IEEE Commun Mag 39(8):118–126
60. Kong X, Feng R (2001) Watermarking medical signals for telemedicine. IEEE Trans Inf Technol Biomed 5(3):195–201

61. Engin M, Çidam O, Engin E (2005) Wavelet transformation based watermarking technique for human electrocardiogram (ECG). J Med Syst 29(6):589–594
62. He X, Tseng K, Huang H, Chen S, Tu S, Zeng F, Pan J (2012) Wavelet-based quantization watermarking for ECG Signals. In: 2012 International conference on computing, measurement, control and sensor network (CMCSN), Taiyuan, China, July 7–9, pp 233–236
63. Dey N, Biswas S, Das P, Das A, Chaudhuri SS (2012) Feature analysis for the reversible watermarked electrooculography signal using low distortion prediction-error expansion. In: 2012 International conference communications, devices and intelligent systems (CODIS), pp 624–627
64. Dey N, Maji P, Das P, Biswas S, Das A, Chaudhuri S (2012) Embedding of blink frequency in electrooculography signal using difference expansion based reversible watermarking technique. Buletinul Ştiinţific Al Universităţii "Politehnica" Din Timişoara, Seria Electronică Şi Telecomunicaţii Transactions On Electronics And Communications 57(71)
65. Nambakhsh M, Ahmadian A, Ghavami M, Dilmaghani R, Karimi-Fard S (2006) A novel blind watermarking of ECG signals on medical images using EZW algorithm. In: Proceedings of the 28th annual international conference of the IEEE engineering in medicine and biology society (Embs'06), New York, USA, 30 Aug 3 Sept 1:3274–3277
66. Kaur S, Farooq O, Singhal R, Ahuja B (2010) Digital watermarking of ECG data for secure wireless communication. In: Proceedings of the 2010 international conference on recent trends in information, telecommunication and computing (Itc2010), Kochi, Kerala, India, 12–13 March, pp 140–144
67. Ibaida A, Khalil I, Van Dhiah A (2010) Embedding patients confidential data in ecg signal for healthcare information system. In: Proceeding of the 32nd annual international conference of the IEEE Embs, Buenos Aires, Argentina, 31 Aug 4 Sept., pp 3891–3894
68. Ibaida A, Khalil I, Van Schyndel R (2011) A low complexity high capacity ECG signal watermark for wearable sensor-net health monitoring system. In: Proceedings of the computing in cardiology (Cinc), Hangzhou, China, 18–21 Sept, pp 393–396
69. Yina G, Dawei Z (2012) Single channel surface electromyography blind recognition model based on watermarking. J Vib Control 18(1):42–47
70. Tseng K, He X, Kung W, Chen S, Liao M, Huang H (2014) Wavelet-based watermarking and compression for ECG signals with verification evaluation. Sensors (Basel) 14(2):3721–3736
71. Dey N, Biswas D, Roy A, Das A (2012) DWT-DCT-SVD based blind watermarking technique of gray image in electrooculogram signal. In: 2012 12th International conference on intelligent systems design and applications (ISDA), pp 680, 685
72. Dey N, Mukhopadhyay S, Das A, Chaudhuri S (2012) Analysis of P-QRS-T components modified by blind watermarking technique within the electrocardiogram signal for authentication in wireless telecardiology using DWT. Int J Image Graphics Signal Process (IJIGSP) 7:33–46
73. Chen S, Guo Y, Huang H, Kung W, Tseng K, Tu S (2014) Hiding patients confidential datainthe ECG signal via a transform-domain quantization scheme. J Med Syst 38(6):1–8
74. Sikander B, Ishtiaq M, Jaffar M, Tariq M, Mirza A (2010) Adaptive digital watermarking of images using genetic algorithm. In: 2010 International conference on information science and applications (ICISA), pp 1–8
75. Kosgharghory S (2011) Hybrid of particle swarm optimization with evolutionary operators to fragile image watermarking based DCT. Int J Comput Sci Inf Technol (ijcsit) 3(3)
76. Soliman M, Hassanien A, Ghali N, Onsi H (2012) An adaptive watermarking approach for medical imaging using swarm intelligent. Int J Smart Home 6(1)
77. Lei B, Tan E, Chen S, Ni D, Wang T, Lei H (2014) Reversible watermarking scheme for medical image based on differential evolution. Expert Syst Appl 41(7):3178–3188
78. Naheed T, Usman I, Khan T, Dar A, Shafique M (2014) Intelligent reversible watermarking technique in medical images using GA and PSO. Optik Int J Light Electron Optics 125(11)
79. Reddy V, Siddaiah P (2015) Hybrid LWT-SVD watermarking optimized using metaheuristic algorithms along with encryption for medical image security. Int J Signal Image Process 6 (1):75–95

80. Memon N, Gilani S, Ali A (2009) Watermarking of CT scan medical images for content authentication. In: International conference on information and communication technologies, 2009 (ICICT '09), pp 175, 180

81. Liu J, He X (2005) A review study on digital watermarking. In: First international conference on information and communication technologies (ICICT 2005), pp 337–341

82. Zkou X, Huang H, Lou S (2001) Authenticity and integrity of digital mammography images. IEEE Trans Med Imaging 20(8):784–791

83. Sebe F, Ferrer T, Herrera J (2000) Spatial domain image watermarking robust against compression, filtering, cropping and scaling. Springer Image Comput 2:44–53

84. Nikolaidis N, Pitas I (1998) Robust image watermarking in spatial domain. IEEE Signal Process Trans 66(3):385–403

85. Nandi S, Roy S, Dansana J, Karaa WB, Ray R, Chowdhury SR, Chakraborty S, Dey N (2014) Cellular automata based encrypted ECG-hash code generation: an application in inter human biometric authentication system. Int J Comput Netw Inf Secur 6(11):1

86. Cao F, Huang H, Zhou X (2003) Medical image security in a HIPAA mandated PACS environment. Computerized medical imaging and graphics. IEEE Trans Image Process 27 (2–3):185–196

Pixel Repetition Technique: A High Capacity and Reversible Data Hiding Method for E-Healthcare Applications

S.A. Parah, F. Ahad, J.A. Sheikh, N.A. Loan and G.M. Bhat

Abstract Electronic healthcare is changing the world worldwide scenario of the conventional healthcare. Transmission of medical images in e-healthcare system is of greater importance, because of the valuable information carried out by them for the purpose of tele-diagnosis and tele-medicine. For successful tele-diagnosis, it is imperative that the medical images should not be tampered with during transit. In this chapter, a high capacity and reversible data hiding system utilizing Pixel Repetition Technique (PRT) has been proposed. The usage of PRT for preprocessing reduces the computational complexity drastically. The proposed system is capable of any sort of tamper detection. This has been achieved by embedding a fragile watermark (computed from checksum of the cover medium blocks) in addition to electronic patient record (EPR). Block-wise division of cover image and intermediate significant bit (ISB) substitution have been used for data embedding. A detailed experimentation carried out shows that besides supporting high payload and reversibility our technique is capable of detecting any tamper caused due to signal processing and geometric attacks. Further, less computational complexity makes the proposed scheme an ideal candidate for real-time electronic healthcare applications.

Keywords Electronic patient's record · Pixel Repetition Technique · Fragility · Checksum · Reversibility · Tamper detection

S.A. Parah (✉) · F. Ahad · J.A. Sheikh · N.A. Loan · G.M. Bhat
Department of Electronics and Instrumentation Technology,
University of Kashmir, Hazratbal, Srinagar, J&K 190006, India
e-mail: shabireltr@gmail.com

F. Ahad
e-mail: bhatfarhana@gmail.com

J.A. Sheikh
e-mail: sjavaid_29ku@yahoo.co.in

N.A. Loan
e-mail: nazirloan786@gmail.com

G.M. Bhat
e-mail: drgmbhat@gmail.com

© Springer International Publishing Switzerland 2017
N. Dey and V. Santhi (eds.), *Intelligent Techniques in Signal Processing for Multimedia Security*, Studies in Computational Intelligence 660, DOI 10.1007/978-3-319-44790-2_17

1   Introduction

With the advancement of digital technology and networked infrastructure, electronics is creeping into every sector. The conventional way of commerce, banking, and healthcare has been replaced by e-commerce, e-banking, and e-healthcare. e-healthcare system involves tele-diagnosis and tele-medicine [1]. For the purpose of tele-diagnosis, digital medical data such as patient's case history, hospital logos, and medical images are transferred over Internet so that the doctor at a remote place can get this data and accordingly diagnose the patient [2]. The exponential growth of Internet, however, has made it possible for unauthorized users to copy and modify the available data and transfer it again [3, 4]. Since medical data carry critical information about the patient and as such it must reach to the receiver (remotely available doctor) in its exact form as an erroneous data may lead to misdiagnosis of a patient [5]. Thus for correct diagnosis at the receiving end a due care needs to be taken in terms of content authentication and copyright protection of the data [6]. There are many techniques which ensure authentication, security, and copyright protection of the medical information but digital watermarking has been found to be one of the best tools for such applications [7–10].

Digital watermarking is the process of embedding special kind of data known as watermark which serves as a basis for authentication, ownership verification, and even for copyright protection [11, 12]. Digital watermarking techniques can be robust or fragile depending upon whether it is required for copyright protection or content authentication. If a watermark is to be used for ownership verification or copyright protection, then such watermark must be robust and resist most of the intentional and non-intentional attacks while as fragile watermark gets completely distorted even when the watermarked media is tampered slightly [13, 14]. In either case of embedding a digital watermark, the cover image quality is deteriorated. This deterioration is unacceptable especially in the case of medical images [15–17]. Thus, medical image watermarking needs to be different from conventional image watermarking schemes. This is because of the fact that the medical images contain the critical information from the diagnosis point of view. In order to secure the diagnostic information, two major watermarking models are followed while embedding the digital watermark in medical images [18]. First method is dividing a medical image in two regions, region of interest (ROI) and region of non-interest (RONI). Since ROI contains the diagnostic region, therefore this region is kept away from embedding process. The digital watermark is usually embedded in RONI as this region contains the least information about the patient's problem. The other method involves reversibility. Reversibility ensures the received image retains its pre-embedding form after data are extracted [19–21]. A lot of work has been carried out to achieve the reversibility. Various interpolation-based hiding techniques have been used to achieve the scaled-up image from the original (small sized) image to ensure reversibility [30, 33]. However, use of interpolation for obtaining cover image from original one results in increase in computational complexity. Thus, the conventional way of achieving the reversibility using interpolation has been replaced by using Pixel Repetition Technique in this work. For obtaining high-quality watermarked medical images, this technique has been found to be suitable for a fairly high payload.

Digital watermarking along with the reversibility is well suited for e-healthcare applications. This chapter presents a reversible, high capacity data hiding technique wherein, besides embedding electronic health record of the patient, a fragile watermark is embedded for authentication of electronic health record (EHR) at the receiving end. Any intentional/unintentional attack on the watermarked medical image results in non-recovery of embedded watermark at the receiving end, informing the receiver that EHR has been altered during transit. Further usage of checksum in the proposed technique makes it capable of detecting the tampered blocks. Rest of the chapter is organized as follows: Sect. 2 presents related work. Proposed work is elaborated in detail in Sect. 3. Data embedding and extraction are presented in Sect. 4. Experimental results and discussion are presented in Sect. 5. The paper concludes in Sect. 6.

2 Related Work

The field of e-healthcare is expanding rapidly due to exponential growth of Internet. Some of the prominent challenges faced for setting up a state-of-the-art e-healthcare system are security of electronic health record (EHR) and authentication of the contents received. Digital watermarking has been found to be an important tool in this regard. As such digital watermarking of medical images is one of the hottest research areas nowadays. In medical image watermarking, the EHR and watermark is embedded in cover medical image which leads to deterioration of its perceptual quality. Since medical images contain critical information from diagnostic point of view, conventional embedding techniques are not employed for medical images. Medical images thus use the concept of region of interest (ROI) and region of non-interest (RONI) separation for saving the quality of the critical information region for diagnostic purpose. Various medical image watermarking techniques based on ROI and RONI have been reported in [22–25]. The separation of the ROI and RONI has a drawback that the payload gets reduced because of the fact that the ROI portion cannot be used for embedding purpose. A medical image watermarking besides being imperceptible must be able to detect the tamper. A lot of work has been reported in this direction. For tamper detection, Sumalatha et al. in [26] have proposed a block-based watermarking technique. The proposed technique detects tamper by computing checksum of non-overlapping blocks of size 4×4. The computed checksum is then embedded in top left 2×2 blocks of the same block. The proposed technique besides being able to detect tamper has been also used for general image authentication. Another algorithm has been used by Guo and Zhuang [27] to attain integrity and to verify tamper. The authors have proposed user-specific region-based lossless watermarking algorithm for tamper detection, in which the region of interest has been kept away from the embedding process. The embedding in the region of interest degrades the image quality and can lead to wrong diagnosis. Abhilasha and Malay [28] have proposed a blind fragile watermarking technique for tamper detection. The final watermark is the combination of mean of ROI and patients' case history. Medical images of size 512×512 have been used for the study. The reported payload is 35,000 bits. Embedding has been done in 8th bit plane of every image using spatial domain watermarking technique. In [29], Moniruzzaman et al. have reported Discrete Wavelet transform-based watermarking system for different modalities of Digital Imaging and Communication in Medicine (DICOM). The low-frequency sub-band has been divided into non-overlapping blocks of size 3×3 in order to embed the watermark of size 85×85. In order to increase the security, the watermark has been encrypted using a logistic map. The average peak signal-to-noise ratio (PSNR) reported is more than 45 dB for computed tomography (CT) scan and magnetic resonance imaging (MRI) images and for X-ray images average PSNR reported is more than 48 dB.

Reversible data hiding algorithms are well suited for medical applications. This is because a reversible digital watermarking technique is able to retrieve the actual cover image in its exact form after data extraction at the receiving end. A lot of research has been reported on reversible image watermarking techniques. In [30], Lee and Huang have proposed a high capacity and reversible hiding algorithm. The reversibility has been achieved by interpolating the neighboring pixels, using maximum difference values of these pixels. In [31], Tang et al. have proposed a high capacity reversible data hiding technique for general standard test images. The reported payload varies from 1.20 bits per pixel (bpp) to 2.45 bits per pixel (bpp) for different set of image database. Hu and Li [32] have put forth a technique for reversible data hiding based on extended image interpolation using maximum difference value between neighboring pixels. This technique, however, suffers from low perceptual quality of watermarked images. Kamran et al. [33] have proposed a high capacity reversible watermarking algorithm based on histogram processing. Naheed et al. [34] have studied an algorithm based on additive interpolation-error expansion for medical images. The authors report two algorithms genetic algorithm (GA) and particle swarm optimization (PSO) to achieve the reversible watermarking and high payload. The authors report an average PSNR of 49 dB for average payload of 38,545 bits. A reversible watermarking algorithm for medical images has been presented by Arsalan et al. [35]. The system utilizes genetic algorithm and companding techniques to ensure reversibility. The reported average PSNR for payload of 0.5 bpp is 45 dB. The work has been carried out using various payloads to check the perceptual quality. Maity and Maity [36] have worked on different modalities and varying sizes of DICOM images for performance evaluation. The authors have replaced conventional reversible watermarking (RW) by Colcut's Reversible Contrast Mapping (RCM) assuring low computational complexity. For capacity of almost 0.5 bpp, PSNR obtained is 44 dB. Based on histogram shifting, Tsai et al. [37] have proposed a reversible data hiding scheme for medical images. The reported high payload has been achieved by using the residual histogram prediction technique of the neighboring pixels in the image.

High capacity, reversibility and tamper detection ability are of greater importance while dealing with medical images. To best of our knowledge, very little research has been reported in this direction. For diagnostic purpose, reversibility plays a significant role in medical images. This chapter presents a high capacity and reversible data hiding algorithm for medical images. The capability to detect any tampered block makes the proposed system an ideal candidate to be used for electronic healthcare applications.

3 Proposed Work

The proposed reversible data hiding scheme for medical images has been discussed in this section. The embedding algorithm has been described in detail with the necessary diagrams, equations, and examples. Reversibility is achieved by using certain preprocessing and transformation of original medical image including scale-down to scale-up of the cover image. The preprocessing includes image resizing and Pixel Repetition Technique (PRT) has been used for such a purpose. The preprocessing has been discussed in Sect. 3.1. Further, checksum generation and tamper detection is discussed in Sect. 3.2 and data embedding along with its explanation is discussed in Sect. 4.

3.1 Pixel Repetition Technique

The input image in which EPR and watermark is to be embedded has been subjected to a few preprocessing steps which result in its scaled-up version. Consider an input image '$I_{M \times N}$,' the image is resized in such a way that the resultant image '$C_{m \times n}$' has the dimension of $M/2 \times N/2$ using Bilinear Transformation. Then with the help of the Pixel Repetition Technique (PRT), the final cover image '$O_{M \times N}$' is obtained using Eqs. (1)–(4).

If $I(i \times j)$ and $O(i \times j)$ represent arbitrary pixels of input image and resultant image, then following equations have been used for obtaining scaled-up version of input image.

$$O_{(i,j)} = C_{(i,j)} \tag{1}$$

$$O_{(i,j+1)} = I_{(i,j+1)} \tag{2}$$

$$O_{(i+1,j)} = I_{(i+1,j)} \tag{3}$$

$$O_{(i+1,j+1)} = I_{(i+1,j+1)} \tag{4}$$

'$O(m \times n)$' is the interpolated image which is twice the size of the cover image.

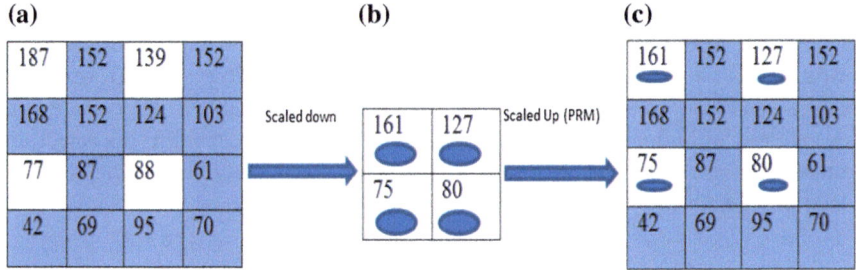

Fig. 1 Pictorial explanation of PRT. **a** Input image (T). **b** Resized image (C). **c** Cover image (O)

The concept of resizing the input image to its scaled-up version can be understood from Fig. 1.

Consider an arbitrary 4×4 block of the original image (I) as depicted in Fig. 1a. The pixels of the scaled-down image are generated using Bilinear Transformation as shown in Fig. 1b. The final scaled-up block of the cover image '$O_{(m \times n)}$' is shown in Fig. 1c.

The concepts of ceiling and floor have been used for rounding to get integer values out of numbers with decimals. It is evident from Fig. 1 that the scaled-up version 'O' contains all the pixels of input image.

3.2 Tamper Detection

For the purpose of the checksum generation and tamper detection, the interpolated image 'O,' i.e., the cover image, is divided into non-overlapping blocks of size $n \times n$. Equation 5 has been used for generating the block sum and the block sum is then reduced 'n' binary bits using Eqs. 6 and 7. The checksum bits are used for tamper detection and these bits should not get changed even after data embedding takes place. Therefore, before actual calculation of the checksum bits, two least significant bits (LSB) of each pixel are set to zero. The following equations are used for checksum generation:

Let $O_{(p,q)}$ be the arbitrary (p, q)th block, then the block sum is calculated as:

$$\text{Blocksum} = \sum_{i=1}^{n} \sum_{i=1}^{n} O_{(p,q)}(i,j) \tag{5}$$

Here we set two LSBs of $O_{(p,q)}$ (i, j) to zero to ensure accurate tamper detection. The block sum obtained from the above equation is a 'K' digit number. If 'K' digit number consists of 'n' subdigits such as $k1, k2, k3\ldots kn$. The subdigits of number 'K' are added to get 'Sum,' as shown in Eq. (6).

$$\text{Sum} = k1 + k2 + k3 + \cdots + kn \tag{6}$$

where n represents total number of digits obtained from Eq. 5.

We obtain binary equivalent of the computed SUM and final checksum is obtained by performing the XOR of every consecutive bit. The checksum is obtained as in Eq. 7.

$$\text{Checksum} = \text{XOR}(b1, b2), \text{XOR}(b3, b4), \ldots, \text{XOR}(bm - 1, bm) \tag{7}$$

where m is the number of bits used to represent binary equivalent of 'Sum.' We have explained the procedure of computing checksum for a 4×4 block in Example 1.

Example 1 Consider a 4×4 block of cover image. Firstly, we set two LSBs of each pixel to zero. The resultant pixel block is shown in Fig. 2.

$$\text{Blocksum} = \sum_{i=1}^{4} \sum_{i=1}^{4} O_{(p,q)}(i,j)$$

$$\text{Blocksum} = 161 + 152 + 125 + 152 + 168 + 152 + 124 + 101 + 73 + 85$$
$$+ 80 + 61 + 40 + 69 + 93 + 68$$
$$= 1704$$

$$\text{Sum} = 1 + 7 + 0 + 4$$
$$= 12$$

$$\text{Binary equivalent of Sum} = 00\,00\,11\,00$$

$$\text{Checksum} = \text{XOR}(0,0), \text{XOR}(0,0), \text{XOR}(1,1), \text{XOR}(0,0)$$
$$= [0\,0\,0\,0]$$

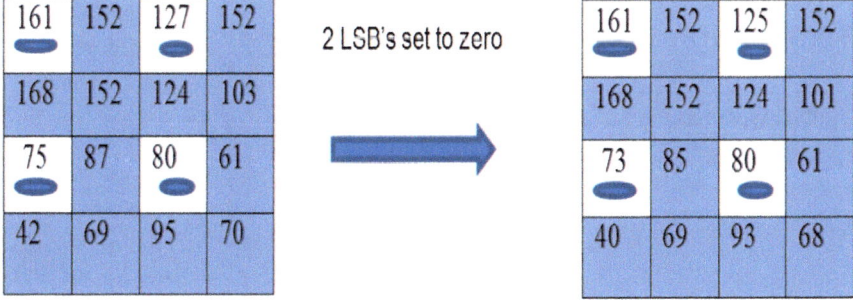

Fig. 2 Block preparation for checksum calculation

The proposed scheme has been carried out using block size 4 × 4 and 8 × 8 and thus, the checksum bits have been reduced to the size of 4 bit for 4 × 4 block while as for an 8 × 8 block an 8 bit checksum has been generated.

4 Data Embedding

Electronic patient record (EPR) and the watermark data besides the checksum bits are embedded in cover media using following steps:

1. Prepare cover from original input image as per description in Sect. 3.1.
2. Compute checksum as per Sect. 3.2.
3. Divide cover image into non-overlapping blocks of size $n \times n$.
4. Leave pivotal pixel of every $n \times n$ block as such to ensure reversibility.
5. Embed EPR, checksum, and watermark bits in remaining $\{(b \times b) - 1\}$, the remaining elements of every block.
6. Concatenate EPR and watermark data to form a single data vector. However, represent checksum bits for every block as a separate 'b' bit vector.
7. Embed EPR and watermark using ISB technique, in first $(n - 1)$ rows of every block except seed pixel and the checksum data in the last (nth) row of each block as shown in Fig. 3.

4.1 Data Embedding Explanation

A typical scenario of embedding data in a cover image block has been presented in Fig. 4. Suppose watermark bit vector, EPR bit vector, and checksum bit vector are, respectively, represented as Wb = [1 0 1 1], EPRb = [1 0 0 1], and CSMb [0 0 0 0].

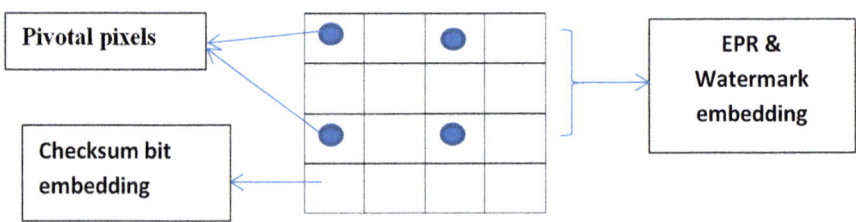

Fig. 3 Embedding procedure for a typical 4 × 4 block

(a)

161	152	125	152
168	152	124	101
73	85	80	61
40	69	93	68

(b)

1010 0001	1001 1000	0111 1101	1001 1000
1010 1000	1001 1000	0111 1100	0110 0101
0100 1001	0101 0101	0101 0000	0011 1101
0010 1000	0100 0101	0101 1101	0100 0100

(c)

1010 0001	1001 1010	0111 1101	1001 1000
1010 1010	1001 1010	0111 1110	0110 0101
0100 1001	0101 0101	0101 0000	0011 1111
0010 1000	0100 0101	0101 1101	0100 0100

Fig. 4 **a** A 4 × 4 block of cover image, **b** binary equivalent of block, **c** ISB embedding

Concatenated watermark and EPR bit vector is represented by [1 0 1 1 1 0 0 1]. The first intermediate significant bits of $n - 1$ rows of cover image block have been used for watermark and EPR embedding while checksum bits have been embedded in nth row(4th in this case). The hidden EPR, watermark, and checksum bits are highlighted in Fig. 4c. The pixel block with embedded watermark, EPR, and checksum data is shown in Fig. 4c.

4.2 Extraction Process

The proposed technique besides being reversible supports blind detection. The data extraction process from the received image can be described in the following steps:

1. The extracted bits are divided into watermark bits, EPR, and checksum bits.
2. For tamper detection, the two LSBs are again set to zero and then the image is again divided into block size of 4 × 4/8 × 8.
3. The checksum of each block is calculated and reduced to four bits or eight bits depending on the block size used at transmitting end.
4. The checksum bits extracted by step 1 are compared with the checksum bits calculated by step 3. With the help of this comparison, it is easy to detect the number of tampered blocks and the block which has been tampered.

5   Experimental Results

The experimentation has been carried out using MATLAB R2014a platform for different CT scan images. The digital watermark used for authentication is of the size 64×64. The test medical images used for evaluating the performance of the presented scheme are shown in Fig. 6. Most of the used images have been obtained from the SKIMS database while as others are commonly used test medical images.

The image quality is established on objective analysis by calculating peak signal-to-noise ratio (PSNR) and structural similarity measure index (SSIM) between original image and 'watermarked and attacked' image. The authentication of the proposed scheme is evaluated by calculating bit error rate (BER %) and normalized cross-correlation (NCC) between embedded 'watermark and EPR' and extracted 'watermark and EPR' for various attacks. Equation (15) is used for calculating BER (%) and Eq. (16) for NCC.

$$\text{MSE} = \frac{1}{MN} \sum_{i=1}^{M} \sum_{j=1}^{N} \left(x_{i,j} - x'_{i,j} \right)^2 \tag{8}$$

$$\text{PSNR} = 10 \log \frac{(2^n - 1)^2}{\text{MSE}} = 10 \log \frac{(255)^2}{\text{MSE}} \text{dB} \tag{9}$$

In the above formulae: M, N are the dimensions of the original image and the watermarked image; $x(i, j)$ is the (i, j)th pixel value of original image, and $(x'i,j)$ is the (i, j)th pixel intensity value of watermarked image.

The structural similarity index (SSIM) is based on the calculations of three terms, namely the luminance, contrast, and structure. The overall index is a given by:

$$\text{SSIM}(x, y) = [l(x, y)^\alpha] \cdot [c(x, y)]^\beta \cdot [s(x, y)]^\gamma \tag{10}$$

where

$$l(x, y) = \frac{2\mu_x \mu_y + C1}{\mu_x^2 + \mu_y^2 + C1} \tag{11}$$

$$c(x, y) = \frac{2\sigma_x \sigma_y + C2}{\sigma_x^2 + \sigma_y^2 + C2} \tag{12}$$

$$s(x, y) = \frac{\sigma_{xy} + C3}{\sigma_x \sigma_y + C3} \tag{13}$$

where μx, μy, σx, σy, and σxy are the local means, standard deviations, and cross-covariance for images x, y. For default exponents and default selections of $C3$, the expression is given by:

$$\text{SSIM}(x,y) = \frac{(2\mu_x\mu_y + C1)(2\sigma_{xy} + C2)}{\left(\mu_x^2 + \mu_y^2 + C1\right)\left(\sigma_x^2 + \sigma_y^2 + C2\right)} \tag{14}$$

$$\text{BER} = \frac{1}{MN}\left[\sum_{i=1}^{M}\sum_{j=1}^{N} w_m(i,j) \oplus w_{me}(i,j)\right] \times 100 \tag{15}$$

$$\text{NCC} = \frac{\sum_{i=1}^{M}\sum_{j=1}^{N} w_m(i,j) \times w_{me}(i,j)}{\sum_{i=1}^{M}\sum_{j=1}^{N} w_m(i,j)^2} \tag{16}$$

In above equations: M, N are the dimensions of the original logo and extracted logo; $w_m(i,j)$ is the (i,j)th pixel of original watermark, and $w_{me}(i,j)$ is the (i,j)th pixel of the extracted logo.

The perceptual quality and fragility analysis for the proposed algorithms are presented below.

5.1 Imperceptibility Analysis

The perceptual quality of the embedding scheme has been evaluated on the basis of subjective and objective analyses. Figure 6 depicts the subjective analysis of watermarked images for a payload of 1, 96, 608 bits when a block size of 4 × 4 and 8 × 8 has been used. It is evident from the subjective analysis of the watermarked images that the proposed system is capable of providing high-quality watermarked images irrespective of high payload. The objective analysis has been carried out using objective quality indices such as PSNR and SSIM. The objective metrics resulted for various images when (4 × 4) and (8 × 8) blocks are used are reported in Table 1. Note that the payload for both block division strategies is 1,96,680 bits.

It is worth to mention here that the average PSNR for the reported images is 46.8406 and 46.5588 dB, respectively, for (4 × 4) and (8 × 8) blocks while the average SSIM is 0.9760 and 0.9667. The PSNR values above 46.8406 dB and SSIM values close to unity reveal that the proposed scheme offers watermarked images of high quality. In order to validate the proposed scheme, the quality metrics of the proposed scheme are compared with Abhilasha and Malay [28] and Kamran and Malik [33]. The comparison results for 8 × 8 block size are reported in Table 2.

It is evident from the Table 2 that the proposed scheme provides high-quality watermarked images besides supporting a huge payload. The maximum capacity of the proposed scheme is 1,96,608 bits which is much higher than the techniques under comparison.

Table 1 Perceptual quality metrics

Image	PSNR (dB)	SSIM	PSNR (dB)	SSIM
	Block size (4 × 4)		Block size (8 × 8)	
Image 1	46.6025	0.9814	46.4628	0.9770
Image 2	46.6191	0.9813	46.4448	0.9767
Image 3	46.5966	0.9800	46.4469	0.9758
Image 4	46.5849	0.9817	46.4597	0.9775
Image 5	46.6208	0.9817	46.4713	0.9771
Image 6	46.6319	0.9818	46.4713	0.9774
Image 7	47.3721	0.9641	46.7808	0.9497
Image 8	46.5906	0.9808	46.4549	0.9497
Image 9	46.6083	0.9812	46.4633	0.9771
Image 10	46.6048	0.9806	46.4642	0.9762
Image 11	47.2619	0.9673	46.7328	0.9542
Image 12	47.2519	0.9675	46.7274	0.9540
Image 13	47.3355	0.9654	46.7836	0.9516
Image 14	47.2991	0.9663	46.7488	0.9528
Image 15	46.6290	0.9782	46.4696	0.9736
Average	46.8406	0.9760	46.5588	0.9667

Table 2 Perceptual transparency comparison

Image	As proposed by Abhilasha and Malay [28]		As proposed by Kamran and Malik [33]		Proposed work	
	PSNR (dB)	Payload (bits)	PSNR (dB)	Payload (bits)	PSNR (dB)	Payload (bits)
Image 1	59.93	35,000	–	–	46.4628	1,96,608
Image 2	62.1218	35,000	–	–	46.4448	1,96,608
Image 3	57.34	35,000	–	–	46.4469	1,96,608
Image 4	59.6964	35,000	–	–	46.4597	1,96,608
Image 5	56.5886	35,000	–	–	46.4713	1,96,608
Image 6	56.3438	35,000	–	–	46.4713	1,96,608
Image 7	58.1550	35,000	–	–	46.7808	1,96,608
Image 8	55.5102	35,000	–	–	46.4549	1,96,608
Image 9	56.4880	35,000	–	–	46.4633	1,96,608
Image 10	57.3814	35,000	–	–	46.4642	1,96,608
Image 14	–	–	53.22	600	46.7488	1,96,608
Image 15	–	–	52.20	800	46.4696	1,96,608
Average	57.9555	35,000	52.71	700	46.5115	1,96,608

5.2 Authentication Analysis

For authentication analysis, the watermarked images obtained are subjected to various image processing and geometric operations. The proposed technique aims at fragility and has been thus carried out in spatial domain. BER and NCC have been used to evaluate the authenticity of the content. For the tamper detection, the number of tampered blocks is used to find the efficiency of the proposed technique. The watermarked image is tested for, salt and pepper noise, Gaussian noise, JPEG compression, median filtering, histogram equalization, low pass filtering, sharpening, Wiener filtering, rotation, and cropping. The results obtained are presented and discussed below.

5.2.1 Salt and Pepper Noise Attack

The watermarked medical images have been subjected to salt and pepper noise of noise density 0.1. Table 3 shows authenticity parameters, i.e., BER and NCC. Figure 7 shows the original, corresponding watermarked and attacked images. Figure 8 shows the extracted watermarks of attacked images.

Table 3 and Fig. 8 clearly reveal that the watermark does not survive after this attack. The average number of tampered blocks is above 11,020, showing that the proposed technique is capable of detecting an attack of this type.

To test the fragility performance of the proposed scheme further, various test images such as CT scan, MRI, and X-ray have been distorted with salt pepper noise with noise density 0.05. The original, watermarked, and distorted versions of the images along with the extracted watermarks are shown in Fig. 9. NCC and BER for extracted watermarks and EPR from various images have been reported in Table 4. The NCC values obtained for this attack are compared with [29] which prove that the proposed technique is more fragile as NCC values for the proposed technique are more deteriorated compared to one under comparison.

Table 3 Image quality metrics for salt and pepper attack (density = 0.1)

Image	Number of blocks tampered	Watermark		Data (EPR)		As proposed in Moniruzzaman et al. [28]
		NCC	BER (%)	NCC	BER (%)	NCC
Image 1	11,015	0.9466	5.30	0.9503	5.00	0.0711
Image 2	11,079	0.9492	4.69	0.9498	5.00	0.0332
Image 3	11,018	0.9472	5.10	0.9494	5.08	−0.0153
Image 4	11,004	0.9506	4.96	0.9499	4.95	–
Image 5	10,988	0.9558	4.37	0.9508	4.99	–
Average	11,020.8	0.9499	4.884	0.9500	5.004	0.0297

Table 4 Image quality metrics for salt and pepper attack (density = 0.05)

Image	Number of blocks tampered (block size 4 × 4)	Watermark		Data		As proposed by Moniruzzaman et al. [29]
		NCC	BER (%)	NCC	BER (%)	NCC
CT scan	7294	0.9800	2.12	0.9752	2.52	0.9837
MRI	6529	0.9732	2.73	0.9750	2.43	0.9788
X-ray	7875	0.9743	2.47	0.9749	2.46	0.9806
Average	7233	0.9758	2.44	0.9750	2.47	0.9810

5.2.2 Addition of White Gaussian Noise

The performance of the proposed scheme has been also evaluated for Gaussian noise. The watermarked images have been distorted with additive white Gaussian noise (AWGN) with mean zero and variance = 0.02. The original, watermarked, and distorted Image 1 is shown in Fig. 10 while the watermark extracted from distorted Image 1 is shown in Fig. 11. The NCC and BER of the extracted watermarks and EPR are reported in Table 5.

For this type of attack, the average number of blocks tampered is 15,178. It is evident from the Table 5 and Fig. 11 that the proposed technique is highly fragile as BER is above 50 %.

The performance of the proposed technique against AWGN is also examined for other watermarked images as shown in Fig. 12. NCC and BER of the extracted watermark and EPR are reported in Table 6. From the Table 6, it clear that our scheme is more fragile as BER is very high and NCC is very low. Further this attack can be easily detected as is evident from the number of tampered blocks.

Table 6 shows that in comparison with [29] the proposed technique is fragile as is revealed by the comparison of average NCC. Because of the Gaussian noise of variance 0.02, the average number of blocks tampered is 15,089.

Table 5 Image quality metrics for AWGN (variance = 0.02)

Image	Number of blocks tampered (block size 4 × 4)	Watermark		Data (EPR)	
		NCC	BER (%)	NCC	BER (%)
Image 1	15,160	0.2962	64.04	0.4634	49.63
Image 2	15,226	0.3031	63.75	0.4628	49.52
Image 3	15,201	0.2731	66.28	0.4582	49.92
Image 4	15,149	0.2979	64.18	0.4639	49.83
Image 5	15,155	0.2837	65.75	0.4651	49.59
Average	15,178	0.2908	64.80	0.4627	49.70

Table 6 Image quality metrics for AWGN (v = 0.02)

Image	Number of blocks tampered (block size 4 × 4)	Watermark		Data		As proposed in Moniruzzaman et al. [29]
		NCC	BER (%)	NCC	BER (%)	NCC
CT scan	15,017	0.2900	64.58	0.4030	49.53	0.9625
MRI	14,813	0.2546	67.65	0.3914	49.53	0.9657
X-ray	15,438	0.5086	49.98	0.5560	49.66	0.9681
Average	15,089	0.3517	60.74	0.4501	49.57	0.9654

5.2.3   JPEG Attack

JPEG is one of the standard compression techniques used to compress the images in order to lower the storage space and the bandwidth required for transmission. The watermarked images have been subjected to JPEG compression with quality factor of 90. The watermarks extracted from various compressed images are shown in Fig. 13. Table 7 shows the performance parameters for JPEG compression for various images for a quality factor of 90.

Table 7 and Fig. 13 clearly state that the proposed technique is highly sensitive to JPEG compression as can be decided from extracted watermarks and their BER and NCC. For quality factor of 90, the average BER (%) and NCC for the extracted watermarks are 49.62 and 0.4599, respectively. A comparison of results for JPEG compression of quality factor 80 is made with [29] to strengthen our argument. The comparison results are reported in Table 8.

It is evident from the Table 8 that the proposed technique is highly fragile to JPEG compression compared to one under comparison [29].

5.2.4   Histogram Equalization Attack

The performance of the proposed system has been also evaluated for histogram equalization where the watermarked images have been subjected to histogram equalization. The performance parameters in terms of BER and NCC for this type of image manipulation are reported in Table 9.

Table 7 Authentication parameters against JPEG attack

Image	Number of blocks tampered (block size 4 × 4)	Watermark		Data	
		NCC	BER (%)	NCC	BER (%)
Image 1	12,386	0.4600	49.49	0.4409	48.75
Image 2	12,478	0.4561	50.07	0.4387	48.85
Image 3	12,394	0.4600	49.61	0.4385	48.98
Image 4	12,477	0.4589	49.80	0.4413	48.82
Image 5	12,430	0.4646	49.19	0.4401	48.67
Average	12,433	0.4599	49.62	0.4399	48.81

Table 8 Authentication parameters against JPEG attack (Q = 80)

Image	Number of blocks tampered (block size 4 × 4)	Watermark		Data		As proposed in Moniruzzaman et al. [29]
		NCC	BER (%)	NCC	BER (%)	NCC
CT scan	11,948	0.4663	48.27	0.4171	48.86	0.9812
MRI	7746	0.4495	48.66	0.2699	49.46	0.9769
X-ray	15,126	0.5314	47.66	0.5156	48.60	0.9853
Average	11,606.67	0.4824	48.20	0.4009	48.97	0.9811

Table 9 Authentication parameters against histogram equalization attack

Image	Number of blocks tampered (block size 4 × 4)	Watermark		Data	
		NCC	BER (%)	NCC	BER (%)
Image 1	15,244	0.0000	85.55	0.5266	48.91
Image 2	15,308	0.0000	85.55	0.5390	53.90
Image 3	15,193	0.0000	85.55	0.5223	49.47
Image 4	15,179	0.0000	85.55	0.5255	49.08
Image 5	15,324	0.0000	85.55	0.5202	49.90
Average	15,249.6	0.0000	85.55	0.5267	50.25

The proposed scheme is highly fragile to this attack as is evident from high BER equal to 85.55 % and very low NCC.

5.2.5 Median Filtering

Sometimes the median filtering is used to mitigate the effect of noise such as salt and pepper noise in order to improve the image quality. The proposed system is tested for this type of manipulation with kernel size 3 × 3. The cover image, watermarked image, and its filtered version are shown in Fig. 14. The performance parameters for watermarks and medical data extracted from various filtered images are presented in Table 10. The average BER of 44.18 % and the average NCC of

Table 10 Image quality metrics for median filtering

Image	Number of blocks tampered (block size 4 × 4)	Watermark		Data	
		NCC	BER (%)	NCC	BER (%)
Image 1	12,102	0.5197	43.97	0.4630	44.96
Image 2	12,137	0.5111	44.43	0.4628	45.06
Image 3	12,094	0.5254	43.14	0.4676	45.01
Image 4	12,093	0.5094	44.63	0.4591	45.25
Image 5	12,115	0.5108	44.75	0.4611	44.85
Average	12,108	0.5153	44.18	0.4627	44.97

0.5153 show that the watermarks get destroyed after this type of attack. The proposed algorithm is also able to detect tamper using this attack.

The fragility of the system can also be proved from the subjective analysis of watermarks as is depicted from Fig. 15. Further, the average number of tampered blocks with this attack is 12,108 which indicate that the proposed algorithm can detect such image manipulation as well.

5.2.6 Low Pass Filtering

A low pass filter is the basis for most image smoothing methods. An image is smoothed by reducing the difference between pixel values by averaging nearby pixels. The watermarked images are passed through a low pass filter of 3×3 kernel. The low pass filter tends to retain the low-frequency information while reducing high-frequency information. The filtered images and their corresponding extracted logos are shown in Fig. 16 while the performance parameters are reported in Table 11 to check the fragility of the system.

The watermarks extracted from the filtered images are not visible as is depicted by Fig. 16 which proves the fragility of the proposed watermarking system. The average number of blocks tampered because of low pass filtering is approximately 12,380.

5.2.7 Wiener Filtering

Wiener filter is often used for removal of blur from images. With an aim to prove the fragility of the proposed system for this attack, the watermarked images are passed through this filter. The objective quality metrics obtained for the watermarks and data extracted from various filtered images are reported in Table 12. The watermarked images after Wiener filtering and the watermarks extracted from them are shown in Fig. 17. From Table 12 and Fig. 17, it is clear that the proposed system is fragile to this attack as well.

Table 11 Authentication parameters against low pass filtering

Image	Number of blocks tampered (block size 4×4)	Watermark		Data	
		NCC	BER (%)	NCC	BER (%)
Image 1	12,414	0.2825	64.18	0.4266	49.81
Image 2	12,336	0.2634	65.41	0.4276	49.86
Image 3	12,408	0.2868	63.75	0.4366	49.67
Image 4	12,344	0.2811	64.26	0.4246	49.73
Image 5	12,401	0.2800	64.40	0.4173	49.92
Average	12,380.6	0.2787	64.4	0.4265	49.80

Table 12 Authentication parameters against Wiener filtering

Image	Number of blocks tampered (block size 4 × 4)	Watermark		Data	
		NCC	BER (%)	NCC	BER (%)
Image 1	12,426	0.2825	64.21	0.4223	50.18
Image 2	12,371	0.2654	65.21	0.4253	50.14
Image 3	12,385	0.2837	63.94	0.4362	49.74
Image 4	12,365	0.2800	64.21	0.4212	50.09
Image 5	12,408	0.2820	64.11	0.4176	49.93
Average	12,391	0.2787	64.33	0.4245	50.02

Table 13 Authentication parameters against sharpening attack

Image	Number of blocks tampered (block size 4 × 4)	Watermark		Data	
		NCC	BER (%)	NCC	BER (%)
Image 1	15,384	0.5805	37.50	0.5326	44.35
Image 2	15,298	0.5842	37.28	0.5308	44.24
Image 3	15,360	0.5882	36.99	0.5340	44.20
Image 4	15,344	0.5830	37.11	0.5442	44.04
Image 5	15,357	0.5791	37.45	0.5341	44.14
Average	15,348.6	0.5830	37.27	0.5351	44.19

5.2.8 Sharpening Attack

The watermarked images have been subjected to sharpening attack. The objective quality metrics obtained for the watermarks and EPRs extracted from various sharpened images are reported in Table 13. The watermarked images obtained after sharpening attack and the corresponding extracted watermarks are presented in Fig. 18.

When the watermarked image is subjected to sharpening attack, the average number of blocks tempered is 15,348. Further high BER and low NCC show that the proposed technique is able to check the authenticity of the content. Figure 18 justifies the argument by depicting subjective quality of the extracted watermarks.

5.2.9 Rotation Attack

We have subjected the watermarked images provided by our system to rotation attack. The watermarked images obtained after data embedding are rotated by 5°. The various rotated watermarked images and the corresponding watermarks are shown in Fig. 19. Table 14 shows the performance parameters for the watermark and EPR extracted after rotation attack. It is evident from the Table 14 that the proposed watermarking system is highly fragile to this attack as BER is of the order of about 72 %. Figure 19 clearly depicts that the proposed technique is very much fragile.

Table 14 Objective quality metrics for rotation attack of 5°

Image	Number of blocks tampered (block size 4 × 4)	Watermark		Data	
		NCC	BER (%)	NCC	BER (%)
Image 1	12,984	0.1906	72.24	0.4223	50.15
Image 2	12,989	0.1929	72.00	0.4208	50.12
Image 3	12,977	0.1881	72.71	0.4230	49.91
Image 4	13,019	0.1875	72.63	0.4245	49.99
Image 5	12,949	0.1952	72.00	0.4229	49.94
Average	12,983.6	0.1909	72.32	0.4227	50.02

Table 15 Objective quality metrics obtained for cropping attack (top left corner)

Image	Number of blocks tampered (block size 4 × 4)	Watermark		Data	
		NCC	BER (%)	NCC	BER (%)
Image 1	4096	1.0000	8.23	1.0000	12.09
Image 2	4096	1.0000	8.23	1.0000	12.19
Image 3	4096	1.0000	8.23	1.0000	12.11
Image 4	4096	1.0000	8.23	1.0000	12.17
Image 5	4096	1.0000	8.23	1.0000	12.02
Average	4096	1.0000	8.23	1.0000	12.12

5.2.10 Cropping Attack

The watermarked images are subjected to cropping attack where different regions of the watermarked images cropped. We have cropped 25 % of the watermarked images at top left corner, top right corner, and from the center of the image. Table 15 shows objective quality metrics obtained after cropping top left corner of the image. Similar observations were also made for cropping at other mentioned cropping positions.

Figure 20 depicts that the proposed algorithm is fragile as the extracted watermarks are not clear enough. Although from above table data, the average number of blocks tampered is 4096.

Table 16 shows the different parameters for watermarked images which are attacked by the cropping attack at the center. The average number of blocks tampered by this type of cropping is approximately 4215. Figure 21 shows the attacked images along with the extracted watermarks.

5.3 Advantages and Weaknesses of Proposed Scheme

The subjective and objective quality parameters of the proposed scheme as observed from Figs. 5, 6, 7, 8, 9, 10, 11, 12, 13, 14, 15, 16, 17, 18, 19, 20 and 21

Table 16 Authentication parameters against cropping attack (center)

Image	Number of blocks tampered (block size 4 × 4)	Watermark		Data	
		NCC	BER (%)	NCC	BER (%)
Image 1	4200	1.0000	0	1.0000	12.96
Image 2	4218	1.0000	0	1.0000	12.98
Image 3	4219	1.0000	0	1.0000	13.03
Image 4	4222	1.0000	0	1.0000	12.97
Image 5	4217	1.0000	0	1.0000	12.97
Average	4215.2	1.0000	0	1.0000	12.98

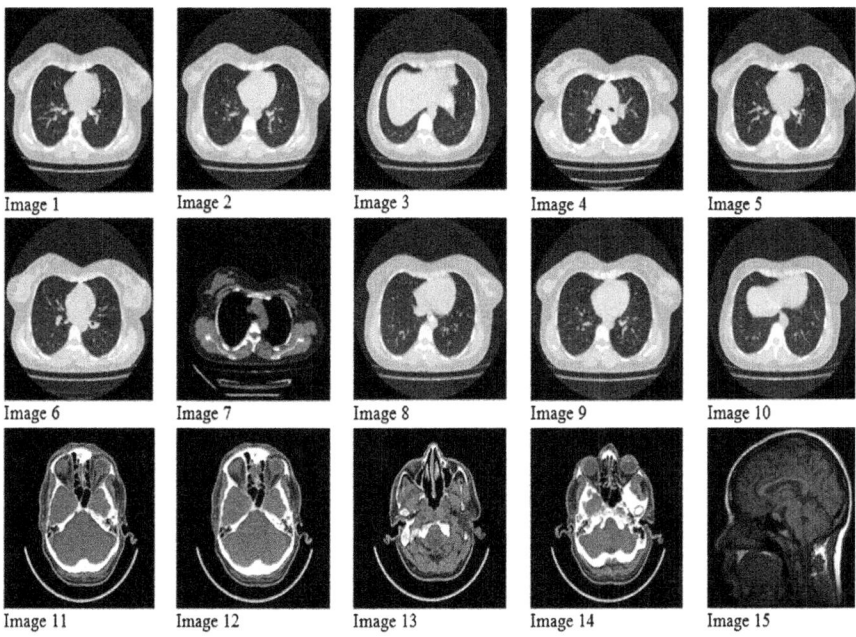

Fig. 5 Original medical images used to test the proposed algorithm

and Tables 1, 2, 3, 4, 5, 6, 7, 8, 9, 10, 11, 12, 13, 14, 15 and 16 show that our scheme has following advantages: (i) good imperceptibility, (ii) reversibility, (iii) tamper detection ability, and (iv) content authentication ability. One of the important weaknesses of proposed data hiding scheme is that the hidden data (EHR) are not robust to various signal processing and geometric attacks. In future, an attempt will be made to make the algorithm robust to various attacks.

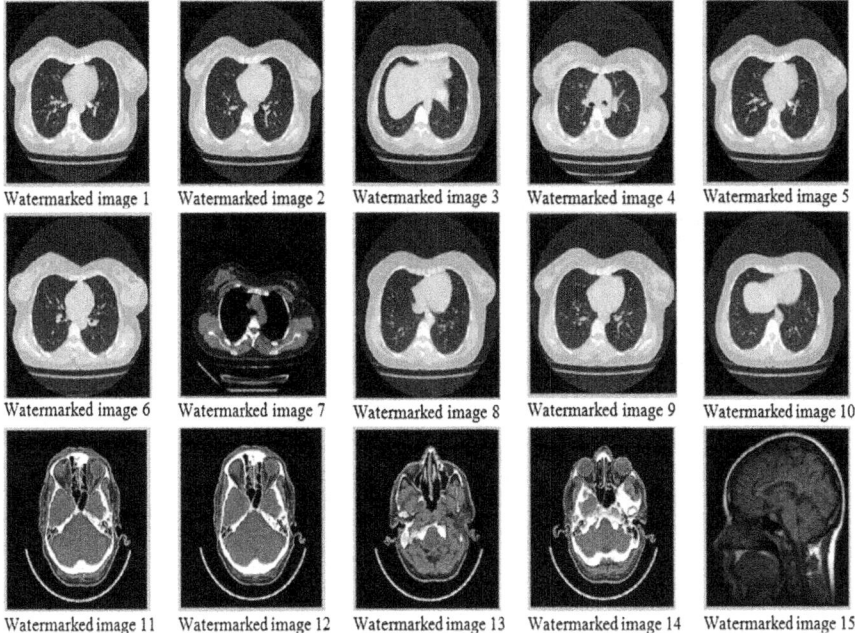

Fig. 6 Watermarked medical images

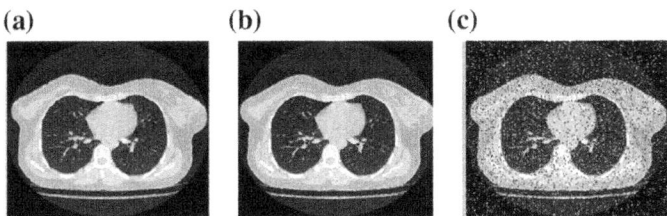

Fig. 7 **a** Original, **b** watermarked, and **c** 'salt and pepper'-attacked Image 1

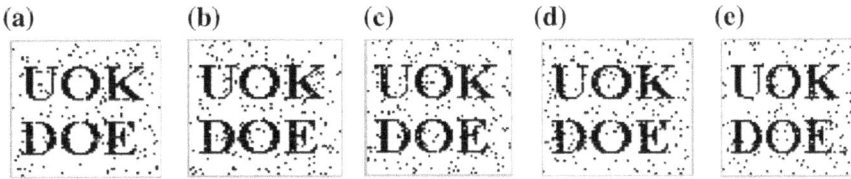

Fig. 8 Extracted watermarks after salt and pepper attack: **a** extracted from Image 1, **b** extracted from Image 2, **c** extracted from Image 3, **d** extracted from Image 4, and **e** extracted from Image 5

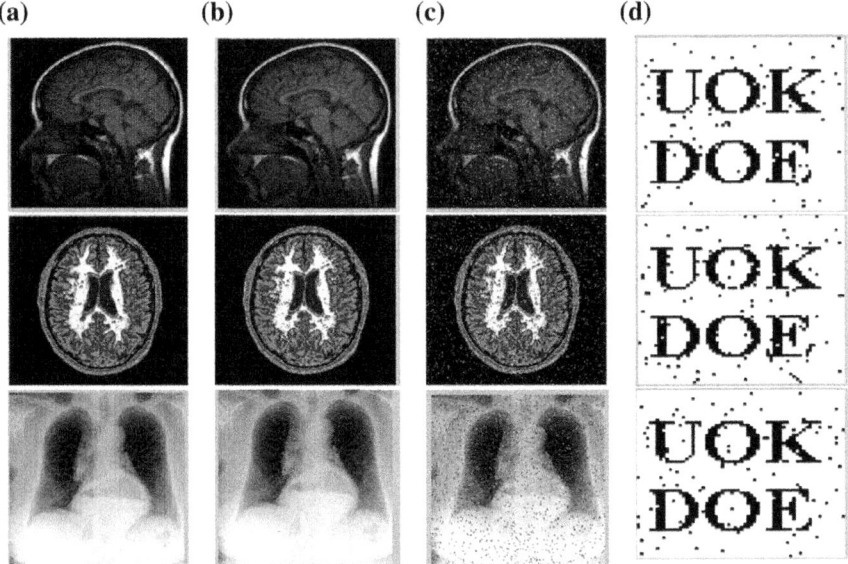

Fig. 9 **a** Original images, **b** watermarked images, **c** 'salt and pepper'-attacked images, and **d** extracted watermarks

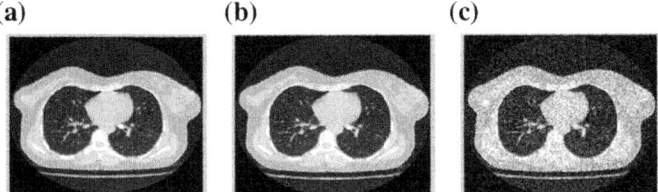

Fig. 10 **a** Original, **b** watermarked and **c** AWGN-attacked image

Fig. 11 Extracted watermark from attacked Image 1

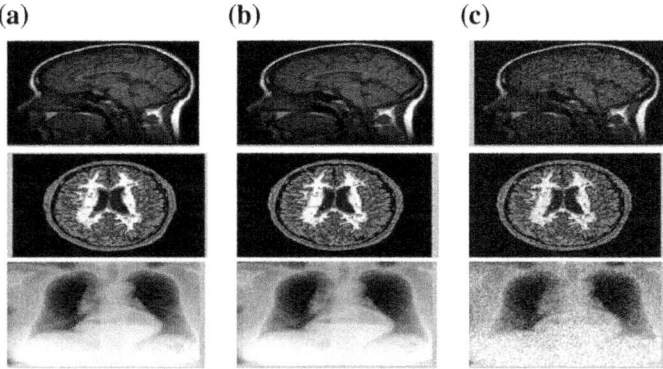

Fig. 12 **a** Original images, **b** watermarked images, and **c** AWGN-attacked images of CT scan, MRI, X-ray, respectively

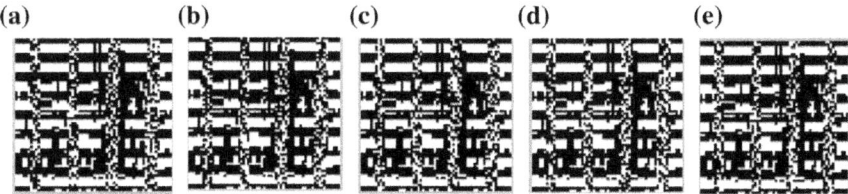

Fig. 13 Extracted watermarks after JPEG compression attack: **a** extracted from Image 1, **b** extracted from Image 2, **c** extracted from Image 3, **d** extracted from Image 4, and **e** extracted from Image 5

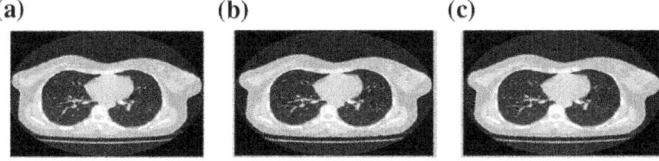

Fig. 14 **a** Original image, **b** watermarked image, and **c** median-filtered image

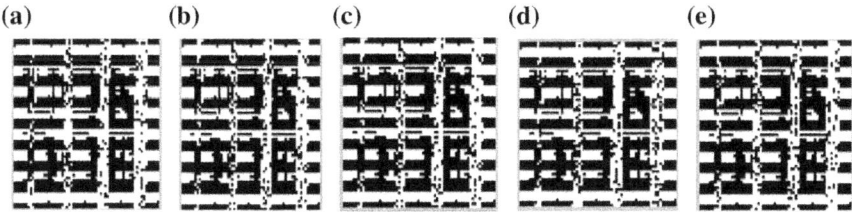

Fig. 15 Extracted watermarks after median filtering attack: **a** extracted from Image 1, **b** extracted from Image 2, **c** extracted from Image 3, **d** extracted from Image 4, and **e** extracted from Image 5

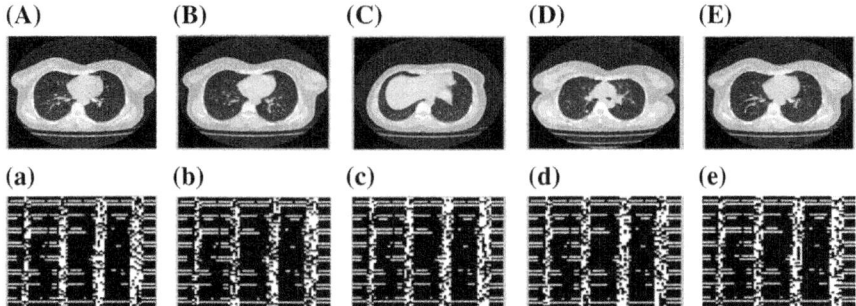

Fig. 16 **A–E** 'Low pass filtering'-attacked images and **a–e** respective extracted watermarks

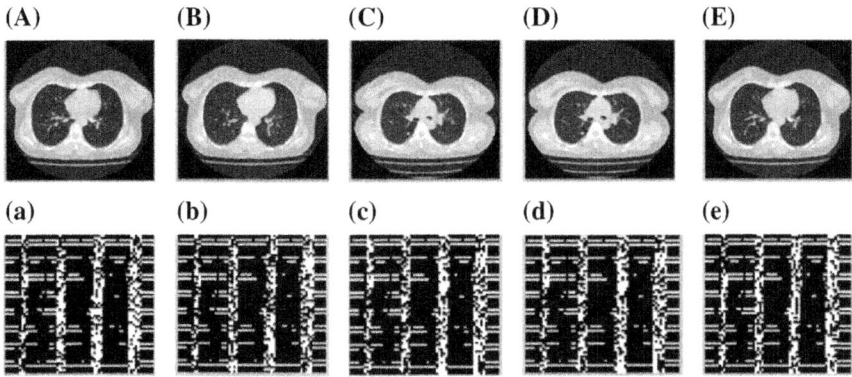

Fig. 17 **A–E** 'Wiener-filtered' images and **a–e** respective extracted watermarks

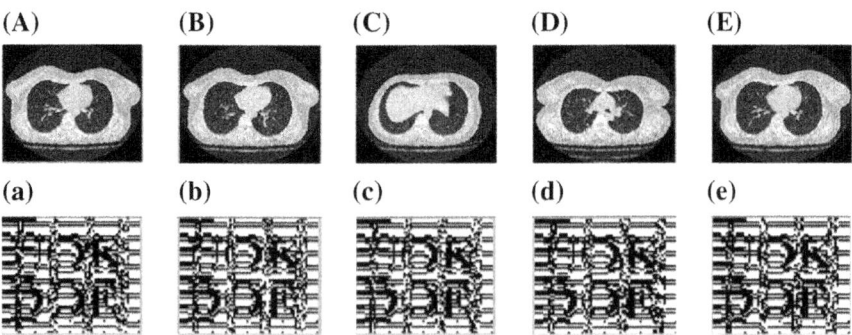

Fig. 18 **A–E** 'Sharpened' images and **a–e** respective extracted watermarks

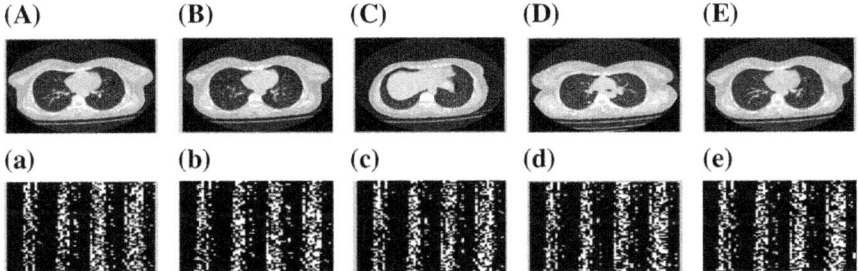

Fig. 19 A–E 'Rotated' images and a–e respective extracted watermarks

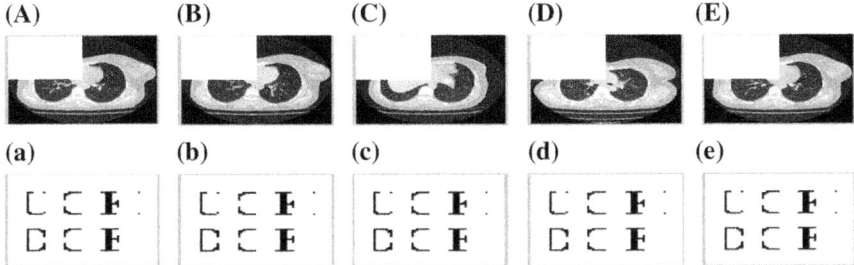

Fig. 20 A–E Left corner-cropped images and a–e respective extracted watermarks

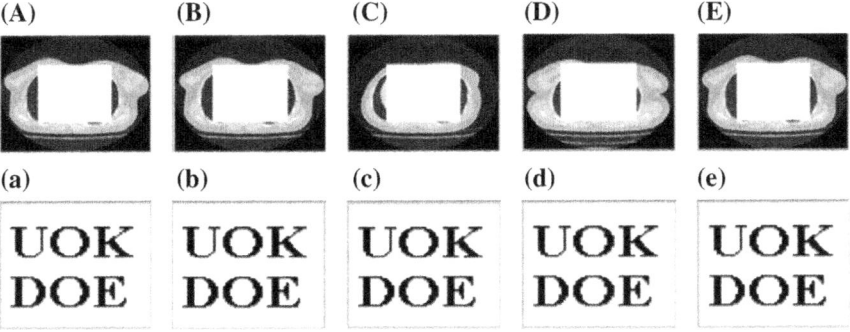

Fig. 21 A–E Center-cropped images and a–e respective extracted watermarks

6 Conclusion

This chapter presents a reversible and high capacity data hiding technique for medical images using Pixel Repetition Technique (PRT). For ensuring detection of tampered blocks, a fragile watermark has been embedded. We have utilized concept of Block Checksum Embedding to determine which of the blocks have been tampered with. Extensive testing has been used for checking efficacy of proposed algorithm. Experimental investigations reveal that the proposed scheme is capable of providing high-quality watermarked images for an extremely high payload. This is substantiated by observed average PSNR value of above 46 dB for a payload of 1,96,608 bits. The algorithm has been further tested for various image processing and geometric attacks to check its content authentication ability. The results reveal the proposed scheme is able to detect any tamper caused due to filtering, compression, noise addition, rotation, and cropping. A comparison of the proposed scheme with few state-of-the-art techniques shows that the proposed algorithm performs better on both counts of tamper detection and image imperceptibility besides being reversible.

References

1. Dey N, Samanta S, Yang XS, Das A, Chaudhuri SS (2013) Optimisation of scaling factors in electrocardiogram signal watermarking using cuckoo search. Int J Bio-Inspired Comput 5(5):315–326
2. Shabir AP, Frahana A, Javaid AS, Bhat GM (2015) On the realization of robust watermarking system for medical images. In: 12th IEEE India international conference (INDICON) on electronics, energy, environment, communication, computers, control (E3-C3), 17–20 December 2015, Jamia Millia Islamia, New Delhi
3. Shabir AP, Javaid AS, Mohiuddin GB (2015) Hiding in encrypted images: a three tier security data hiding system. Multidimension Syst Signal Process. doi:10.1007/s11045-015-0358-z
4. Dey N, Das P, Roy AB, Das A, Chaudhuri SS (2012) DWT-DCT-SVD based intravascular ultrasound video watermarking. In: 2012 world congress on information and communication technologies (WICT), IEEE, pp. 224–229
5. Shabir AP, Javaid AS, Mohiuddin GB (2014) Data hiding in scrambled images: a new double layer security data hiding technique. Comput Electr Eng 40(1):70–82
6. Mohamed MA (2013) A proposed security technique based on watermarking and encryption for digital imaging and communications in medicine. Egypt Informat J 14:1–13
7. Chakraborty S, Samanta S, Biswas D, Dey N, Chaudhuri SS (2013) Particle swarm optimization based parameter optimization technique in medical information hiding. In: 2013 IEEE international conference on computational intelligence and computing research (ICCIC), IEEE, pp 1–6
8. Shabir AP, Javaid AS, Mohiuddin GB (2012) High capacity data embedding using joint intermediate significant bit and least significant bit technique. Int J Inf Eng Appl 2(11). ISSN 2224-5782(print) ISSN 2225-0506 (online)
9. Bhat GM, Shabir AP, Javaid AS (2010) FPGA implementation of novel complex PN code generator based data scrambler and descrambler. Int J Sci Technol Thail 4(01):125–135
10. Shabir AP, Shazia A (2015) Robustness analysis of a digital image watermarking technique for various frequency bands in DCT domain. In: Proceedings of IEEE international conference iNiS, 21–23 December, Indore, India
11. Shabir AP, Javaid AS, Mohiuddin GB (2014) A secure and robust information hiding technique for covert communication. Int J Electron 102(8):1253–1266. doi:10.1080/00207217.2014.954635
12. Shabir AP, Jahangir AA, Javaid AS, Nazir AL, Farhana A, Bhat GM (2015) A high capacity data hiding scheme based on edge detection and even-odd plane separation. In: 12th IEEE India international conference (INDICON) on electronics, energy, environment, communication, computers, control (E3-C3), 17–20 December 2015, Jamia Millia Islamia, New Delhi

13. Acharjee S, Chakraborty S, Samanta S, Azar AT, Hassanien AE, Dey N (2014) Highly secured multilayered motion vector watermarking. In: Advanced machine learning technologies and applications, Springer International Publishing, pp 121–134
14. Dey N, Samanta S, Chakraborty S, Das A, Chaudhuri SS, Suri JS (2014) Firefly algorithm for optimization of scaling factors during embedding of manifold medical information: an application in ophthalmology imaging. J Med Imaging Health Inform 4(3):384–394
15. Shabir AP, Javaid AS, Mohiuddin GB (2014) A secure and efficient spatial domain data hiding technique based on pixel adjustment. Am J Eng Technol Res 14(2):38–44
16. Chakraborty S, Maji P, Pal AK, Biswas D, Dey N (2014) Reversible color image watermarking using trigonometric functions. In: 2014 international conference on electronic systems, signal processing and computing technologies (ICESC), IEEE, pp 105–110
17. Dey N, Maji P, Das P, Biswas S, Das A, Chaudhuri SS (2013) An edge based blind watermarking technique of medical images without devalorizing diagnostic parameters. In: 2013 international conference on advances in technology and engineering (ICATE), IEEE, pp 1–5
18. Shabir AP, Javaid AS, Farhana A, Nazir AL, Bhat GM (2015) Information hiding in medical images: a robust medical image watermarking system for E-healthcare. Multimedia Tools Appl. doi:10.1007/s11042-015-3127-y
19. Baiying L, Ee-Leng T, Siping C, Dong N, Tianfu W, Haijun L (2014) Reversible watermarking scheme for medical image based on differential evolution. Expert Syst Appl 41:3178–3188
20. Arijit KP, Nilanjan D, Sourav S, Achintya D, Sheli SC (2013) A hybrid reversible watermarking technique for color biomedical images. In: IEEE international conference on computational intelligence and computing research, IEEE, 978-1-4799-1597-2/13
21. Pal AK, Das P, Dey N (2013) Odd-even embedding scheme based modified reversible watermarking technique using Blueprint. arXiv preprint arXiv:1303.5972
22. Rahimi F, Rabbani H (2011) A dual adaptive watermarking scheme in contourlet domain for DICOM images. Biomedical Engineering Online
23. Malay KK, Sudeb D (2010) Lossless ROI medical image watermarking technique with enhanced security and high payload embedding. In: IEEE international conference on pattern recognition, pp 1457–1460
24. Jianfeng L, Meng W, Junping D, Qianru H, Li L, Chin-Chen C (2015) Multiple watermark scheme based on DWT-DCT quantization for medical images. J Inf Hiding Multimedia Signal Process, 6. ISSN 2073-4212
25. Solanki N, Malik SK (2014) ROI based medical image watermarking with zero distortion and enhanced security. Int J Educ Comput Sci 10:40–48
26. Sumalatha L, Rosline KG, Vijaya VK (2012) A simple block based content watermarking scheme for image authentication and tamper detection. Int J Soft Comput Eng (IJSCE) 2
27. Guo X, Zhuan T (2009) Lossless watermarking for verifying the integrity of medical images with tamper localization. J Digit Imaging 22:620–628
28. Abhilasha S, Malay KD (2014) A blind and fragile watermarking scheme for tamper detection of medical images preserving ROI. In: International conference on medical imaging, m-health and emerging communication systems (MedCom), IEEE
29. Moniruzzaman M, Hawlader MAK, Hossain MF (2014) Wavelet based watermarking approach of hiding patient information in medical image for medical image authentication. In: 17th international conference on computer and information technology, Daffodil International University, Dhaka, Bangladesh, IEEE
30. Lee C-F, Huang Y-L (2012) An efficient image interpolation increasing payload in reversible data hiding. Expert Syst Appl 39:6712–6719
31. Tang M, Hu J, Song W (2014) A high capacity image steganography using multi-layer embedding. Optik 125:3972–3976
32. Hu J, Li T (2015) Reversible steganography using extended image interpolation technique. Comput Electr Eng. doi:10.1016/j.compeleceng.2015.04.014

33. Kamran AK, Malik SA (2014) A high capacity reversible watermarking approach for authenticating images: exploiting down-sampling, histogram processing, and block selection. Inf Sci 256:162–183
34. Naheed T, Usman I, Khan TM, Dar AH, Shafique MF (2013) Intelligent reversible watermarking technique in medical images using GA and PSO. Optik. doi:10.1016/j.ijleo.2013.10.2014
35. Arsalan M, Sana AM, Asifullah K (2012) Intelligent reversible watermarking in integer wavelet domain for medical images. J Syst Softw 85:883–894
36. HK Maity, SP Maity (2012) Joint robust and reversible watermarking for medical images. In: 2nd international conference on communication, computing and security [ICCCS-2012], Elsevier, Procedia Technology, vol 6, pp 275–282
37. Tsai P, Yu CH, Hsiu LY (2009) Reversible image hiding scheme using predictive coding and histogram shifting. Sig Process 89:1129–1143

Part IV
Security for Various Multimedia Contents

A New Method of Haar and Db10 Based Secured Compressed Data Transmission Over GSM Voice Channel

Javaid Ahmad Sheikh, Sakeena Akhtar, Shabir Ahmad Parah and G.M. Bhat

Abstract In this chapter, a successful attempt for transmission of secured compressed data over GSM voice channel has been made. In this work, given text of any length is converted to real-time speech signal and then compressed with discrete wavelet transform, encoded with gold code sequence, modulated using quadrature phase-shift keying and sent over the GSM voice channel. The aim of this paper was to present a secured digital data compression and modulation for robust data transmission in terms of voice. The main objective was to achieve higher data rates, lower bit error rate, and less utilization of the bandwidth. The performance of the proposed technique is compared to an already existing technique for data transmission over voice channel. It has been observed that the proposed technique shows much better results in terms of bit error rate (BER), mean square error (MSE), and peak signal-to-noise ratio (PSNR) as compared to the already existing techniques. The proposed scheme has a great significance in GSM systems where data security and quality of service (QoS) are two main issues.

Keywords Text-to-speech (TTS) synthesis · Speech compression · Gold code · Haar wavelet · Daubechies wavelet · Thresholding

J.A. Sheikh (✉) · S. Akhtar · S.A. Parah · G.M. Bhat
PG Department of Electronics and Instrumentation Technology, University of Kashmir, Hazratbal, Srinagar 190006, India
e-mail: sjavaid_29ku@yahoo.co.in

S. Akhtar
e-mail: mirsakina77@gmail.com

S.A. Parah
e-mail: shabireltr@gmail.com

G.M. Bhat
e-mail: drgmbhat@gmail.com

© Springer International Publishing Switzerland 2017
N. Dey and V. Santhi (eds.), *Intelligent Techniques in Signal Processing for Multimedia Security*, Studies in Computational Intelligence 660,
DOI 10.1007/978-3-319-44790-2_18

1 Introduction

We live in a world where there is continuous rise in demand for all types of wireless services such as voice, data, and multimedia. This has led to a need for higher capacity and a lower bit error rate in wireless networks with efficiency of transmitted data. Over the past few years, the studies have revealed that there is an increase in data transmission over the voice channel for real-time monitoring systems and secure voice communications, with all of them focusing on the GSM's voice codes: high rate (HR), full rate (FR), and enhanced full rate (EFR) [1–3]. We thus find that there is a need to review the data transmission performance over the GSM voice channel or data transmission in terms of voice. Hence, in this work, an attempt has been made to overcome the drawbacks or disadvantages of data transmission over the GSM voice channel. But before proceeding further, an important question needs to be answered, why to transmit data in terms of voice? We know that there are a number of data transmission techniques [4] available such as near-field communication (NFC), General Packet Radio Service (GPRS), High Speed Downlink Protocol Access (HSDPA), Wireless-Fidelity (Wi-Fi), short message service (SMS), and a lot more that are used for high speech data transmission. In contrast to the above-mentioned techniques, there are some important advantages of data transmission in terms of speech/voice:

- The data rate requirement for voice transmission is low as compared to the data transmission which requires a higher data rate because voice transmission has real-time delay constraints while data transmission has less stringent delay constraints.
- As the voice has higher tolerance for bit error rate, voice bits are allowed to use higher-order modulation than the data bits.
- Due to the higher-order modulation, the voice bits have larger symbol duration which is subsequently reduced to fewer bits due to compression.
- The compressed voice bits when sent over the channel require less bandwidth and thus saving the overall bandwidth.

One more question to be answered is why not to use GSM encoders (HR, FR, EFR) for transmission of data in terms of voice? In order to transmit data over the existing voice channel, there are certain data transmission methods:

- The data symbols need to be mapped into three voice characteristics.
- One needs to derive a pre-optimized codebook for speech like symbols.
- Selection of accurate modulation techniques for GSM network.

These transmission methods require a lot of efforts to convert the data into voice, but are still inadequate to convert data into voice fully. Instead, these methods try to introduce the temporal characteristics of speech into the data. Moreover, due to the characteristic of the GSM encoders, the modulated signal should mimic the properties of the speech signal (formants, higher harmonics, pitch frequency, etc.) for better transmission. Due to this reason, not all the digital modulation techniques are suitable for transmitting the data over the GSM voice channel as the GSM speech coders do not assure the phase-shifting preservation. Thus, the coherent and phase-based modulation techniques are not used for modulated data transmission over the voice channel [5]. Hence, the remaining options are ASK and FSK. In the proposed work, so as to overcome above-mentioned problems, the given text is converted to a speech signal and we know human ears are insensitive to phase. Instead of mimic of temporal characteristics of a speech signal, the text is converted to a real-time speech signal using text-to-speech conversion system. When compared with the reference paper [6], it was observed that the proposed technique produces much less value of bit error rate. Although the method used in our proposed technique is totally different from that taken as a reference, the motive is same: transmitting secure data as voice with less bit error rate and at higher data rates. Both these parameters have better results in our proposed technique as compared to the existing one. Thus, the proposed technique would be beneficial in various applications such as telemetry, secure communications, wireless automatic teller machines, and faxes we can say in general—virtually anywhere—where the regular transmission of low data rate burst is required.

The proposed scheme used in this chapter is particularly for wireless communication which is anywhere any time in nature. Besides fading of a signal in a communication system, the major challenges before the design engineers is to combat the effect of data corruption and check the security issues needed to avoid the corruption of data. In this chapter, a multilayer encryption technique has been used to incorporate the security for the transmission of compressed data over wireless GSM channels. The prime focus on the security was accomplished by authenticity, integrity, and confidentiality of the transmitted data. Since the wireless channel is very hostile and is prone to many attacks, it was imperative to check the efficacy of the proposed scheme using QoS.

In this chapter, Daubechies wavelets have been used for compression and denoising purposes. Among the Daubechies family of wavelets, only the results of Haar and Db10 wavelets have been shown. The Haar wavelet is not chosen as a comparison with Db10 wavelet. The motive is to find the best suitable wavelet to compress and denoise a speech signal using Daubechies wavelet family. The Haar wavelet is the only orthogonal wavelet with linear phase and having only one vanishing moment. Among the two wavelets used, it has been observed that Db10 shows better results than Haar wavelet.

The rest of the chapter is organized as follows. Section 2 will provide a brief literature survey of various implemented techniques for data transmission over the voice channel. A basic text-to-speech synthesis will be discussed in Sect. 3. In Sect. 4, an overview of wavelet transform including choice of wavelet will be presented. Followed by Sects. 5, 6, and 7 where signal decomposition, signal compression, and thresholding selection will be elaborated, respectively. Section 8 will give a brief description about the gold code sequence used for the coding. QPSK modulator/demodulator will be presented in Sects. 9 and 10, respectively.

How to reconstruct the speech signal back from the demodulated signal will be shown in Sect. 11. Section 12 presents the simulation results of the proposed work. Finally, the paper will be concluded in Sect. 13.

2 Literature Survey

Sebastian Ciornei et al. [7] proposed highly secure end-to-end voice calls over existing voice channels. In contrast to the normal data channels, the audio channels provide high security due to the peer-to-peer nature of the connections. They resolved the problem for transmitting data over existing and upcoming vocoders by studying the sensitivity analysis of QAM modulation parameters over the AMR-WB vocoder. It was observed that with AMR-WB, the data transfer rates and bit error rates show better performance.

Mahapatra et al. [8] have implemented an integrated voice and data transmission technique, where the symbol duration of voice is minimized with the information of voice BER using adaptive modulation so that same voice information is sent with the compressed symbol duration. Consequently, some additional bits are stored for data transmission, which effectively increases the transmission rate. Hence, with this scheme, the overall throughput in a wireless broadband network can be improved, provided the receiver CNR remains within the desired range. Also, the shrinking of the symbol period will increase the system complexity at the cost of the increased data rates.

Kotnik et al. [6] presented a novel digital data modulation/demodulation method for data transmission over the voice channel. Voiced speech signal characteristics are analyzed using the autoregressive modeling for speech production, and LPC coefficients are calculated. The coefficients are converted to linear spectrum frequencies (LSF) and are applied to modulation synthesis filter. The filter is excited by a signal having the characteristics to that of the human vocal tract. A speech like modulated signal is produced and sent over the voice channel. The technique is compared with the FSK modulation which is taken as a reference, and comparative results are obtained. It has been shown that the ARDMA technique used shows better results at higher bit rates than the FSK modulation technique.

LaDue et al. [2] have described a new approach to transmit data through GSM voice channel. A set of predefined signals with finite bandwidth were used. The data were encoded into symbols, and symbols were voice-coded, GSM–modulated, and sent over the voice channel. This method again allows data transmission over the existing voice channels.

The above referred techniques [6–9] suffer from a number of drawbacks. All of these techniques have focused only on the GSM voice channel and GSM encoders (HR, FR, EFR). None of them have given any review on data transmission in terms of voice over the GSM voice channel. Moreover, none of the above-mentioned techniques used any other compression technique than those provided by the GSM encoders, although there are better compression techniques available such as wavelet based as is used in our work. Technique [6] stresses on not to use phase-based and noncoherent modulation techniques with the GSM encoders. In our proposed work, both the practical implementation and transmission of the compressed data in terms of speech using the phase-based modulation technique (QPSK) have been realized. An attempt has been made to overcome all these drawbacks, so that a better quality of compressed data transmission in terms of speech with low bit rate is obtained.

3 A Text-to-Speech (TTS) Synthesis

A text-to-speech system (Fig. 1) converts the desired text into speech. It has two main parts: the front end and the back end. The front end does two functions. The first of which is to convert raw text containing symbols, such as numbers and abbreviations into the equivalent words. This process is termed as normalization or tokenization. The second function is to assign phonetic transcriptions to each word. The front end converts each phonetic unit of the text into units such as phrases, clauses, and sentences. This process of assigning phonetic transcriptions to words is called text-to-phoneme conversion. Phonetic transcriptions and prosody information together will form the symbolic linguistic representations. The back end, that is the synthesizer, uses this linguistic information and converts it into sound using signal processing. The back end will also include the computation of the target prosody such as pitch contour and phoneme durations that are then imposed on the speech output [10].

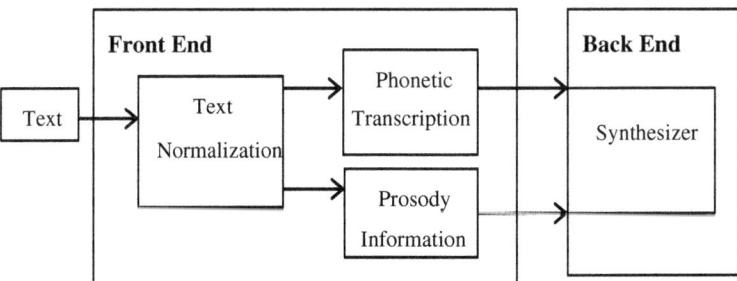

Fig. 1 Block schematic for a text-to-speech system

4 Wavelet Transform: Choice of Wavelet

Wavelet is a new technique for examining and comparing a speech signal, and it is a more advantageous technique because it holds both time and frequency aspect of a signal. Besides application to speech signals, wavelets play an important role in image as well as video applications in various different fields [11, 12]. Wavelet breaks speech signal into different coefficients. Some of the coefficients having small value are treated as insignificant during data compression and are hence discarded [13]. Wavelets are obtained by a single mother wavelet by delay and shifting.

$$\psi_{a,b} = \frac{1}{\sqrt{a}} \psi \frac{(t-b)}{a} \tag{1}$$

where 'a' and 'b' are called as scaling and shifting parameters. In a high-quality speech coder, choosing a mother wavelet is of key importance. DWT due to its orthogonal properties does not produce any redundant information. The scaling and shifting parameters are thus responsible for multiresolution analysis algorithm, which decomposes a signal into scales having different time and frequency resolutions. Various mother wavelet (e.g., Haar, Daubechies, Coiflets, Symlet, and Biorthogonal) functions are differentiated upon these scaling functions; thus, the choice of wavelet decides the final waveform shape [14]. In our proposed work, Daubechies wavelets 'Haar' and 'Db10' have been used.

The use of wavelet transform for removing common noise from a speech signal or simply denoising a speech signal has been found much effective as compared to the Fourier transform. In telecommunication, noise is the main element that limits the capacity of the systems, and also in signal measurement systems, the accuracy of the results is also affected by the noise. Thus, removing noise from a signal is of core importance mainly in communication and signal processing systems [15].

5 Signal Decomposition

In DWT, a signal ($S(t)$) to be examined is passed through an analysis filter bank followed by a particular decomposition level. At each decomposition level, the analysis filter bank consists of a high-pass and low-pass filter. The signal is passed through a series of such high-pass and low-pass filters. The output of high-pass filter is known as detail coefficients (Cd) and contains the valuable information of the signal while the output of the low-pass filter is known as approximation filters (Ca) and contains least information of the signal. Detail coefficients are low-scaled high-frequency components while approximation coefficients are high-scaled low-frequency components. The frequency components that are not very

Fig. 2 Decomposition of
DWT coefficients

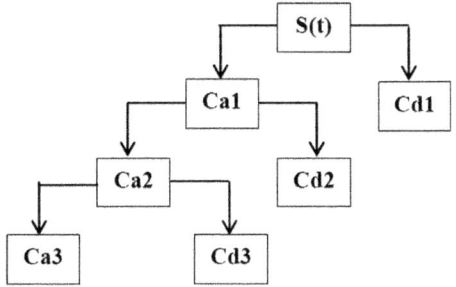

prominent in the original signal will have very low amplitude, and this part of the
DWT signal can be discarded without loss of any valuable information, allowing
data compression at higher data rates. During the decomposition procedure, a signal
is broken down into many lower-resolution components [16, 17]. This is called the
wavelet decomposition tree as shown in Fig. 2 given below. Theoretically, the
decomposition level can be infinite, but for human voice, level 5 decomposition is
the best. Wavelet transformation is thus an important tool for analyzing signals of
different kinds [18]. The most important application has been found in the
biomedical field [19–21].

6 Signal Compression

Speech compression is to remove redundancy from the speech signal to achieve
compression and to reduce transmission (for voice over transmission) and storage
costs (for speech recording). Speech compression techniques are mainly focused on
removing short-term correlation (in the order of 1 ms) among speech samples and
long-term correlation (in the order of 5–10 ms) among pitch patterns. Speech
coding is a main issue in digital speech processing. Speech coding is the phe-
nomenon of converting the speech signal to a more compressed form, which can
then be transmitted with a much smaller memory. The compression of speech has
found a great importance in communication systems as in wireless communication,
bandwidth is a major resource that is taken into consideration and the service
providers frequently come across with the challenge of accommodating more
number of users within a limited bandwidth. Compression of speech can overcome
this challenge by lowering the bit rate and providing a good speech quality. By
using compression techniques, the program execution time and storage of processor
are reduced to a greater extent. Compression reduces the data transfer rate and
bandwidth requirement with security of data [22].

7 Threshold Selection

Wavelet thresholding is a signal approximation technique that uses the capabilities of wavelet transform for signal denoising. Using a small value of threshold for denoising a signal may result the output signal close to that of the input, but cannot denoise the signal fully. A large threshold, on the other hand, yields a signal with a large number of zero coefficients. Thus, the selection of threshold for denoising purposes needs to be done carefully. Thresholding operation is basically a nonlinear operation. In signal processing, thresholding is performed on coefficients in the frequency domain at a time. Each coefficient is compared with the threshold: If the coefficient is smaller than threshold, it is set to zero, otherwise it is kept or modified. There are two types of thresholding algorithms exploited in this work: hard thresholding and soft thresholding.

7.1 Hard Thresholding

In this method, input is compared against the threshold: If input is greater than the threshold, it is kept, else it is set to zero. Mathematically, hard thresholding is given by

$$
\begin{aligned}
Y &= T(X, Y) = X; \quad &\text{for } |X| > \lambda \\
Y &= 0; \quad &\text{for } |X| \leq \lambda
\end{aligned}
\tag{2}
$$

where λ is the threshold limit.

It is clear from Eq. (2) that input is compared with the threshold: If input is greater than the threshold, input is equal to the input or equal to zero if less than the threshold. The hard thresholding process removes noise by thresholding the detail coefficients only, while keeping the approximation coefficients unaltered.

7.2 Soft Thresholding

In this method, the input shrinks toward zero by the threshold T and hence this method is also called as shrinkage function. Mathematically, it is given as follows:

$$
\begin{aligned}
Y &= T(X, Y) = \text{sign}\{X\} \, (|X| - 1); \quad &\text{for } |X| > \lambda \\
Y &= 0; \quad &\text{for } |X| \leq \lambda
\end{aligned}
\tag{3}
$$

The soft thresholding scheme is basically modified form of hard thresholding. In this method, the output is equal to zero if the input X is less than or equal to threshold λ. If the input is greater than λ, the output takes the value $|X - 1|$. Hard thresholding shows some discontinuities at λ, while soft thresholding does not show such discontinuities. Therefore, it is clear that soft thresholding is more stable than hard thresholding against the noise for efficient signal denoising in signal processing. Wavelet thresholding has been widely used, and among the various capabilities of wavelet transform, thresholding can denoise a signal and also will approximate the signal to the actual one, but cannot denoise the signal fully.

8 Spreading Sequences

Spreading sequences are the class of sequences having typical properties to that of random signal (like noise). These sequences have been used for spreading in spread spectrum modulation, securing message communication, and also synchronization of transmitter and receiver.

8.1 Pseudo Noise (PN) Sequences

The term 'pseudo noise or pseudo random' is used particularly to mean random in appearance, but reproducible by deterministic means. A pseudo noise (PN) sequence is well defined as a coded sequence of 1s and 0s with certain statistical attributes. These sequences are generally generated by means of a feedback shift register. Broadly speaking, PN sequences are of two classes: (1) periodic sequence and (2) aperiodic sequence.

8.1.1 Aperiodic Sequences

An aperiodic sequence is defined logically by a sequence of N plus or minus ones and is represented by

$$a1, a2, a3, \ldots aN \quad ai = \pm 1$$

The sequence is said to be true PN sequence, if it retains certain properties. The autocorrelation property is one of the properties given as

$$C(K) = \sum_{n=1}^{N-k} a_n a_{n+K} \tag{4}$$

where $K = 0, 1, \ldots N - 1$.

8.1.2 Periodic Sequences

In periodic sequences, the sequence of 0s and 1s (or −1s and +1s) repeats itself exactly with a known period of N. (N is also defined as the length of the period.) This is illustrated by

$$a1, a2, a3, \ldots aN - 1, aN, a1, a2 \quad ai = \pm 1$$

Such a sequence is said to be random if the following conditions are satisfied:

Balance property: Good balance requires that in each period of the sequence, the number of binary 1s (+1s) differs from the number of binary 0s (−1s) by at most one digit (should be ideally equal).

To be exact,

$$\text{No of 1s} = \frac{2^n}{2} = 2^{n-1} \tag{5}$$

and

$$\text{No of 0s} = \frac{2^n}{2} - 1 = 2^{n-1} - 1 \tag{6}$$

where n is the length of the shift register.

i. Run property: By the term run, we mean a series of consecutive 1s (+1s) or 0s (1s) grouped consecutively. In every period, half of the runs of the same sign have the length 1 and one-fourth have the length 2. One-eighth have length 3 and so on. Also, the number of positive runs equals the number of negative runs.

ii. Correlation property: The autocorrelation function of a sequence is periodic and two valued if it can be described by

$$C(K) = \sum_{n=1}^{N-k} a_n a_{n+K} \tag{7}$$

where N is the length of the sequence and K is the lag (shift) of the autocorrelation sequence.

$$C(K) = N \quad K = 0, N, 2N \ldots \text{ i.e., } K = 1N \tag{8}$$

and

$$C(K) = -1 \quad K = 1, 2, 3 \ldots \text{ i.e., } K \neq 1N \tag{9}$$

where 1 is any integer. When N is infinitely large, the autocorrelation sequence C (K) approaches that of completely random binary sequence.

Correlation is one of the important properties of the pseudo random codes. Various techniques and theories have been proposed to have good autocorrelation and cross-correlation properties. Robert Gold in 1967 proposed that modulo +2 addition of two maximal length sequences having length of $(2^n - 1)$ results a code known as gold code with good correlation properties.

9   QPSK Modulator

Implementation of a QPSK modulator directly results from the interpretation of a QPSK signal as the sum of two BPSK signals with carriers having phase quadrature. Figure 3 shows a simplified block diagram of a QPSK modulator employing two BPSK modulators with orthogonal carriers. The serial data are converted to 2-bit symbols at half of the rate. Quadrature component (Q) and in-phase component (I) are changed to NRZ pattern in a parallel manner. The outputs of the I and Q modulators are linearly combined to obtain QPSK signal.

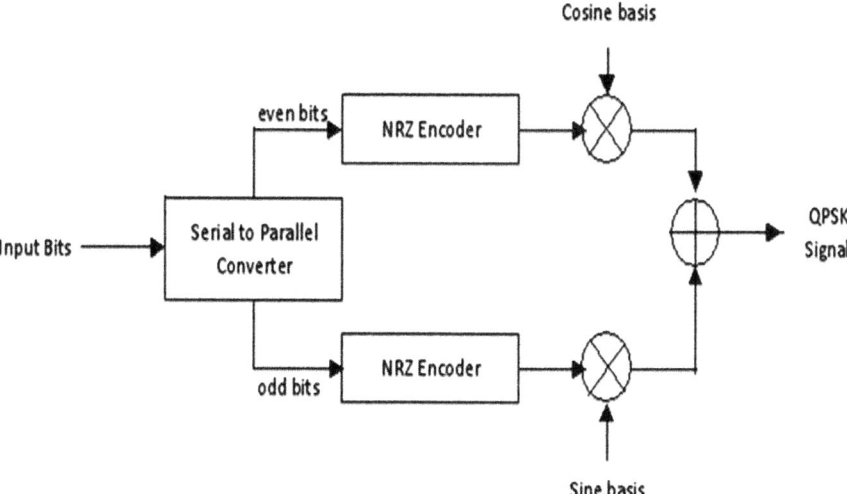

Fig. 3 QPSK modulator

10   QPSK Demodulator

In QPSK demodulator, a coherent demodulator in which the carrier frequency and phase are known to the receiver is used as an example. A coherent demodulator uses a phase lock loop (PLL) at the receiver. A PLL basically locks to the incoming carrier frequency and traces the variations in frequency and phase. During demodulation, received signal is multiplied by $\cos(\omega t)$ and $\sin(\omega t)$ which acts as reference frequency generators on in-phase component (I) and quadrature component (Q). On each component/arm, an integrator is used which integrates the output over one bit period and the threshold detector makes the desired decision based on the value of the threshold. The in-phase and the quadrature bits are finally remapped to detect the QPSK demodulated signal. Figure 4 given below is the in-phase arm detector while for the quadrature arm detector $\sin(\omega t)$ basis function must be used.

11   Signal Reconstruction

After the thresholding step, the signal is encoded with gold code sequence, modulated using QPSK modulation technique, and is sent over the AWGN channel. At the receiving end, the received signal is first decoded using same gold code sequence and demodulated. The signal is reconstructed by applying inverse wavelet transform on original approximation coefficients and modified detail coefficients which gives us the concept of compression, i.e., signal is reconstructed from original approximation coefficients and some detail coefficients.

Figure 5 given below depicts the whole procedure/flowchart of the proposed work.

From the flowchart shown in Fig. 5 given above, a desired text of any length is first converted into speech using text-to-speech conversion system. The text-to-speech conversion system is already available with the MATLAB tool, we need not to devise it separately, and we use the inbuilt text-to-speech conversion system of MATLAB. The converted speech signal is stored as .wav and stored in MATLAB directory. The speech thus acts as a test recorded speech signal which is decomposed using discrete wavelet transform. There are a number of wavelet families, but in our proposed technique Haar and Db10 wavelets are used as

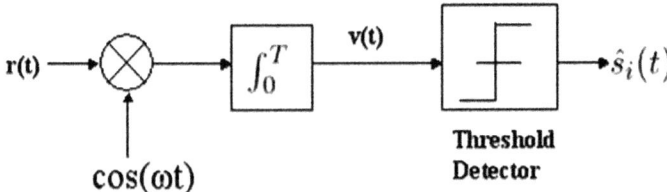

Fig. 4 QPSK demodulator

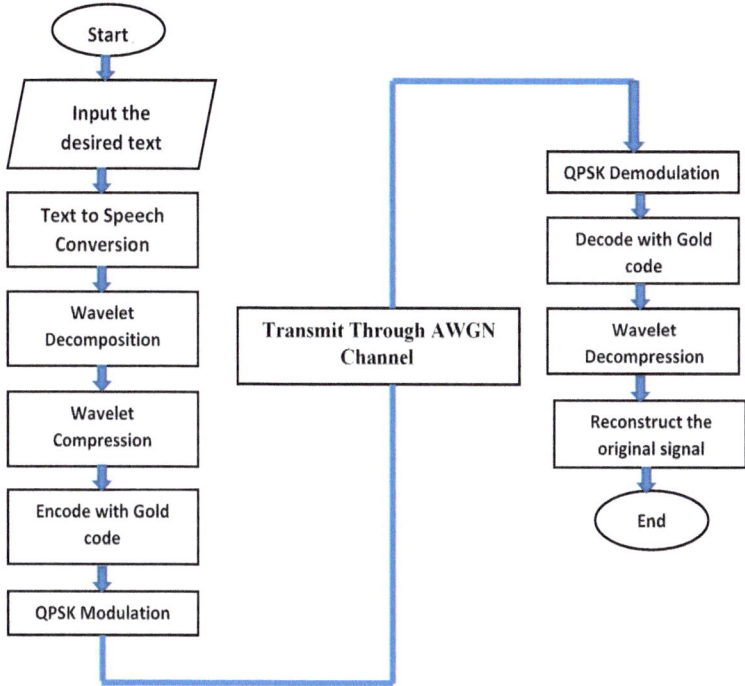

Fig. 5 Block diagram of the proposed technique

compression methods. For denoising purposes, both hard and soft thresholding algorithms are used. After thresholding, the signal is encoded using gold code sequence for security reasons, modulated using QPSK modulator, and sent over the AWGN channel. At the receiving end, the signal is first decoded, demodulated using QPSK demodulator, and decompressed using the inverse wavelet. The speech samples are thus reconstructed depending upon these decompressed parameters to obtain back the original speech signal. Due to the AWGN channel, some noise is incorporated in the reconstructed signal, and in order to avoid such noise, an adaptive FIR filter is used, so that some high noise frequency components are avoided. It has also been observed that the speech signals without encoding are more prone to some background/channel noise as compared to the speech signals which are first coded with gold code sequence and then transmitted. So, use of a filter for filtering such a signal (encoded signal) is of least importance because no such high noise frequency components are present, but it definitely increases the sound quality of the reconstructed speech signal as is done in this work. For the purpose of comparison, Figs. 6 and 7 given below show the effect of channel noise on a signal which is transmitted without coding. The speech signal waveforms with encryption using gold code sequence are shown in Fig. 10 onward.

Figures 6 and 7 represent the original and reconstructed speech signals compressed with Haar and Db10 wavelets and transmitted through an AWGN channel

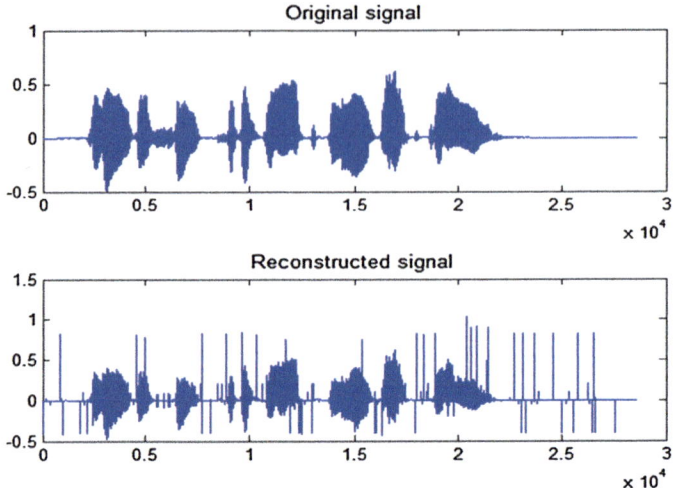

Fig. 6 Original and reconstructed speech signals without coding using Haar wavelet

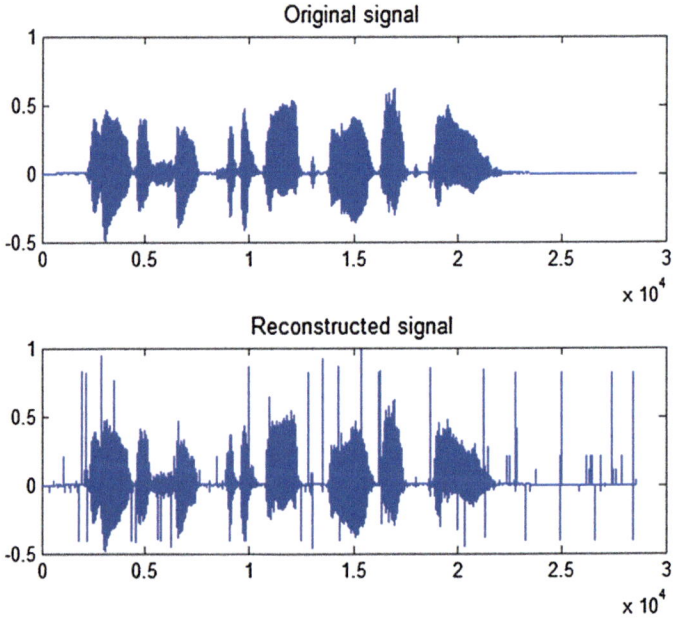

Fig. 7 Original and reconstructed speech signals without coding using Db10 wavelet

but without encryption. There is no coding technique used to encode the transmitted data. The figures clearly show the effect of noise on the received/reconstructed speech signals.

12 Simulation Results

The proposed technique used in this chapter has been verified on a converted test recorded speech signal. To evaluate the performance of the implemented technique, various performance parameters related to a speech signal are calculated. These include % of zero coefficients (PERF0), retained signal energy (PERF2), mean square error (MSE), peak signal-to-noise ratio (PSNR), compression ratio (CR), and bit error rate (BER). The above quantities are calculated using the following formats:

$$PERF2 = \frac{100\|XC\|^2}{\|x(n)\|^2} \tag{10}$$

where XC is the compressed signal.

$$MSE = \left\{\sum err^\wedge 2\right\}/N \tag{11}$$

where err is the error signal.
N is the size of original signal.

$$PSNR = 10\log10\left\{\max(A)/MSE\right\} \tag{12}$$

where A is the original signal.

$$CR = x(n)/y(n) \tag{13}$$

where x(n) is the length of original signal.
y(n) is the length of compressed signal.

Tables 1 and 2 given below show performance measures of speech signal using Haar and Db10 wavelets for compression and denoising techniques.

From the above-obtained results (Tables 1 and 2), it is clear that among the four techniques implemented, Db10 with hard thresholding gives the best results with 98 % retained signal energy, 87 peak signal-to-noise ratio and less bit error rate.

Table 1 Performance measures of signal S1 (3.58 s long) with hard thresholding

Wavelet used	PERF0	PERF2	PSNR	MSE	CR	BER
Hard thresholding without coding						
Haar	96.93	96.61	81.40	4.71e-04	1	0.0117
Db10	95.29	99.00	85.36	1.88e-04	1	0.0072
Hard thresholding with coding						
Haar	97.03	95.35	86.97	1.30e-04	1	0.0047
Db10	95.35	98.16	87.76	1.0e-04	1	0.0039

Table 2 Performance measures of signal S1 (3.58 s long) with soft thresholding

Wavelet used	PERF0	PERF2	PSNR	MSE	CR	BER
Soft thresholding without coding						
Haar	96.93	80.35	76.26	0.0015	1	0.0204
Db10	95.29	87.94	79.42	7.42e-04	1	0.0143
Soft thresholding with coding						
Haar	97.03	78.16	82.52	3.6e-04	1	0.0081
Db10	95.35	81.07	83.99	2.59e-04	1	0.0065

A signal having high PSNR and less MSE is considered a better reconstructed signal. The BER and SNR comparison of the proposed technique (Haar and Db10 wavelets) with and without coding is shown in Figs. 8 and 9, respectively. From the graphs, it is clear that the performance of the coded signal is better than the signal which is not coded using gold code sequence. The coded speech signal is having higher PSNR and less BER. Figure 10 shows a comparison of the proposed technique with already existing techniques of data transmission over wireless GSM channel.

The time domain waveforms and power spectral density of the original and reconstructed signals using Haar and Db10 wavelets encoded with gold code sequence are shown in Figs. 11, 12, 13, 14, 15, 16, 17, 18, 19, 20, 21, and 22, respectively.

Figure 8 shows the BER comparison of the two techniques that have been implemented. The comparison has been made between the speech signals: one coded with gold code sequence (shown in blue) and other without coding (shown in orange). Also the results of hard and soft thresholding algorithms have been made for Haar and Db10 wavelets, respectively. From the graph obtained, it is clear that transmission of secured compressed speech signals (encoded with gold code sequence) has less BER as compared to one without coding and hard thresholding is preferable over soft thresholding.

Figure 9 shows the PSNR comparison of the proposed techniques. It is observed that the PSNR of the speech signal coded with gold code and denoised with Db10 wavelet (hard thresholding) has high value than the signal without coding.

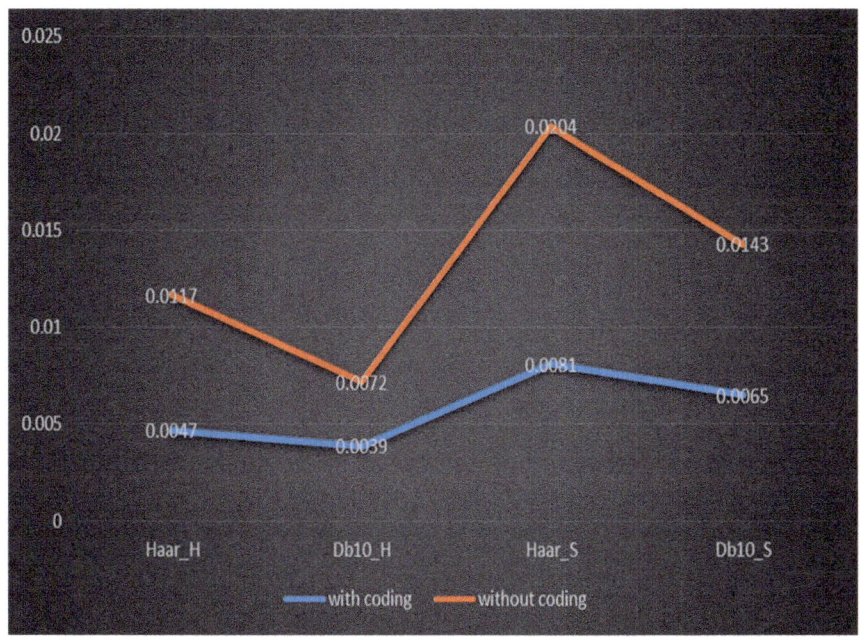

Fig. 8 BER (bit error rate) comparison of the proposed techniques

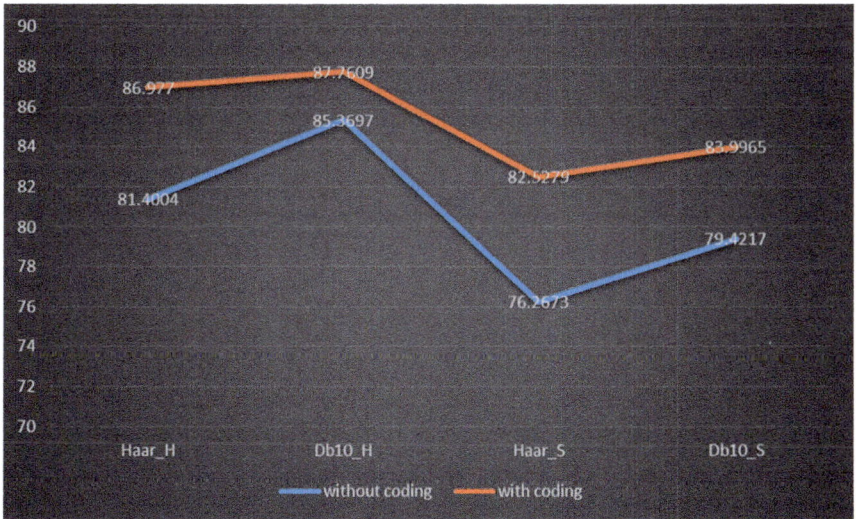

Fig. 9 Peak signal-to-noise ratio (PSNR) comparison of the proposed techniques

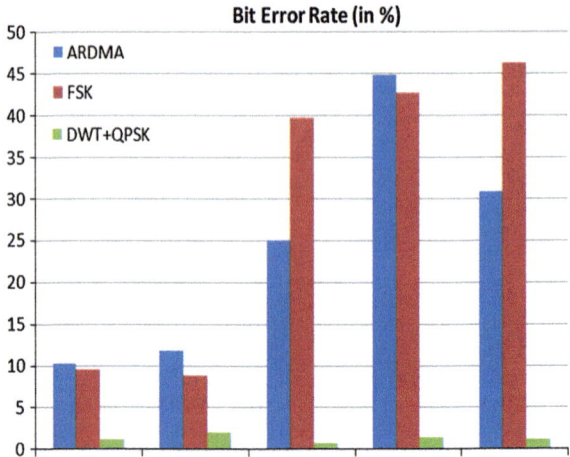

Fig. 10 BER comparison of existing (Bojan Kotnik et al.: ARDMA, FSK) and proposed work (DWT QPSK)

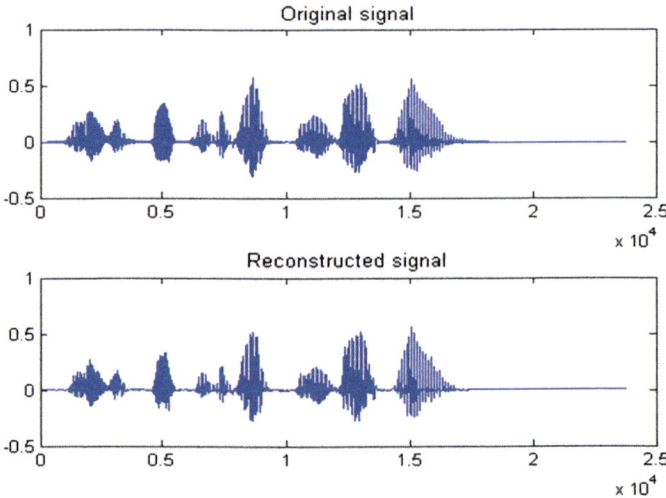

Fig. 11 Original and reconstructed speech signal using Haar wavelet with hard thresholding

Figure 10 gives the BER comparison of the proposed technique with the existing technique of data transmission over GSM voice channel. The motive of the two techniques is same, i.e., to transmit data over GSM voice channel. The existing technique has used neither any compression technique nor encoding technique for data security and transmission. In contrast, the proposed technique has met with the objectives of compression and secured data transmission simultaneously. The

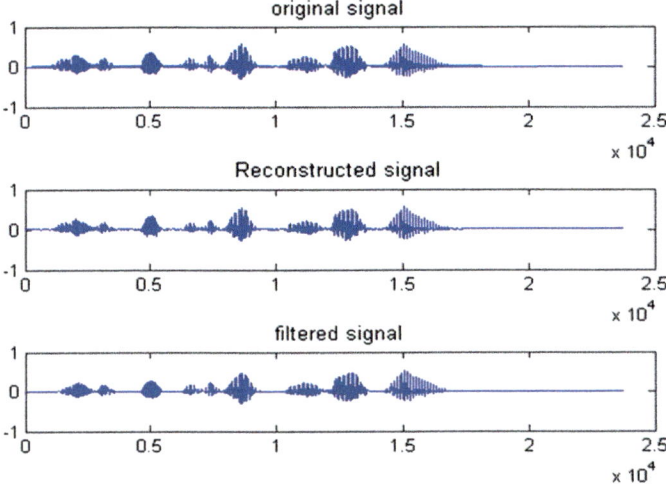

Fig. 12 Original, reconstructed, and filtered speech signals using Haar wavelet with hard thresholding

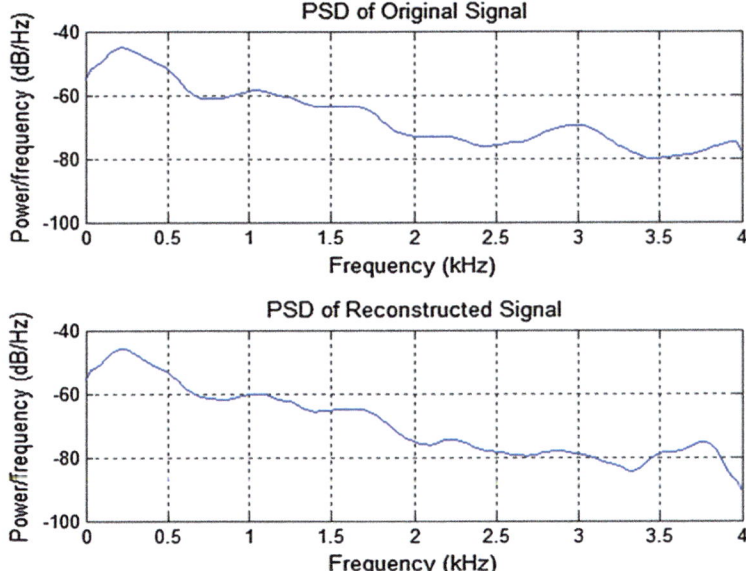

Fig. 13 PSD of speech signal using Haar wavelet with hard thresholding

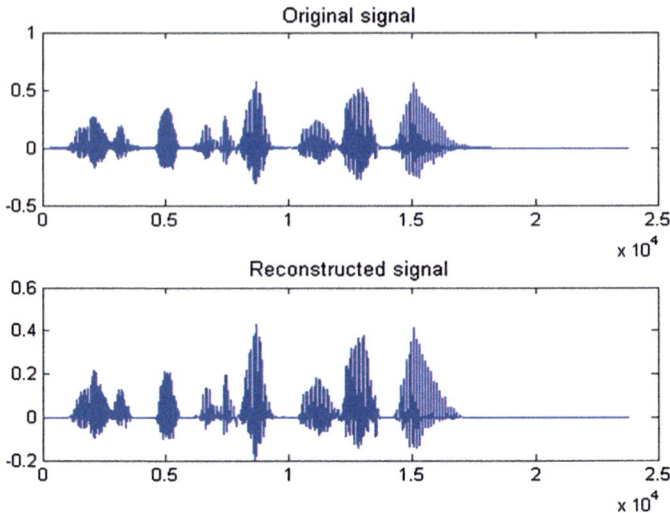

Fig. 14 Original and reconstructed speech signal using Haar wavelet with soft thresholding

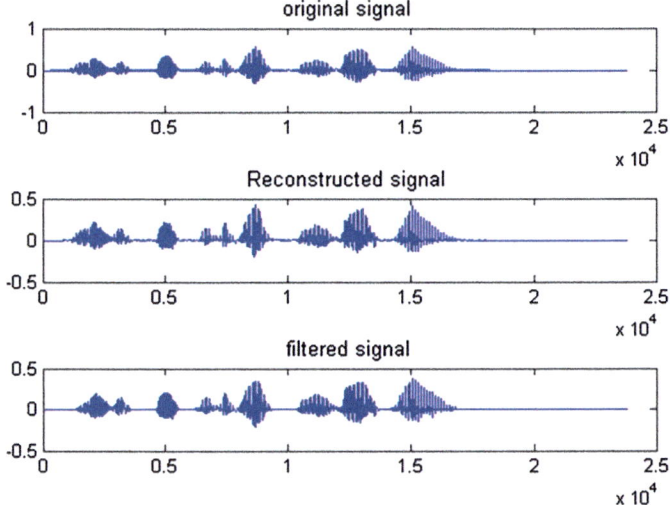

Fig. 15 Original, reconstructed, and filtered speech signal using Haar wavelet with soft thresholding

proposed technique (DWT with QPSK modulation) shown in green has the least BER (in %) when compared to the already existing techniques.

Figures 11, 12, 13, 14, 15, 16, 17, 18, 19, 20, 21, and 22 show the various time domain waveforms of the original, reconstructed, and filtered speech signals and the power spectral density of the reconstructed/received speech signals using the

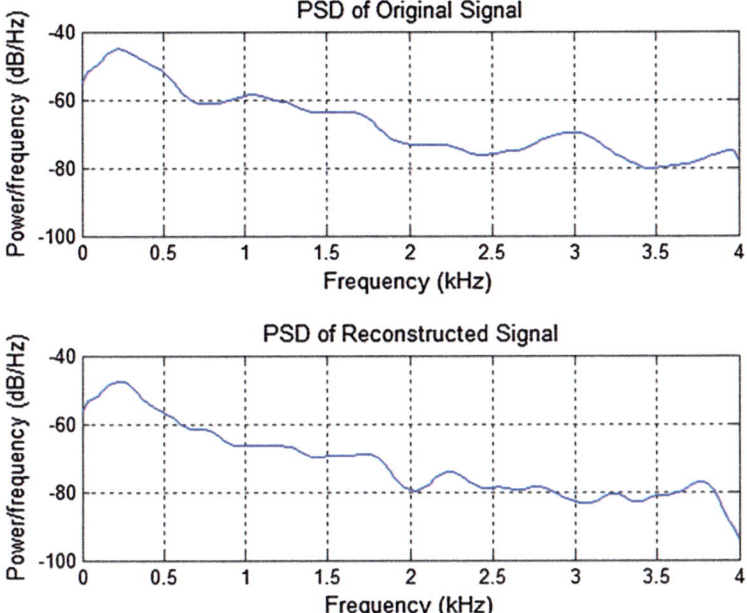

Fig. 16 PSD of speech signal using Haar wavelet with soft thresholding

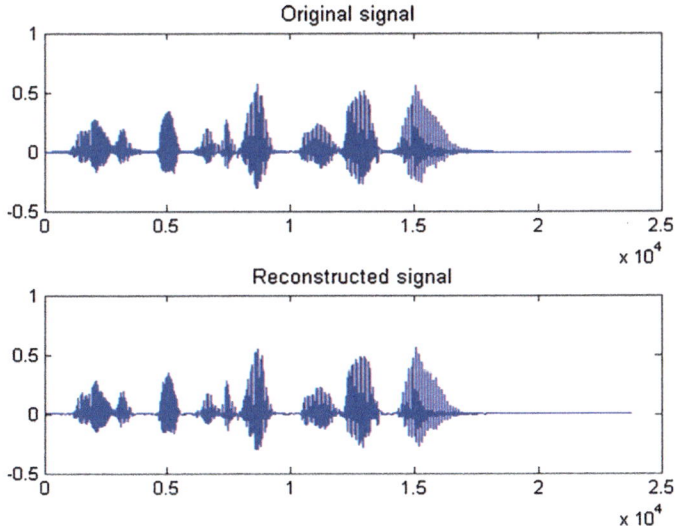

Fig. 17 Original and reconstructed speech signal using Db10 wavelet with hard thresholding

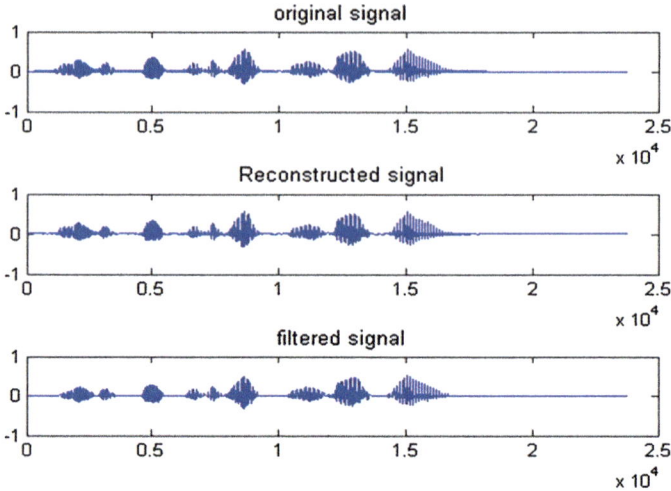

Fig. 18 Original reconstructed and filtered speech signal using Db10 wavelet with hard thresholding

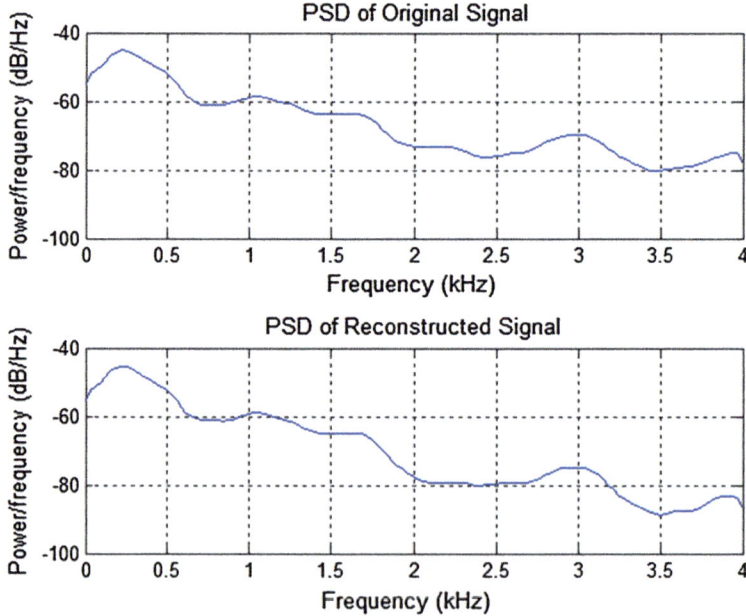

Fig. 19 PSD of speech signal using Db10 wavelet with hard thresholding

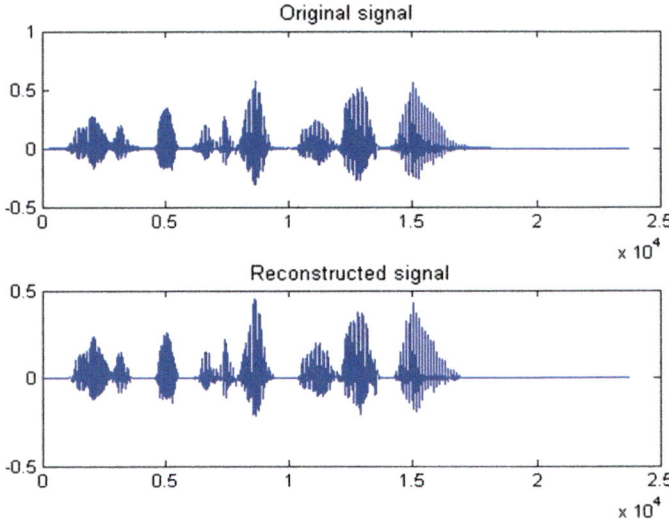

Fig. 20 Original and reconstructed speech signal using Db10 wavelet with soft thresholding

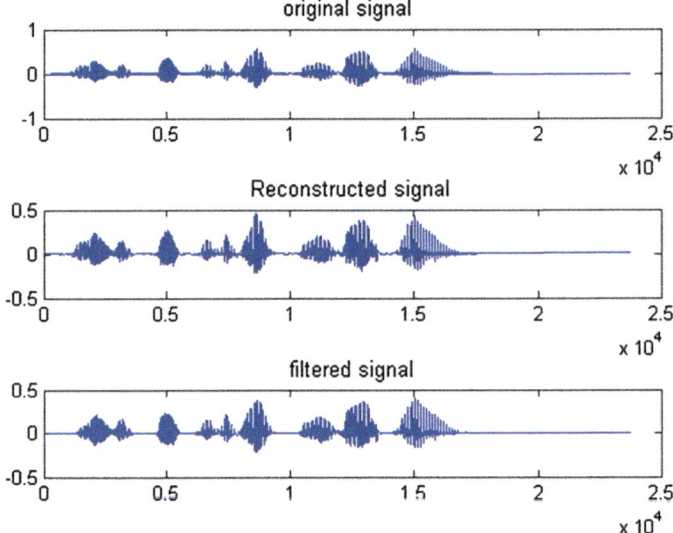

Fig. 21 Original, reconstructed and filtered speech signal using Db10 wavelet with soft thresholding

proposed methods. The waveforms of the speech signals obtained using Haar and Db10 wavelets with hard and soft thresholding algorithms have a slight difference if observed keenly. The speech signals compressed with Db10 wavelet with hard thresholding have a better sound quality than the one compressed with Haar wavelet.

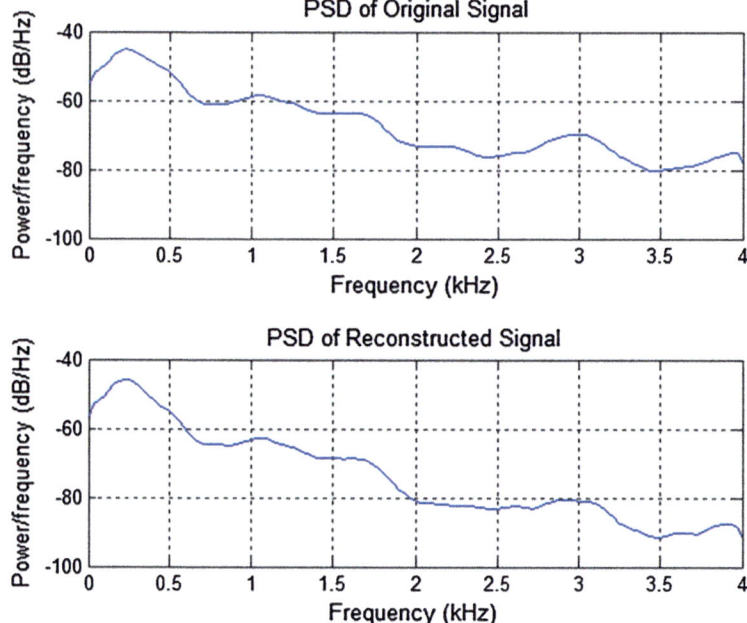

Fig. 22 PSD of speech signal using Db10 wavelet with soft thresholding

13 Conclusion

This paper presents a new idea of data transmission over the GSM voice channel. Due to the specific characteristics of the GSM coders which do not allow the usage of PSK-based and ASK-based digital modulation techniques, a new method of data transmission with phase-based modulation technique is realized in this paper. The proposed work is based on the wavelet-based compression technique which is a better technique than that provided by the GSM coders. Hence, the data can be compressed to a greater extent leading to less utilization of the available bandwidth. Further using the combination of DWT with QPSK, the reconstructed signal shows less bit error rate with high peak signal-to-noise ratio (PSNR). Therefore, for data transmission in terms of voice, the proposed work shows the best performance as compared to the already implemented techniques. A great advantage of this type of technique is that it can be used to deploy a data service where no cabling is required. The proposed technique has the capability to use only voice channel for transmission of data which results to save various logical channels such as paging and broadcasting channels. Hence, a lot of bandwidth and radio engineering resources can be saved.

References

1. Katugampala N, Al-Naimi V, Villette S, Kondoz A (2005) Real time end to end secure voice communications over GSM voice channel. In: 13th European signal processing conference, Antalya, Turkey, pp 1–4
2. LaDue CK, Sapozhnykov VV, Fienberg KS (2008) A data modem for GSM voice channel. IEEE Trans Veh Technol 57(4):2205–2218
3. Boloursaz M, Hadavi AH, Kazemi R, Behnia F (2013) A data modem for GSM adaptive multi rate voice channel. In: East-west design and test symposium, pp 1–4, 27–30
4. Ehmayssani T, Baudoin G (2008) Data transmission over voice dedicated channels using digital modulations. In: Radio electronika 18th international conference, IEEE, pp 1–4, ISBN: 978-1-4244-2087-2
5. Yarman BS, Guz U, Gurkan H (2006) On the comparative results of SYMPES: a new method of speech modeling. Int J Electron Commun (AEU) 60:421–427
6. Kotnik B, Mezgec Z, Svečko J, Chowdhury A (2009) Data transmission over GSM voice channel using digital modulation technique based on autoregressive modeling of speech production. Digit Signal Process 19:612–627
7. Ciornei S, Bogdan I, Scripcariu L, Popa M (2014) Sensitivity analysis of quadrature amplitude modulation over AMR-WB voice code. In: Proceedings of the second international conference on advances in computing, communication and information technology—CCIT 2014. ISBN: 978-1-63248-051-4
8. Mahapatra R, Dhar AS, Datta D (2009) Integrated voice and data transmission employing adaptive modulation in wireless networks. Int J Electron Commun 63:1012–1025
9. Swetha N, Anuradha K (2013) Text to speech conversion. Int J Adv Trends Comput Sci Eng 2 (6):269–278 ISSN: 2278-3091
10. Patil MV, Gupta A, Varma A, Salil S (2013) Audio and speech compression using DCT & DWT techniques. Int J Innov Res Sci Eng Technol, 2(5). ISSN: 2319-8753
11. Dey N, Biswas S, Das P, Das A, Chaudhuri SS (2012) Lifting wavelet transformation based blind watermarking technique of photoplethysmographic signals in wireless telecardiology. In: 2012 world congress on information and communication technologies (WICT), IEEE, pp 230–235
12. Dey N, Pal M, Das A (2012) A session based watermarking technique within the NROI of retinal fundus images for authencation using DWT, spread spectrum and Harris corner detection. Int J Mod Eng Res 2(3):749–757
13. Sheikh JA, Parah SA, Akhtar S, Bhat GM (2015) Performance evaluation and comparison of speech compression using linear predictive coding and discrete wavelet transform. In: Commune 2015 international conference on advances in computers, communication and electronic engineering, 16–18 March 2015, University of Kashmir, pp 352–355
14. Kumar S, Singh OP, Mishra GR, Mishra SK, Trivedi A (2012) Speech compression and enhancement using wavelet coders. Int J Electron Commun Comput Eng 3(6). ISSN (Online): 2249–071X, ISSN (Print): 2278–4209
15. Joseph SM, Auto B (2012) Speech coding based on orthogonal and biorthogonal wavelet. In: 2nd international conference on communication, computing and security, procedia technology, Elsevier, vol 6, pp 397–404
16. Jagtap SK, Mulye MS, Uplane MD (2015) Speech coding techniques. Procedia Comput Sci 49:253–263
17. Sheikh JA, Akhtar S, Parah SA, Bhat GM (2015) On the design and performance evaluation of compressed speech transmission over wireless channel. In: 12th IEEE India international conference (INDICON) on electronics, energy, environment, communication, computers, control (E3-C3), 17–20 December 2015, Jamia Millia Islamia, New Delhi

18. Dey N, Roy AB, Das A, Chaudhuri SS (2012) Stationary wavelet transformation based self-recovery of blind-watermark from electrocardiogram signal in wireless telecardiology. In: Recent trends in computer networks and distributed systems security, Springer, Berlin Heidelberg, pp 347–357
19. Dey N, Dey M, Mahata SK, Das A, Chaudhuri SS (2015) Tamper detection of electrocardiographic signal using watermarked bio-hash code in wireless cardiology. Int J Signal Imaging Syst Eng 8(1–2):46–58
20. Dey N, Mishra G, Nandi B, Pal M, Das A, Chaudhuri SS (2012) Wavelet based watermarked normal and abnormal heart sound identification using spectrogram analysis. In: 2012 IEEE international conference on computational intelligence and computing research (ICCIC), IEEE, pp 1–7
21. Dey N, Acharjee S, Biswas D, Das A, Chaudhuri SS (2013) Medical information embedding in compressed watermarked intravascular ultrasound video. arXiv preprint arXiv:1303.2211
22. Ahmad J, Akhtar S, Majeed S, Ahmad S (2016) On the design and performance evaluation of DWT based compressed speech transmission with convolutional coding. Commun Appl Electron 4(9):36–40. doi:10.5120/cae2016652173

StegNmark: A Joint Stego-Watermark Approach for Early Tamper Detection

S.A. Parah, J.A. Sheikh and G.M. Bhat

1 Introduction

The advancement in the field of computer networks and digital signal processing have resulted in easy availability of digital multimedia content over the Internet nowadays. The ugly side of this advancement manifests itself in terms of the information security issues that creep in during multimedia transmission [1, 2]. Information security is necessary sue to the fact that transmitted digital data on the network is at a risk of being completely copied or intercepted illegally. Secrecy of data being transmitted over a network is of greater importance especially if the data pertains to national security issues, classified information, trade secrets, banking transactions etc. [3]. To protect the secrecy of the transmitted data, cryptographic techniques are generally used to encrypt the data at the transmitter before transmission. However, exponential growth in the computing powers of modern computers has resulted in the security of the data yielded by these techniques being threatened. Further, as the cryptographic techniques encrypt secret messages into unrecognizable forms before transmission, the disguised nature of the encrypted message can easily arouse suspicion among eavesdroppers and bring an unexpected attack from hackers [4, 5]. Data hiding of late has flourished as a supplementary solution for secret communication and protecting digital media. In data hiding paradigm, many parameters of conflicting nature contend with each other, they are capacity, imperceptibility, robustness, and security. On the basis of these

S.A. Parah (✉) · J.A. Sheikh · G.M. Bhat
Post Graduate Department of Electronics and Instrumentation Technology,
University of Kashmir, Hazratbal, Srinagar, J&K 190006, India
e-mail: shabireltr@gmail.com

J.A. Sheikh
e-mail: sjavaid_29ku@yahoo.co.in

G.M. Bhat
e-mail: drgmbhat@gmail.com

© Springer International Publishing Switzerland 2017
N. Dey and V. Santhi (eds.), *Intelligent Techniques in Signal Processing for Multimedia Security*, Studies in Computational Intelligence 660, DOI 10.1007/978-3-319-44790-2_19

paradigms, data hiding is broadly classified in two main fields, steganography and digital watermarking [6–8].

Steganography refers to a data hiding technique for hiding a secret message within publicly available data. The word steganography means "covered writing," which gives a fair indication of the fact, that it is a technique where the emphasis is laid on the hiding or covering up what is being communicated [9–11]. Although steganography is used in order to send a secret or covert message, the sender does not want anyone to know the information is being sent. This makes the sender to hide the secret message within a host (cover) medium. The cover medium, or overt message, is publicly available data that is used to hide the covert message. The data is hidden in the cover medium in such a manner that an adversary only sees the content of the overt file and not the actual secret information that is being sent. The main aim of a steganography is to achieve high payload coupled with perceptual transparency and at the same time provide a fair degree of security to the secure data being carried out by the cover medium [12–14].

Digital watermarking refers to the embedding of the special watermark which can be annotations, corporate business logos, digital signature etc. Digital watermarking is used for copyright protection of media, fingerprinting, broadcast monitoring and covert communication between parties, etc. [15, 16]. Depending upon perception of watermark, watermarking techniques are further classified as visible and invisible watermarks [12, 17]. Further, depending on the intent for which watermark is used, watermarking techniques are also classified as robust and fragile watermarking techniques [18]. Robust techniques are used for copyright protection, on the other hand fragile watermark techniques are used for tamper detection and content authentication. The data size of a watermark embedded in a cover medium is generally very small compared to the data hidden in a steganographic system [19, 20].

This chapter has been based on the fusion of steganography and fragile watermarking [21, 22] and presents a technique called StegNmark, which is capable of content authentication at the receiver utilizing early tamper detection ability of proposed scheme. The early detection could be quite useful in situations where time factor plays an important role, an example being secure data extraction indicating a warfare plan to attack the enemy. Rest of the chapter is organized as follows. Section 2 provides detailed survey of literature. The proposed system has been described in Sect. 3. Experimental results and discussions are presented in Sect. 4. Section 5 describes the advantages and limitations of the proposed technique. The chapter concludes in Sect. 6.

2 Related Work

Data hiding schemes have been used as a potent tool for establishing covert channels for secure communication and solving digital rights management (DRM) issues throughout the last decade. Data hiding encompasses two major schemes called as steganography and watermarking. Steganography strives for establishing covert communication without giving a clue to the adversary that secret communication is taking place. While as the main goals of a watermarking algorithm are copyright protection, tamper detection, and content authentication. Lot of literature is available that describes various high capacity steganographic techniques and robust/fragile watermarking techniques. In either of the schemes, data is hidden into the cover media either in spatial domain or transform domain. Spatial domain techniques are characterized by high payload and less computational complexity while as transform domain techniques are robust to various image processing and geometric attacks. However, transform domain techniques are computationally complex. Some of the important spatial-domain data hiding techniques pertaining to image data include [23–27]. A few early watermarking techniques include [28, 29]. Here, m-sequences are embedded into the least significant bit (LSB) of the cover media for providing secure and an effective transparent embedding technique. Use of m-sequences helps in watermark detection at the receiver as they have good correlation properties. Furthermore, these techniques are computationally inexpensive to implement. In [27], the authors reshape the m-sequence into two-dimensional watermark blocks which are added and detected on a block-by-block basis. The block-based method, referred to as variable-w-two-dimensional watermark (VW2D) is shown to be robust to JPEG compression. This technique has also been shown to be an effective fragile watermarking scheme which can detect image alterations on a block basis [30]. Several spatial-domain data hiding techniques in images are proposed in. One such technique consists of embedding a texture-based watermark into a portion of the image with similar texture. The idea being that due to the similarity in texture, it will be difficult to perceive the watermark. The watermark is detected using a correlation detector.

Wu and Tsai reported a data hiding method in images by using the concept of pixel value differencing (PVD) [31]. Cover image is partitioned into non overlapping blocks of two consecutive pixels say p_i and $p_i + 1$. From each block a difference di is obtained by subtracting $p_i + 1$ from p_i. The PVD method embeds data in the edge area to achieve good perceptual qualities of the stego-image. PVD suffers from a drawback, which is that histogram analysis of the pixel differencing reveals the presence of secret message. Zang and Wang [32] proposed a modified scheme to enhance the security and avoid unusual steps seen in the pixel difference histogram while preserving the advantage of low visual distortion of PVD. Hiding capacity or payload is one of the important aspects of a data hiding system. Many studies [33–35] have tried to improve the hiding capacity by using various approaches. Parah et al. [36] have reported high capacity data hiding technique in which the data to be secured has been embedded in intermediate significant bits in addition to least significant bits of cover image. To improve the quality of the stego-image, the data to be embedded is broken down in data blocks of variable length and each block is embedded in the cover media in such a way that highest length data vector is embedded in lower order bit plane and the lower length data vector is embedded into a higher order bit plane. The security parameter has been taken care of by embedding data using a secret key [37]. Transform domain techniques are generally used for watermarking applications. In transform domain techniques, the data to be hidden or watermark are embedded in the coefficients of the transformed version of the host image. The commonly used transform domain techniques used for hiding watermark in the frequency domain are discrete Fourier transform (DFT), discrete cosine transform (DCT), and discrete wavelet transform (DWT). The most commonly used transform for embedding the watermark in digital images is DCT. In the reported technique the DCT has been obtained on 8 × 8 blocks of host image, a pseudo-random subset of blocks has been chosen and midrange frequencies are slightly altered to encode a binary sequence. Various watermarking schemes use more elaborate model of human visual system (HVS) and use the concept of image adaptive watermark to improve imperceptibility. Some of such techniques are described in [38–40]. DWT has gained a lot of momentum in the current research going on in the field of data hiding. This is because of the fact DWT provides a powerful insight into an image's spatial and frequency attributes [41–43]. Kundur et al. proposed a DWT-based robust watermarking technique [44] referred to as digital image watermarking using image-based fusion. This technique employs multiresolution wavelet decomposition of both the host image and watermark. The main issue with DCT and DWT when used for data hiding purpose is that such transforms usually render low relatively low payload hiding systems. Data hiding in image edges of late has got attention of research community for high capacity embedding. In [45], a high capacity hiding scheme utilizing LSB substitution method and hybrid edge detection has been reported. In [46], Mashallah et al. have proposed a steganography algorithm based on comparison of a pixel with its eight neighbours. By selecting a threshold the embedding has been done using LSB substitution method. In [47] Marghny et al. have proposed a simple LSB substitution method for embedding data. The authors have reported a PSNR of 48 dB for maximum payload of 65,536 bits. In [49] Wei Sun et al. have reported a reversible data hiding technique using block truncation code (BTC). The capacity of the proposed technique in bits is 64,008. In [50], Ou and Sun proposed a steganographic technique based on absolute moment block truncation code (AMBTC). For achieving high payload the authors have divided the image into blocks based on a threshold. The data is embedded in blocks which are termed as complex blocks and smooth blocks. The reported payload is different for different test images, for Lena the payload is 226,474 bits, for baboon it is 124,969 bits, for peppers it is 232,819 bits, and for boat it is 206,629 bits. Ying Hsuan et al. [51] have proposed an improved version of [50]. The authors have divided the blocks in complex and smooth blocks using a predefined adjustable threshold and have used both the blocks for embedding purpose using the difference between the two mean values of the block. The payload reported varies from 50,000 to 300,000 bits.

A thorough literature survey of the data hiding area reveals that huge amount of work is being done in the fields of steganography and watermarking with an aim to improve payload, imperceptibility, robustness, and security and to reduce computational complexity. The work in the areas of steganography and watermarking somehow seems to be isolated. To best of our knowledge, no technique has been put forth wherein both the concepts of steganography and watermarking have been used for ensuring high payload covert communication of fairly high imperceptibility and capability of early tamper detection. The proposed chapter presents a technique in this regard wherein both the steganography and watermarking (Fragile) have been used to provide what we call as StegNmark. The proposed technique is capable of providing high-quality stego-images for a fairly high payload and early tamper detection.

3 Proposed StegNmark System

The proposed StegNmark system for early tamper detection of secure data is shown in Fig. 1. The cover medium which is an image is firstly subjected to fragile watermark embedding in the watermark embedder. The watermark embedding is carried out in spatial domain using ISB embedding under the control of watermark key that has been generated by the master key generator [37]. The embedding of watermark in ISB plane results into the watermark being fragile in nature. The resultant watermarked image is passed on to second data embedder that embeds secret data under the control of stego-key generated by master key generator. It is pertinent to mention here that the master key generator generates both keys in such a way that ensures the interleaving of the watermark bits within the secret data bits before the embedding takes place. The second embedder outputs the resultant watermarked and stego-image referred to as StegNmark. The StegNmark is received at the receiver, where same watermark key is used to extract watermark. Since the embedded watermark being fragile is highly sensitive to any signal processing attack carried out on the StegNmark. Therefore any possible attack on the StegNmark during its transit from transmitter to receiver results in false detection of the watermark, and as such informs the receiver that the secure data has been tampered with during the transit. Based on detection or false detection of the watermark from the StegNmark, decision about secure data extraction from StegNmark is made. The case of false detection prompts the receiver, to skip data extraction and as such saves precious time of the processor. It is to be noted here that time required to extract the fragile watermark is relatively very small compared to time required to extract all the secure data from the StegNmark. The saved time slot can also be used for to send some sort of automatic request to the transmitter for retransmission. The mathematical model for embedding and extraction is as follows:

3.1 Data Embedding and Extraction

Let "C" be the original 8-bit grayscale cover image of $M_c \times N_c$ pixels. It can be therefore represented as

$$C = \left\{ x_{ij} | 0 \leq i < M_c, \ 0 \leq j < N_c, \ x_{ij} \in \left\{ 0, \ldots, 2^k - 1 \right\} \right\} \tag{1}$$

with $k = 8$, the pixel values is bound to lie in the range of 0–255.

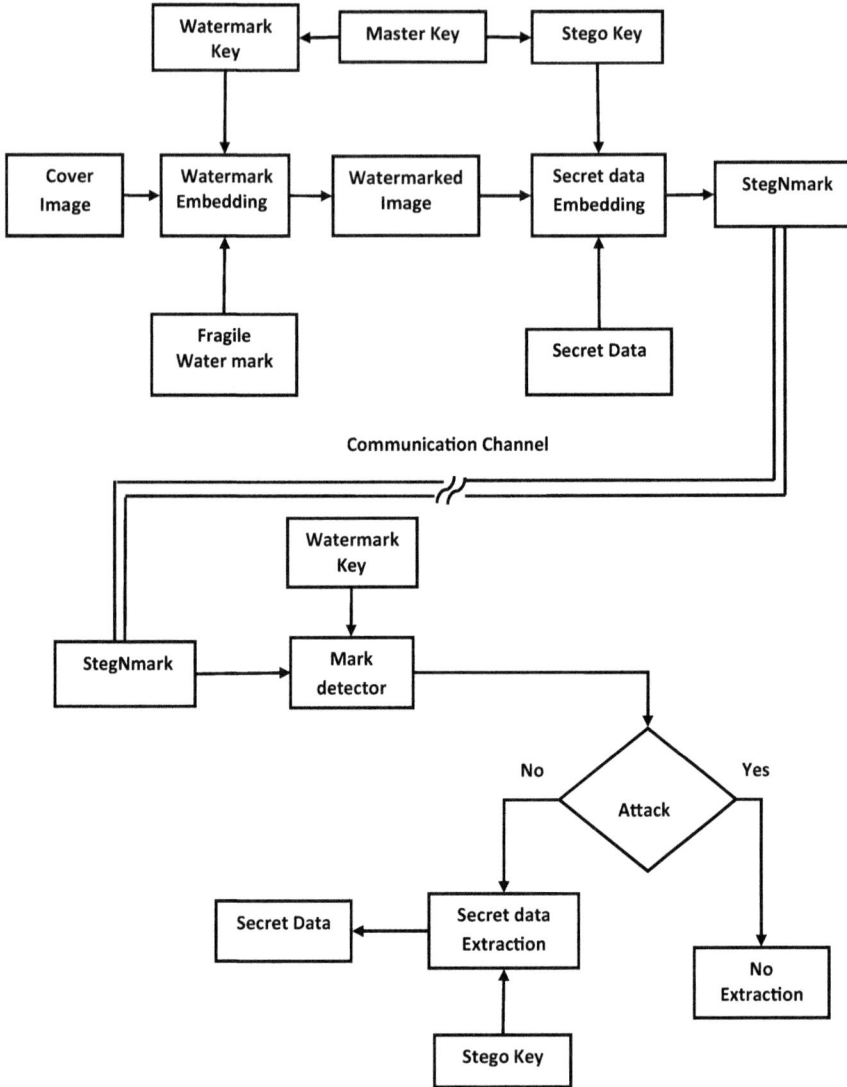

Fig. 1 Proposed StegNmark system

If M is the n_1-bit secret message vector and W is n_2 bit watermark, vector such that:

$$M = \{m_i | 0 \leq i < n_1, \ m_i \in \{0, 1\}\} \tag{2}$$

$$W = \{w_i | 0 \leq i < n_2, \ w_i \in \{0, 1\}\} \tag{3}$$

Let D represent joint data vector containing secret data and watermark data. If n is the length of such a data vector, it is given by $n = n_1 + n_2$. The total data vector D containing secret data and watermark information is given by

$$D = \{d_i | 0 \leq i < n,\ d_i \in \{0, 1\}\} \tag{4}$$

It is pertinent to mention here that $n = M_c \times N_c$ as well.

Corresponding to this data vector master key is used to generate a Pseudorandom Address Vector (PAV) (which serves as watermark and stego-key), of exactly same no of elements as data vector and is given by:

$$PAV = \{A_i | 0 \leq i < n,\ A_i \in \{0, 1, 2 \ldots n\}\} \tag{5}$$

It is pertinent to mention here that first n_2 addresses (watermark key) as pointed by PAV are used to embed watermark while as last n_1 addresses (stego-key) are used for secret data embedding. In ISB approach, data are embedded in a kth plane by embedding ith data bit into the location pointed to by PAV value at ith iteration. In our case, we have embedded data into first ISB. For extraction at the receiver, master key is used to generate PAV which in turn leads to generation of watermark and stego-keys. The respective keys are used to extract data from the kth ISB plane in which data is embedded at the transmitter side. It is worth noting that secret data is extracted only if watermark data received is unaltered signifying no attack has been carried out on StegNmark. To have a clear idea of the efficacy of the proposed scheme we define StegNmark ratio R as:

$$\text{StegNmark ratio } (R) = \text{Secret data bits/watermark bits}$$

As long as $R \leq 1$ the time taken to extract the secure data is less or equal to that taken to extract logo and as such efficacy of the proposed scheme is poor in this region. However, in general when data hiding is used for covert communication the secret data size is very large compared to that of watermark size, i.e. $R \gg 1$. For this scenario the proposed scheme performs better, in other words greater the value of StegNmark ratio R, better the efficacy of the proposed scheme.

4   Experimental Results and Discussions

This section presents the experimental results and discussions in detail. The system has been tested for a number of standard grayscale test images (512×512). The test images have been obtained from standard image gray scale image data base and are available at www.decsai.ugr.es. The proposed system has been evaluated in terms of various objective image quality metrics such as peak signal-to-noise ratio (PSNR), normalized cross correlation (NCC), structural similarity measure index (SSIM), average difference (AD), and normalized absolute error (NAE). The data

Fig. 2 Cover images

Fig. 3 Watermark

size hidden in each test image has been varied from 1024 bits to 258,048 bits. It is to be noted that a binary fragile watermark (logo) of size 64 × 64 = 4096 bits is inserted in each test image in addition to the variable length data vector inserted in the image. Figure 2 shows the various original cover images used for testing the proposed scheme. While as the watermark embedded is shown in Fig. 3.

Various Image quality metrics used for performance analysis of scheme have been defined as:

$$\text{PSNR} = 10 \ \log \frac{(2^n - 1)^2}{\text{MSE}} = 10 \ \log \frac{(255)^2}{\text{MSE}} \tag{6}$$

$$\text{MSE} = \frac{1}{MN} \sum_{j=1}^{M} \sum_{k=1}^{N} \left(x_{j,k} - x'_{j,k} \right)^2 \tag{7}$$

$$\text{NAE} = \frac{\sum_{j=1}^{M} \sum_{k=1}^{N} \left| x_{j,k} - x'_{j,k} \right|}{\sum_{j=1}^{M} \sum_{k=1}^{N} \left| x_{j,k} \right|} \tag{8}$$

$$\text{AD} = \frac{1}{MN} \sum_{j=1}^{M} \sum_{k=1}^{N} \left(x_{j,k} - x'_{j,k} \right) \tag{9}$$

Here M, N are the dimensions of the cover image and the stegomarked image; $x \ (i, j)$ is the (i, j)th pixel value of original image and $x' \ (i, j)$ is the (i, j)th pixel intensity value of stegomarked image.

The structural similarity index (SSIM) is based on the calculations of three terms, namely the luminance, contrast, and structure. The overall index is a given by:

$$\text{SSIM}(x, y) = [l(x, y)]^{\alpha} \cdot [c(x, y)]^{\beta} \cdot [s(x, y)]^{\gamma} \tag{10}$$

where;

$$l(x, y) = \frac{2\mu_x\mu_y + C1}{\mu_x^2 + \mu_y^2 + C1} \tag{11}$$

$$c(x, y) = \frac{2\sigma_x\sigma_y + C2}{\sigma_x^2 + \sigma_y^2 + C2} \tag{12}$$

$$s(x, y) = \frac{\sigma_{xy} + C3}{\sigma_x\sigma_y + C3} \tag{13}$$

where μ_x, μ_y, σ_x, σ_y, and σ_{xy} are the local means, standard deviations, and cross-covariance for images x, y. The expression for default exponents and default selections of $C3$ is given by:

$$\text{SSIM}(x, y) = \frac{(2\mu_x\mu_y + C1)(2\sigma_{xy} + C2)}{(\mu_x^2 + \mu_y^2 + C1)(\sigma_x^2 + \sigma_y^2 + C2)} \tag{14}$$

$$\text{BER} = \frac{1}{MN}\left[\sum_{i=1}^{M}\sum_{j=1}^{N} w_m(i,j) \oplus w_{\text{me}}(i,j)\right] \times 100 \tag{15}$$

$$\text{NCC} = \frac{\sum_{i=1}^{M}\sum_{j=1}^{N} w_m(i,j) \times w_{\text{me}}(i,j)}{\sum_{i=1}^{M}\sum_{j=1}^{N} w_m(i,j)^2} \tag{16}$$

Here M, N are the dimensions of the original logo and extracted logo; $w(i, j)$ is the (i, j)th pixel of original watermark and $w_{\text{me}}(i, j)$ is the (i, j)th pixel of the extracted logo.

4.1 Imperceptibility Analysis

Imperceptibility is one of the chief goals of a data hiding system. A data hiding system is considered as better if human visual system cannot distinguish between cover and stego-media for a fairly large payload. A highly imperceptible data hiding system is accompanied by high value of PSNR (usually above 40 dB), NCC, and SSIM very close to unity and near zero values of NAE and AD. The results

Lena	Peppers	Baboon	Boats	Bridge	Barbara

Fig. 4 Stegomarked images

obtained for the various test images, when subjected to varying nature of payload from 1024 to 262144 bits have been shown in various tables below. The stego-images for a payload of one bit per pixel or 262144 bits corresponding to cover images shown in Fig. 2 have been shown in Fig. 4. It is clear from the subjective analysis that the system is capable of providing high-quality watermark images.

Tables 1, 2, 3, 4, 5, and 6 show various objective quality indices for test images, Lena, Pepper, Baboon, Boats, Bridge, and Barbara, respectively. It is evident from the different tables that the PSNR and SSIM decrease as data size varies.

It is evident from Tables 1, 2, 3, 4, 5, and 6 that the proposed system is capable of providing good quality stego-images even for a high payload of one bit per pixel. This is substantiated by the fact that average PSNR for all the used test images for a payload of 262144 bits is above 49 dB. Further Unity value of NCC coupled with near zero NAE values justifies our argument. Subjective quality analysis is also in tune with objective quality analysis. In this context, Fig. 5 shows various stego-images of Lena for different quantum of embedded data. It is evident that human eye hardly perceives any degradation of stego-image for various payloads.

A comparison of proposed technique with some state-of-the-art techniques for a payload of 262144 bits has been undertaken. Figure 6 shows the comparison results

Table 1 Image indices of test image Lena for varied range of embedded data bits

Secret data bits	Watermark bits	PSNR (dB)	SSIM	AD	NAE	NCC
1024	4096	56.9554	0.9990	0.0277	0.0003	1.0000
2048	4096	56.8236	0.9990	0.0279	0.0003	1.0000
4096	4096	56.6565	0.9989	0.0281	0.0003	1.0000
8192	4096	56.3615	0.9988	0.0280	0.0004	1.0000
16384	4096	55.7652	0.9986	0.0278	0.0005	1.0000
32768	4096	54.8076	0.9981	0.0282	0.0008	1.0000
65536	4096	53.3254	0.9974	0.0277	0.0014	1.0000
131072	4096	51.3593	0.9961	0.0278	0.0025	1.0000
258048	4096	49.0594	0.9938	0.0263	0.0047	1.0000

Table 2 Image indices of test image Peppers for varied range of embedded data bits

Secret data bits	Watermark bits	PSNR in dB	SC	AD	NAE	NCC
1024	4096	57.1702	0.9997	0.0262	0.0003	1.0000
2048	4096	57.0737	0.9997	0.0263	0.0003	1.0000
4096	4096	56.9306	0.9997	0.0260	0.0003	1.0000
8192	4096	56.5670	0.9996	0.0260	0.0004	1.0000
16384	4096	55.9436	0.9994	0.0258	0.0005	1.0000
32768	4096	54.9351	0.9991	0.0259	0.0008	1.0000
65536	4096	53.4512	0.9984	0.0269	0.0014	1.0000
131072	4096	51.4360	0.9973	0.0253	0.0026	1.0000
258048	4096	49.1055	0.9947	0.0269	0.0048	1.0000

Table 3 Image indices of test image Baboon for varied range of embedded data bits

Secret data bits	Watermark bits	PSNR in dB	SC	AD	NAE	NCC
1024	4096	57.1107	1.0000	0.0268	0.0003	1.0000
2048	4096	57.0273	0.9999	0.0267	0.0003	1.0000
4096	4096	56.8356	0.9999	0.0267	0.0003	1.0000
8192	4096	56.4968	0.9999	0.0267	0.0004	1.0000
16384	4096	55.9264	0.9999	0.0266	0.0005	1.0000
32768	4096	54.9059	0.9998	0.0267	0.0008	1.0000
65536	4096	53.4377	0.9997	0.0263	0.0013	1.0000
131072	4096	51.4301	0.9990	0.0267	0.0024	1.0000
258048	4096	49.0896	0.9978	0.0251	0.0046	1.0000

Table 4 Image indices of test image Boats for varied range of embedded data bits

Secret data bits	Watermark bits	PSNR in dB	SSIM	AD	NAE	NCC
1024	4096	57.2185	0.9991	0.0255	0.0002	1.0000
2048	4096	57.1259	0.9991	0.0256	0.0003	1.0000
4096	4096	56.9188	0.9990	0.0256	0.0003	1.0000
8192	4096	56.5926	0.9989	0.0259	0.0003	1.0000
16384	4096	55.9696	0.9988	0.0242	0.0005	1.0000
32768	4096	55.0011	0.9984	0.0237	0.0007	1.0000
65536	4096	53.4943	0.9978	0.0227	0.0013	1.0000
131072	4096	51.4549	0.9973	0.0193	0.0024	1.0000
258048	4096	49.1137	0.9955	0.0112	0.0045	1.0000

Table 5 Image indices of test image Bridge for varied range of embedded data bits

Secret data bits	Watermark bits	PSNR in dB	SSIM	AD	NAE	NCC
1024	4096	56.8417	0.9999	0.0285	0.0003	1.0000
2048	4096	56.7691	0.9999	0.0285	0.0003	1.0000
4096	4096	56.6517	0.9999	0.0279	0.0003	1.0000
8192	4096	56.3305	0.9999	0.0279	0.0004	1.0000
16384	4096	55.7687	0.9999	0.0261	0.0006	1.0000
32768	4096	54.8261	0.9998	0.0172	0.0009	1.0000
65536	4096	53.3621	0.9997	0.0013	0.0015	1.0000
131072	4096	51.3746	0.9994	0.0168	0.0027	1.0000
258048	4096	49.0595	0.9987	0.0027	0.0051	1.0000

Table 6 Image indices of test image Barbara for varied range of embedded data bits

Secret data bits	Watermark bits	PSNR in dB	SSIM	AD	NAE	NCC
1024	4096	57.1552	0.9996	0.0265	0.0003	1.0000
2048	4096	57.1005	0.9996	0.0265	0.0003	1.0000
4096	4096	56.8598	0.9996	0.0267	0.0003	1.0000
8192	4096	56.5030	0.9995	0.0265	0.0004	1.0000
16384	4096	55.9506	0.9995	0.0264	0.0005	1.0000
32768	4096	54.9343	0.9991	0.0262	0.0008	1.0000
65536	4096	53.4650	0.9986	0.0259	0.0014	1.0000
131072	4096	51.4216	0.9977	0.0247	0.0026	1.0000
258048	4096	49.0930	0.9958	0.0265	0.0050	1.0000

of our technique when compared to those in [48–51]. It is evident from the comparison results that our technique performs better.

4.2 Authentication Analysis

The stego-images obtained using the proposed method have been subjected to various image processing and geometric attacks like median filtering, Low Pass Filtering, Histogram equalization, additive White Gaussian Noise, Salt and Pepper noise, JPEG compression, and rotation to evaluate tamper detection and hence content authentication ability of the scheme [43, 53–55, 57]. All the attacks are performed on stegomarked images containing a payload of 262144 bits (258048 secret data bits and 4096 watermark bits) i.e., 12.5 % of the cover image are shown below. The results show that the time required for the extraction of the watermark and stego-bits from the StegNmark besides showing Bit Error Rate and Normalized

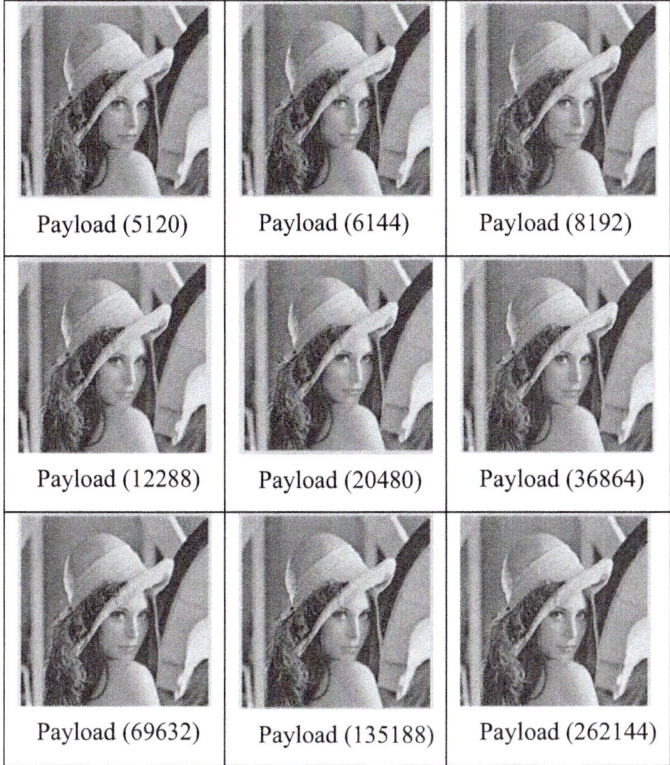

Fig. 5 Stego-images for different embedded data size

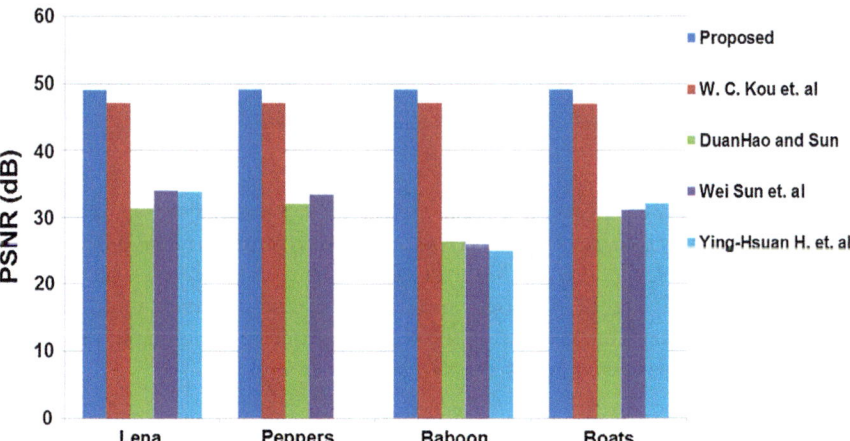

Fig. 6 Comparison of proposed StegNmark technique

Table 7 Performance parameters of StegNmark system for no attack

Image	Bit error rate (BER)	PSNR in dB	Watermark extraction time in seconds	Secret data extraction time in seconds
Lena	0	49.0697	0.0936	9.4381
Peppers	0	49.0928	0.0468	9.6253
Baboon	0	49.1141	0.0468	9.4225
Boats	0	49.1147	0.0624	9.4069
Bridge	0	49.0647	0.0624	9.1885
Barbara	0	49.0871	0.0468	9.4849
Average	0	49.0879	0.0598	9.4260

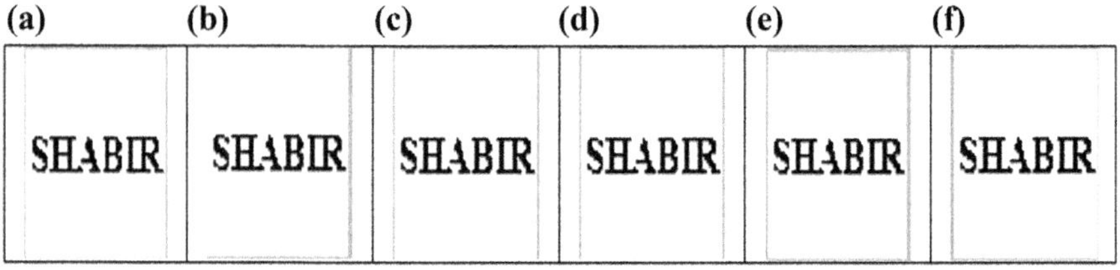

(a)   (b)   (c)   (d)   (e)   (f)

SHABIR   SHABIR   SHABIR   SHABIR   SHABIR   SHABIR

Fig. 7 Extracted watermarks from **a** Lena. **b** Peppers. **c** Baboon. **d** Boats. **e** Bridge. **f** Barbara

Cross Correlation corresponding to every attack. BER and NCC have been calculated using Eqs. 15 and 16, respectively. Following results show performance of the proposed technique to various attacks:

(a) No Attack

Table 7 shows the performance parameters of the proposed system under ideal condition i.e., no attack. It is clear that BER in this case is zero and as such we are able to extract the logo completely at the receiving end as shown in Fig. 7, indicating that StegNmark has not been attacked during transit.

(b) Median Filtering

The stegomarked images have been subjected to median filtering. The results obtained in terms of various subjective and objective parameters have been shown in Fig. 8 and Table 8. Average bit error rate of 40 % is observed which is large enough to get a highly distorted watermark at the receiver informing that StegNmark has been attacked during transit. The extracted watermarks for such an attack from various test images have been shown in Fig. 9.

(c) Low Pass Filtering

The stegomarked images have been subjected to Low pass filtering attack for filter Kernel of size (3 × 3). The average BER calculated after low pass filtering is more than 0.49 i.e. 49 %. The attacked stegomarked images have been shown in

Fig. 8 Stegomarked and then median filtered images

Table 8 Performance parameters of StegNmark system for median filtering

Image	Bit error rate (BER)	PSNR in dB	Watermark extraction time in seconds	Secret data extraction time in seconds
Lena	0.3980	35.3469	0.0624	9.1573
Peppers	0.4141	35.0668	0.0468	9.5941
Baboon	0.4205	25.0554	0.0468	9.4849
Boats	0.4094	30.9318	0.0624	9.3133
Bridge	0.4085	26.7260	0.0468	9.4225
Barbara	0.3966	25.4274	0.0468	9.4069
Average	0.4011	29.8213	0.052	9.3958

(a)	(b)	(c)	(d)	(e)	(f)

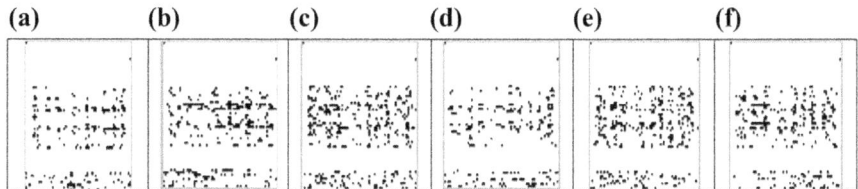

Fig. 9 Extracted watermarks from different median filtered images. **a** Lena. **b** Peppers. **c** Baboon. **d** Boats. **e** Bridge. **f** Barbara

Fig. 10 and image quality metrics reported in Table 9 while as extracted watermarks have been depicted in Fig. 11. It is evident that distorted watermarks are extracted at the receiving end thus enabling us to detect the tampering.

(d) Histogram Equalization

Table 10 shows the performance parameters of the used test images when subjected to the histogram equalization attack. The average BER as calculated is more than 50 %. Figure 12 shows the attacked stegomarked image while as Fig. 13 shows the extracted watermarks. It is evident that our system is also able to detect this attack.

(e) Additive White Gaussian Noise (AWGN)

The stegomarked images were subjected to Additive White Gaussian Noise (AWGN) of variance 0.0002. Table 11 shows the objective quality parameters of the extracted watermarks while as Fig. 14 shows stegomarked images due to this

Fig. 10 Stegomarked and then low pass filtered images

Table 9 Performance parameters of StegNmark system for low pass filtering

Image	Bit error rate (BER)	PSNR in dB	Watermark extraction time in seconds	Secret data extraction time in seconds
Lena	0.4986	31.9411	0.0624	9.1417
Peppers	0.4978	31.5296	0.0486	9.4225
Baboon	0.5000	24.5464	0.0468	9.3445
Boats	0.4993	28.8655	0.0312	9.3445
Bridge	0.5013	25.8400	0.0312	9.4693
Barbara	0.5002	25.0195	0.0624	9.5161
Average	0.4998	27.5231	0.0471	9.3685

(a)	(b)	(c)	(d)	(e)	(f)

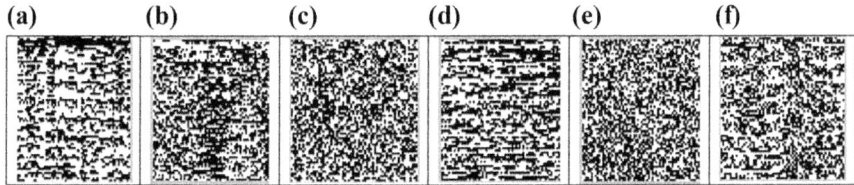

Fig. 11 Extracted watermarks from different low pass filtered images. **a** Lena. **b** Peppers. **c** Baboon. **d** Boats. **e** Bridge. **f** Barbara

Table 10 Performance parameters of StegNmark system for histogram equalization

Image	Bit error rate (BER)	PSNR in dB	Watermark extraction time in seconds	Secret data extraction time in seconds
Lena	0.5038	19.1171	0.0936	9.3757
Peppers	0.5022	20.5897	0.0936	9.7033
Baboon	0.4976	16.2418	0.0624	9.6253
Boats	0.5024	16.9374	0.0780	9.6672
Bridge	0.5029	19.3388	0.0936	9.5941
Barbara	0.4930	20.8493	0.1248	9.7033
Average	0.5003	18.8456	0.0910	9.6323

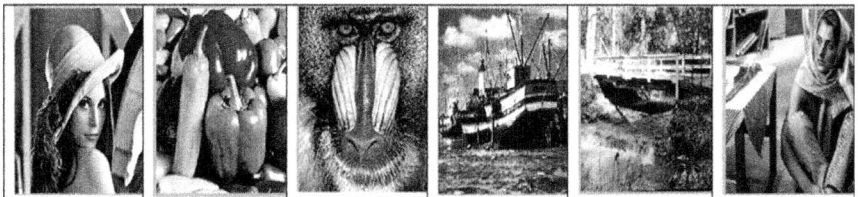

Fig. 12 StegNmarked and then histogram equalized images

(a)   (b)   (c)   (d)   (e)   (f)

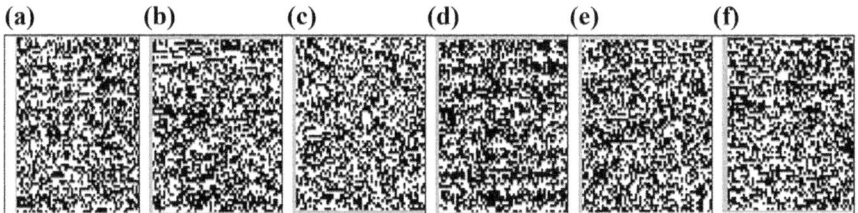

Fig. 13 Extracted watermarks from different stegomarked test images. **a** Lena. **b** Peppers. **c** Baboon. **d** Boats. **e** Bridge. **f** Barbara

Table 11 Performance parameters of StegNmark system for Gaussian noise ($v = 0.0002$)

Image	Bit error rate (BER)	PSNR in dB	Watermark extraction time in seconds	Secret data extraction time in seconds
Lena	0.5010	36.7153	0.0624	9.2197
Peppers	0.4994	36.7122	0.0468	9.5629
Baboon	0.5001	36.6897	0.0312	9.6721
Boats	0.5004	36.6925	0.0468	9.4537
Bridge	0.5009	36.7342	0.0468	9.4381
Barbara	0.4991	36.6964	0.0468	9.3913
Average	0.5001	36.7067	0.0488	9.4563

attack. The average PSNR of the attacked images is more than 36 dB while as BER of the extracted watermarks is almost 50 %. Figure 15 reveals that the proposed watermarking technique is fragile and hence one can detect the tamper caused due to this attack as well.

(f) Salt and Pepper Noise

The stegomarked images have been tested for Salt and Pepper noise, of noise density 0.1. The average BER as can be seen from the Table 12 is 0.05 (5 % error). Figure 17 shows the extracted watermarks after a carrying out this attack while as Fig. 16 shows Stegomarked test images (Fig. 17).

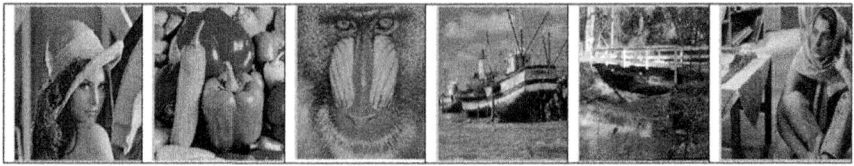

Fig. 14 Stegomarked images attacked by additive white Gaussian noise

(a)   (b)   (c)   (d)   (e)   (f)

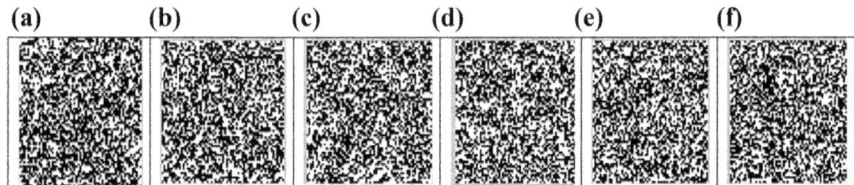

Fig. 15 Extracted watermarks from different stegomarked test images. **a** Lena. **b** Peppers. **c** Baboon. **d** Boats **e** Bridge. **f** Barbara

Table 12 Performance parameters of StegNmark system for Salt and Pepper (density = 0.1)

Image	Bit error rate (BER)	PSNR in dB	Watermark extraction time in seconds	Secret data extraction time in seconds
Lena	0.0502	25.4243	0.0624	9.1573
Peppers	0.0500	25.2373	0.0312	9.3289
Baboon	0.0503	25.6660	0.0468	9.4537
Boats	0.0502	25.5064	0.0468	9.4693
Bridge	0.0493	25.3004	0.0624	9.4069
Barbara	0.0503	25.2745	0.0468	9.3913
Average	0.0501	25.3456	0.0494	9.3738

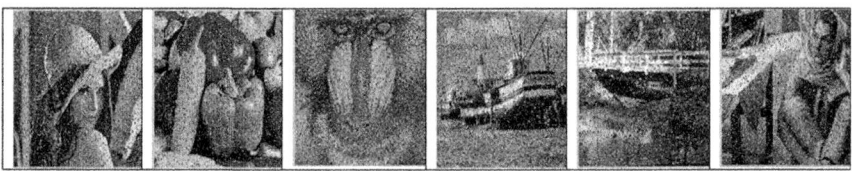

Fig. 16 Stegomarked images and then attacked by salt and pepper noise

(a)   (b)   (c)   (d)   (e)   (f)

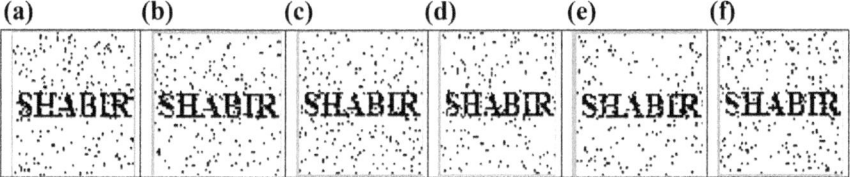

Fig. 17 Extracted watermarks from different stegomarked test images. **a** Lena. **b** Peppers. **c** Baboon. **d** Boats. **e** Bridge. **f** Barbara

Table 13 Performance parameters of Stegomarked system for JPEG (Q.F. = 90)

Image	Bit error rate (BER)	PSNR in dB	Watermark extraction time in seconds	Secret data extraction time in seconds
Lena	0.5001	40.3948	0.0780	9.1885
Peppers	0.4997	38.5491	0.0468	9.7969
Baboon	0.4991	37.0992	0.0624	9.5629
Boats	0.5015	38.8475	0.0468	9.4069
Bridge	0.5012	37.3631	0.0468	9.4225
Barbara	0.5002	39.8791	0.0624	9.5629
Average	0.5008	38.6888	0.0572	9.4890

(g) JPEG Compression

JPEG compression is one of the mostly used unintentional attacks on images. We have subjected the stegomarked images obtained from the proposed system to JPEG compression of quality factor of 90. The performance indices obtained have been listed in Table 13. It is seen that, even for high values of quality factor the scheme has a BER of about 50 % indicating the proposed scheme is fragile to this attack as well. Figure 18 shows compressed stegomarked images. Figure 19 which shows watermarks extracted from various compressed stegomarked images justifies fragility of the proposed algorithm.

(h) Rotation

The stegomarked images have been subjected to rotation with an aim to find if this tamper caused due to rotation could be detected by our scheme. Figure 20 shows various stegomarked images rotated by 5. From Table 14, it is clear that after 5 rotation the average BER is 0.50 which is large enough to detect that tampering has occurred during transit. Figure 20 shows rotated versions of various stego-marked images. While as Fig. 21 shows extracted watermarks. It is clear from the visual quality of extracted watermarks that our scheme is able to detect this attack as well.

Fig. 18 Stegomarked images and then compressed by JPEG

(a)   (b)   (c)   (d)   (e)   (f)

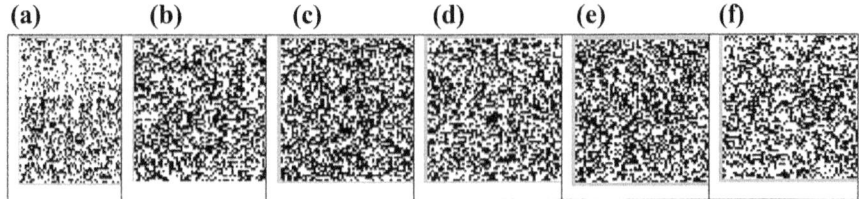

Fig. 19 Extracted watermarks from JPEG compressed images. **a** Lena. **b** Peppers. **c** Baboon. **d** Boats. **e** Bridge. **f** Barbara

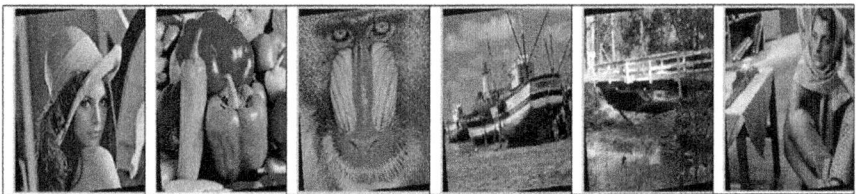

Fig. 20 Stegomarked images and then 5° rotated image

Table 14 Performance parameters of StegNmark system for 5 rotation

Image	Bit error rate (BER)	PSNR in dB	Watermark extraction time in seconds	Secret data extraction time in seconds
Lena	0.5012	14.6598	0.0312	9.6253
Peppers	0.5012	13.5546	0.0312	10.1245
Baboon	0.4981	15.6102	0.0468	10.608
Boats	0.5000	14.4022	0.0468	9.7189
Bridge	0.5014	13.5637	0.0468	10.3273
Barbara	0.4998	13.8485	0.0312	9.7189
Average	0.5005	14.2731	0.0390	10.0205

(a)   (b)   (c)   (d)   (e)   (f)

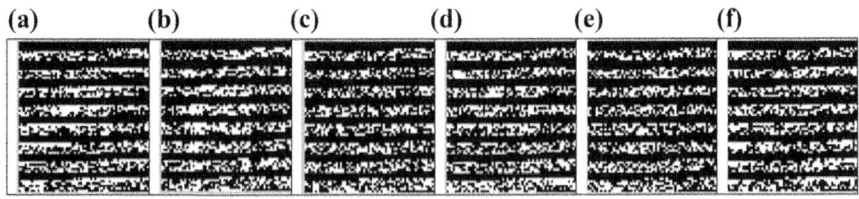

Fig. 21 Extracted watermarks from rotated images. **a** Lena. **b** Peppers. **c** Baboon. **d** Boats. **e** Bridge. **f** Barbara

Table 15 Performance parameters of StegNmark system for sharpening

Image	Bit error rate (BER)	PSNR in dB	Watermark extraction time in seconds	Secret data extraction time in seconds
Lena	0.4964	23.4502	0.1248	9.9529
Peppers	0.4955	22.4029	0.0936	9.8437
Baboon	0.4951	15.5887	0.0624	9.7184
Boats	0.4990	20.5193	0.0936	9.7813
Bridge	0.4722	17.0188	0.0936	9.7813
Barbara	0.4987	17.6173	0.0936	9.8905
Average	0.4928	19.4328	0.0936	9.8280

Fig. 22 Stegomarked images and then sharpened image

(i) Sharpening

We have subjected the stegomarked images to Sharpening attack. The image quality indices obtained have been reported in Table 15. Figure 22 shows stegomarked images subjected to sharpening. From Table 15 it is clear that after sharpening the stegomarked images the average BER is 0.49 which is high enough to detect the tamper. Figure 23 shows the distorted, extracted watermarks which clearly reveal the fragility and hence content authentication ability of the proposed algorithm.

It is evident from the results as presented in Tables 7, 8, 9, 10, 11, 12, 13, 14, and 15 that the proposed scheme is highly fragile to all the attacks carried over it. This fragility of watermark can be used for content authentication at the receiver. Further as observed in the quoted tables, the time required to extract watermark

(a)   **(b)**   **(c)**   **(d)**   **(e)**   **(f)**

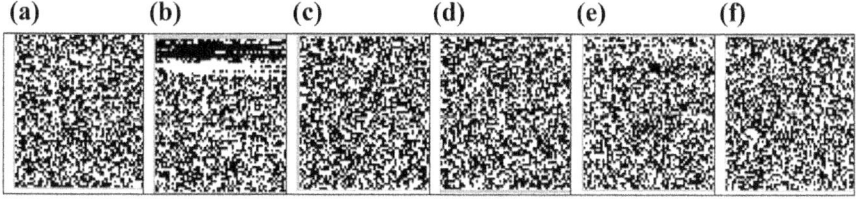

Fig. 23 Extracted watermarks from sharpened images. **a** Lena. **b** Peppers. **c** Baboon. **d** Boats. **e** Bridge. **f** Barbara

Table 16 Average extraction time for watermark (logo) and secret data for varying size of secret data

Secret data bits	Watermark bits	StegNmark ratio R = secret data bits/watermark bits	Watermark extraction time in seconds	Secret data extraction time in seconds
1024	4096	0.25	0.0580	0.0156
2048	4096	0.5	0.0587	0.0468
4096	4096	1.0	0.0597	0.0579
8192	4096	2.0	0.0680	0.1872
16384	4096	4.0	0.0580	0.4212
32768	4096	8.0	0.0578	0.8424
65536	4096	16.0	0.0588	2.2932
131072	4096	32.0	0.0528	5.9212
258048	4096	63	0.0597	9.5544

Watermark and secret data extraction times represent the average values calculated for six above mentioned test images used

from the stegomarked image is very small compared to the secret data extraction time. The average extraction time for a payload of 262,144 bits (4096 watermark bits and 258,048 secret data bits) comes out to be 0.0597 s for watermark and 9.5544 s for secret data. It is clear from the mentioned results that we can easily detect tamper early by extracting first watermark only. The decision about extraction of secret data watermark would strictly depend upon exactness of watermark extracted. Table 16 shows the average logo extraction time and data extraction for all the test images used for variable secret data and fixed logo size. It is obvious from the Table 16 that efficiency of our scheme is better for StegNmark ratio of greater than unity.

5 Advantages and Limitations of Proposes System

StegNmark, a joint stego and watermarking scheme has been put forward in this chapter. The proposed scheme is based on the well-known fact that watermark data is always less compared to stego-data. This is due to the fact that watermarking is mainly used for copyright protection or content authentication, while as main purpose of steganography is covert communication. The proposed system has been found to produce high-quality stegomarked images for a fairly high payload. Besides this, the system is capable of early tamper detection, thus ensuring that precious time is not lost in extraction of tampered data during a critical situation. Instead the system sends retransmission request once it detects the tamper. One of the disadvantages of scheme is that the hidden data is fragile to various signal processing and geometric attacks.

The scheme has been tested for general standard images only. The future work aims at extending it so that it can be used for electronic healthcare applications. Further we aim to implement this scheme on field-programmable gate array (FPGA) for developing a state-of-the-art real-time system.

6 Conclusion

An effective technique combining the features of digital image watermarking and steganography called as StegNmark has been presented in this chapter. The use of fragile watermark facilitates early tamper detection and informs the receiver that the secure data has been attacked during transit. This technique has been conceived based on the general fact related to data hiding that watermark data embedded in a cover image for content authentication is generally less compared to when data hiding is used for covert communications. The proposed technique has been tested for various test images and results compared with some existing state-of-the-art techniques. It has been observed that the proposed technique provides a better imperceptibility in addition to early tamper detection ability. Given the merits of the scheme, it can find huge application in critical applications pertaining to real-time national security issues.

References

1. Chamlawi R, Khan A, Usman I (2010) Authentication and recovery of images using multiple watermarks. Comput Electr Eng 36:578–584
2. Elshoura SM, Megherbi DB (2012) A secure high capacity full-gray-scale-level multi-image information hiding and secret image authentication scheme via Tchebichef moments. Signal Process: Image Commun. doi:10.1016/j.image.2012.12
3. Chang C, Chuang J (2002) An image intellectual property protection scheme for gray level images using visual secret sharing strategy. Pattern Recognit Lett 23:931–941
4. Dey N, Samanta S, Yang XS, Das A, Chaudhuri SS (2013) Optimisation of scaling factors in electrocardiogram signal watermarking using cuckoo search. Int J Bio-Inspired Comput 5(5):315–326
5. Lou DC, Liu JL (2002) Steganographic method for secure communications. Comput Secur 21:449–460
6. Wu M, Liu B (2004) Data hiding in binary image for authentication and annotation. IEEE Trans Multimed 6:528–538
7. Shabir AP, Javaid AS, Bhat GM (2015) Hiding in encrypted images: a three tier security data hiding system. Multidimens Syst Signal Process, Springer. doi:10.1007/s11045-015-0358-z
8. Cox I, Miller M, Bloom J (2001) Digital watermarking: principles and practice. Morgan Kaufman, Los Altos
9. Fan L, Gao T, Cao Y (2012) Improving the embedding efficiency of weight matrix-based steganography for grayscale images. Comput Electr Eng. doi:10.1016/j.compeleceng.2012.06.014

10. Dey N, Samanta S, Chakraborty S, Das A, Chaudhuri SS, Suri JS (2014) Firefly algorithm for optimization of scaling factors during embedding of manifold medical information: an application in ophthalmology imaging. J Med Imaging Health Inform 4(3):384–394

11. Cox I, (2009), Information hiding, watermarking and steganography. Public Seminar, Intelligent Systems Research Centre (ISRC), University of Ulster at Magee, Northern Ireland

12. Hussain F (2012) A survey of digital watermarking techniques for multimedia data. MIT Int J Electron Commun Eng 2:37–43

13. Chakraborty S, Maji P, Pal AK, Biswas D, Dey N (2014) Reversible color image watermarking using trigonometric functions. In: 2014 International conference on electronic systems, signal processing and computing technologies (ICESC), IEEE, pp 105–110

14. Hartung F, Kutter M (1999) Multimedia watermarking techniques. Proc IEEE 87(7):1079–1107

15. Shabir AP, Javaid AS, Bhat GM (2014) Data hiding in scrambled images: a new double layer security data hiding technique. Comput Electr Eng Elsevier 40(1):70–82

16. Dey N, Maji P, Das P, Biswas S, Das A, Chaudhuri SS (2013) An edge based blind watermarking technique of medical images without devalorizing diagnostic parameters. In: 2013 International conference on advances in technology and engineering (ICATE), IEEE, pp 1–5

17. Chopra D, Gupta P, Sanjay G, Gupta A (2012) LSB based digital image watermarking for grayscale image. IOSR J Comput Eng 6:36–41

18. Chakraborty S, Samanta S, Biswas D, Dey N, Chaudhuri SS (2013) Particle swarm optimization based parameter optimization technique in medical information hiding. In: 2013 IEEE international conference on computational intelligence and computing research (ICCIC), IEEE, pp 1–6

19. Bender W, Gruhl D, Morimoto N, Lu A (1996) Techniques for data hiding. IBM Syst J 35 (3&4):313–336

20. Acharjee S, Chakraborty S, Samanta S, Azar AT, Hassanien AE, Dey N (2014) Highly secured multilayered motion vector watermarking. Advanced machine learning technologies and applications. Springer, Berlin, pp 121–134

21. Shabir AP, Javaid AS, Nazir AL, Bhat GM (2016) Robust and blind watermarking technique in DCT domain using inter-block coefficient differencing. Digit Signal Process, Elsevier. doi:10.1016/j.dsp.2016.02.005

22. Shabir AP, Javaid AS, Farhana A, Nazir AL, Bhat GM (2015) Information hiding in medical images: a robust medical image watermarking system for E-healthcare. Multimed Tools Appl, Springer. doi:10.1007/s11042-015-3127-y

23. Schyndel RG, Tirkel A, Osborne CF (1994) A digital watermark. In: Proceedings of IEEE international conference on image processing, (ICIP'1994), pp 86–90

24. Caronni G (1995) Assuring ownership rights for digital images. In: Proceedings of the reliable IT systems, (VIS'95'), pp 251–263

25. Tanaka K, Nakamura Y, Matsui K (1990) Embedding secret information into a dithered multi-level image. In: IEEE Proceedings of the military communications conference'90, pp 216–220

26. Pitas I (1996) A method for signature casting on digital images. In: IEEE proceedings of the international conference on image processing, Lausanne, Switzerland, vol 3, pp 215–218

27. Wolfgang RB, Delp EJ (1996) A watermark for digital images. In: IEEE proceedings of the international conference on image processing, Lausanne, Switzerland, vol 3, pp 219–222

28. Bhat GM, Shabir AP, Javaid AS (2010) FPGA implementation of novel complex PN code generator based data scrambler and descrambler. Int J Sci Technol, Thailand 4(01):125–135

29. Shabir AP, Javaid AS, Mohiuddin GB (2012) High capacity data embedding using joint intermediate significant bit and least significant technique. Int J Inf Eng Appl 2(11), ISSN: 2224-5782(print) ISSN: 2225-0506 (online)

30. Wolfgang RB and Delp EJ (1999) Fragile watermarking using the VW2D watermark. In: Proceedings of electronic imaging, San Jose, CA, 25–27 Jan 1999, vol 3657, pp 204–213

31. Wu DC, Tsai WH (2003) A stegnographic method for images by pixel value differencing. Pattern Recognit Lett 24(9–10):1613–1626
32. Zang X, Wang S (2004) Vulnerability of pixel value differencing steganography to histogram analysis and modification for enhanced security. Pattern Recognit Lett 25:331–339
33. Hamood AK, Jalab HA, Kasirun ZM, Zidain BB, Zadain AA (2010) On capacity and security of stegnographic approaches: an overview. J Appl Sci 10:1825–1833
34. Qi K, Zang DF, Xie D (2010) A high capacity stegnographic scheme for 3D point cloud models. Inf Technol J 9:4121–4421
35. Ahmad AL, Kahiah ML, Zaidain BB, Zaidain AA (2010) A novel embedding method to increase capacity of low bit encoding audio steganography technique using noise gate software algorithm. J Appl Sci 10:59–64
36. Shabir AP, Javaid AS, Ghulam MB (2015) A secure and efficient spatial domain data hiding technique based on pixel adjustment. Am J Eng Technol Res, US Library congress, (USA), 14 (2):38–44, 2014
37. Parah SA, Sheikh JA, Bhat GM (2012) High capacity data embedding using joint intermediate significant bit and least significant technique. Int J Inf Eng Appl 2(11):1–12
38. Podilchuk C, Zeng W (1998) Image-adaptive watermarking using visual models. IEEE J Select Areas Commun 16:525–539
39. Wolfgang RB, Podilchuk CI, Delp EJ (1999) Perceptual watermarks for digital images and video. In: Proceedings of the IEEE, vol 7, pp 1108–1126
40. Swanson MD, Zhu B, Tewfik AH (1996) Transparent robust image watermarking. In: Proceedings of the international conference on image processing, Lausannae, Switzerland, pp 211–214
41. Jahne B (2002) Digital image processing, 5th edn. Springer 1991 revised and extended edition, New York
42. Gonzalez RC, Woods RE, Eddins SL (2009) Digital image processing using MATLAB, 2nd edn. Gatesmark Publishing, Knoxville
43. Shih FY (2008) Digital watermarking and steganography: fundamentals and techniques. CRC Press, Boca Raton
44. Kundur D, Hatzinakos D (1997) A robust digital image watermarking technique using wavelet based fusion. In Proceedings of the international conference on image processing, (ICIP) vol 1, pp 544–547
45. Wen-Jan C, Chin-Chen C, Hoang NTL (2010) High payload steganography mechanism using hybrid edge detector. Expert Syst Appl, Elsevier 37:3292–3301
46. Mashallah AD, Sajad N, Seyed EA (2013) A new image steganography method based on pixel neighbors and 6 most significant bit (MSB) compare. ACSIJ Adv Comput Sci, 2(5), No. 6, Nov 2013, ISSN: 2322–5157
47. Marghny HM, Naziha MA, Mohamed AB (2012), Data hiding technique based on LSB matching towards high imperceptibility. MIS Review National Chengchi University and Airiti Press Inc., vol 18, No 1, pp 57–69
48. Kuo WC, Chang JC, Wang CC (2011) Data hiding method with high embedding capacity character. Int J Image Process 6:310–317
49. Sun W, Zhe-Ming L, Yu-Chun W, Fa-Xin Y, Rong-Jun S (2013) High performance reversible data hiding for block truncation coding compressed images. SIViP, Springer 7:297–306. doi:10.1007/s11760-011-0238-4
50. Ou D, Sun W (2015) High payload image steganography with minimum distortion based on absolute moment block truncation coding. Multimed Tools Appl, Springer 74:9117–9139. doi:10.1007/s11042-014-2059-2
51. Ying-Hsuan H, Ching-Chun C, Yi-Hui C (2015) Hybrid secret hiding schemes based on absolute moment block truncation coding. Multimed Tools Appl, Springer. doi:10.1007/s11042-015-3208-y
52. Jain AK (2013) Fundamentals of digital image processing, I edn. Prentice-Hall, Upper Saddle River

53. Shabir AP, Javaid AS, Bhat GM (2012) On the realization of a secure, high capacity data embedding technique using joint top-down and down-top embedding approach. Elixir Comput Sci Eng 49:10141–10146
54. Shabir AP, Javaid AS, Bhat GM (2012) High capacity data embedding using joint intermediate significant bit and least significant technique. Int J Inf Eng Appl 2(11):1–11
55. Shabir AP, Javaid AS, Bhat GM (2013) On the realization of a spatial domain data hiding technique based on intermediate significant bit plane embedding (ISBPE) and post embedding pixel adjustment (PEPA). In: Proceedings of IEEE international conference on multimedia signal processing and communication technologies (IMPACT'2013) (AMU, Aligargh, 23–25 Nov 2013) pp 51–55

Adaptive Color Image Watermarking Scheme Using Weibull Distribution

V. Santhi

Abstract Digital watermarking is considered to be an important technique for protecting copyrights of digital images. In this paper, a new adaptive watermarking algorithm using Weibull distribution is proposed. In order to embed a digital watermark, the cover image is transformed into frequency components and it is modulated. This proposal uses discrete wavelet transformation and discrete cosines transformation techniques to obtain frequency components. The selected robust frequency components are modulated for embedding a watermark. In addition, the quality of watermarked images is preserved through adaptive calculation of scaling and embedding parameters. The efficiency of the proposed algorithm is tested through various attacks. The obtained similarity measure values of Gaussian low-pass filter, compression, and Poison attacks are 98, 96, and 98 %, respectively. The calculated values for other attacks are also tabulated and efficiency of the proposed algorithm is proved.

Keywords Digital watermarking · Discrete wavelet transform (DWT) · Discrete cosine transform (DCT) · Singular value decomposition (SVD) · Copyright protection · Adaptive watermarking · Weibull distribution

1 Introduction

Due to the advancements in technology and tremendous growth in Internet, the end users are facilitated to share huge amounts of data [1, 2]. Data to be shared over Internet include text, images, video sequences, and audio clips. During transmission, these multimedia data have also subjected to various malicious attacks either intentionally or unintentionally. In order to protect digital images from malicious attacks and unauthorized manipulation, a promising technique is required.

V. Santhi (✉)
School of Computer Science and Engineering, VIT University,
Vellore 632014, India
e-mail: vsanthinathan@gmail.com

© Springer International Publishing Switzerland 2017   453
N. Dey and V. Santhi (eds.), *Intelligent Techniques in Signal Processing
for Multimedia Security*, Studies in Computational Intelligence 660,
DOI 10.1007/978-3-319-44790-2_20

In particular, protecting copyrights of digital images becomes cumbersome. Digital watermarking is considered as promising techniques for protecting copyright of digital images. It is a technique which embeds a piece of secret data called digital watermark into cover data. The embedded data could be used to prove copyright holder, and thereby, it helps to avoid copyright violation and ownership misuse [3, 4]. The objective of digital watermarking is not only to protect copyrights of digital images but also to preserve the quality of underlying digital content of cover images. However, many content creators or owners are reluctant to incorporate copyright information in digital images before publishing it online. So it is extremely necessary to provide utmost security for digital content by embedding a piece of digital data. Digital Watermarking is of two types: visible and invisible. Since, the vulnerability of invisible watermark is usually quite high, a new adaptive invisible watermarking scheme is proposed in this chapter. Digital Watermark is embedded either in spatial domain or in frequency domain [5–7]. In spatial domain, pixel intensity values are modified to insert watermark whereas in transform domain, frequency components are modified. The most commonly used transformation techniques include discrete Fourier transform (DFT), discrete cosine transform (DCT), and discrete wavelet transform (DWT). The insertion of watermark in transform domain does not make major difference in intensity values and it is diminished to greater extent [8].

This proposal inserts imperceptible watermark using adaptively calculated scaling and embedding factors in transform domain. In order to increase the security level, a key is derived from cover image that could be used by owner exclusively.

The organization of the paper is as follows: Review of related works is given in Sect. 2. Section 3 presents preliminaries of basic techniques briefly. The description of proposed work is presented in Sect. 4. Section 5 discusses about performance evaluation of the proposed work. Concluding remarks are given in Sect. 6.

2 Review of Related Work

Invisible watermarking is also called imperceptible watermarking. In order to protect copyrights of the digital images, imperceptible watermarking can be employed to a greater extent. The early watermarking scheme which hides undetectable electronic watermark using least significant bit (LSB) manipulation is proposed in [9]. A scheme is proposed in [10] to embed robust label into digital images for copyright protection through bit modification method. In [11], invisible watermarking scheme in DCT domain is proposed. In this scheme, watermark is inserted using bidirectional approach. If the region is identified as robust region, then more number of bits are embedded by modifying significant coefficients. In [12], the noise sensitivity of each pixel is analyzed based on its local region and each bit of the watermark is spread with pseudo-noise sequence in spatial or DCT domain such that to keep the noise level below the threshold value. Discrete wavelet transform is another very useful way to analyze signal in frequency domain. The wavelet domain could be used to spread watermark across the whole image by exploiting the hierarchical nature of the wavelet representation of an image. Gaussian distributed random vector is considered as watermark and inserted in all high-pass bands as discussed in [13].

In [14], Podilchuk et al. have proposed a digital watermarking technique for images and video signals by exploiting human visual system to provide a transparent and invisible but robust watermark. In this approach, watermarking process is carried out not only based on the frequency response of the human eye but also the properties of the image itself. But the scaling is empirically selected to carry out the process. In [15], a watermarking scheme is proposed in which visually recognizable pattern is embedded into an image by modifying certain middle-frequency part of an image, which could then be proved to be robust to attacks. A novel technique to exploit the multiresolution property of the wavelet transform and to embed a circular self-similar watermark in the first and the second level details of sub band coefficients of wavelet decomposition is discussed in [16]. To provide robustness against the geometrical distortions such as scaling, rotation and translation, a local invariant feature of the image called scale-invariant feature transform (SIFT) feature technique is used for inserting a watermark [17, 18]. A Taylor series approximated locally optimum test (TLOT) detector based on the hidden Markov model (HMM) in the wavelet domain is used for embedding watermark in [19]. This kind of watermarking could be classified as informed watermarking.

In [20], Ali et al. have proposed DWT-SVD domain-based watermarking by adjusting the mutation factor and the crossover rate dynamically in order to insert watermark with a robust adaptive scaling factor. In this approach, authors have used self-adaptive differential evolution algorithm. Ali et al. [21] have applied a similar differential algorithm to identify the best multiple scaling factor to balance a trade-off between imperceptibility and robustness. In [22], Cai et al. introduced a new angle quantization index modulation method which is capable to disperse the interference to the watermark signal from one angle to more angles and thus make the watermark resistible to attacks. Recently, Soliman et al. have introduced a new SVD-based image watermarking technique using a combination of weighted quantum particle swarm optimization and a human visual model in [23]. This adaptive technique has shown the superiority over classical particle swarm optimization (PSO) technique and quantum-behaved particle swarm optimization (QPSO) technique.

Genetic algorithm has been used adaptively for finding the best suitable scaling factor or for inserting watermark in images as discussed in [24, 25]. In [24], Vahedi et al. have used wavelet transformation technique by incorporating HSV model for determining the maximum tolerable scaling factor. Similarly in [25], Horng et al. have combined DCT and SVD techniques with luminance masking and HVS model to insert watermark for protection of document images.

Based on the study made in similar area, it is observed that most of the existing watermarking algorithm uses empirically selected values as scaling parameter.

In addition, it is also found that the robustness test has not been carried out extensively on watermarked images. The above-mentioned issues are addressed in this chapter, and the obtained results are compared with the results of Horng et al.'s work and Vahedi et al.' work.

3 Preliminaries

3.1 Discrete Wavelet Transform (DWT)

It is a tool that convert signal from spatial domain to frequency domain. The major advantage of DWT over Fourier transform is that it provides temporal resolution contains both frequency and location information. Discrete wavelet transformation technique decomposes cover images into various frequency bands. Low-frequency band represents average information and middle-frequency band represents both vertical and horizontal details whereas high-frequency band represents both vertical and horizontal details together [26].

3.2 Discrete Cosine Transform (DCT)

It is also a mathematical tool used to convert signal from spatial domain to frequency domain. As DCT is very good in energy compaction, it represents energy of the entire image with few significant frequency components. These significant components are robust components and it is resistive to various attacks. The very first component is called DC component and it is called average signal. The other components are called AC components [27, 28]. The DCT technique is predominantly used in data compression application. The following equations comprise DCT tool. The signal is converted from spatial domain to frequency domain using (1). Similarly, inverse transforms are accomplished by (2):

$$C(u,v) = \frac{2}{\sqrt{MN}} \sum_{x=0}^{m-1} \sum_{y=0}^{n-1} f(x,y) \cos \frac{(2x+1)u\pi}{2M} \cos \frac{(2y+1)v\pi}{2N} \quad (1)$$

where $u = 1\ldots M - 1$ and $v = 1\ldots N - 1$

$$f(x,y) = \frac{2}{\sqrt{MN}} \sum_{x=0}^{m-1} \sum_{y=0}^{n-1} c(u,v) \cos \frac{(2x+1)u\pi}{2M} \cos \frac{(2y+1)v\pi}{2N} \quad (2)$$

3.3 Singular Value Decomposition

It is a mathematical tool which decomposes a given matrix $A \in R^{m \times n}$ into unitary matrices and a singular matrix. It is represented mathematically as shown in (3).

$$A = US(V^T),$$ (3)

where $U \in R^{m \times n}$, $V \in R^{m \times n}$ are the orthogonal matrices and $S \in R^{m \times n}$, is a singular matrix. The orthogonal matrices should satisfy the properties given in (4) and (5).

$$UU^T = U^T U = I_m$$ (4)

$$VV^T = V^T V = I_n$$ (5)

The singular matrix consists of nonzero elements along the diagonal called singular values and zero-valued elements elsewhere. These singular values are the representation of intrinsic algebraic image properties. If any modification is carried out in singular values, it makes a small difference in intensity values of images [21, 29, 30].

3.4 Weibull Cumulative Distribution

It is a continuous probability distribution, and it has a wide range of applications [31]. The expression for Weibull distribution is given in (6).

$$f(x) = \frac{\alpha}{\beta} \frac{(x-\alpha)^{\delta-1}}{\beta} e^{\frac{-(x-\alpha)}{\beta}}; \quad \alpha < x < \infty, \quad \beta > 0, \quad \delta > 0$$ (6)

Weibull distribution is governed by parameters such as location (α), shape $(\delta > 0)$, and scale $(\beta > 0)$. The shape (δ) and scale (β) parameters play a vital role in deciding the value of location parameter. The value of location parameter is negligible and it is assumed as zero in this work. The simplified version of Weibull cumulative distribution function is given (7).

$$F(x) = 1 - e^{\left(\frac{x-\alpha}{\beta}\right)^{\delta}}; \alpha < x < \infty, \quad \beta > 0, \quad \delta > 0$$ (7)

The graphical representations of Weibull cumulative distribution with various values of scale and shape parameters are shown in Fig. 1. The value of scale and shape parameter could be considered as a key for deriving scaling parameter. As the

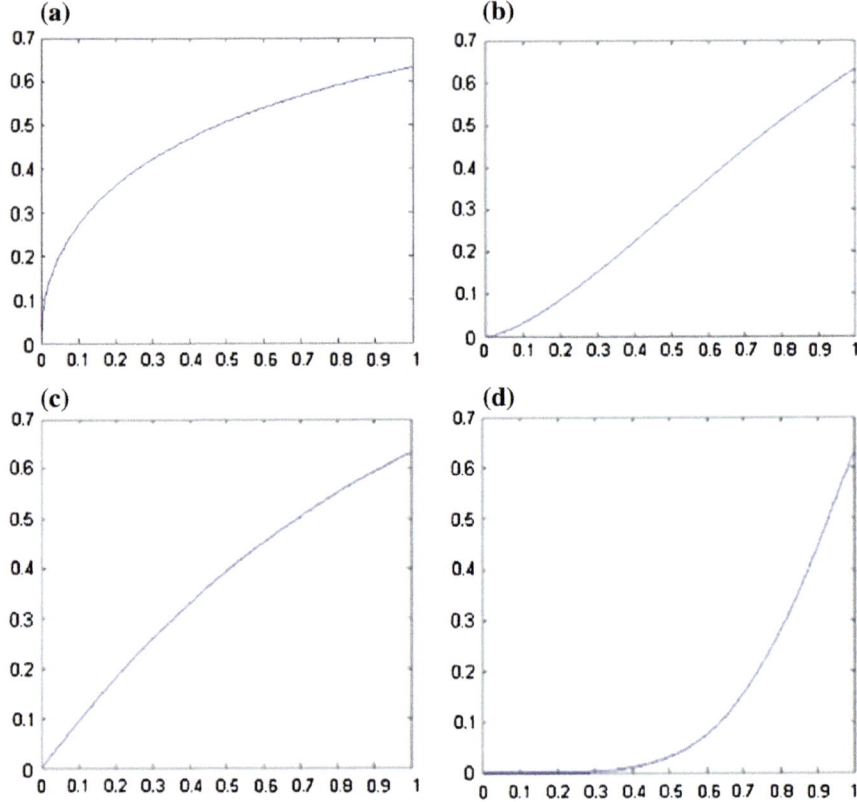

Fig. 1 Weibull cumulative distribution graphs for various scale and shape parameters. **a** Scale-1 shape-0.5, **b** scale-1 shape-1.5, **c** scale-1 shape-1, **d** scale-1 shape-5

key value is provided only with content creator, the security level of the proposed system is increased. The assumed values of parameters are $\beta = 1$ and $\delta = 1.5$. The value obtained by Weibull cumulative distribution with the above-selected parameters would not exceed 0.6321. Thus, it makes an appropriate choice for calculating scaling and embedding parameter for watermark insertion.

4 Proposed Work

In this section, a robust watermarking scheme is presented in transformation domain using SVD technique. The watermarking process consists of two modules, namely embedding and extraction. In order to insert a watermark in frequency components, the scaling and embedding parameters need to be calculated from the content of the cover image itself.

During embedding process, the cover image is converted from RGB color space into YIQ color space to extract luminance components Y. The component Y is decomposed into various frequency bands using DWT. The middle-frequency band is selected and block transformation technique is applied on each block of size 8×8. In each DCT-transformed block, there would be one DC component, all DC components are collected from various blocks, and DC matrix is constructed. This DC matrix is SVD decomposed to obtain singular values for inserting watermark. Similarly, the reverse procedure is applied to extract watermark from watermarked image. In the following section, the adaptive procedure for calculating scaling and embedding factors is presented.

4.1 Adaptive Calculation of Scaling and Embedding Parameters

The values of scaling and embedding parameters are derived from contents of cover images. In order to calculate scaling factor, the absolute values of middle-frequency components are stored as Z_{LH} and to bring scaling parameter (α) value between the range of $0 < \alpha < 1$, the normalization process is carried out on each element of Z_{LH} using (8), and the obtained results are stored in Z_1.

$$Z_1(i,j) = \frac{\text{element}(Z_{LH}) - \min(z_{LH})}{\max(z_{LH}) - \min(z_{LH})}, \tag{8}$$

where $i = 0, 1, 2,...N - 1$ and $j = 0, 1, 2, 3...M - 1$.

The obtained matrix is given as input to Weibull cumulative distribution by choosing appropriate shape and scale parameters, and it is shown in (9).

$$Z_2 = \text{Wblcdf}(Z_1, 1, 1.5) \tag{9}$$

The value of scale parameter β is set as 1, and the shape parameter δ is set as 1.5. The mean value of Z_2 is calculated using (10), and to avoid zero mean, parameter β_1 with minimum value of 0.001 is added as given in (11).

$$\mu_1 = \frac{1}{mn} \sum_{i=1}^{m} \sum_{j=1}^{n} Z_2(i,j) \tag{10}$$

$$\alpha_1 = \beta_1 + \mu_1 \tag{11}$$

The same procedure is applied to calculate the value of α_2 from other middle-frequency band of cover image. The normalization technique is applied to each element of the matrix Z_{HL} using (12) and normalized matrix Z_3 is obtained.

$$Z_3(i,j) = \frac{\text{element}(Z_{HL}) - \min(z_{HL})}{\max(z_{HL}) - \min(z_{HL})}, \tag{12}$$

where $i = 0, 1, 2, \ldots, N - 1$ and $j = 0, 1, 2, 3, \ldots .M - 1$.

The obtained matrix is given as an input to Weibull cumulative distribution by choosing appropriate shape and scale parameters and it is shown in (3).

$$Z_4 = \text{Wblcdf}(Z_3, 1, 1.5) \tag{13}$$

The mean value of Z_4 is calculated using (14), and to avoid zero mean, the parameter β_1 is added with minimum value of 0.001, as shown in (14).

$$\mu_2 = \frac{1}{mn} \sum_{i=1}^{m} \sum_{j=1}^{n} Z_4(i,j) \tag{14}$$

$$\alpha_2 = \beta_1 + \mu_2 \tag{15}$$

$$\alpha = \frac{1}{2}(\alpha_1 + \alpha_2) \tag{16}$$

The scaling factor is obtained by taking average of both α_1 and α_2 as given in (16), and embedding factor is calculated by subtracting α from 1.

4.2 Embedding Process

In this proposal, a color image of size 512×512 and a watermark of similar size are considered for testing it. Both images are converted into grayscale images using YIQ components [32]. The images in gray scale are converted into frequency components using discrete wavelet transformation technique. The selected frequency bands are DCT converted followed by SVD technique. The algorithm for watermark embedding process is given below:

Embedding Algorithm:	
Input : A, W	// A cover image, W watermark Image
Output: A"	//A" watermarked Image

1) $Read(A, W)$ — // Read Cover image and the Watermark image
2) $[Y1\ I1\ Q1] \leftarrow RGB2YIQ(A)$ — // Convert Cover image to YIQ components
3) $[Y2\ I2\ Q2] \leftarrow RGB2YIQ(W)$ — // Convert Watermark image to YIQ components
4) $[LL1\ LH1\ HL1\ HH1] \leftarrow DWT(Y1)$ — // Apply DWT to Y_1 component of cover image
5) $[LL2\ LH2\ HL2\ HH2] \leftarrow DWT(Y2)$ — // Apply DWT to Y_2 component of watermark
6) $[LL1'\ LH1'\ HL1'\ HH1'] \leftarrow DCT(LL1\ LH1\ HL1\ HH1)$ — // Apply DCT to each band of cover image
7) $[LL2'\ LH2'\ HL2'\ HH2'] \leftarrow DCT([LL2\ LH2\ HL2\ HH2])$ — //Apply DCT to each band of watermark image
8) $[\ U1\ S1\ V1T\ ] \leftarrow [LL1'\ LH1'\ HL1'\ HH1']$ — // Apply SVD to each band of cover image
9) $[\ U2\ S2\ V2T\ ] \leftarrow [LL2'\ LH2'\ HL2'\ HH2']$ — // Apply SVD to each band of watermark image
10) $Snew = S1 + \alpha S2$ — // Watermark insertion using key
11) $Y'' \leftarrow iDWT + iDCT + iSVD(Snew)$ — // Extract new Y" from the image
12) $A'' \leftarrow YIQ2RGB([Y''\ I1\ Q1])$ — // Reconstruct watermarked image Y"

The architectural diagram for watermark embedding process is shown in Fig. 2.

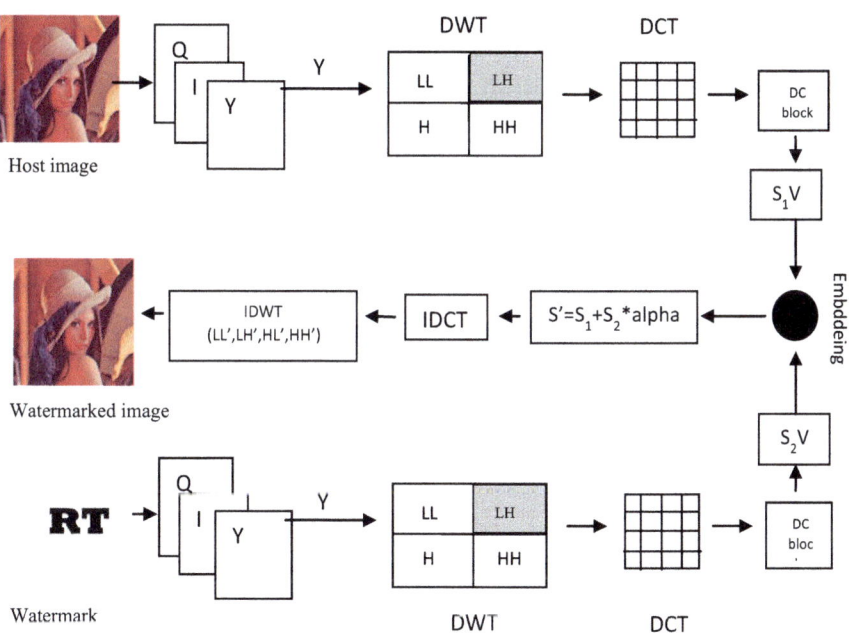

Fig. 2 Watermark embedding process

4.3 Watermark Extraction Process

During the watermark extraction process, both the watermarked image and original image are converted into grayscale images using YIQ color model. For the conversion of image into frequency components, DWT technique is applied. The algorithm for watermark extraction process is given below.

Extraction Algorithm:
Input : A'', A, W *//Watermarked Image, Original Image and watermark*
Output : W'' *// Extracted Watermark*

1) $Read(A'', A, W)$ // Read watermarked image, Cover image and Watermark
2) $[Yw\ Iw\ Qw] \leftarrow RGB2YIQ(A'')$ //Convert Watermarked image to YIQ components
3) $[LLw\ LHw\ HLw\ HHw] \leftarrow DWT(Yw)$ //Apply DWT to Y component of Watermarked image
4) $[LL'w\ LH'w\ HL'w\ HH'w] \leftarrow DCT([LLw\ LHw\ HLw\ HHw])$ //Apply DCT to each band
5) $[USw\ VwT] = [LL'w\ LH'w\ HL'w\ HH'w]$ // Apply SVD to each band
6) $S''w = (Snew - S1) / \alpha$ // Calculation for extracting Watermark
7) $Y''w = iDWT + iDCT + iSVD(S''w)$ // Extract Y''_w from the image
8) $W'' = YIQ2RGB([Y''w\ I2\ Q2])$ // Reconstruct Watermark using Y''_w

---

5 Performance Evaluation

The performance analysis of proposed algorithm is carried out with four color images, namely pepper, lake, Lena, and house images of size 512×512. The proposed algorithm is simulated and tested using a MATLAB tool of version R2013b. The test images and watermark used are shown in Fig. 3.

The scaling parameter is calculated using adaptive calculation procedure mentioned in Sect. 4.1. The calculated scaling factors are varying from 0.01 to 0.05 and are shown in Table 1.

(a)   **(b)**   **(c)**   **(d)**   **(e)**

Fig. 3 Host images and watermark. **a** Pepper, **b** Lake, **c** Lena, **d** House, **d** Watermark

Table 1 Calculated scaling factor

Images	Scaling factor
Pepper	0.01781
Lake	0.02366
House	0.02483
Lena	0.02529

In order to carry out the analysis, objective metrics peak signal-to-noise ratio (PSNR) and normalized correlation coefficient (NCC) are used. The metric PSNR calculates noise present in any image; similarly, NCC measures similarity between any two images [26, 27, 33]. The value of PSNR is calculated using (17) and (18).

$$MSE = \frac{1}{mn}\sum_{i=1}^{m}\sum_{j=1}^{n}[H(i,j) - H'(i,j)]^2 \tag{17}$$

where m and n represent the number of rows and columns of the cover image H, and MSE represents mean square error.

$$PSNR = 10\log 10\left(\frac{255^2}{MSE}\right) \tag{18}$$

The similarity measure is calculated through normalized correlation coefficient (NCC) [27, 33]. In this proposal, it is measured between original and extracted watermark as shown in [19].

$$NCC = \frac{\sum_{i=1}^{m}\sum_{j=1}^{n}W(i,j)W'(i,j)}{\sqrt{\sum_{i=1}^{m}\sum_{j=1}^{n}W(i,j)}\sqrt{\sum_{i=1}^{m}\sum_{j=1}^{n}W'(i,j)}} \tag{19}$$

If the obtained value of NCC is 1 which shows maximum similarly and value 0 indicates that there is no similarity between images. The calculated PSNR values for various test images and the corresponding watermarked images are shown in Table 2.

In order to test the robustness of the watermark, various attacks are implemented on watermarked images. The attacks considered for testing are noise addition, sharpening, median filtering, compression, cropping, Gaussian blur, resizing,

Table 2 Measured PSNR values of watermarked images

Name of watermarked images	PSNR in dB
Pepper	35.4678853
Lake	32.9375932
House	32.52741559
Lena	32.40484709

rotation, and intensity adjustments. The extracted watermark images from various test images are shown in Fig. 4 with calculated NCC values. Similarly, obtained PSNR values and NCC values after salt and pepper noise attack are shown Fig. 5.

The experimental results of salt and pepper attack and Gaussian attack are shown in Tables 3 and 4, respectively.

The strength of the algorithm can easily be visualized through the obtained NCC values after implementing various attacks. It is observed that the similarity of the

(a)   **(b)**   **(c)**   **(d)**

Fig. 4 Extracted watermark from various test images. **a** NCC = 0.993, **b** NCC = 0.991, **c** NCC = 0.996, **d** NCC = 0.995

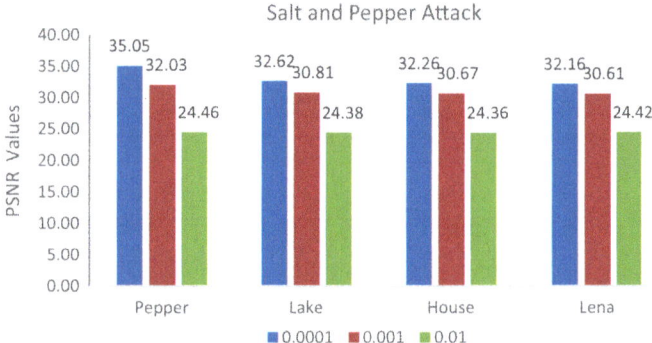

Fig. 5 Calculated PSNR values after salt and pepper noise attack

Table 3 Extracted watermark and its NCC value for various salt and pepper noise values using Lena image	Salt and pepper noise	Extracted watermark	NCC
	0.0001	RT	0.994844
	0.001	RT	0.980098
	0.01	RT	0.946304

Table 4 Extracted watermark and its NCC value for various Gaussian noise attack using Lena image

Gaussian noise	Extracted watermark	NCC value
0.0001		0.95018
0.001		0.88902
0.01		0.48071

extracted watermark and the original watermark is varying from 75 to 100 % for Gaussian low-pass filter, cropping, resize, JPEG compression, Poisson noise attacks. But for rotation attack, it is only 19–45 %. The obtained results and extracted watermark are shown in Table 5.

Table 5 Extracted watermark and calculated NCC values after various attacks using Lena image

Attack	Watermarked image	Extracted watermark	NCC
Sharpening			0.937501
Gaussian blur (SD: 1, size: 5 × 5)			0.986329
Cropping (512 × 30)			0.9879
Cropping (50 × 50)			0.9404
Rotation (2°)			0.372490
Resize (70 %)			0.9340983

(continued)

Table 5 (continued)

Attack	Watermarked image	Extracted watermark	NCC
Median filtering		RT	0.9878491
Compression (60 %)		RT	0.9637
Poisson attack		RT	0.9731926
Intensity adjustment			0.7399688
Speckle			0.9677

Table 6 Comparison of NCC values for Lena image with existing works

Attack type	Parameter chosen	Vahedi et al.'s work	Horng et al.'s work	Proposed algorithm
Gaussian noise	Mean = 0, variance = 0.5	–	0.6041	0.5623
	Mean = 0, variance = 1	–	0.6363	0.5227
Salt and pepper noise	0.01	–	0.6119	0.9356
	0.05	0.9149	–	0.8745
	0.08	–	0.5532	0.8489
	0.15	–	0.5451	0.7839
	0.1	–	0.5499	0.8324
Median filtering	2 × 2	–	0.7277	0.9676
	3 × 3	0.9785	0.7709	0.9878
	5 × 5	0.9029	0.8302	0.9743
Mean filtering	2 × 2	–	0.6621	0.9763
	3 × 3	0.6577	0.5826	0.9874
	5 × 5	0.6577	0.5627	0.9824
JPEG compression	10	–	0.6875	0.9643
	30	–	0.6741	0.9677
	50	–	0.6906	0.9750
Rotation	1°	0.8757	–	0.9624
Scaling	0.8	0.9478	–	0.9644
	1.5	0.9046	–	0.9705

The obtained results of the proposed algorithm are compared with existing algorithm and are shown in Table 6.

In order to prove the superiority of the proposed algorithm, the obtained results are compared with results of Vahedi et al.' s work [24] and Horng et al.'s work [25]. It is observed from the comparison Table 6 that the proposed algorithm has proved its efficiency through higher values of NCC.

6   Conclusion

A new adaptive invisible watermarking algorithm in transform domain using Weibull distribution function is proposed in this paper. The major contribution of the proposed method is that it uses adaptive model for calculating the scaling factor. The obtained results of proposed invisible watermarking algorithm are compared with the results of Vahedi et al.'s work and Horng et al.'s work. It is proved that the performance of the proposed adaptive invisible watermarking scheme is better than the results of above-said works. Thus, the performance of proposed adaptive watermarking model for invisible watermarking scheme confirmed its efficiency through experimental analysis.

References

1. Mistry D (2010) Comparison of digital water marking methods. Int J Comput Sci Eng 2(09):2905–2909
2. Yeung MM (1998) Digital watermarking. Commun ACM 41(7):31–33
3. Kwok SH (2003) Watermark-based copyright protection system security. Commun ACM 46(10):98–101
4. Su JK, Hartung F, Girod B (1998) Digital watermarking of text, image, and video documents. Comput Graph 22(6):687–695
5. Zhao J, Koch E (1998) A generic digital watermarking model. Comput Graph 22(4):397–403
6. Wang B, Ding J, Wen Q, Liao X, Liu C (2009) An image watermarking algorithm based on DWT DCT and SVD. In: IEEE international conference on network infrastructure and digital content, pp 1034–1038
7. Yang W, Zhao X (2011) A digital watermarking algorithm using singular value decomposition in wavelet domain. In: International conference on multimedia technology, pp 2829–2832
8. Barni M, Bartolini F, Cappellini V, Piva A (1998) Copyright protection of digital images by embedded unperceivable marks. Image Vis Comput 16(12):897–906
9. Van Schyndel RG, Tirkel AZ, Osborne CF (1994) A digital watermark. In: Proceedings of IEEE international conference on image processing, pp 86–90
10. Zhao J, Koch E (1995) Embedding robust labels into images for copyright protection. In: Proceedings of the conference on intellectual property rights and new technologies, pp 242–251
11. Ruanaidh JÓ, Dowling WJ, Boland FM (1996) Watermarking digital images for copyright protection. In: IEEE proceedings on vision, image and signal processing, vol 143, issue 4, pp 250–256

12. Kankanhalli MS, Ramakrishnan KR (1998) Content based watermarking of images. In: Proceedings of the sixth ACM international conference on multimedia, pp 61–70
13. Zhu W, Xiong Z, Zhang YQ (1999) Multiresolution watermarking for images and video. IEEE Trans Circuits Syst Video Technol 9(4):545–550
14. Podilchuk CI, Zeng W (1998) Image-adaptive watermarking using visual models. IEEE J Sel Areas Commun 16(4):525–539
15. Hsu CT, Wu JL (1999) Hidden digital watermarks in images. IEEE Trans Image Process 8 (1):58–68
16. Tsekeridou S, Pitas I (2000) Embedding self-similar watermarks in the wavelet domain. IEEE international conference on acoustics, speech, and signal processing, vol 6, Istanbul, pp 1967–1970
17. Lee HY, Kim H, Lee HK (2006) Robust image watermarking using local invariant features. Opt Eng 45(3):037002:1–037002:11
18. Lin YT, Huang CY, Lee GC (2011) Rotation, scaling, and translation resilient watermarking for images. IET Image Proc 5(4):328–340
19. Wang C, Ni J, Huang J (2012) An informed watermarking scheme using hidden markov model in the wavelet domain. IEEE Trans Inf Forensics Secur 7(3):853–867
20. Ali M, Ahn CW (2014) An optimized watermarking technique based on self-adaptive DE in DWT–SVD transform domain. Sig Process 94:545–556
21. Ali M, Ahn CW, Pant M (2014) A robust image watermarking technique using SVD and differential evolution in DCT domain. Optik Int J Light Electron Opt 125(1):428–434
22. Cai N, Zhu N, Weng S, Ling BWK (2015) Difference angle quantization index modulation scheme for image watermarking. Sig Process Image Commun 34:52–60
23. Soliman MM, Hassanien AE, Onsi HM (2015) An adaptive watermarking approach based on weighted quantum particle swarm optimization. Neural Comput Appl 1–13
24. Vahedi E, Zoroofi RA, Shiva M (2012) Toward a new wavelet-based watermarking approach for color images using bio-inspired optimization principles. Digit Signal Proc 22(1):153–162
25. Horng SJ, Rosiyadi D et al (2014) An adaptive watermarking scheme for e-government document images. Multimed Tools Appl 72(3):3085–3103
26. Dubolia R, Singh R, Bhadoria SS, Gupta R (2011) Digital image watermarking by using discrete wavelet transform and discrete cosine transform and comparison based on PSNR. In: International conference on communication systems and network technologies, pp 593–596
27. Santhi V, Thangavelu DA (2011) DC coefficients based watermarking technique for color images using singular value decomposition. Int J Comput Electr Eng 3(1):1793–8163
28. Lusson F, Bailey K, Leeney M, Curran K (2013) A novel approach to digital watermarking, exploiting colour spaces. Sig Process 93(5):1268–1294
29. Liu RZ, Tan TN (2001) SVD based digital watermarking method. Acta Electron Sin 29(2):168–171
30. Run RS, Horng SJ, Lai JL et al (2012) An improved SVD-based watermarking technique for copyright protection. Expert Syst Appl 39(1):673–689
31. Asgharzadeh A, Valiollahi R, Raqab MZ (2011) Stress-strength reliability of Weibull distribution based on progressively censored samples. SORT 35(2):103–124
32. Liu Z, Liu C (2008) Fusion of the complementary Discrete Cosine Features in the YIQ color space for face recognition. Comput Vis Image Underst 111(3):249–262
33. Su Q, Niu Y, Zou H, Liu X (2013) A blind dual color images watermarking based on singular value decomposition. Appl Math Comput 219(16):8455–8466

Multi-fingerprint Unimodel-based Biometric Authentication Supporting Cloud Computing

P. Rajeswari, S. Viswanadha Raju, Amira S. Ashour and Nilanjan Dey

Abstract Cloud computing is one of the emerging technologies that transfers network users to the next level. Security is one of the critical challenges faced by cloud computing. Biometrics proves its efficiency to achieve secured authentication. A new attribute should be created to handle authentication information in the infrastructure. The proposed system model presented a new idea for biometric security system based on fingerprint recognition. It automated the verification method to match between two human fingerprints, where fingerprints are considered a commonly used biometrics to identify an individual and to verify their identity. The proposed system presented a new model of a security system, where the users were asked to provide multiple [two] biometric fingerprints during the registration for a service. These templates are stored at the cloud providers' end. The users are authenticated based on these fingerprint templates which have to be provided in the order of random numbers that are generated every time. Both fingerprint templates and images were provided whenever encrypted for enhanced security. The proposed multi-fingerprint system achieved superior accuracy of 98 % compared to the single-fingerprint and manual-based attendance management systems.

P. Rajeswari
Department of CSE, KL University, Guntur, India
e-mail: rajilikhitha@gmail.com

S. Viswanadha Raju
Department of CSE, JNTUH, Hyderabad, India
e-mail: svraju.jntu@gmail.com

A.S. Ashour
Department of Electronics and Electrical Communications Engineering,
Faculty of Engineering, Tanta University, Tanta, Egypt
e-mail: amirasashour@yahoo.com

N. Dey (✉)
Department of Information Technology, Techno India College of Technology,
Kolkata, India
e-mail: neelanjan.dey@gmail.com

© Springer International Publishing Switzerland 2017
N. Dey and V. Santhi (eds.), *Intelligent Techniques in Signal Processing for Multimedia Security*, Studies in Computational Intelligence 660,
DOI 10.1007/978-3-319-44790-2_21

Keywords Cloud computing · Biometrics · Encryption · Decryption · Fingerprints · Authentication

1 Introduction

Cloud computing is the most promising and evolving network trend that provides opportunities to use infrastructure, application, or hardware as a service. Several organizations consider cloud computing to be secure, cost-effective, and suitable for their needs for sharing distributed resource services with the aid of Internet [1]. Multitenancy, flexibility, massive scalability, reliability, and elasticity are the basic characteristics of cloud technology [2]. However, the cloud computations are operational on unfixed nodes in a network. Operations are often carried out without trusted nodes, leading to data loss and privacy issues. The main challenges for the cloud computing include the security, data privacy, and the personal data protection from the third party. Consequently, cloud computing is predominantly interesting with the biometric recognition to achieve scalability, accessibility, and availability.

Biometric technology is involved in a great number of applications including security, appropriate identity identification, access control, authentication, and surveillance [3–9]. Furthermore, it is used in businesses and governments worldwide, such as in justice and law enforcement, financial, and transactional control. Biometrics refers to an authentication scheme that measures automatically unique physical/human characteristics, such as fingerprint, voice, ear, eye color, iris, retina, and palmprint. Biometric security stores unique biological characteristics in digital form and then uses this template for matching. These inherent attributes cannot be definitely shared and are unique biometric attributes as its person has its own.

One of the significant biometrics is the fingerprint as it is unique and consistent over the time. Thus, it can be used for identification and verification over a century. Fingerprint recognition has forensic applications such as criminal investigation and missing children as well as governmental applications including social security, border control, driving license, and passport control. Moreover, it can be applied in commercial applications such as e-commerce, Internet access, automated teller machine (ATM), and credit card [10, 11]. The process of fingerprint recognition is becoming automated and results in many automatic fingerprint identification systems (AFIS). Fingerprint usually appears as a series of dark lines that represent the ridges, while the valleys between these ridges appear as white space.

Using more than one finger provides high reliability. The number of the used fingerprints is directly proportion to the reliability of the biometric system. The main advantage is no extra hardware is required; however, more memory is required. Meanwhile, the memory problem is insignificant as the storage devices' cost is cheap. Subsequently, the main contribution of the current work is to solve the security issues of cloud computing based on two-fingerprint recognition with supporting proof of private matching identification to resolve security issues of cloud computing's credibility efficiently. Since, using more than one finger provides high reliability, where this number is directly proportional to the biometric system reliability. The present work proposed a novel security model based on multiple fingerprints combined with encryption and random numbers as authentication tools. The users are authenticated based on their fingerprint templates which have to be provided in the order of random numbers that are generated every time. Both fingerprint templates and images are provided consistently and are encrypted for enhanced security. The authenticated user registers with two fingerprints, assigning single-digit values for each of the two fingers. Finger impressions are provided in the order of random number generated by a random number generator. The access to the user is granted only if the fingerprint matches the user-provided fingerprints with the random numbers generated by a random number generator.

Since, encryption is the conversion of data into a form that cannot be easily understood by unauthorized people. Decryption is the process of converting encrypted data back into its original form. For enhanced security, the biometric images from both user's and service provider's end can be encrypted. There are number of encryption algorithms that are used for the fingerprint images. One such algorithm is elliptic encryption algorithm that is adopted in this paper.

The structure of the remaining sections is as follows. Section 2 included the related literatures to the cloud computing and the biometric approaches. Section 3 represented the methodologies used through the current work followed by the proposed system in Sect. 4. In addition, the results along with the discussion are included in Sect. 4. Finally, the conclusion is conducted in Sect. 5.

2 Related Work

Biometrics refers to the use of unique physiological characteristics to identify an individual. A number of biometric traits have been developed and are used to authenticate the person's identity. The idea is to use the special characteristics of a person to identify an individual. Biometric authentication uses human traits like fingerprints, tongue impressions, iris, and face recognitions that are unique to each individual and thus differentiating the users. Combining biometric techniques and cloud computing for the purpose of a secure cloud computation is a new technology. Since, the operations are carried out beyond trusted boundaries; cloud is more vulnerable to hacking and security breaches. When using biometrics as an authentication tool, a cloud user, during enrollment for a service, registers with the unique traits (fingerprints, tongue, face, and iris). These get stored as templates at the cloud service provider's end. Whenever, an access is required, the cloud user is prompted to provide the enrolled trait that is compared against the template and authenticated accordingly. Though these biometric traits are unique, problems arise when eavesdroppers gain access to these stored fingerprint templates.

Cryptography is one of the most active ways for enhancing information by using biometric features to encrypt the data. Tiwari and Saklani [12] proposed a biometric-based encryption and decryption system as many random keys were generated using pseudorandom generator. The biometric key designed from the sender's and the receiver's fingerprints can neither be forgotten nor shared. This approach provided a novel way for authentication in different approaches.

Sabri et al. [13] proposed a scheme using bio-hash function for biometric security to improve the security performance in cloud as per various perspectives of cloud customers. The experiments were performed using the CASIA benchmark of the fingerprint-V5 datasets. It was established that using bio-hash function was more efficient for protecting the biometric template compared to cryptobiometric authentication technique, and the achieved error rate was minimized by 25 %.

Liang et al. [14] proposed a biometric fingerprint encryption approach based on threshold. This approach did not require template storage, and its biometric key was relatively long. Omar et al. [15] suggested a biometric encryption technique to enhance the confidentiality in cloud computing for biometric data. The authors discussed the cloud computing virtualization as well as the biometric encryption. Masala et al. [16] designed a novel cloud platform to support an elementary Web applications shared by companies. The proposed platform guaranteed secure access for multiple users related to different companies. The user authentication was based on a biometric approach through integrated face and finger modalities.

On space-limited tokens such as driver license and smart cards, Khan et al. [17] proposed an efficient multimodal fingerprint and face biometrics authentication system. Fingerprint templates were encrypted and encoded/embedded into the face images. The experimental results established that the accuracy based on the decoded images was similar to that with the original images. The proposed approach proved its efficiency without degrading the overall decoding and matching performance of the biometric system. Moreover, the proposed scheme can be applied to cope with the unimodal biometric systems' problems. Emersic et al. [18] evaluated various schemes that combined existing cloud-based uni-modal biometric into a multi-biometric cloud service. From the perspective of training complexity, performance gains, and resource consumption, several fusion strategies were analyzed. The evaluation results were conducted based on two biometric cloud services, namely the fingerprint recognition service and the face recognition service. The authors concluded that existing biometric cloud services can be integrated with minimal effort and guarantee improved security for potential client applications.

Combining multi-input data proves to be a very promising trend, both in experiments and in real-life biometric authentication applications. Multimodal biometric systems can overcome some of the limitations of unimodal systems. The problem of non-universality is addressed since multiple traits can ensure sufficient population coverage. Also, multimodal biometric systems make it difficult for an intruder to simultaneously spoof the multiple biometric traits of a registered user. From the preceding literatures, it is concluded that using multimodal biometrics provided efficient security performance for the cloud computing services.

3   Methodology

A general approach in a biometric system is to store all captured biometric images in the enrollment phase, and then the authentication is performed using a matching process. However, this scheme suffers from security weaknesses [19]. Typically, security is interested in overcoming nonstandard formats and missing data to provide information, where the device system is agnostic. In addition, the security techniques should be flexible with the biometric matchers, implement anti-spoofing technologies, provide flexibility in expected response, and search speed. Vulnerable storage may lead to an attacker stealing biometric templates and impersonating the legitimate user. The stolen biometric information may compromise other systems [20]. A cloud private matching algorithm is proposed in [21]. Two encrypted images are compared under double encrypted conditions, from the client and from cloud storage. Several techniques have been proposed for biometric template protection, such as the cancellable biometrics [22]. This approach satisfies: (i) un-recoverability of the original biometric data from the stored biometric template and (ii) the issue of a new biometric template when an existing template is compromised.

In the current work, three phases are involved in the proposed new model, namely (i) the registration phase that enrolls with the two fingerprints, assigning single-digit values for each of two fingers; (ii) the access phase, where the finger impressions are provided in order of random number generated by a random number generator; and (iii) the matching phase that matches the user-provided fingerprints with the random numbers generated by a random number generator and encrypted image stored at the database as illustrated in Fig. 1.

A detailed description for the proposed phases in Fig. 1 is as follows.

3.1   Registration Phase

The registration phase is required by any user who enrolls for a service as illustrated in Fig. 2. The registration is performed using any two finger traits.

Afterward, the user assigns any selectable two single-digit numbers. All the two inputs (fingerprint images), two single-digit numbers, and the mapping of numbers to fingers are all encrypted and stored at the service provider's end as demonstrated in Fig. 3.

Fig. 1 Proposed model phases

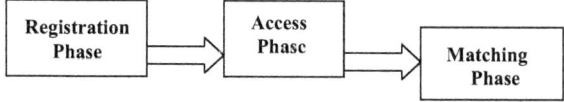

Fig. 2 User registering the biometric traits with the cloud service provider

(a)

(b)

Fig. 3 Inputs given by the user stored at the cloud, where **a** inputs from the user, and **b** cloud service provider

Since, the biometric encryption refers to the process of secure key management to complement existing cipher systems. Thus, it does not directly provide a mechanism for the encryption/decryption of data; nevertheless, it provides a replacement to typical passcode key-protection protocols. Although, the biometric encryption process can be applied to any biometric image, the initial implementation in the current system is achieved using fingerprint images. Additionally, instead of comparing the biometric data directly, a key is derived from these data, and subsequently knowledge of this key is proved. The goal of a biometric encryption system is to embed a secret into a biometric template in a way that can only be decrypted with a biometric image from the enrolled person. Cryptography plays an important task in accomplishing information security. It has many public key cryptography schemes that for encrypting or signing data at the source before transmission, and then decrypting or validating the signature of the received message at the destination. The potential for using the discrete logarithm problem in public-key cryptosystems has been recognized in [23]. There are several public key cryptographies, such as RSA, El-Gamal, and elliptic curve cryptography. The elliptic curve cryptography (ECC) is considered to be more suitable for limited resources applications than other public key cryptography algorithms because of its small key size. Therefore, ECC was chosen in this work because of its advantages over other public key cryptographies. Generally, a random generator is used to produce private keys and elliptic curve cryptography domain parameters. It uses a randomly generated seed to produce the random number, where the cryptanalysts may exploit it. The elliptic curve is considered an effective cryptosystem [24, 25]. Nowadays, the ECC is adopted by many standardizing institutes, such as the ANSI [26–29].

Consequently, in the proposed system, multiple inputs are involved to provide efficient high security system. This new proposed model applied encryption algorithms for the two fingerprint inputs, namely the elliptic curve algorithm for the biometric images and the Rivest–Shamir–Adleman (RSA) algorithm for the numbers and mappings.

3.1.1 Elliptic Curve Cryptography

The elliptic curve cryptography (ECC) is the algorithm used for encrypting the biometric traits. The steps for the elliptic encryption algorithm are as follows [30–35].

Elliptic Encryption Algorithm

The user will do the following:
    Selects an elliptic curve $Ep(a,b)$, $y_2=x_3+ax+b$ *(modp)*
    Obtain a point on the elliptic curve as point G
    Selects a private key k
    Generates public key $K=KG$
    Sends $Ep(a,b)$ and point K,G to B
When cloud receives the information, it will do the following:
    Encoded to be transmitted to the point M on $Ep(a,b)$
    Generates a random integer $r(r<n)$
    Calculates points $C_1= M+ rK$; $C_2=r$
    Passes C_1 and C_2 to the user
After receiving the information, the user will:
    Calculates C_1-kC_2
    Output result is the point M

Since, $C_1 - kC_2 = M + rrKk(rG) = M + rK - r(rkG) = M$, then point M can be explicitly decoded.

3.1.2 The Rivest-Shamir-Adleman Algorithm

Typically, the biometric cryptosystems combine both the cryptography and biometrics to gain the strengths of both methods. Cryptography provides high and adjustable security levels, while the biometrics provides non-repudiation without the necessity to remember the passwords or to carry tokens. In biometric cryptosystems, a cryptographic public key is generated from the biometric template of a user. This key is stored in the database with the necessity to a biometric authentication to be revealed successfully.

Subsequently, the RSA algorithm for public-key cryptography that is based on the presumed difficulty of factoring large integers is used. A user of the RSA creates and then publishes the product of two large prime numbers along with an auxiliary value as their public key. The prime factors must be kept secret. Anyone can use the public key to encrypt a message; however, if the public key is large enough, it will be encrypted only with knowledge of the prime factors that can feasibly decode the message. Generally, the RSA algorithm involves three steps, namely key generation, encryption, and decryption and is explained as follows.

Two-digit chosen numbers are encrypted using RSA algorithm. The RSA involves a public key and a private key, where the public key can be known to everyone for encrypting the messages. Messages encrypted with the public key can only be decrypted in a reasonable amount of time using the private key. The keys for the RSA algorithm are generated as follows [36–39]:

Key Generation for the RSA Algorithm

Choose two distinct prime numbers p and q

Find the modulus $n = p * q$

Find the totient $\phi(n)=(p\text{-}1)*(q\text{-}1)$

Choose e such that $1 <e<\phi(n)$, and $gcd(\phi(n),e)=1$

Determine d(using modular arithmetic) such that $d =e\text{-}1$ (mod $\phi(n)$)

Obtain the Public Key=$[e,n]$ and the Private Key=$[d,n]$

Furthermore, d can be picked such that $de - 1$ can be evenly divided by $(p - 1)$ $(q - 1)$. This is often computed using the extended Euclidean algorithm, since e and $\phi(n)$ are relatively prime and d is the modular multiplicative inverse of e, where d is kept as the private key exponent. The public key has modulus n and the public (or encryption) exponent e. The private key has modulus n and the private (or decryption) exponent d, which is kept secret.

Meanwhile, the encryption steps are as follows:

Encryption Procedure

User transmits the public key (modulus n and exponent e) to Cloud

Keep this private key secret

When Cloud sends a number "M" to user:

Convert M to an integer such that $0 <m<n$ by using agreed upon reversible protocol known as a padding scheme

Cloud computes with the user's public key information, the cipher text c corresponding to $c\equiv me$ (mod n)

Cloud sends message "M" in cipher text, or c to the user

Moreover, the decryption procedure is as follows:

Decryption Procedure
    User recovers m from c by using the private key exponent
'd' by the computation $m \equiv cd \pmod{n}$
    Given m
    User can recover the original message "M" by reversing the
padding scheme

This procedure is applied since $c \equiv me \pmod{n}$, $cd \equiv (me)d \pmod{n}$, and $cd \equiv mde \pmod{n}$. By the symmetry property of the mods, it is found that $mde \equiv mde \pmod{n}$. Since $de = 1 + k\phi(n)$, thus $mde \equiv m_1 + k\phi(n) \pmod{n}$, $mde \equiv m(mk)\phi(n) \pmod{n}$, and $mde \equiv m \pmod{n}$. This is valid for all m and the original message $cd \equiv m \pmod{n}$ is obtained based on Euler's theorem and the Chinese remainder theorem. Thus, these are employed in the proposed security model.

3.2 Access Phase

In order to access the cloud, the user provides finger impressions of the two registered fingerprints. The order of the impressions is based on the generated two-digit random number. The generation of the random numbers is done where the proposed security model has an edge over other models that provide single-fingerprint system. Hence, once an intruder gains access to a fingerprint template, it is required to claim being an authenticated user. However, in a multiple fingerprint system, even if an intruder manages to lacerate a stored template, still number tagged to each of the finger remains hidden. For authentication purpose, a random number generator (RNG) is used to generate a three-digit random number (with repetition) every 20th second. The three digits constitute the numbers that are chosen by the user during enrollment phase. User now provides the finger impressions in the order of the generated random number. Figure 4 illustrates the proposed system during the access phase.

3.3 Matching Phase

In this phase, a legitimate user is validated and an eavesdropper is invalidated. It is required to provide the impressions varies with the generated random number, even if the stored templates are hacked. Thus, by means of trial and error, if a hacker tries with different permutations, access will be denied after three consecutive wrong attempts. The user has to reset the numbering that was earlier assigned. This phase also includes a method of reassignment of a biometric template along with numbers

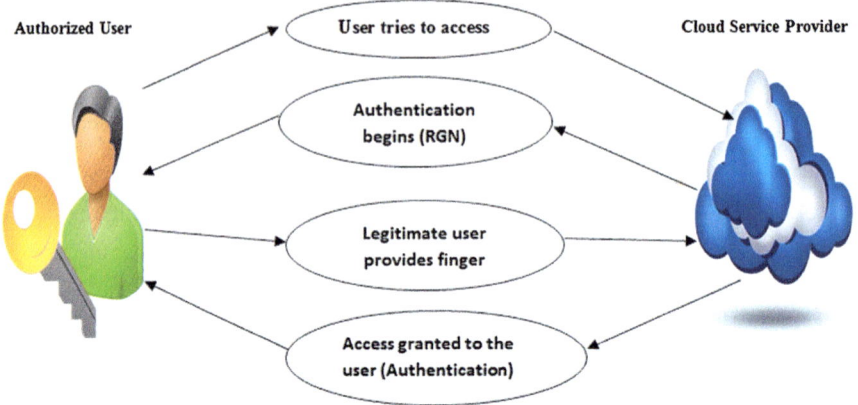

Fig. 4 User access process with the cloud service provider

and mappings when the existing one assumed to be compromised after three consecutive wrong attempts.

Consequently, the proposed system based on two-fingerprint biometrics is given in the next section. The proposed multi-input unimodal system is tested on the attendance management system database, which consists of records corresponding to an authorized person that has access to the system. Each record may contain the minutiae templates of the person's fingerprint and user name of the person or other information to the template. The database design for the system implements relational data model, which is a collections of tables that stored data. The SQL Server is fast and easy, and it can store very large records and requires little configuration. Thus, the database is implemented in Microsoft SQL Server database (Sql Server, 2005).

4 Proposed System

The attacks in the realm of a fingerprint are intended to either circumvent the security afforded by the system or to deter the normal functioning of the system. Protecting the template is a challenging task due to variability in the acquired biometric traits [40]. Biometric template and multi-input protection schemes are more viable than unimodal in terms of security, revocability, and impact on matching accuracy [41]. In particular, the biometric unimodal fingerprint system is prone to various types of security attacks, such as attack at the scanner, where the attacker can physically destroy the recognition scanner and cause a denial of service. A fake biometric trait such as an artificial finger can also be created by attacker to bypass the recognition system [42].

Biometric unimodal fingerprint system is susceptible to various types of threats including (i) the denial of service, where the adversary overwhelms computer and network resources to the point that legitimate users can no longer access the resources [43], (ii) the circumvention, where an adversary gains access to data or computer resources that he may not be authorized to access, (iii) the repudiation, where a legitimate user accesses the resources offered by an application and then claim that an intruder had circumvented the system, (iv) the covert acquisition, where an adversary compromises and abuses the means of identification without the knowledge of a legitimate user [44], and (v) the collusion, where in any system there are different user privileges. Users with super-user privileges have access to all of the system's resources. Collusion occurs when a user with super-user privileges abuses his privileges and modifies the system's parameters to permit incursions by an intruder.

However, the security attacks and threats are minimized by using multi-input fingerprint biometric system. In the realm of template transformation, the so-called biometric cryptosystems are gaining popularity. These systems combine biometrics and cryptography at a level that allows biometric matching to effectively take place in the cryptographic domain, hence exploiting the associated higher security. Uludag et al. [44] converted the fingerprint templates (minutiae data) into point lists in two-dimensional (2D) space, which implicitly hide a given secret (e.g., a 128-bit key). The list does not reveal the template data, since it was augmented with chaff points to increase security. The template data was identified only when matching minutiae data from an input fingerprint is available. Furthermore, the biometric cryptosystems can contribute to template security by supporting biometric matching in secure cryptographic domains.

Consequently, in the proposed system, multi-fingerprint security model is applied, where the users during registration use two finger templates of their choice and assign a single-digit number for each of these two fingers. These recorded images are encrypted using elliptical algorithm and stored at the service provider's end as demonstrated in Fig. 5.

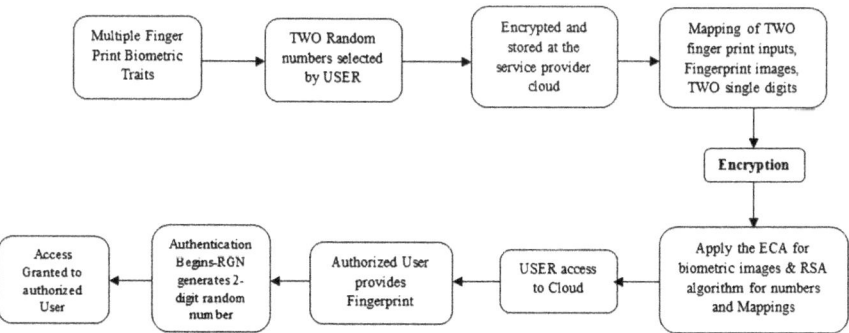

Fig. 5 Proposed system block diagram for multiple fingerprint biometric traits

As deployed in Fig. 5, the encryption algorithm is applied at three levels:

1. Fingerprint images are collected from the end user.
2. Two single-digit numbers are assigned to two fingerprints of the end user.
3. Two single-digit numbers are mapped to the two fingerprint images.

Given a fingerprint matcher, its accuracy and speed performance in a realistic setting are evaluated to measure the system performance. Unlike passwords and cryptographic keys, biometric templates have high uncertainty. There is considerable variation between biometric samples of the same user taken at different instances of time. Therefore, the match is always done probabilistically, which in contrast to exact match required password- and token-based approaches. The inexact matching leads to two forms of errors, namely the false (impostor) acceptance rate (FAR) and false (genuine individual) rejection rate (FRR). The FAR/FRR ratios depend mainly on the difficulty of the algorithms used for the fingerprint extraction. Usually, algorithms with high/medium complexity lead to acceptable low FRR/FAR [27]. The proposed RSA algorithm has outperformed unimodal fingerprint in terms of false accept, false reject, and failure to register.

An impostor may sometime be accepted as a genuine user, if the similarity with the template used falls within the intra-user variation of the genuine user. The FAR normally defined the probability of someone else matching as the main user either in a percentage or a fraction. The FAR formula is given by:

$$FAR = \frac{FA}{N} \times 100 \tag{1}$$

where FA is the number of false accept and N is the total number of verification. Moreover, when the acquired biometric signal has poor quality, even a genuine user may be rejected during authentication. This error is labeled as a 'false reject.' If the user fails to match against the owned template, then it is considered to be falsely rejected. The probability of this case is referred to as the false rejection rate (FRR). Thus, the higher the probability of false rejection, the greater the likelihood the user will be rejected. The FRR is defined by as follows:

$$FRR = \frac{FR}{N} \times 100 \tag{2}$$

where FR is the number of false reject and N is the total number of verification. Furthermore, the system may also have other less frequent form of error, such as the failure to enroll (FTE). The FTE assumes that nearly 4 % of the population has illegible fingerprints. This consists of senior population, laborers who use their hands a lot, and injured individuals. Due to the poor ridge structure present in such individuals, these users cannot be enrolled into the database and therefore cannot be subsequently authenticated. The FTE provides the possibility of someone failing to enroll in a system. It can be given in a percentage or a fraction. The FTE expression is given as follows:

$$FTE = \frac{FE}{N} \times 100 \tag{3}$$

where FE is the total number of failure enroll, and N is the total number of verification.

For performance analysis, the application developed was tested using the bio-data and fingerprints collected from one hundred (100) users of which 25 were staff and 75 were students using a live-scan method. The fingerprints were taken from thumb and forefinger of a respective member of the group in which each person must remember the exact finger that was used for the purpose of verification. Changing the finger will lead to failure in the enrollment due to the poor ridge structure present in such fingers. The minutiae data were extracted from the fingerprint images and stored in a database as a template for the subject along the user's ID.

4.1 Problem Solution

During authentication, the biometric of the user is captured again and minutiae data are extracted to form the test template that is matched against the already stored template in the database. In each case, if the matching score is less than the threshold, the person is rejected, otherwise the person is accepted using Eqs. 1–3. The respective values for the false acceptance rate (FAR) and false rejection rate (FRR) are 0.0 and 2.00 %, respectively. This implies an accuracy of 98.0 % considering the genuine acceptance rate. During the test, the FAR of zero value refers to no cases of false acceptance, i.e., a person that was not preregistered was not falsely enrolled for attendance. There was a FRR during the test in which the system failed to identify some preregistered users. These could be attributed to improper placement of the finger on the scanner and fingers that have been slightly scarred due to injuries. Table 1 reports the detailed evaluation of the current system with successful and unsuccessful verification of faculty and students.

Table 1 illustrates that in total a number of faculty are 25 and the students are 75; the successful verification of faculty and the students are 25 and 71, respectively. However, the unsuccessful rate is 0 % for the faculty, while four for the students. Furthermore, a comparative study is performed to compare the single-fingerprint evaluation system to the multi-fingerprint system as demonstrated in Table 2.

Table 1 Evaluation for the successful and unsuccessful verification

	Successful verification	Unsuccessful verification
Faculty	25	0
Students	71	4
Total	96	4

Table 2 Comparison of single- and multi-fingerprint evaluation systems

	Number of attendees	Successful verification	Unsuccessful verification	Accuracy (%)
Single fingerprint	100	96	4	96
Multi-fingerprint	100	96	4	**98**

Bold value indicates the significant high accuracy values that achieved by the proposed system

Table 3 Comparative study in terms of the execution time and the accuracy for 100 attendees

System type	Total time (s)	Total time (min)	Average Execution time (s)	Accuracy
Manual attendance system	2100.55	28.2	14.42	94
Single-fingerprint-based attendance system	455.41	7.30	3.20	96
Multi-fingerprint-based attendance system (proposed)	900	15.0	6.00	**98**

Bold value indicates the significant high accuracy values that achieved by the proposed system

Table 2 deploys that the multi-fingerprint proposed system achieved superior accuracy compared to the single-fingerprint system. The reported accuracies are 98 and 96 % for the multi-fingerprint system and the single-fingerprint system, respectively, in the case of 100 attendees. The time taken for the verification process is reported in Table 3 for the single and the proposed multi-fingerprint, and the manual attendance systems. The comparison of the manual system with the fingerprint-based attendance management system is performed.

Table 3 reports the comparison between the proposed and developed fingerprint-based attendance management systems with the existing manual attendance system (use of paper sheet/attendance register) as well as the single-fingerprint system in terms of the time. The manual attendance system's average execution time for one hundred attendees is approximately 14.42 s as against 3.20 s for the single-fingerprint-based attendance management and 6.00 s for multi-fingerprint model system.

Though the time complexity of the proposed multi-fingerprint system is more, where every individual has to give 2 fingerprints, however, the achieved accuracy rate is superior to value of 98 %. Hence, the proposed system is highly recommended as it outperforms single-fingerprint system in terms of high security. The authenticated result presented in this proposed model should be evaluated using other public multimodal real-user databases. Especially, it must be necessary to measure the performance of the suggested approaches with a large dataset, containing more individuals.

The proposed two-fingerprint biometric multimodal system provides high security by storing the templates at the cloud providers' end. The users are authenticated based on these finger templates which have to be provided in the order of random numbers that are generated every time. Both fingerprint templates and images are provided consistently and encrypted for enhanced security. In the future, it is recommended to involve large dataset to support and evaluate the proposed biometric two-fingerprint-based multiple input unimodal system has adverse implications for real-time problems. In addition, the proposed system can be used with different applications. Since, the systems' security requirements depend on the applications requirements and cost-benefit analysis. Thus, elaborate spoofing attacks may defeat a practical biometric system. Such case can be studied as a future scope. Development of novel encryption algorithms that can improve the proposed system can be applied in the future.

5 Conclusions

Cloud computing is concerned with using maximum remote services through an Internet/network using numerous minimum resources. It delivers these resources to the users through the Internet. There are various critical problems that the cloud computing faces including the reliability, data privacy, and security. However, the security issues are the most critical between these problems.

As the biometric systems are playing a key role in government and commercial applications which are outsourced to the cloud, providing security and privacy of user is the biggest concern to be considered. Subsequently, the present work brings about a novel security model where two fingerprints constitute an authentication. The proposed system is an efficient technique of providing high-level security by multiple fingerprints. It can be considered as a new security model of authentication, which provides high security from the intruders in the cloud. This novel and hybrid approach is theoretically portrayed to provide security for the biometric templates which enhances the privacy of the user.

The experimental results proved 98 % accuracy of the proposed system, which outperforms both the single-fingerprint and the manual system with the fingerprint-based attendance management systems.

References

1. Kuo AM-H (2011) Opportunities and challenges of cloud computing to improve health care services. J Med Internet Res
2. Peelukhana R, Bala PS, Aghila G (2011) Securing virtual images using blind authentication protocol. Int J Eng Sci Technol (IJEST)
3. Nandi S, Roy S, Dansana J, Karaa WBA, Ray R, Chowdhury SR, Chakraborty S, Dey N (2014). Cellular automata based encrypted ECG-hash code generation: an application in inter human biometric authentication system. Int J Comput Netw Inf Secur 6(11)
4. Biswas S, Roy AB, Ghosh K, Dey N (2012) A Biometric authentication based secured atm banking system. Int J Adv Res Comput Sci Softw Eng

5. Dey M, Dey N, Mahata SK, Chakraborty S, Acharjee S, Das A (2014) Electrocardiogram feature based inter-human biometric authentication system. In: 2014 International conference on electronic systems, signal processing and computing technologies (ICESC), pp 300–304

6. Dey N, Nandi B, Das P, Das A, Chaudhuri SS (2013) Retention of electrocardiogram features insignificantly devalorized as an effect of watermarking for a multi-modal biometric authentication system, published by Advances in Biometrics for Secure Human Authentication and Recognition, CRC Press, ISBN-9781466582422, pp. 450

7. Bose S, Chowdhury SR, Sen C, Chakraborty S, Redha T, Dey N (2014) Multi-thread video watermarking: a biomedical application. In: International conference on circuits, communication, control and computing (I4C), pp 242–246

8. Dey N, Pal M, Das A (2012) A session based watermarking technique within the NROI of retinal fundus images for authencation using DWT, spread spectrum and Harris corner detection. Int J Mod Eng Res 2:749–757

9. Field W, Hellman ME (1976) New directions in cryptography by MEMBER. IEEE Trans Inf Theory 22:6

10. Davide M, Dario M, Jain AK, Salil P (2005) Handbook of fingerprint recognition, 2nd edn. Springer, New York

11. Sasirekha K, Mary CI (2012) Biometric based network video security system with RSA implementation. Int J Biom Bioinform

12. Tiwari P, Saklani A (2013) Role of biometric cryptogra phy in cloud computing. Int J Comput Appl 70–79

13. Sabri HM, Ghany KKA, Hefny HA, Elkhameesy N (2014) Biometrics template security on cloud computing. In: 2014 International conference on advances in computing, communications and informatics (ICACCI), pp 672–676

14. Liang B, Wu Z, You L (2014) A novel fingerprint- based biometric encryption. In: 2014 ninth international conference on P2P, Parallel, Grid, Cloud and internet computing (3PGCIC), pp 146–150

15. Omar MN, Salleh M, Bakhtiari M (2014) Biometric encryption to enhance confidentiality in Cloud computing. In: 2014 International symposium on biometrics and security technologies (ISBAST), pp 45–50

16. Masala GL, Ruiu P, Brunetti A, Terzo O, Grosso E (2015) Biometric authentication and data security in cloud computing. In: Proceedings of the international conference on security and management (SAM) (p 9). The steering committee of the world congress in computer science, computer engineering and applied computing (WorldComp)

17. Khan MK, Zhang J (2008) Multimodal face and finger print biometrics authentication on space-limited tokens. Neuro Comput 71(13):3026–3031

18. Emersic Z, Bule J, Zganec-Gros J, Struc V, Peer P (2014) A case study on multi-modal biometrics in the cloud/Multi-modal. ElektrotehniskiVestnik 81(3):74

19. Goh TA, Ngo D (2006) Random multispace quantization as an analytic mechanism for bio hashing of biometric and random identity inputs. IEEE Trans Pattern Anal Mach Intell 1892–1901

20. Feng YC, Yuen PC, Jain AK (2010) A hybrid approach for generating secure and discriminating face template. IEEE Trans Inf Forensics Secur 03–117

21. Kanade S, Petrovska-Delacretaz D, Zi BD (2009) Cancelable iris biometrics and using error correcting. IEEE Conference computer vision and pattern recognition. CVPR 2009, Miami, pp 120–127

22. Gudavalli M, Raju VS, Adari R (2012) A novel approach for multi biometric template protection. IEEE international conference on computing and security (ICCS'12), Ulaanbaatar, Mongolia

23. Singh K, Verma R, Chehal R (2012) Modified prime number factorization algorithm (MPFA) for RSA public key encryption. Int J Soft Comput and Eng (IJSCE) 4(2):204–206

24. Miller VS (1986) Use of elliptic curves in cryptography. Adv Cryptol CRYPTO 85:417–426

25. Koblitz N (1987) Elliptic curve cryptosystems. Math Comput 203–209

26. IEEE-P1363–2000 (2000) IEEE standard specifications for public-key cryptography. IEEE Computer Society Press, Silver Spring, MD, USA

27. ISO/IEC15946 (2002) Information technology -security techniques: cryptographic techniques based on elliptic curves. International Organization for Standardization, Switzer land

28. NSIX9.62 (1999) Public key cryptography for the financial services industry: the elliptic curve digital signature algorithm (ECDSA). American National Standards Institute, New York, USA

29. NIST. FIPS 186–2 (2000) Digital signature standard (DSS). National Institute for Standards and Technology, Gaithersburg

30. Bos JW, Halderman JA, Heninger N, Moore J, Naehrig M, Wustrow E (2014) Elliptic curve cryptography in practice. In: Financial cryptography and data security. Springer, Berlin, pp 157–175

31. Lenstra AK, Verheul ER (2001) Selecting cryptographic key sizes. J Cryptol 14(4):1–255

32. Michaelis K, Meyer C, Schwenk J (2013) Randomly failed The state of randomness in current Java implementations. In Dawson E (ed) CT-RSA, 7779:129–144 of LNCS, Springer

33. Miers I, Garman C, Green M, Zerocoin RAD (2013) Anonymous distributed E-Cash from Bitcoin. In: IEEE symposium on security and privacy, pp 397–411

34. Miller VS (1986) Use of elliptic curves in cryptography. In: Williams HC (ed) CRYPTO, volume 218 of LNCS, Springer, pp 417–426

35. Juels A, Weis SA (2005) Authenticating pervasive devices with human protocols. In: Advances in cryptology–CRYPTO 2005. Springer, Heidelberg, pp 293–308

36. Singh K, Verma R, Chehal R (2012) Modified prime number factorization algorithm (MPFA) for RSA public key encryption. Int J Soft Comput Eng(IJSCE) 2:204–206

37. Sharma S, Yadav JS, Sharma P (2012) Modified RSA public key cryptosystem using short range natural number algorithm. Int J Adv Res Comput Sci Softw Eng 2:134–138

38. Wiener M (1990) Cryptanalysis of short RSA secret expo Nents. IEEE Trans Inf Theory 36:553–558

39. Latha U, Rameshkumar K (2013) A study on attacks and security against fingerprint template database 2(5)

40. Jain AK, Uludag U (2004) Attacks on biometric systems: a case study in fingerprints. In: Proceedings of SPIE-EI 2004, security, seganography and watermarking of multimedia contents VI, San Jose, CA, pp 622–633

41. Jain AK, Ross A, Uludag U (2005) Biometric template security: challenges and solutions. In: Proceedings of the European signal processing conference (EUSIPCO '05), Antalya, Turkey

42. Moore A, Ellison RJ, Linger R, (2001) Attack modeling for information security and survivability, technical report, Carnegie Mellon University

43. Abdullayeva F, Imamverdiyev Y, Musayev V, Man WJ (2008) Analysis of security vulnerabilities in biometric systems. In: The second international conference: problems of cybernetics and informatics

44. Uludag U, Pankanti S, Prabhakar S, Jain AK (2004) Biometric cryptosystems: issues and challenges. Proc IEEE 92:948–960

Printed by Printforce, the Netherlands